CHEMICAL ENGINEERING PRACTICE

Under the General Editorship of
HERBERT W. CREMER
C.B.E., M.Sc., F.R.I.C., M.I.Chem.E.
M.Inst.F., M.Cons.E.

Managing Editor
TREFOR DAVIES
B.Sc., F.R.I.C., M.I.Chem.E.

IN TWELVE VOLUMES

VOLUME 3
SOLID SYSTEMS

NEW YORK
ACADEMIC PRESS INC., PUBLISHERS
LONDON
BUTTERWORTHS SCIENTIFIC PUBLICATIONS
1957

BUTTERWORTHS PUBLICATIONS LTD.
88 KINGSWAY, LONDON, W.C.2

U.S.A. Edition published by
ACADEMIC PRESS INC., PUBLISHERS
111 FIFTH AVENUE
NEW YORK 3, NEW YORK

Set in Monotype Baskerville type
Made and printed in Great Britain by William Clowes and Sons, Limited
London and Beccles

CONTENTS

SIZE REDUCTION

Principles of Crushing and Grinding 1
   HAROLD HEYWOOD, D.Sc.(Eng.), Ph.D., Wh.Sch., M.I.Mech.E., M.I.Chem.E., M.Inst.F.
      Reader in Mechanical Engineering, Department of Mechanical Engineering, Imperial College of Science and Technology, University of London

Methods of Sizing Analysis 24
   HAROLD HEYWOOD, D.Sc.(Eng.), Ph.D., Wh.Sch., M.I.Mech.E., M.I.Chem.E., M.Inst.F.
      Reader in Mechanical Engineering, Department of Mechanical Engineering, Imperial College of Science and Technology, University of London

Crushing and Grinding Equipment 48
   J. C. FARRANT, M.I.Chem.E., M.A.I.Mech.E. (Senior Member)
      Retired—formerly Manager of the Grinding, Screening and Filtering Division, International Combustion Products Limited, Derby
   R. NORTH, A.M.I.Mech.E.
      Technical Department, International Combustion Products Limited, Derby

The Mechanics of Pulverizers 97
   HAROLD HEYWOOD, D.Sc.(Eng.), Ph.D., Wh.Sch., M.I.Mech.E., M.I.Chem.E., M.Inst.F.
      Reader in Mechanical Engineering, Department of Mechanical Engineering, Imperial College of Science and Technology, University of London

Special Applications of Grinding Machines 109
   R. A. SCOTT, Ph.D., F.Inst.P.
      Head of Research Department, Henry Simon Ltd., Stockport

SCREENING, GRADING AND CLASSIFYING 128
   R. A. SCOTT, Ph.D., F.Inst.P.
      Head of Research Department, Henry Simon Ltd., Stockport

Tabling and Jigging 172
   G. H. HIGGINBOTHAM, B.Sc., Ph.D.
      Development Engineer, Simon-Carves Limited, Stockport

Flotation 209
   J. E. FELSTEAD, B.Sc., A.M.I.Min.E.
      Development Engineer, Simon-Carves Limited, Stockport

Sedimentation 248
   R. FORBES STEWART, M.C., A.R.T.C., F.R.I.C., M.I.Chem.E.
      Director, Dorr-Oliver Co. Ltd., London

Wet Classification 287
   R. FORBES STEWART, M.C., A.R.T.C., F.R.I.C., M.I.Chem.E.
      Director, Dorr-Oliver Co. Ltd., London

CONTENTS

	PAGE
Dense Medium Coal Washing	312

R. SYMINGTON, B.Sc., A.R.I.C., A.M.Inst.F.
Contract Engineer, Simon-Carves Limited, Stockport

Air Flow Selection 337
R. A. SCOTT, Ph.D., F.Inst.P.
Head of Research Department, Henry Simon Ltd., Stockport

MIXING OF SOLIDS 362
R. A. SCOTT, Ph.D., F.Inst.P.
Head of Research Department, Henry Simon Ltd., Stockport

STORAGE AND HANDLING OF SOLIDS 380
J. A. W. HUGGILL, M.A., D.Phil., A.Inst.P.
A.E.I.—John Thompson Nuclear Energy Company Limited, Cheshire

Sampling, Measuring and Gauging of Solids 438
E. J. SEBESTYEN, Dipl. Ing.
Development Engineer, Simon Handling Engineers Ltd., Stockport

CLEANING GASEOUS MEDIA

Cyclones 464
DR. ING. R. F. HEINRICH
Chief Engineer, Precipitator Division, Chemical Plant Department, Simon-Carves Limited, Stockport
J. R. ANDERSON, A.M.I.E.E.
Deputy Chief Engineer, Precipitator Division, Chemical Plant Department, Simon-Carves Limited, Stockport

Electro-Precipitation 484
DR. ING. R. F. HEINRICH
Chief Engineer, Precipitator Division, Chemical Plant Department, Simon-Carves Limited, Stockport
J. R. ANDERSON, A.M.I.E.E.
Deputy Chief Engineer, Precipitator Division, Chemical Plant Department, Simon-Carves Limited, Stockport

1

PRINCIPLES OF CRUSHING AND GRINDING

HAROLD HEYWOOD

THEORIES OF THE MECHANISM OF CRUSHING AND GRINDING

THE theories deduced by Rittinger and by Kick in the late nineteenth century regarding the mechanism of fracture are still the subject of controversy. Rittinger's law stated that the energy necessary for reduction in particle size was directly proportional to the increase in surface. Kick's law, in a slightly abbreviated form, was that the energy required for producing analogous changes of configuration of geometrically similar bodies varied as the volumes of these bodies. A simple illustration will show that these laws are incompatible. If we consider two cubes of material, one having twice the dimensions of the other, then the relative volumes will be in the ratio 8 to 1, and if they both break down in a similar manner, say into eight cubes each having a dimension of one-half the original size, then by Kick's law the ratio of the work required to fracture the cubes will also be in the ratio 8 to 1. The increase in surface area is in the ratio 4 to 1, and so by Rittinger's law this is also the ratio of the work required to fracture the two cubes. Thus, according to Kick's law the work required varies as the cube of the initial particle dimension, whereas according to Rittinger's law it varies as the square of the dimension.

This reasoning may also be applied to size reduction in a number of stages, in each of which a cube of material is broken into eight cubes of half the initial dimensions. Denoting the ratio of size reduction in each stage by r, then the overall size reduction for n stages will be r^n. Let D be the size of the initial cube and d that of the final cubes after n stages of reduction, then

$$r^n = \frac{D}{d}, \text{ or } n \log r = \log \frac{D}{d}$$

Since the volume of material remains constant, the energy, E, for each stage by Kick's law will be constant, and the total energy required will be nE; but

$$n = \frac{\log D/d}{\log r} \text{ and total energy } nE = \frac{E}{\log r} . \log D/d.$$

Since E and r are constant for each stage of reduction, the total energy for a series of stages is proportional to $\log D/d$, which is another form in which Kick's law may be stated.

By Rittinger's law the energy for reduction would be proportional to the increase in specific surface, *i.e.* to $(1/d - 1/D)$, but

$$\frac{1}{d} - \frac{1}{D} = \frac{1}{D}\left(\frac{D}{d} - 1\right) = \frac{1}{D}(r^n - 1)$$

If the ratio of reduction r per stage is 2 and the initial size of cube is 1, then the relative energies for a number of stages of reduction are shown by *Table 1*.

These figures show that by Rittinger's law the energy per stage increases with the degree of size reduction. OWEN[1, 2] developed this method of comparison very fully, and it is apparent that such simple basic assumptions are inadequate to explain the fundamental principles of the process of crushing.

Table 1. Relative Energies for Size Reduction

No. of stages of reduction	$\dfrac{D}{d}=r^n$	Kick's law, energy proportional to $\log D/d$	Rittinger's law, energy proportional to $(r^n - 1)$
1	2	1	1
2	4	2	3
3	8	3	7
4	16	4	15

In reality, both laws are far too greatly simplified and depend upon ideal processes that are not fulfilled in practice. Thus, if a cube of perfectly homogeneous and isotropic material is steadily and uniformly loaded up to the limit of elasticity, then the work done in compression equals the energy stored elastically and will be proportional to the volume of the cube. If the material is brittle the modulus of elasticity will remain approximately constant up to the fracturing load and at this point the work done in compression equals a quantity termed the limiting strain energy. In this respect the work done is in agreement with Kick's law, but we have still to consider the fracture pattern or size distribution of the fragments that are produced when the energy applied exceeds the limiting strain energy. The size distribution of these fragments will not be of the simple form assumed in the above hypothetical illustration, and without a knowledge of the fracture pattern no deduction can be made as to whether the surface increase is also proportional to the work done in crushing. Another factor which has a profound influence on the problem is that minerals are not homogeneous or isotropic but contain planes of weakness or incipient fracture at which stresses become concentrated. If it is assumed that all the strain energy is concentrated at these planes of weakness, then the total fracture energy would be proportional to the new surface produced. Hence the relationship between fracture energy and surface produced will in practice be exceedingly complex and is linked essentially with the size distribution of the products of fracture. An analogy which may be helpful in visualizing this process is that of the impact test on steels; this is done with square bars in which a V-notch is cut along the edge at which a tensile stress is developed by impact of a weight. As a consequence, almost all the energy applied to the specimen is concentrated in a plane through this V-notch and even ductile metals are broken cleanly into two pieces. If the specimen were tested without the V-notch then the energy would be more uniformly distributed throughout the volume of the specimen and there would not be a clean fracture. A number of research workers, starting with the postulation of a random distribution of fissures or planes of weakness, have made theoretical deductions regarding the size distribution of the fragments resulting from crushing and of the energy required for this process. These theories will be discussed further in a later section and experimental procedures and results will be described in order to present a visual picture of the phenomena concerned.

Compression and Impact Tests

Free-crushing conditions are attained when the crushing force ceases to be applied immediately after the material fractures. Packed crushing occurs if the crushing force is maintained after the initial breakage and produces further size reduction of the fragments formed by the initial

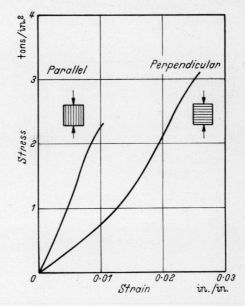

Figure 1. Stress–strain curves for Barnsley soft coal. (By courtesy of the Journal of the Institute of Fuel)

crushing. Energy is dissipated by friction between the particles in a bed subjected to packed crushing so that free crushing is the most efficient process for size reduction. Free-crushing conditions are attained by compression tests on cubes or pieces of material, or by impact tests in which a weight is allowed to fall on the material, provided that a stop is placed so that the force applied is restrained from following up the fractured material. Applying these principles to industrial machines for size reduction, free-crushing conditions are fulfilled in jaw crushers where there is a definite minimum gap between the jaws, or in crushing rolls if these are set to have a definite clearance. Packed crushing occurs in mills of the normal ring-roll type where the rolls are pressed against the grinding ring by gravitational or centrifugal force, or in ball and tube mills.

The results of compression tests on cubes of coal are shown in *Figure 1*, in which the stress–strain curves are plotted for compression, parallel and perpendicular to the bedding planes respectively. The curves approximate to straight lines and the limiting strain energy per unit volume at fracture is determined from the area below the curves. These show that the coal is much stronger and has a greater limiting strain energy when compressed in a direction perpendicular to the bedding planes. The energy applied up to fracture cannot be varied in the compression tests, since this is controlled by the dimensions and elastic properties of the material, but by impact tests the energy applied can be varied at will, and may be either greater or less than the limiting strain energy. If the applied energy were less than the limiting strain energy determined by compression test, and provided that this energy were uniformly distributed throughout the volume of the specimen, there would be no fracture and no surface increase. If the applied energy were greater than the limiting strain energy, then free crushing would normally be followed by packed crushing, and although there would be an additional increase in surface the efficiency of crushing

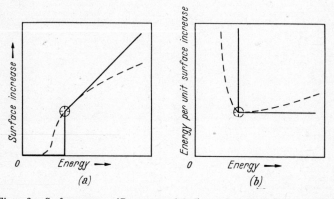

Figure 2. Surface energy. (*By courtesy of the* Journal of the Institute of Fuel)

would be reduced because of the energy lost by friction. When the energy applied by impact is the same as the limiting strain energy, then conditions should be identical with the compression test, and this has been verified by practical experiments. Experiments made by the writer[3] showed that for brittle materials the ratio of the energy to the surface increase is almost identical for compression and impact tests made under conditions of equal applied energy.

The idealized case is illustrated graphically in *Figure 2*. The surface increase is plotted against the energy applied in diagram (*a*), the relative values for the compression test being shown by the circle. An impact test in which the applied energy was less than the limiting strain energy would not produce any fracture or external surface increase, the energy being absorbed elastically or by the development of internal fissures. The same surface increase would be obtained when the two energies were equal, but when the impact energy exceeded the limiting strain energy, there would be a greater surface increase. Assuming for the moment that Rittinger's law applied, the relationship between surface increase and energy would be a straight line directed towards the origin, but in fact terminating at the point representing the compression test. Diagram (*b*) shows the relationship between the ratio energy to surface as a function of the applied energy and is derived from the corresponding lines in diagram (*a*). The experimental relationships are shown by the dotted lines in these diagrams; a slight surface increase is obtained with impact energies less than the limiting strain energy because of planes of weakness in the material and the probability that the energy will be concentrated on a projecting corner of the specimen and will not be uniformly distributed. Impact energies greater than the limiting strain energy do not produce as much surface increase as in the ideal case because of the energy loss in friction and the kinetic energy imparted to the particles.

Steady compression tests on cubes of brittle materials almost invariably exhibit increasing strength with decreasing size, showing that planes of weakness are more predominant in the larger pieces of material. The limiting strain energy per cubic inch at fracture is also greatest for the small cubes and so is the ratio of energy to surface increase. Data obtained by the writer with compression tests on cubes of coal is given in *Table 2*, and similar effects for other materials are shown in researches by AXELSON and PIRET[4]. A method of compression testing that can be applied to particles of about 1 mm in size is described by CAREY and STAIRMAND[5].

Table 2. *Compression Tests on Cubes of Coal* (S 276)[34]

Size of cube in.	Breaking stress lb./in.^{2}	Limiting strain energy		Surface of fragments		Ratio of energy to surface increase ft.Lb./ft.^{2}
		ft.Lb.	ft.Lb./in.^{3}	ft.^{2}	ft.^{2}/in.^{3}	
4·00	440	11·65	0·182	18·65	0·293	0·648
2·57	1,015	8·39	0·492	12·81	0·752	0·670
1·72	1,195	1·90	0·372	2·83	0·554	0·701
1·66	1,410	2·21	0·480	3·35	0·730	0·684
0·74	1,370	0·485	1·180	0·364	1·016	1·228

Impact tests by means of a falling weight are described by GROSS and ZIMMERLEY[6], WILSON[7], PRENTICE[8] and FAIRS[9]. Small pieces of ductile metal wire have, in some cases, been placed below the anvil supporting the material on which the impact was received, and measurement of the deformation of these wires enables a correction to be made for the impact energy of the falling weight not absorbed by the crushing process. Impact crushers in the form of a ballistic pendulum have been devised by BOND[10], GAUDIN and HUKKI[11,12], and these have the advantage that the small proportion of the applied energy which is not absorbed by the crushing process can be determined directly from the rebound of the pendulum. Much may be learnt about the characteristics of materials by compression and impact tests, but it is doubtful whether these can be related directly to the performance of industrial pulverizers in which the product of comminution is a fine powder.

Crushing Roll and Ball Mill Tests

Higher intensities of comminution may be applied to materials by using experimental crushing rolls and ball mills. Free-crushing conditions are attained if the crushing rolls are set to have a fixed separation, provided that the feed size is not more than about three times the roll gap. The experimental rolls used in research by the writer are shown in *Figure 3*. These rolls were $2\frac{3}{4}$ in. diameter, mounted in ball-bearings and geared together

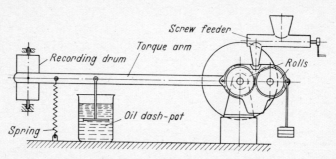

Figure 3. Experimental crushing rolls. (*By courtesy of the* Journal of the Institute of Fuel)

to run at an equal speed of 500 r.p.m. The frame carrying the rolls was pivoted to form a torque arm, so that the reaction torque produced when particles were being crushed could be measured directly by the displacement of the arm against a restraining spring. The reaction torque due to bearing friction with the rollers running empty could also be measured, hence from

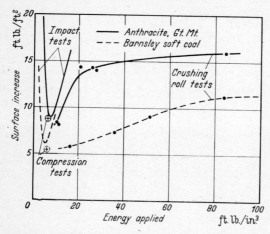

Figure 4. Compression, impact and crushing roll tests. (*By courtesy of the* Journal of the Institute of Fuel)

the difference an accurate determination could be made of the energy put into the crushing process. An autographic record of the position of the torque arm during crushing was made by passing a high tension electric spark between a point on the end of the torque arm and a rotating drum, which punctured a sheet of paper wrapped round the drum. The gap between the rolls was adjustable over a small range, but was fixed rigidly during any particular test. Tests were made with decreasing roll gaps and a fixed size of feed particle until a stage was reached at which packing occurred. This departure from free-crushing conditions was apparent from the very large increase in the torque on the rollers.

Table 3. Crushing Roll Tests on Coal (S 276)[34]

Roll gap in.	Energy applied ft.Lb./in.^{3}	Final surface ft.^{2}/in.^{3}	Ratio of energy to surface increase ft.Lb./ft.^{2}
0·045	7·5	2·66	4·27
0·040	8·1	2·93	4·07
0·035	12·3	4·68	3·29
0·030	19·0	6·70	3·29
0·025	43·0	11·55	4·04
0·020	162·5	19·45	8·76

A comparison of compression, impact and crushing roll tests is shown in *Figure 4*. These curves show how the impact test coincides approximately with the compression tests when the impact energy is equal to the limiting strain energy of the compression tests. The crushing roll tests also agree with the compression test when the reduction ratio is low, but the ratio of energy to new surface increases with the intensity of crushing.

Table 4. 8 in. Ball Mill Tests on Coal (S 276)[34]

Revolutions of mill	Energy applied ft.Lb./in.^{3}	Final surface ft.^{2}/in.^{3}	Ratio of energy to surface increase ft.Lb./ft.^{2}
113	137	29·7	5·04
450	545	74·4	7·58
900	1,090	94·0	11·92
1,800	2,180	106·0	21·07

Additional tests were made on coals by grinding in a ball mill of 8 in. diameter fitted with lifter bars and constructed to the U.S. Bureau of Mines specification described in a later subsection. This mill was fitted with a torsion dynamometer so that the net work used in lifting the ball charge could be measured accurately, and the surface increase of the ground product was calculated from the measured size distribution. The results of tests in crushing rolls and the ball mill are shown in *Table 3* and *Table 4*.

Effect of Reduction Ratio on Energy per Unit Surface Increase

The test results given in *Table 3* and *Table 4* show that the energy required to produce unit area of new surface increases with the reduction ratio. The final specific surface is a measure of the extent of degradation of the product and Rosin has suggested that this is analogous to entropy in thermodynamics. The final specific surface is therefore used as the independent variable against which to plot the energy surface characteristics of the material. The results of compression, crushing roll and ball mill tests on coal (S 276) are plotted in *Figure 5*, showing the variation in foot-pounds per square foot of surface increase with the final specific surface in square feet per pound. All these curves exhibit a minimum value for the energy to surface ratio, but these minima increase with the intensity of comminution and show that the early stages of grinding by any particular process are more efficient. The conclusions deduced from these and other tests may be summarized as follows:

(a) For any given type of crushing or grinding mill there are optimum conditions of comminution for which the energy to surface ratio is a minimum. If finer grinding is attempted in such a machine the ratio of energy to surface will be increased.
(b) The minimum value of the energy to surface ratio for each type of machine increases with the fineness of grinding.

This increase in the energy to surface ratio could be due to a reduction in the efficiency of the grinding process, *i.e.* that the mechanical forces applied were less effectively utilized in the processes of finer comminution, or it could be due to a change in the properties of the material with the

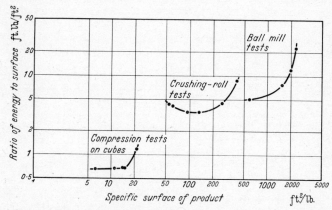

Figure 5. Variation of energy to surface ratio with fineness of product; tests on South Wales coal. (By courtesy of the Journal of the Institute of Fuel)

reduction in particle size. Consideration of the internal structure of materials and the presence of internal fissures, which are known to exert an important influence on the properties of larger pieces, would suggest that the latter effect has the predominating influence.

SIZE DISTRIBUTION LAWS

The study of crushing and grinding processes is inseparable from that of the size distribution of the fragments of ground product, and graphical methods of expressing the size distribution of a powdered material by various curves plotted to a base of particle size are as follows:

(a) The percentage by weight smaller than a given particle size, namely, the cumulative undersize curve.
(b) The percentage by weight larger than a given particle size, namely, the cumulative oversize curve or residue curve.

(c) The percentage by weight in a hypothetical fraction limited between particle sizes differing by one unit of length, plotted against the mean size of this hypothetical fraction.

This latter curve is the frequency or size distribution curve, and is the differential curve of (a), or of (b) if algebraic sign is ignored. These three curves are plotted in *Figure 6*, in which W is the percentage by weight of material smaller than a given particle size. Hence the undersize curve is

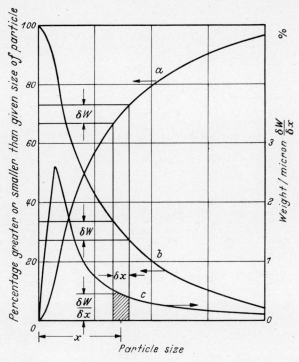

Figure 6. (a) *Cumulative undersize curve*; (b) *Cumulative oversize or residue curve*; (c) *Size frequency curve*. (By courtesy of The Institution of Mechanical Engineers)

W plotted against x, the oversize or residue curve is $100 - W$ plotted against x, and the frequency or size distribution curve is dW/dx plotted against x. Considering a small fraction of particles of mean size x and with a size range dx, the weight of these particles is dW and is represented by the area shaded in *Figure 6*. The total area enclosed by the size distribution curve is equal to the summation of all such areas and equals 100 per cent, hence the two cumulative curves (a) and (b) may be regarded as the integral or summation curves of the frequency curve.

These curves can also be plotted to a logarithmic scale of particle size in order to accommodate the very large range encountered in practice, but this will alter the mathematical relationships. Schuhmann[13] has also plotted the undersize percentages to a logarithmic scale, and the resulting graphs approximate to straight lines over a considerable proportion of the finer size range. It can be shown mathematically that there must be a limit to the range of particle size over which the straight line relationship occurs. The law for the straight line section is:

$$\log W = \log C + n \log x,$$

where W is the undersize percentage by weight, x the particle size, C and n are constants. Hence

$$W = Cx^n \text{ and } \frac{dW}{dx} = nCx^{n-1}$$

If n is greater than 1, then dW/dx increases with x. Since dW/dx must become zero at some finite value of x, it follows that the straight line cannot be prolonged to the maximum particle size, but that the slope of the graph must decrease with increasing particle size.

Considering the surface area of particles, since the surface area dS of a small weight of particles dW which is composed of particles of size x is proportional to dW/x, it follows that

$$\frac{dS}{dx} = \frac{1}{x} \cdot \frac{dW}{dx} = ncx^{n-2}$$

and the total surface
$$= \int dS = \int \frac{dW}{x} = nC \int x^{n-2} \, dx$$
$$= \frac{n}{n-1} \cdot C \left[x^{n-1} \right]_{x_1}^{x_2}$$

where x_2 and x_1 are the particle size limits.

Now if x_1 is 0 and n is greater than 1 this integral has a finite value, but if x_1 is 0 and n is less than 1 the integral, and consequently the surface area, is infinite. The integral has finite values for all values of n if the lower limit of size is finite, and Gaudin and associated workers have suggested that in the many practical cases where the observed value of n is approximately 1, the lower size limit should be that of the unit crystal, about 5 Å or 0·0005 μ[11, 12]. Gaudin's theorem is that if the particles are divided into grades each limited by sizes with a 2 to 1 or other constant ratio, then the surface per grade is constant for multiple fracturing. Examination of the published curves of the characteristics of ground materials shows, however, that there is a tendency for n to increase towards the finer limit of particle size measurement, and in the writer's opinion this effect may modify Gaudin's theorem.

ROSIN and RAMMLER[14] deduced that the fineness characteristic curve of residue plotted against particle size could be represented by the law $R = 100e^{-bx^n}$, where R is the percentage residue corresponding to particle size x. This equation may be transposed to the logarithmic form, thus

$$\frac{100}{R} = e^{bx^n}; \quad \log \frac{100}{R} = bx^n$$

and
$$\log \log \frac{100}{R} = \log b + n \log x$$

Graphs of the cumulative residues plotted to a scale of log log $100/R$ against log x should therefore be in the form of straight lines, the slope n characterizing the type of distribution and the constant b the mean size. The Rosin equation has been modified slightly by Bennett to the form

$$R = 100\mathrm{e}^{-(x/x_m)^n}$$

in which x_m is a direct measure of the mean particle size of the system. This equation has been used by BENNETT[15] and co-workers[16] in a study of the breakage of coal.

The Rosin form of plotting has been used in many investigations on the crushing and grinding of various materials and an approximately straight line relationship between the above variables has been obtained. Nevertheless, there are frequent exceptions to the law, and it is not valid to assume that even if a straight line is obtained over the range of measured particle sizes this may be extrapolated. The same criticism that applies to the Gaudin equation already mentioned also applies to the Rosin equation, namely, that if the slope n is unity, as frequently is found in practice, then the calculated specific surface is infinite. This feature is shown as follows:

Differentiating the Rosin equation gives

$$\frac{\mathrm{d}R}{\mathrm{d}x} = \frac{\mathrm{d}W}{\mathrm{d}x} = 100\, nbx^{n-1}\mathrm{e}^{-bx^n}$$

and

$$\frac{\mathrm{d}S}{\mathrm{d}x} = 100\, nbx^{n-2}\mathrm{e}^{-bx^n}.$$

Hence

$$S = \int \mathrm{d}S = 100\sqrt[n]{b}\, \Gamma\!\left(\frac{n-1}{n}\right)$$

where Γ is the gamma function. If $n=1$, $(n-1)/n$ is 0 and $\Gamma(0)$ is infinite: hence the specific surface is infinite. If n is less than 1 $(n-1)/n$ is negative, and the gamma function is also negative. Thus a real value for the specific surface is only calculable if n is greater than 1.

The writer has suggested[17] that the size frequency curve can be represented by the equation:

$$\frac{\mathrm{d}N}{\mathrm{d}x} = a\mathrm{e}^{-bx^n},$$

where $\mathrm{d}N$ is the number of particles in the size range $\mathrm{d}x$. Hence $\mathrm{d}W/\mathrm{d}x = ax^3\mathrm{e}^{-bx^n}$. The integration of this equation is somewhat complex, but it fulfils the mathematical requirements that $\mathrm{d}W/\mathrm{d}x$ is zero at $x=0$ and tends to zero at a maximum value of x. The calculated specific surface is finite for all values of n, either greater or less than 1. The four different methods of plotting described above are compared in *Figure 7*, which shows the size characteristics of the same powder to the different scales.

The size distributions of particles produced by different methods and intensities of comminution as shown by the Rosin and Rammler method of plotting are illustrated in *Figure 8*. Curve (*a*) is the characteristic curve for fragments produced by a compression test on a 4 in. cube of coal; the

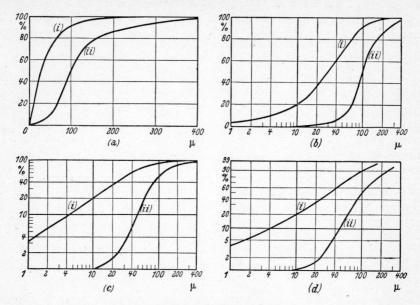

Figure 7. Undersize percentage against particle size in microns. Curve (i) Finely ground silica. Curve (ii) Fine sand
(a) Undersize to arithmetic scale; Particle size to arithmetic scale
(b) Undersize to arithmetic scale; Particle size to logarithmic scale
(c) Undersize to logarithmic scale; Particle size to logarithmic scale
(d) Undersize to Rosin–Rammler scale of log log 100/100-u; Particle size to logarithmic scale

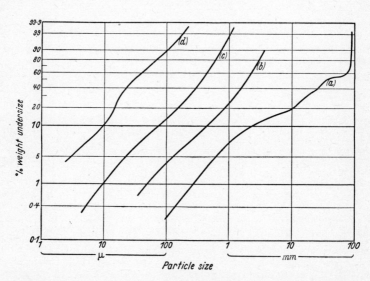

Figure 8. Size distribution curves for coal breakage and grinding; undersize to Rosin–Rammler scale, particle size to logarithmic scale. (a) Compression test on 4 in. cube; (b) Jaw crusher test at ⅛ in. gap; (c) Crushing roll test at 0·025 in. gap; (d) Ball mill test at 900 revs

vertical part of the curve at the right shows that a single large fragment represents some 20 per cent by weight of the original cube and this, together with several slightly smaller pieces, represents some 45 per cent of the original weight. The horizontal part of the curve shows that there are very few particles between 4 cm and 7 cm size, and the remainder of the curve represents the small particles and fine dust which is invariably produced by the crushing. Curve (a) is typical for the free crushing of a single lump; if the original cube were smaller than 4 in., the vertical part of the curve would be at a correspondingly smaller particle size, but the slope of the section representing the fine particles would be the same as for the example shown.

If the fragments from the first crushing are now subjected to further size reduction the larger pieces will be broken down in a similar manner and each fracture will contribute to the amount of fines. In this way the discontinuity of the characteristic curve for a single crushing is eliminated and a more nearly uniform distribution of particle sizes is obtained. This effect is shown by the curve (b) which represents the products after passing through a jaw crusher having 0·025 in. minimum gap, and by the curve (c) which represents the products after passing through crushing rolls with 0·025 in. gap. Both these processes are free-crushing processes. The effect of grinding the crushed material in a ball mill for 900 revolutions is shown by curve (d). The same process of continuous size reduction is evident, but there is a certain selective action with a tendency to produce preferentially a particular range of particle size as shown by the increased slope of part of the curve.

Such curves are evidence of some definite pattern of fracture occurring throughout all these processes, as shown by the curves being approximately parallel in the region of small particle size. Research has been conducted by Carey and Stairmand[5] in which the associated energy of the material is deduced from a series of such parallel curves, and is used to assess the efficiency of pulverizers.

THE COMPONENT THEORY OF COMMINUTION

The frequency curve representing the size distribution of finely ground materials is normally of the asymmetrical Gaussian form, as illustrated in *Figure 6*. The curve for a closely size-graded sample, *e.g.* a sieved fraction, would be of the symmetrical Gaussian form, but the transition from this to the asymmetrical form does not take place by simple displacement of the mode or peak towards the axis of zero particle size.

An examination of the curves representing the preliminary crushing of coarse material in *Figure 8* shows that these have a double inflection. The differential curves, therefore, consist of two components, each mono-modal. This bi-modal type of distribution curve is a characteristic of the early stages of crushing and, indeed, is almost invariable. The frequency or size distribution curves for the particles produced in a series of jaw crusher tests are shown in *Figure 9*[34]. The feed material was Barnsley soft coal in lumps of size equivalent to a 2 in. cube, and the minimum gap of the jaw crusher was set to various apertures. The size distribution curves for this material are bi-modal, the primary components consist mainly of particles that are larger than the minimum jaw gap, and the secondary components of particles that are smaller than the gap.

The sequence of curves shows that as the intensity of crushing increases, both the weight of the primary component and also its modal size (the size of maximum frequency) decrease; the weight of the secondary component increases but the modal size remains approximately constant. It may be deduced that the characteristics of the primary component are controlled by the setting of the machine and therefore this component is termed

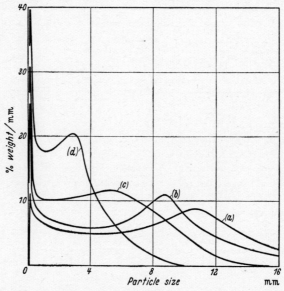

Figure 9. *Size frequency curves for coal from jaw crusher.* (a) *Jaw gap* 9·9 *mm;* (b) *jaw gap* 7·4 *mm;* (c) *jaw gap* 5·3 *mm;* (d) *jaw gap* 2·8 *mm*

transient, but that the characteristics of the secondary and tertiary components which are mainly of smaller particle size than the minimum gap of the jaw crusher, and therefore not subjected to recrushing, are controlled by the structure of the material and are termed persistent.

The process of size reduction may be visualized in the following way: single lumps break down into a primary transient component and a secondary persistent component, a second stage of crushing acts preferentially on the larger particles so that the primary component is reduced in weight and modal size. The product of this second crushing has characteristics similar to that from the first crushing and hence the secondary component is increased in weight but unchanged in modal size. Continued crushing may eliminate the primary component completely, and if the material is transferred to a different machine with a more intense crushing action, the former secondary component becomes crushed and a tertiary component will develop. When the action of this second machine has progressed as far as possible, the material must be transferred to a fine grinding machine which is capable of crushing the particles produced by the preceding one. Components of still smaller modal size could be developed in this stage, but it is more probable that the particle sizes will blend and produce a mono-modal distribution curve as the final state of the powder. This sequence of stages is illustrated by the series of size distribution curves in *Figure 10*,

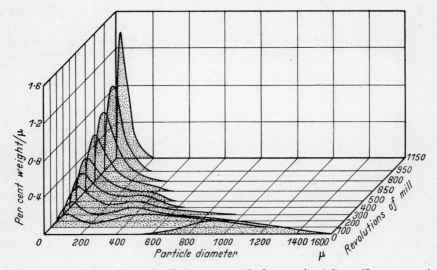

Figure 10. Variation of size distribution curves with fineness of grinding. (By courtesy of The Institution of Mechanical Engineers)

shown grouped isometrically. The curves are based on grinding tests with anthracite, the original feed being a screened fraction of mono-modal size distribution. The development by a bi-modal distribution is shown with the gradual elimination of the primary component and a reversion to the mono-modal size distribution of the fine particles.

LABORATORY GRINDABILITY TESTS USING SMALL MILLS

These tests are made to enable the output of industrial mills to be predicted for the different materials that are to be processed in practice. This is particularly important in the pulverized coal industry because of the great variation in the grinding characteristics of coals from different parts of the world, but has also application to mineral dressing processes. The essential features of a laboratory grindability test are that it should simulate the large-scale processes as nearly as possible, *i.e.* the final product of the test should be the same order of fineness as that from the industrial mill, and selective grinding effects must be eliminated.

Small ball mills have been used for grindability tests and an attempt has been made to reproduce an open circuit grinding process which is comparable with industrial practice, instead of using a simple batch grinding process. In open circuit grinding the throughput of the mill is several times the rate of new feed addition, and is classified into an oversize and undersize fraction, the former being returned to the mill with the new feed. Under conditions of equilibrium the weight of undersize product is equal to the feed rate of new material. This condition is simulated in the small mills by sieving the ground material after a certain time of grinding, returning the oversize fraction to the mill, plus a sufficient weight of feed to bring the mill charge back to the original weight. In the test devised by BALTZER and HUDSON[18] of the Canadian Fuel Research Laboratories, a 500 g sample of coal was ground for 1,000 revolutions and then screened on 100-mesh sieve; the residue on the sieve was returned to the mill with additional new feed to make up the 500 g charge, and the milling repeated for a second cycle. Screening with the addition of new feed was repeated and a third cycle of grinding performed, and the final weight of material passing the 100-mesh sieve was used as a measure of the grindability of the whole. MAXSON, CADENA and BOND[19] used an even more elaborate method of grinding in which the amount of material circulating was two-and-a-half times the rate of feed. Starting with a 700 cm^3 sample of ore, the material passing the 100 mesh was removed from the mill charge after each stage of grinding and replaced by new feed. The process of grinding and screening was repeated until the number of grammes ground to pass 100-mesh sieve per revolution of the mill became constant. As a result of very thorough investigations by various laboratories in the U.S.A., the following two methods were proposed by the A.S.T.M.[20] for grindability tests on coals, though there is no reason why these tests should not be applied to other minerals.

(1) The U.S. Bureau of Mines ball mill test in which the coal is ground to a fineness of 80 per cent through 200-mesh sieve in a number of cycles as described later.
(2) The Hardgrove mill test in which a constant amount of energy is applied to the grinding process, and the increase in surface is taken as a measure of the grindability of the coal.

United States Bureau of Mines Ball Mill Test

The ball mill is constructed from steel tube 8 in. diameter and 8 in. long, and is fitted with three lifter bars each $\frac{7}{16}$ in. square, running the full length of the mill and spaced at 120° intervals. The ball charge consists of 100 steel ball bearings, each 1 in. diameter and the mill is rotated in a horizontal position about its cylindrical axis at 40 r.p.m. The feed consists of 500 g of coal graded between 12- and 200-mesh sieves. A preliminary grinding cycle consists of 50 revolutions of the mill, after which the coal is screened on a 200-mesh sieve and the weight of undersize determined. The number of revolutions which would be required to produce 50 g, *i.e.* 10 per cent of undersize, is then calculated by direct proportion, and this number of revolutions constitutes a cycle. Both the oversize and undersize material are then returned to the mill and the additional number of revolutions given to complete the first cycle; after this the coal is again screened on 200-mesh sieve, but the oversize only is returned to the mill for a second cycle of grinding of the same number of revolutions. These cycles of grinding followed by sieving and return of the oversize to the mill are continued until more than 80 per cent of the original sample has been ground finer than 200-mesh sieve. The exact number of revolutions required to grind 80 per cent of the sample through 200 mesh is determined by plotting the results and interpolating from the curve, and the grindability is expressed as 72,000 divided by this number of revolutions.

The writer has conducted a number of experiments with such a mill and has found that if the number of revolutions per cycle is maintained constant the percentage ground to pass 200-mesh sieve decreases in successive cycles. This is evidently due to a selective grinding action, for as grinding proceeds, the harder constituents of the coal become concentrated in the oversize

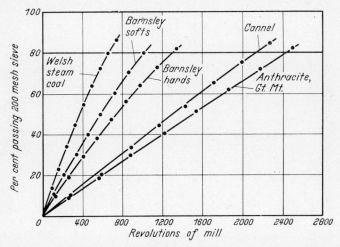

Figure 11. Ball mill tests, U.S. Bureau of Mines method. (*By courtesy of the Journal of the Institute of Fuel*)

material. Nevertheless, the process of eliminating the fine dust at each stage of grinding does ensure that eventually these harder components are subjected to the grinding process. Typical test results on some British coals are shown in *Figure 11*.

The Hardgrove Machine Method

The Hardgrove machine is not a ball mill in the normal interpretation, but consists of eight 1 in. diameter steel balls which roll on a stationary ring and are driven by a rotating ring above them; it may be compared in construction with a ball thrust-bearing, and the action is one of crushing between the balls and the rings. A load of 64 lb. is applied to the balls and the test is made by rotating the mill for 60 revolutions. The sample of coal weighs 50 g and is graded between 16- and 30-mesh sieves. The ground material, after the single grinding cycle of 60 revolutions, is sieved on a set of sieves in square root two progression from 16 to 325 mesh and the surface area is calculated from the weights retained between each pair of sieves and the mean sieve size. The original surface is deducted from the final surface and the grindability is expressed as the new surface produced divided by the new surface which would be produced by a standard coal taken as a 100 grindability.

Comparison of Grindability Tests—Comparative tests by various methods were made by YANCEY and GEER [21, 22] on a large number of coals, and the relationship between the Hardgrove grindability per cent and the U.S. Bureau of Mines ball mill test expressed as 72,000 divided by the number of revolutions for 80 per cent through 200-mesh sieve, is shown by the curve in *Figure 12*. The number 72,000 was chosen to give approximate agreement between the grindability value obtained by the two methods, but the curve shows that the relationship is not strictly linear. The line curve (b) in *Figure 12* shows the results of tests made in this country. The

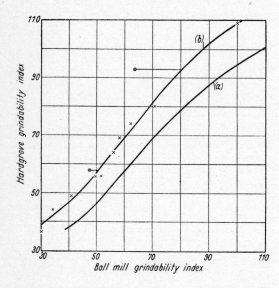

Figure 12. Grindability comparisons. Curve (a) relationship deduced by Yancey and Geer. Curve (b) experimental results by Heywood

shape of the curve is similar to that for United States coals, and the shift in position is due to a difference in the procedure for calculating the surface increase obtained by means of the Hardgrove mill. The two experimental points with circles are for coals with a very high shale content; the effect of selective grinding of the softer component is apparent, and the Hardgrove grindability figures are predominantly affected by the grindability value of the clean coal. Tests conducted by Yancey and Geer on a mixture of soft coal and anthracite showed very marked selective grinding with the Hardgrove mill but hardly any in the final stages of grinding by the ball mill.

Summarizing the characteristics of these two methods, the Hardgrove mill test is made by constant applied energy, and the ball mill test at constant fineness of product. The fineness of the product from the 8 in. ball mill is more in accordance with that of industrial pulverized fuel, and in general this would seem to be a more satisfactory method of test, especially for high ash coals. Grindability tests on minerals other than coal could be conducted with this mill, but the sample weight should be adjusted proportionally to the specific gravity, so as to equal the volume of 500 g of coal. This is necessary to ensure that the volume of the sample is always the same percentage of the volume of the ball charge.

THE EFFICIENCY OF CRUSHING AND GRINDING MACHINES

The effective output of a crushing or grinding machine may be measured in terms of the new surface produced, or per unit weight of output by the specific surface of the product in square feet per pound. The energy input to the machine may be expressed in foot-pounds per pound of output. Hence the effectiveness of the machine is shown by the ratio of energy to surface produced expressed in foot-pounds per square foot of surface, though other units may be used if preferred. The relative performance of the machine may then be assessed by comparing this figure with the foot-pounds per square foot required in some ideal process of size reduction for the same material. The problem is to devise an ideal process for the basis of comparison. Pioneer research workers in this subject, MARTIN[23], WHITE[24] and

Table 5. *Relative Efficiency of Crushing and Grinding Machines*

Author	Material used	Type of machine	Relative efficiency per cent	Basis of comparison
Gross and Zimerley	Quartz	Ball mills	45	Impact tests
White	Quartz	Crushers Battery stamps Tube mills	4·4 10·5 18·5	Impact tests
Wilson	Clinker	Tube mill	20–40	Impact test
Prentice	Quartz	Gyratory crusher Stamp mill Tube mills	33 35 23	Impact, compression and other tests
Fairs	Limestone, barytes and anhydrite	Unit impact crusher Batch ball mill Swing hammer mills (*a*) coarse (*b*) fine	20–27 10–14 20–30 4–10	Compression test
Heywood	Coal	Swing hammer mill Ball mill Ring-roll mill	20 10 30	Compression and impact tests

others, used the lattice surface energy of the crystal structure as such a basis, and estimated the relative efficiencies of ball mills to be only a small fraction of 1 per cent. Such a comparison is not now considered to be realistic, for it does not take into account all the complex features of the crushing process. A more valid basis for comparison is a free-crushing process in which all the energy applied is used to fracture the feed material, and which is therefore the best that can be attained under practical conditions of operation. Gross and Zimmerley[6], Wilson[7], Prentice[8], Fairs[9] and others have used the results of compression and impact tests as a basis for assessing the relative efficiency of industrial mills and have obtained values varying from 10–60 per cent. Although this is probably a more realistic approach, it is still not free from defects. Since the fineness of the product in the ideal process is generally much less than in the industrial one, it is not certain how much of the increased energy required by the latter is due to a change with varying size of the properties of the material. The ideal process should, therefore, have the same intensity of comminution as the industrial process, but practical difficulties have so far prevented the construction of a device fulfilling this requirement. Typical figures for relative efficiencies as determined by various workers are given in *Table 5*.

The order of relative efficiencies obtained shows that in spite of the empirical design of many crushing and grinding machines, these operate with a reasonable degree of relative efficiency. Though there is scope for further improvements it does not seem probable that any revolutionary change can be effected which will greatly reduce the energy required for fine grinding.

SUMMARY OF THE FUNDAMENTAL PRINCIPLES OF CRUSHING AND GRINDING

The preceding subsections have given an outline of the various theories of crushing and grinding, and of practical experiments that have been made to substantiate them. The derivation of a satisfactory theory that will apply to all conditions of crushing and grinding is still unsolved. A summary of the present position is given in this subsection.

Kick's law applies to perfectly elastic solids, Rittinger's law to perfectly brittle or rigid solids, and both apply to homogeneous bodies. TAPLIN[25] formulated a theory that Kick's law referred to the energy of preparation, *i.e.* the energy required to stress the material to the point of fracture, and that Rittinger's law concerned the additional energy for the separation into fragments. It was shown that the ratio of the energy of fracture to the energy of preparation varied inversely as the particle size; consequently, with finer grinding, Rittinger's law would become predominant. Statistical laws describing the method of breakdown of a solid into fragments have been developed theoretically by ANDREASEN[26], BROWN[27], BENNETT[15, 16] and LIENAU[28], in which is postulated a random system of fissures or planes of weakness within the material. There is little doubt that the process of crushing and grinding is determined by such a structure, rather than from consideration of homogeneous materials. Bennett has expressed this effect by the theorem that when work is done on a brittle material the energy appears in part as a new surface of fragmented products and in part as the creation of fresh inner weakness, and that the size distribution of the products, if there is no preferential distribution of the breakage forces, will be such that in the broken product from the complete fracture of a single lump the amount of material smaller than a given size bears a simple exponential relation to the size. This can be expressed mathematically as

$$W_x = (1 - e^{-x/x_m})$$

where W_x is the fraction of the material smaller than size x and x_m is a mean size for the system of particles.

PONCELET[29] has made a very complete theoretical and experimental investigation of the formation of cracks in a brittle material, and has formulated the following stages in the mechanism:

(1) Deformation of the solid by the application of outside forces results in tensile stresses.
(2) Development of one or more cracks as a direct result of these stresses.
(3) The formation of compressive and transverse pulses caused by these breaks, which travel through the solid. The latter pulses cause the cracks to progress mainly at the initial crack velocity, which is a constant for the material.
(4) The generation of tension and more transverse pulses by reflection of the compression pulses at the boundaries of the solid, causing offspring cracks to form and progress preferentially in the smaller fragments liberated by the parent cracks. Some of the larger fragments liberated may fracture no further, and remain as residual particles with internal cracks.
(5) The formation of residual particles of smaller and smaller sizes as the process is repeated, until the whole solid is reduced to a collection of residual particles.

A number of size distribution laws have been proposed, usually of an exponential form, but none of these has been proved to apply to the whole range of sizes of a fragmented product, but only within a limited range of sizes.

Examination of the shape of the size distribution curves for particles produced by various methods of comminution indicates that size reduction occurs by the formation of components. The primary and secondary components due to coarse crushing are widely separated; as comminution proceeds the primary component is reduced both in magnitude and particle size, and eventually merges into the secondary component. Further size reduction in another machine having a higher intensity of action will cause repetition of this process.

Practical tests to measure the grindability of coal and other minerals enable the output of industrial mills to be predicted. The procedure for coals has been standardized, but there is, at the moment, no standard test for other minerals. In this connection reference should be made to the recent work of BOND[30, 31], in which a third theory of comminution has been developed. In this theory it is stated that if W represents the work in kilowatt-hour per ton required to reduce from a feed with 80 per cent passing a diameter of $F\ \mu$, to a product with 80 per cent passing $P\ \mu$, then

$$W_i = W\left(\frac{\sqrt{F}}{\sqrt{F}-\sqrt{P}}\right)\sqrt{\frac{P}{100}}$$

where W_i is the 'Work Index', or kilowatt-hour per ton required to reduce from a theoretically infinite particle size to 80 per cent passing 100 μ.

Although the efficiency of industrial mills based on the theoretical surface energy of the lattice structure may be extremely small, the practical relative efficiency based on crushing or impact tests for the same material may be of the order of 30 per cent in many cases. One of the difficulties in assessing the relative efficiency of mills is the inconsistency of laboratory determinations

Table 6. Surface Energies of Quartz, ergs/cm²

Author	Basis	Ergs/cm²
Edser	Theoretical	920
Gross & Zimmerley	Impact tests	56,000
Piret	Impact tests	174,000 (permeability)
		89,000 (adsorption)
Schellinger	Ball mill	107,000

of the surface energy of the material. SCHELLINGER[32] has conducted experiments with a ball mill immersed in a water-jacket to act as a calorimeter, and quotes the values for the surface energy of quartz given in *Table 6*.

The literature on theoretical and experimental investigations of the mechanism of fracture covers a vast extent of research work; future research should be concentrated on the application of this knowledge to improvements in the efficiency of industrial machines, and a reduction in the energy used for size reduction processes.

REFERENCES

[1] OWEN, J. S., *Trans. Instn Min. Metall., Lond.*, 42 (1933) 407
[2] OWEN, J. S., *Trans. Instn Min. Metall., Lond.*, 44 (1935) 421
[3] HEYWOOD, H., *J. Inst. Fuel*, 9 (1935) 94
[4] AXELSON, J. W., and PIRET, E. L., *Industr. Engng Chem.*, 42 (1950) 665
[5] CAREY, W. F., and STAIRMAND, C. J., *Instn Min. Metall., Lond.*, Symposium on Mineral Dressing (1952) 117
[6] GROSS, J., and ZIMMERLEY, S. R., *Trans Amer. Inst. min. (metall.) Engrs*, 87 (1930) 7, 27, 35
[7] WILSON, R., *Amer. Inst. min. (metall.) Engrs, Tech. pap.* No. 810 (1937)
[8] PRENTICE, T. K., *Trans. Inst. Min. Metall., Lond.*, 55 (1946) 427
[9] FAIRS, G. L., *Trans. Inst. Min. Metall., Lond.*, 63 (1954) 211
[10] BOND, F. C., *Trans. Amer. Inst. min. (metall.) Engrs*, 169 (1946) 58
[11] GAUDIN, A. M., and HUKKI, R. T., *Trans. Amer. Inst. min. (metall.) Engrs*, 169 (1946) 67
[12] GAUDIN, A. M., and YAVASCA, S. SUPHI, *Trans. Amer. Inst. min. (metall.) Engrs*, 169 (1946) 88
[13] SCHUHMANN, R., *Trans. Amer. Inst. min. (metall.) Engrs, Tech. pap.* No. 1189 (1940)
[14] ROSIN, P., and RAMMLER, E., *J. Inst. Fuel*, 7 (1933) 29
[15] BENNETT, J. G., *J. Inst. Fuel*, 10 (1936) 22
[16] BENNETT, J. G., BROWN, R. L., and CRONE, H. G., *J. Inst. Fuel*, 14 (1941) 111

[17] HEYWOOD, H., *Proc. Instn mech. Engrs, Lond.*, 125 (1933) 383
[18] BALTZER, C. E., and HUDSON, H. P., *Rep. Dep. Min., Can.*, No. 737–1 (1933)
[19] MAXSON, W. L., CADENA, F., and BOND, F. C., *Trans. Amer. Inst. min. (metall.) Engrs*, 112 (1934) 130
[20] *A.S.T.M. Stand.*, D408–37T ; D409–51
[21] YANCEY, H. F., and GEER, M. R., *Trans. Amer. Inst. min. (metall.) Engrs*, 119 (1936) 353
[22] BLACK, C. G., *Trans. Amer. Inst. min. (metall.) Engrs*, 119 (1936) 330
[23] MARTIN, G., *Trans. ceram. Soc.*, 25 (1926) 63; 26 (1927) 34
[24] WHITE, H. A., *J. chem. Soc. S. Afr.*, 30 (1929) 1
[25] TAPLIN, T. J., *Min. Mag.*, 50 (1934) 18, 87
[26] ANDREASEN, A. H. M., *IngenVidensk. Skr.* (Denmark) No. 3 (1939)
[27] BROWN, R. L., *Research, Lond.*, 1 (1947) 93
[28] LIENAU, C. C., *J. Franklin Inst.*, 221 (1936) 485, 673, 769
[29] PONCELET, E. F., *Trans. Amer. Inst. min. (metall.) Engrs*, 169 (1946) 37
[30] BOND, F. C., *Trans. Amer. Inst. min. (metall.) Engrs*, 193 (1952) 484
[31] BOND, F. C., *Instn Min. Metall., Lond.*, Symposium on Mineral Dressing (1952) 101
[32] SCHELLINGER, A. K., *Trans. Amer. Inst. min. (metall.) Engrs*, 193 (1952) 369
[33] HEYWOOD, H., *J. Inst. Fuel*, 'Conference on Pulverized Fuel' (1947) 594
[34] *Reps Fuel Res. Bd, Lond.*, (1936) 49; (1937) 42 ; (1938) 71

2

METHODS OF SIZING ANALYSIS

HAROLD HEYWOOD

SIZING analyses are processes for determining the relative proportions of powdered materials that are within certain defined particle size ranges. Such analyses are necessary for studying the characteristics of crushing and grinding machines and are frequently required for the routine control of the quality of the product. The purpose of this section is to discuss the fundamental properties of particles that affect sizing, and to describe those methods of analysis which have been found by practical experience to be the most effective.

DEFINITIONS

A particle is a discrete piece of material. As only material particles are considered, the lower limit is a molecule, but there is no real maximum limit to the size. Particulate material consists of a large number of individual particles, and a particulate system refers to the particulate material and the ambient fluid, either gaseous or liquid. The essential characteristic of a particulate system is that the particle size shall be small in relation to the volume occupied by the system. A particulate system may be static or in motion, it may be closely packed so that the particles are in contact though not joined, or the particles may be dispersed and widely separated. Although a particle is generally a solid, many of the laws relating to such particles also apply to drops of liquid dispersed in a gas, or to small bubbles of gas dispersed in a liquid. The above definitions are fundamental, but there are also conventional definitions which apply to special types of particulate material; *e.g.* a powder generally consists of particles which have been produced by artificial comminution and are below 1 mm in size. Dust implies natural occurrence, or that the particles are an unintentional product which is a nuisance or danger to health; it also implies that the particles are, or have been, airborne. Grit consists of relatively large particles, hard and angular in shape, which are separately visible to the unaided eye. Definite size limits cannot be set to any of these products and various interpretations of the terms apply to different industries.

PRINCIPLES OF SIZING ANALYSIS

A powder is normally composed of particles extending over a range of sizes. The essential process in a sizing analysis is to split the complete range into smaller subdivisions and to determine either the number or the weight of particles within the size limits of each subdivision. This subdivision into groups may be real or virtual; in the process of sieving the grouping is real, since the particles are separated by a series of screens with different mesh apertures. The separation is virtual in size analysis by microscopical measurement, as the analysis is conducted by counting the numbers of particles between certain size limits as observed in a sample mounted on a glass slide. Sizing analyses may also be effected by utilizing the motion of particles in a fluid, either by normal gravitational settlement or under the action of centrifugal force. Elutriation consists essentially of separation by an upward flowing column of fluid in a vertical tube; according to the velocity and physical properties of the fluid a certain size of particle will be carried to the top of the tube and away from the apparatus, while larger particles will remain in the lower portion of the tube; this is a case of real separation. In sedimentation methods the particles are uniformly dispersed in a liquid and allowed to settle, the relative weights between various size limits are determined from the rate of settlement, and in the normal methods by which this analysis is conducted the separation is virtual. Practical methods of sizing analysis thus comprise sieving, microscopical measurement (including the use of the electron microscope), and settlement or relative motion in a fluid. The characteristics and range of application of these processes are shown by the following summary.

Geometrical Similarity

Controlling factors are dimensions and shape of particles. Independent of particle density.

Sieves. Normal range B.S. 5 mesh to B.S. 350 mesh. (3353 to 44 μ). Coarser sieves or punched plates used for larger particles.

Optical Microscope. Normal range 100 μ to theoretical limit of resolution of 0·2 μ. Measurements below 1 μ are subject to diffraction errors.

Electron Microscope. 0·01 μ is normal limit of resolution.

Hydrodynamical Similarity

Controlling factors are dimensions, shape and density of particles, viscosity and density of liquid.

Elutriation. Normal range 100 to 10 μ for lighter minerals in water, with lower limit of 5 μ for heavy minerals if water velocity and temperature are carefully controlled.

Sedimentation. Normal range 50 to 2 μ for sedimentation in water, with lower limit of 1 μ if temperature maintained constant. Lower limit may be reduced to 0·1 μ by centrifugal sedimentation. Particles larger than 50 μ may be sedimented in viscous liquids.

Aerodynamical Similarity

Controlling factors are dimensions, shape and density of particles, viscosity of gas. Density of gas usually negligible compared with density of particle.

Elutriation. 100 to 10 μ for normal operation with air, lower limit of 5 μ for heavy minerals if air velocity and temperature are carefully controlled.

Centrifugal air elutriation may be used over a range of 100 to 5 μ.

Sedimentation. A size range of 100 to 3 μ may be measured by this method.

Descriptions of Sizing Analysis Methods

Sieving—The process of sieving is the most convenient method of sizing analysis for particles larger than the apertures of a sieve with 200 meshes per inch. The process may be extended to a lower limit of 350 meshes per inch (British Standard series), or 400 meshes per inch (Tyler series), but the unavoidable errors of weaving for such fine gauzes are relatively great. The accuracy of sieve analyses has been improved by the introduction of various standard dimensions for test sieve gauzes, such as by the Institution of Mining and Metallurgy (I.M.M.) and British Standards Institution (B.S.I.) in this country, and similar standards in the United States and on the Continent. British and American sieves are designated by the number of meshes per lineal inch, but continental sieves are usually designated by the number of apertures per square centimetre. All such test sieves are woven from metallic wire and the standards specify the nominal width of aperture, *i.e.* the average, and give tolerances for the size and number of oversize apertures. Complete details of British Standard and other sieve series are given in B.S. 410:1943. (Test Sieves), amended 1955.

Sieving is a statistical process and there is always an element of chance as to whether a particle of 'near mesh' dimensions will or will not pass through the sieve; thus the 'end point' of a sieving analysis is indefinite and must be fixed arbitrarily, either by a standard time for sieving, or by continuing the process until the weight of the material passing the sieve per minute is less than a fixed percentage; *e.g.* 0·1 per cent per minute. Comparative sieving tests have shown that by the adoption of a standard method of procedure the errors between analyses made in different laboratories may be reduced to one-sixth of those occurring without such control[1]. Whilst it is impossible to specify a single method of sieve analysis to cover all the different types of material used in industry, there are general principles which should be observed, and recommended methods of sieving have been set out in B.S. 1796:1952 (Methods for Use of B.S. Fine Mesh Test Sieves). It is always an advantage to eliminate the fine dust as the first stage of procedure. If the material is to be sieved dry there should be a preliminary sieving period of 5 min on the finest sieve that is to be used, after which the residue may be placed on the coarsest of a nest of sieves and shaking and tapping continued for a standard time or until the specified rate of passing has been attained. Some materials, *e.g.* those containing clay particles, are best sieved in the wet state; in these cases the procedure should be to wash the material first on the finest sieve. The dried residue from this sieve is then placed on the nest of sieves and further separation effected as for dry sieving.

Sieves should be vibrated or lightly tapped during the process of sieving and also given a shaking motion by hand. It is essential to remove the sieves from the nest, invert each sieve individually over a sheet of paper, clean with a soft brush and replace the residues on the appropriate sieves at intervals not greater than 5 min. If this is not done, many of the apertures become blocked by particles and are thereby rendered inoperative. Hand labour is saved by the use of sieving machines which combine the required shaking and vibration, but even when using these it is still necessary to remove and brush the sieves at suitable intervals.

Particles larger than 5 mesh B.S. are graded on coarser gauze sieves, or on plates with punched square holes. Screens with circular holes are often used for industrial grading of materials; the ratio of equivalent sizes of round hole and square aperture, *i.e.* such that the same particle will just pass both apertures, is 1·15 to 1·2 for mineral particles, the higher value applying to those of flatter shape. The method of reporting the sieve analysis and a comparison of particle size as measured by sieving and by other processes are discussed in a later part of this section.

Microscopical Measurement—This method of size analysis is used for small samples of dust extracted from industrial atmospheres, and which are generally insufficient in weight for gravimetric methods; it is also used for the calibration of other methods of size analysis. The sample of dust may be collected directly on a glass cover slip and examined in the dry state, or the dust particles may be dispersed in a suitable mounting fluid such as glycerine jelly or Canada Balsam. In either case effective dispersion of the particles is essential to break up aggregates.

The procedure is to count the number of particles within suitable size limits selected to cover the complete size range of the particles in the sample. The mean projected diameter is measured, and this is defined as the diameter of a circle having an area equal to that of the particle when viewed in a direction perpendicular to the plane of maximum stability. This dimension is an average of the length and breadth of the particle and takes no account of the thickness, which may often be appreciably less than the other two dimensions, but it is possible to make a correction for this factor and the method is explained in a later section. It is not feasible to measure the projected area of the particles individually when hundreds or even thousands of particles are to be examined, and this dimension is determined in effect by statistical methods of measurement. Thus, MARTIN[2] measured the intercept dimension bisecting the area of the particle in a random direction, and FERET[3] measured the distance between two parallel lines tangential to the profile of the particle; these dimensions are illustrated in *Figure 1*. The procedure that is now almost universally adopted is to compare the image of the particle with dimensioned circles or discs on a graticule inserted in the eye-piece of the microscope. The circles should have a constant size ratio, *e.g.* $\sqrt{2}$, and *Figure 2* shows the graticule designed by FAIRS[4]. The microscope is adjusted to produce a known magnification and the particles are assigned to the various size groups and counted as they are examined. The results of such an analysis express the frequency of occurrence of the various size particles on a number basis, but for many purposes it may be necessary to convert this count to a distribution on a weight basis. The difficulty in this conversion arises from the fact that the volume of a particle varies as the cube of the size, assuming a constant shape, and thus the weight of one relatively large particle may equal the combined weight of several thousand of the smaller particles present, *e.g.* if the size range is 100 to 1, the weight range will be 1,000,000 to 1. In order, therefore, to attain statistical accuracy without counting many thousands of particles, a procedure has been devised by Fairs in which the complete size range of the particles is subdivided into, say, three smaller ranges. Separate counts are made at different magnifications on each of the size ranges and the three counts subsequently combined numerically to give the overall distribution. The original paper by Fairs[4] gives full details of the procedure and explains how the statistical accuracy of the count should be controlled.

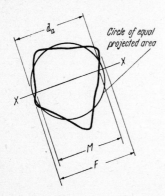

Figure 1. Two-dimensional mean diameters
d_a = mean projected diameter
M = Martin's statistical diameter
F = Feret's statistical diameter

An experiment by the writer showed that experienced operators could obtain measurements within 2 per cent of the correct mean projected diameter by comparison with the circles on the graticule[5]. Martin's statistical diameter measurement was 3 per cent low, and Feret's statistical diameter measurement was 12 per cent high, the reasons for these variations have been described in the above paper.

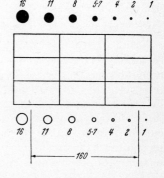

Figure 2. Microscope graticule designed by Fairs[4]. (By courtesy of the Royal Microscopical Society)

Although the theoretical limit of resolution for a particle is about 0·2 of a micron with monochromatic illumination, such small particles can only be measured by experienced operators using high quality equipment which is in perfect adjustment. Even then there is likely to be considerable error due to the diffraction ring which surrounds the outline of the particle and it is now generally considered that visual measurement should be restricted to a lower limit of 1 μ particle size.

Recent developments in automatic particle counters which incorporate electronic scanning devices, enable particles to be sized and counted at rates which may eventually attain several thousand a second. The introduction of such counters not only saves much time and tedious labour, but also eliminates some of the inaccuracies of the present microscope method due to the relatively small number of particles that can be counted visually. A description of automatic particle counters is included in the Proceedings of a conference by the Institute of Physics[6].

Electron Microscope—The electron microscope enables particles of one-hundredth of a micron diameter to be resolved distinctly. In this instrument the light rays of the normal microscope are replaced by an electron beam of very short effective wave length, which is focused electronically after passing through the dust sample. The electron beam is not directly visible, but may be photographically recorded or shown on a fluorescent screen. This extension of the lower limit of visibility has proved invaluable in the study of fine particle structure and many particles which appear to be single under the optical microscope have been found to consist of aggregates of smaller particles. The third dimension and the shape of small particles may be studied by means of a shadowing process, in which a metallic deposit is evaporated in a vacuum on to the particles at a known angle of incidence. The relative dimensions of the particle and shadow indicate the shape and thickness of the particle. Researches by WATSON[7] have shown that particles of coal dust of electron microscope size are very similar in proportions to the larger ones examined under the optical microscope.

Although the electron microscope can contribute much to a knowledge of the shape and structure of very minute particles, the same difficulties that apply to the optical microscope as regards sizing apply to an even greater extent with the electron microscope, and it is not easy to determine the size distribution for such particles unless the size range is restricted, as for example in certain carbon blacks.

Elutriation and Sedimentation—Elutriation is a process for grading particles by means of an upward current of fluid, usually water or air; sedimentation grades the particles according to the velocity of fall in a column of fluid at rest. The two processes are therefore similar in that they both operate by the relative motion between particles and fluid, although there may be a slight difference in practice due to the turbulent motion of an upward current of fluid at high velocities, and to the non-uniform velocity across the section of a rising column of fluid. Elutriation is useful when it is required to subdivide a powdered material into a number of closely sized fractions.

The size of the particles in both these processes is calculated from the Stokes's equation for the terminal or free falling velocity of a particle in a fluid, namely:

$$v_f = \frac{d_f^2}{18} \cdot \frac{(\sigma - \varrho)}{\eta} \cdot g$$

where v_f is the free falling velocity in centimetres per second, d_f is the particle diameter in centimetres, σ is the density of the particle in grammes per cubic centimetre, ϱ is the density of the fluid in grammes per cubic centimetre, η is the absolute viscosity of the fluid in poises, and g is the gravitational acceleration, 981 cm/sec^2.

This equation may be rearranged conveniently for size analysis calculations in the form

$$d_f = 175 \sqrt{\left(\frac{\eta}{(\sigma - \varrho)} v'\right)}$$

where v' is the terminal velocity in centimetres per minute and d_f is in microns.

The theory for Stokes's equation assumes that the particles are spherical and fall in a continuous fluid without interference due to neighbouring particles or the walls of the vessel, and also the conditions are such that the fluid flow round the particles is laminar or streamline. Although mineral particles are usually flat in shape and therefore fall with a smaller velocity than spherical particles of the same volume, Stokes's equation may be applied to the sizing analysis since it determines the diameter of the free falling sphere which is equivalent to the actual particle. Particle interference has a negligible effect if the total volume of suspended particles does not exceed 1 per cent of the volume of the suspension, provided that the particles are effectively dispersed. Particles consisting of siliceous materials are well dispersed in distilled water to which 0·1 per cent by weight of sodium hexametaphosphate has been added, but some materials are more difficult to disperse and the concentration of the suspension may have to be reduced to less than the 1 per cent quoted above.

At a certain velocity of fall, corresponding to a Reynolds number of 0·2, the fluid motion round the particle departs from the streamline form, and Stokes's equation no longer applies accurately. It is possible to make an appropriate correction for such conditions and the figures in *Table 1* show the correct free falling diameters that correspond to diameters calculated from Stokes's equation. The method of using this table is as follows: the velocity of fall of the particle being known, calculate the diameter by Stokes's equation and use the tabulated values to determine the correct size of sphere (for the specific gravity applicable) which has this falling velocity.

Table 1. *Deviations from Stokes's Equation for Particles Falling in Water at 60° F*

Particle diameter calculated from Stokes's equation (microns)	Correct diameter of sphere having same free falling velocity for various specific gravities (microns)				
	1·3	2·65	3·85	8·9	19·3
30	30·0	30·0	30·1	30·2	30·4
40	40·0	40·1	40·3	40·6	41·4
50	50·1	50·3	50·7	51·5	53·3
60	60·1	60·8	61·1	63·1	66·4
70	70·3	71·1	72·3	75·3	80·7
80	80·4	82·2	83·8	88·7	97·3
100	101·2	105·0	108·1	118·6	136
120	122	130	135	155	181
140	144	157	166	196	237
160	167	187	201	243	303

The various designs of practical elutriators are described briefly below.

Elutriation by Means of Water—The apparatus used for this purpose consists essentially of a series of cylindrical glass vessels, the diameter of the vessels increasing in order of sequence; since the volumetric rate of water flow is constant, the upward velocity in succeeding vessels decreases. The sample is placed in the smallest vessel, and all particles smaller than a certain size are washed out of this vessel and into the next one; further separation takes place in the second vessel, and the finest particles are washed into the third vessel. The very finest particles which are washed

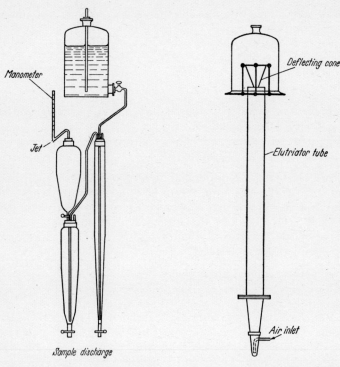

Figure 3. Water Elutriator. (After A. H. M. ANDREASEN[8], by courtesy of The Institution of Mechanical Engineers)

Figure 4. Air Elutriator. (After H. W. GONELL[12], by courtesy of The Institution of Mechanical Engineers)

right through the apparatus may be collected by filtering or they may be allowed to settle. As many as twenty vessels have been placed in series, but usually the number is restricted to three, and a design of this type of apparatus due to ANDREASEN[8] is shown in *Figure 3*. A very convenient design of elutriator is the Andrew's Kinetic Elutriator, which incorporates a circulating system in the larger vessel in order to reduce the time of operation. The essential features of this apparatus have been described by the inventor[9]. Blyth has designed an elutriator which has been found very satisfactory for the preparation of closely graded fractions of minerals[10].

The process of elutriation is very suitable for the grading of comparatively heavy minerals, such as sand and metallic ores, particularly over a size range of 100 to 10μ. The grading of coal dust is not quite so satisfactory on account of the small difference between the densities of coal and water, and any variation in the density of the coal sample has a marked effect on the results of the analysis. Elutriation by liquids other than water can only be accomplished with difficulty, although apparatus has been devised for the elutriation of Portland cement by means of a circulating flow of kerosene.

Elutriation by Means of Air Flow—This is obviously desirable when the powder to be tested is in practice subjected to the grading action of an air flow. Especially is this the case for the grading of flue dust consisting of particles covering a wide range of density, which would result in considerable difference between the results of elutriation by air and by water. The major difficulties involved in air elutriation are to ensure the breaking up of aggregates of particles and to prevent fine particles adhering to the sides of the elutriator tube.

A comprehensive survey of the development of the air elutriator has been made by PEARSON and SLIGH[11]. Briefly summarizing this, it may be stated that there are three main types, namely, the down-blast type, the up-blast type, and the circulating type.

The down-blast type is illustrated in *Figure 4* and was originally developed by GONELL[12]. In this elutriator, the air jet blows downwards into a thimble containing the powdered material. This type of elutriator has been adopted in B.S. 893:1940 for the size analysis of flue dust, but has been modified to the up-blast type. Full description and dimensions are given in the above standard.

The circulating type of elutriator has been developed by ROLLER[13], and is a modification of the up-blast type, but the sample of powder is caused to circulate in a U-tube at the base of the elutriator tube. The method of operating all these types of elutriator is to blow with a known velocity of air current until equilibrium has been attained or until the rate of loss has been reduced to an arbitrary value, and then to weigh the residue of coarse particles that are retained in the apparatus.

The elutriator designed by HAULTAIN[14] is used very largely in the mining industry. This consists of six cylindrical or conical tubes in series, each increasing in diameter in the direction of air flow. Automatic control is provided to regulate the air flow and the powder which is initially placed in the first tube is thereafter divided into a number of fractions each of which has a restricted size range.

A form of apparatus has been devised by which the particles are separated during settlement across a horizontally flowing stream of air[15]. At present this procedure is restricted to particles larger than 20 μ, but it is proving a useful method for the analysis of flue dust. The BAHCO centrifugal dust classifier is a machine in which the particles are separated in radial channels within a rotating disc.

METHODS OF SIZING ANALYSIS

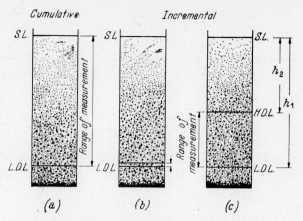

Figure 5. Principles of Sedimentation methods. (*By courtesy of* The Institution of Mining and Metallurgy)

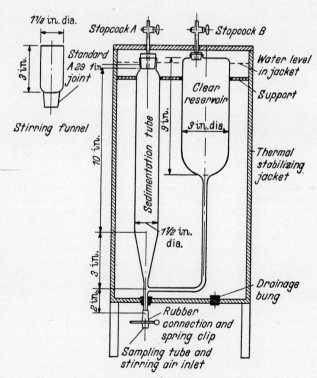

Figure 6. Sedimentation column. (*From* C. J. STAIRMAND[18], *by courtesy of* The Institution of Chemical Engineers)

Sedimentation Analyses—Sedimentation is the process most frequently used for the sizing analysis of particles smaller than the sieving range. In almost all forms of this method of test, initially the particles are uniformly distributed in a column of fluid. After a certain period of settlement there will be a density gradient or variation of concentration over the height of the column. The procedure for the analysis is usually to determine the density of the suspension either over the whole depth of the column or at a certain fixed level which can be called the lower datum level. The former method is termed cumulative and the latter, incremental. These procedures are illustrated in *Figure 5*, which shows the range of measurement for the respective methods[16]. In the ideal application of the incremental method the density or concentration would be measured at the lower datum level (L.D.L.), but in practice the range of measurement must extend over a finite height. In many forms of apparatus this range is very small compared with the height from the lower datum level to the surface level, but in the hydrometer method the range of measurement, h_1–h_2, is quite an appreciable proportion of the total height of the suspension column h_1, so that this particular case, although classed as an incremental method, departs from the theoretical basis.

Cumulative methods, which include the sedimentation balance method, are easy to apply, but have the disadvantage that the curve relating concentration and time of settlement, or in the case of the sedimentation balance, the curve relating weight settled and time of settlement, must be differentiated in order to obtain the size analysis curve relating percentage weight and particle size. Incremental methods have the advantage that the weight-size distribution is given directly by the experimental results, but suffer from a disadvantage in that very small differences in density or in concentration must be measured.

Sedimentation Equipment—Many ingenious forms of equipment have been devised for this form of analysis, and it is only possible to describe here those methods which are in common use. There are two forms of the cumulative method, one being the sedimentation balance, a new form of which has recently been developed by BOSTOCK[17]. This measures, by means of an automatic balance, the weight of particles that have settled from the suspension during various periods of time. The other, shown in *Figure 6*, is a form of sedimentation column designed by STAIRMAND[18], in which the settled material is collected in a small tube which is removed from the column and the particles extracted by centrifuging. The most commonly used of the incremental methods is the pipette method, in which a small sample of the suspension is extracted from the sedimentation column after various time intervals of settlement. The liquid is evaporated and the weight of suspended particles determined. The apparatus designed by Andreasen[8] is very convenient for this method of analysis and *Figure 7* shows the author's arrangement of the pipette in a casing which very greatly reduces temperature changes. A known weight of the powder is shaken with the dispersing fluid, which may be distilled water with the addition of a suitable dispersing agent, alcohol or any other liquid that can be evaporated conveniently. The samples are withdrawn normally at a depth of 20 cm below the surface, through a two-way cock which is then turned to discharge the sample into the evaporating dish. The liquid is evaporated and the weight of particles plus any dispersing agent determined. The maximum size of particle present in each sample is calculated by Stokes's equation and hence the weight–size distribution is obtained.

In the hydrometer method the density of the suspension is measured directly by the depth of immersion of the hydrometer stem. The disadvantage has already been mentioned, namely, that the length of the

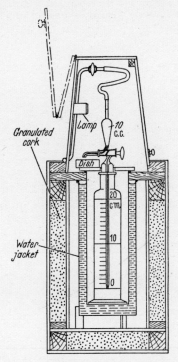

Figure 7. Sedimentation pipette. (After A. H. M. ANDREASEN[8], by courtesy of The Institution of Mechanical Engineers)

hydrometer bulb is an appreciable proportion of the total depth of immersion, but it may be assumed with sufficient accuracy for many purposes that the reading of the instrument refers to the density at the centre of buoyancy of the hydrometer bulb. Relative weights of the particles in suspension are proportional to the densities of the suspension minus the density of the clear liquid, and these may be expressed as a percentage of the initial weight of particles present. The procedure is simple and rapid in operation and is frequently used for the size analysis of soil samples; a complete description of the method of use is given in B.S. 1377:1948, Methods of Test for Soil Classification and Compaction.

Efforts have been made to design sedimentation equipment which would obviate the withdrawal of samples and the somewhat tedious system of weighing. One such method is to pass a narrow beam of light through a suspension of the particles contained in a parallel-sided transparent vessel, the beam being horizontal and at a known depth below the surface of the suspension. During the process of settlement the intensity of the transmitted light increases and may conveniently be measured by means of a photoelectric cell.

If it is assumed that the shadow area of the particles is equal to that of the particle, then the optical density of the suspension, which is the logarithm of the ratio of intensities of the incident and transmitted light, is proportional to the surface area of the particles in the light beam. The relationship is expressed by the equation:

$$S = \frac{9212}{cL} \cdot \log_{10} \frac{I_c}{I_s}$$

where S is the specific surface of the powder in square centimetres per gramme, I_c is the light intensity through clear liquid, I_s the light intensity through the suspension, c the concentration of powder in grammes per litre of suspension, and L the length of the light path through the suspension in centimetres. The optical properties of the suspension of fine particles are, however, very complex, and scattering and diffraction affect the relationship between optical density and particle surface in an irregular manner. Hence the above equation does not apply to particles less than about 20 μ in size. Nevertheless this type of equipment is rapid to use and the recording can be performed automatically, so that in spite of the theoretical difficulties it may be used for the routine control of manufactured powder products where an absolute measurement of particle size is not essential.

Further details of the theory and the mathematics of sedimentation, and of comparative tests on various types of sedimentation equipment have been described in other papers by the writer[16, 19, 20].

Centrifugal Sedimentation in Liquids—Centrifugal separation avoids the long time required for gravitational settlement of particles below 5 μ in size, and also reduces the tendency for agglomeration to occur during settlement and the effect of Brownian movement. Research investigations have been made using parallel-sided and sector-shaped centrifugal sedimentation cells, but a form of centrifuge recently developed by DONOGHUE and BOSTOCK[21], and shown in *Figure 8*, enables the percentage weights corresponding to six different particle sizes to be determined simultaneously. This reference also quotes the researches of other workers and gives an outline of the mathematics of the settlement process.

The procedure is briefly as follows: The dilute suspension of particles is introduced into the stepped cone which rotates at a constant speed, normally 480 r.p.m. After a certain period of centrifuging, during which the settled particles adhere to the vertical sides of the steps, the remaining liquid suspension is removed by pumping. Thin metallic strips are attached over half the circumference of the steps so that the deposit may be more easily removed for weighing. This is a cumulative method of analysis as described in the previous section on sedimentation, and the sizing analysis is derived by differentiating the curve relating p, the total percentage deposited, plotted against the radius of the step, R. Thus the percentage by weight above a certain particle size is given by the equation:

$$W_x^\infty = p + R \frac{(R^2 - S^2)}{2S^2} \cdot \frac{dp}{dR}$$

S being the radius of the inner surface of the rotating centrifuge. The size of particle, x, corresponding to the above weight, and centrifuging time t in seconds is given by the relationship $\ln (R/S) = \dfrac{(\sigma - \varrho)\omega^2 t}{18\eta} \cdot x^2$
where ω is the rate of rotation in radians per second.

This equipment has been used to determine the sizing analysis of fine powders to a lower limit of considerably less than 1 μ diameter.

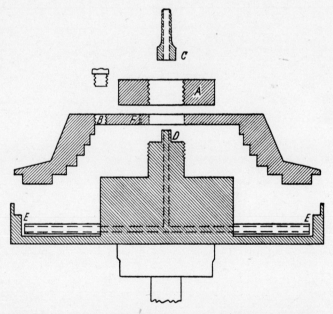

Figure 8. Cross section of centrifuge rotor. (From J. K. Donoghue and W. Bostock[21], *by courtesy of* The Institution of Chemical Engineers)

EQUIVALENT DIMENSIONS OF IRREGULARLY SHAPED PARTICLES

The previous subsections have described the various methods by which the sizes of irregularly shaped particles may be measured, *i.e.* by sieving, by microscopical measurement and by motion in a fluid. Each of these processes employs a different principle of measurement and consequently would give a different equivalent size for the same particle. As it is frequently necessary in practice to use two of these methods of analysis to cover the whole range of particle size in a powdered material, further consideration must be given to the relative values of equivalent diameters as determined by these methods of analysis.

If a particle is assumed to be resting on a horizontal plane in the most stable position as shown in *Figure 9*, the limiting dimensions, in increasing magnitude are:

(*i*) the thickness T measured perpendicular to the plane of greatest stability;
(*ii*) the breadth B which is the minimum dimension across the profile in a direction parallel to the above plane;
(*iii*) the length L which is measured perpendicular to the breadth and in the same plane. (*Note*: This may be less than L', the maximum dimension of the particle.)

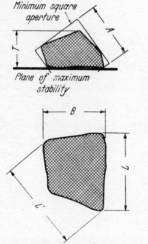

Figure 9. Limiting dimensions of a particle

The proportions of the particle may be defined by the following ratios:

Flatness, $m = B/T$

Elongation, $n = L/B$

The actual form of the particle is described by comparison with geometrical solids, such as a cube, a tetrahedron, or a sphere. The term shape as commonly applied to an irregular particle is, therefore, a combination of these two separate factors, namely, the geometrical form of the particle and the proportions as defined above. These factors enable the shape of particles to be defined numerically, but it is impossible when examining thousands of particles to measure them individually, and it is therefore usual to define the size of a particle in terms of a circle or of a sphere, which has some equivalent property. Thus if the particles are measured through the microscope, the size refers to a circle of equivalent projected area, and the diameter of this circle, termed the mean projected diameter, has already been defined in the section on microscopical measurement. This diameter is a two-dimensional mean, since it represents the average of the two larger limiting dimensions, *i.e.* B and L, assuming that the particle has settled on the microscope slide in the most stable position (see *Figure 1*). The minimum limiting dimension, *i.e.* the thickness T, is not measured by the microscopical method.

Three-dimensional mean diameters represent the size characteristic of the particles more completely, and such equivalents are the diameter of a sphere which has the same volume as the particle, or has the same terminal

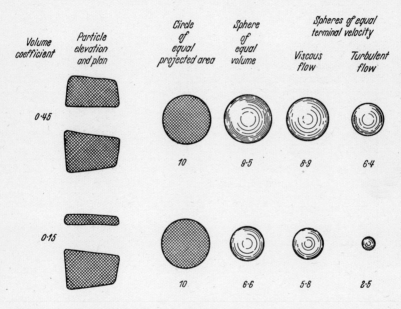

Figure 10. Equivalent diameters of irregular particles. (*By courtesy of* The Institution of Mining and Metallurgy)

velocity in a fluid. The size of square sieve aperture through which the particle will just pass is also an equivalent diameter. These various equivalent diameters can only be identical for spherical particles, and the greater the irregularity, particularly as regards flatness or flakiness, the greater will be the divergence in the equivalent values. Hence a clear definition of the method of measuring particle size is essential when reporting on sizing analyses, or comparing analyses made by the different methods.

These various mean diameters, some of which are shown graphically in *Figure 10*, are summarized overleaf.

Diameter of circle having equivalent projected diameter, or mean projected diameter, d_a
Diameter of sphere of equivalent volume, δ
Diameter of sphere of equivalent surface, Δ
Equivalent sieve aperture, A
Diameter of sphere having equal free falling or terminal velocity in a fluid, d_f

RELATIONSHIP BETWEEN EQUIVALENT DIAMETERS

The mean projected diameter d_a is selected as a basis, as it is the only diameter that can be measured directly for all sizes of particles, ranging

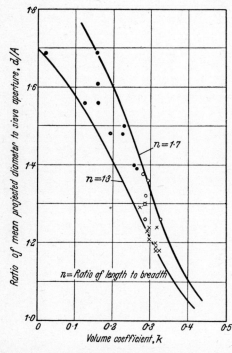

Figure 11. Relationship between mean projected diameter of particle and sieve aperture as affected by particle shape. (*By courtesy of* The Institution of Mechanical Engineers)

● 10 mesh mineral particles
○ 1 in. to 2 in. coal
× ½ in. stone chippings

from those visible by means of the electron microscope, to those of a size which can be handled individually. The following shape coefficients, namely, the volume coefficient k and the surface coefficient f, are defined as:

$$\text{Volume of particle} = kd_a^3$$
$$\text{Surface of particle} = fd_a^2$$

Many thousands of measurements on individual particles have shown that the volume coefficient k is statistically constant for various types of mineral particle. This does not imply that all particles of a given mineral have the same shape, for large variations occur, but if many thousands of particles are considered there is a statistically constant mean shape. The volumes and the mean projected diameters may be measured individually if the particles are sufficiently large, and k calculated from the above equation, but if the particles are too small to be weighed individually, then closely graded fractions of particles are prepared and the number of particles and their mean projected diameter determined for the minimum weighable quantity.

The following example is for stainless steel particles graded between 100 and 120 mesh B.S. sieves. From measurements through the microscope of 600 particles the mean value of d_a^3 was found to be $3 \cdot 22 \times 10^6 \mu^3$; and 12,012 particles weighed $0 \cdot 1145$ g, the density being $8 \cdot 3$ g/c.c.

$$\text{Hence mean weight of particle} = 9 \cdot 53 \times 10^{-6} \text{g}$$
$$\text{and mean volume of particle} = 1 \cdot 15 \times 10^{-6} \text{c.c.}$$
$$= 1 \cdot 15 \times 10^6 \mu^3$$

Therefore
$$k = \frac{1 \cdot 15 \times 10^6}{3 \cdot 22 \times 10^6} = 0 \cdot 36$$

The ratio between the mean projected diameter d_a, as measured by the microscope, and the corresponding sieve aperture A through which the particle will just pass, is affected by the proportions and the geometrical shape of the particles. The aperture A is primarily dependent upon the breadth B of the particle, but is also affected by the thickness T; the length L has relatively little effect, provided this does not exceed say, four times the breadth. As flattened particles can pass through a square aperture in a diagonal position (see *Figure 9*), the diameter d_a may be very much greater than the side dimension of the equivalent square aperture A, and the curves in *Figure 11* show the variation in the ratio d_a/A, as affected by the volume coefficient k for various mineral particles. This ratio is also a function of the elongation n, as shown by the graphs.

Table 2. *Free Falling Diameters Equivalent to Various Sieve Apertures*

B.S. sieve mesh No.	Nominal aperture (microns)	k d_a/A	Diameter of free falling sphere (microns)			
			0·1 1·63	0·2 1·43	0·3 1·24	0·4 1·10
350	44*		36·4	40·0	40·8	40·9
300	53		43·6	48·1	49·1	49·4
240	64*		52·4	57·7	59·0	59·4
200	76		61·8	67·9	69·8	70·3
150	104		83·0	91·8	94·6	95·3
100	152		118	131	135	137
72	211		160	177	183	186
52	295		214	243	254	259
36	422		282	327	355	367
25	599		355	428	476	505
18	853		432	555	634	685
14	1,204		510	728	832	923
10	1,676		600	865	1,062	1,202
7	2,411		724	1,096	1,384	1,578
5	3,353		861	1,380	1,770	2,070

* Amendment to B.S. 410 : 1943, issued in 1955.

The relationships between d_a, δ, Δ, and A, are purely geometrical and unaffected by particle size, but the relationship between d_a and d_f, the diameter of the sphere of equal free falling velocity, depends on the nature of the fluid flow round the falling particle, *i.e.* whether this is stream-line or turbulent. From the results of experiments on the free falling velocities of mineral particles it has been possible to deduce the relationship between the equivalent diameters d_a, d_f, and A. The figures in *Table 2* show this relationship for particles of quartz. The relationship between these equivalents will be slightly different for particles of other densities, but the variation will only be slight for sieve sizes below 100 mesh.

EXPRESSION OF RESULTS OF A SIZING ANALYSIS

The numerical results of a sizing analysis should be recorded in a tabular form such as given below. The particle size intervals should be in approximately constant ratio, *e.g.* 2 or $\sqrt{2}$. The percentages by weight of the sample consisting of particles that are smaller than the various limiting sizes, are termed the cumulative undersize percentages by weight.

Report on Sizing Analysis of Fine Sand Collected by Filter

Sieving analysis on British Standard Test Sieves: conducted by dry sieving method according to B.S. 1796:1952. Sample weight 50 g.

Cumulative percentages undersize

	Weight per cent
Through 100 mesh B.S. test sieve	98·2
Through 150 mesh B.S. test sieve	81·4
Through 200 mesh B.S. test sieve	55·2
Through 300 mesh B.S. test sieve (including loss of 0·4 per cent)	41·0

Sedimentation analysis of sub-sieve particles: conducted by pipette method. Concentration of suspension 1 per cent by weight of powder in distilled water with 0·1 per cent of sodium hexametaphosphate.

Cumulative percentages undersize

Size (microns)	Weight per cent undersize
40	34·0
20	21·6
10	13·2
5	7·5
2	2·6

The following notes indicate items which should be included in the report so that the method of analysis is fully understood. In recording sieving analyses, state the standard sieve series used, the method of shaking the sieves, *i.e.* hand or machine, the duration of sieving, and whether the analysis was conducted in the dry or wet state. The loss in weight by dry sieving should also be recorded, and should be within 0·5 per cent of the weight of sample, which should also be stated. In reporting on sedimentation analyses the concentration of the suspension should be recorded, the sedimentation liquid, and any dispersing agent added.

It is often an advantage to express the results of a sizing analysis graphically and this facilitates the comparison of the results for a number of powdered materials. Standardization of the method of plotting has obvious advantages and it is the usual practice to plot vertically the cumulative per cent undersize by weight to a base of particle size. The range of particle sizes may, however, be so great that a logarithmic scale of particle size is necessary to show the complete characteristic of the powder. There are also alternative scales that can be used to express the per cent undersize. Thus, plotting the undersize on a logarithmic scale straightens the curve in the range of smaller particle sizes, but increases the curvature for the large particles. The normal undersize curve can be reduced to the form of an approximately straight line, by plotting the percentages undersize to a log probability scale, though a warning must be given against extrapolating such an approximately straight line beyond the limits of actual particle size measurement. A method of plotting giving somewhat similar characteristics to the latter method was devised by ROSIN and RAMMLER[22], and based upon a law for the size distribution which they developed. In this form of plotting, the scale for the percentage undersize is expressed in the form $\log \log 100/(100-u)$ where u is the percentage undersize.

DIRECT MEASUREMENT OF SURFACE AREA

The object of producing a material in powdered form is nearly always to increase the surface area of the material, and the purpose of many researches is to correlate the surface area with some other property, for example, the rate of combustion of pulverized coal, the hardening properties of Portland cement, the covering properties of a pigment and rates of chemical reaction in general. The characteristic required is the specific surface of the powder, usually defined as the aggregate surface area of all the particles expressed in square centimetres per gramme. The specific surface of a powder may be calculated from the sizing analysis, but an accurate figure depends on an exact knowledge of the analysis of the finest particles and upon a suitable allowance for the shape[23]. There are methods of powder examination which give directly a value for the specific surface, and these may be employed with advantage if the complete size distribution is not required. Such methods comprise specific surface determination by permeability methods, by photo-extinction and molecular adsorption. As these various methods depend on different properties it follows that they will not all give the same results on a certain sample of powder, but each method will give comparative characteristics as regards surface for a series of powdered materials.

The permeability method has been found very satisfactory for the direct determination of specific surface over a very wide range of powder fineness.

The method is essentially based upon the relationship between pressure drop and surface area for a fluid flowing through a static bed of the particles. A liquid flow may be used for coarser particles, but gas flow is generally preferred for fine powders. The simple theory assumes that the fluid flow in the pore spaces between the particles is laminar or streamline; as in the case of the motion of particles, modifications of the simple theory are necessary if the fluid flow becomes turbulent, or if the pore dimensions are so small that they are comparable in size with the mean free path of

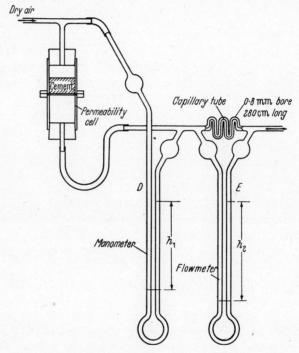

Figure 12. Permeability apparatus. (From F. M. Lea and R. W. Nurse, by courtesy of The Institution of Chemical Engineers. Crown copyright reserved)

molecules in a gas. Assuming streamline flow, the pressure drop with a volumetric steady flow in unit time of Q is given by the following equation:

$$S^2 = \frac{gP}{5\eta\sigma^2 L} \cdot \frac{A}{Q} \cdot \frac{\varepsilon^3}{(1-\varepsilon)^2}$$

where S = specific surface, P = pressure drop, A = cross sectional area of the bed, L = length of bed, η = absolute viscosity of air, σ = density of the powdered material, and ε = porosity of the bed; all factors in consistent units. The porosity of the bed is the ratio of the volume of void space to the volume occupied by the bed. The effect of the porosity of the bed is determined by the expression $\varepsilon^3/(1-\varepsilon)^2$ which indicates the extreme sensitivity of the permeability to the voidage factor. The test is conducted by apparatus such as is illustrated in *Figure 12*, showing the design by LEA and NURSE[24], which has been adopted as the standard method for determining the specific surface of Portland cement. (B.S. 12:1947). The powder is carefully compressed into a bed of known length, cross sectional area and porosity. The rate of gas flow is measured by a capillary tube, and the pressure drop through the bed by a manometer. This gives numerical values for all the factors which may then be inserted in the above equation, and the specific surface calculated. For the somewhat complex corrections for molecular flow with very fine powders the above reference should be consulted. A convenient form of apparatus has been designed by RIGDEN[25,26], which eliminates the flow meter and measures directly the volume of air passed through the particle bed by displacement of liquid in a cylinder. The apparatus is shown diagrammatically in *Figure 13* and the flow rate is determined by timing the change in level of the liquid in the cylinder, usually oil, between two different levels. Flow conditions are not steady in this apparatus, so that the surface is calculated by a modified equation given below:

$$S^2 = \frac{2g\varrho At}{5\eta\sigma^2 aL} \cdot \frac{\varepsilon^3}{(1-\varepsilon)^2} \cdot \frac{1}{\ln h_1/h_2}$$

where ϱ is the density of the oil, t is the time for difference in oil levels in the two limbs of the U-tube to change from h_1 to h_2, a is the cross sectional area of each limb of the U-tube, and other symbols are as in the previous equation for steady flow.

The principle of photo-extinction measurements has been described on page 36. If a measurement of the optical density of the suspension is made whilst the particles are in random motion, *i.e.* when stirred, then the simple theory enables the specific surface of the particles to be calculated. This measurement eliminates the effect of particle shape, as by the theory of CAUCHY[27] the mean projected area of a particle in random motion is one quarter of the surface area of the particle, independent of the shape. The objection to the method is the complex nature of light scattering which renders the simple theory incorrect for fine particles. The method may, however, be used to give comparative specific surfaces, and, if required, a method of correction devised by ROSE[28] may be used to allow for the variation in the light extinction coefficient.

The above methods determine the external surface of the particles; adsorption measures the total area of the solid which is accessible to the molecules of the adsorbed substance. Nearly all solid particles have an internal surface consisting of minute fissures or of very fine pores which may, or may not, be accessible to these molecules. Thus the surface area measured by the adsorption method is considerably greater than the external surface by a factor which varies from 2–10 normally, but can be as high as 100, according to the internal structure of the particles and the size of the adsorbed molecules which are assumed to form a layer of monomolecular thickness on the surface of the particle. Dye molecules in a liquid are relatively large, gas molecules are smaller and able to penetrate the more minute fissures.

The usual procedure for gas adsorption is to measure the volume of adsorbed gas at a temperature near the normal boiling point, *e.g.* if nitrogen is the gas used the powder sample is maintained at the temperature of liquid oxygen. The volume of gas adsorbed is calculated from pressure changes within the system. The theory of adsorption and the method of measurement are described in papers by GREGG[29] and JOY[30].

SUMMARY OF OBSERVATIONS ON SPECIFIC SURFACE MEASUREMENTS

It is not possible to give an absolute value for the specific surface of a powdered material, since this will depend on the assumptions made, *i.e.*

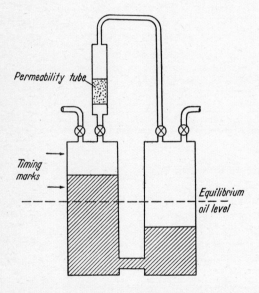

Figure 13. Permeability apparatus. (From P. J. RIGDEN[5], by courtesy of The Society of Chemical Industry)

whether only external surface is required, or whether a certain proportion of the internal fissure surface is to be included. If the object of the analysis is to assess the suitability of the powder as regards rate of reaction in a liquid or a gas, then the appropriate method would be adsorption of a liquid or gas respectively. In research on grinding processes it is usually the external surface that is required and the permeability method is most frequently used for the purpose of determining this.

REFERENCES

[1] HEYWOOD, H., *Trans. Instn Min. Metall., Lond.*, 55 (1946) 373
[2] MARTIN, G., BLYTH, C. E., and TONGUE, H., *Trans. ceram. Soc.*, 25 (1923–4) 61
[3] FERET, R., *Ann. Inst. Bâtim.* (1937) No. 2

[4] FAIRS, G. L., *J. R. micr. Soc.*, 71 (1951) 209
[5] HEYWOOD, H., *Trans. Instn Min. Metall., Lond.*, 55 (1946) 391
[6] 'The Physics of Particle Size Analysis', *Brit. J. appl. Phys. Conf.*, 1954
[7] WATSON, H. H., *Nature, Lond.*, 173 (1954) 362
[8] ANDREASEN, A. H. M., *Kolloidchem. Beih.*, 27 (1928) 349
[9] ANDREWS, L., *Proc. Inst. Engng Inst.*, 1927–8, 25
[10] PRYOR, E. J., BLYTH, H. N., and ELDRIDGE, A., 'Recent developments in Mineral Dressing', *Instn. Min. Metall. Symp.* (1953) 11
[11] PEARSON, J. C., and SLIGH, W. H., *U.S. Bur. Stand. Tech. Pap. No.* 48 (1915)
[12] GONELL, H. W., *Z. Ver. dtsch Ing.*, 72 (1928) 945
[13] ROLLER, P. S., *U.S. Bur. Min. Tech. Pap. No.* 490 (1931)
[14] HAULTAIN, H. E. T., *Trans. Canad. Inst. Min. Metall.*, 40 (1937) 229
[15] LUCAS, D. H., *Engineering*, 177 (1954) 272
[16] HEYWOOD, H., 'Recent developments in Mineral Dressing', *Instn Min. Metall. Symp.* (1953) 31
[17] BOSTOCK, W., *J. sci. Instrum.*, 29 (1952) 209
[18] STAIRMAND, C. J., 'Particle Size Analysis', *Instn chem. Engrs Symp.*, (1947) 128
[19] HEYWOOD, H., *Proc. Instn mech. Engrs*, 140 (1938) 287
[20] JARRETT, B. A., and HEYWOOD, H., 'Physics of Particle Size Analysis', *Brit. J. appl. Phys. Conf.* (1954) 21
[21] DONAGHUE, J. K., and BOSTOCK, W., *Trans. Inst. chem. Engrs*, 33 (1955) 72
[22] ROSIN, P., and RAMMLER, E., *J. Inst. Fuel*, 7 (1933) 29
[23] PIDGEON, F. D., and DODD, C. G., *Analyt. Chem.*, 26 (1954) 1823
[24] LEA, F. M., and NURSE, R. W., 'Particle Size Analysis', *Instn chem. Engrs Symp.* (1947) 47
[25] RIGDEN, P. J., *J. Soc. chem. Ind., Lond.*, 62 (1943) 1; 66 (1947) 130T
[26] FAIRS, G. L., 'Recent developments in Mineral Dressing', *Instn Min. Metall. Symp.* (1953) 59
[27] CAUCHY, A., *C. R. Acad. Sci., Paris*, 13 (1841) 1060
[28] ROSE, H. E., *J. Soc. chem. Ind., Lond.*, 67 (1948) 283; 2 (1952) 217
[29] GREGG, S. J., 'Particle Size Analysis', *Instn chem. Engrs Symp.* (1947) 40
[30] JOY, A. S., *Vacuum*, 3 (1953) 254

3

CRUSHING AND GRINDING EQUIPMENT

J. C. FARRANT AND R. NORTH

CRUSHING EQUIPMENT

MOST hard minerals used in chemical plants have to be reduced to fine particle size to render them suitable for processing and the preliminary reduction from large to relatively small pieces is accomplished by the use of a crusher or a number of different crushers arranged in series. For some purposes, the product from a fine crushing machine is sufficiently small for direct use, but in the majority of cases the material from the crushing plant passes to a mill (or pulverizer) for final reduction to the required size. (The description of grinding equipment is covered in a separate subsection.)

Machines dealt with under the present heading represent main types in general use and their descriptions show the fundamental principles involved. Many modifications of these types exist, but such variations are usually undertaken to meet special conditions, or according to the preference of the particular manufacturer.

The leading types of machines are enumerated below.

(*a*) Jaw crusher.
(*b*) Gyratory and cone crushers.
(*c*) Roll crusher.
(*d*) Swing hammer crusher.
(*e*) Rod mill.

The first three employ the principle of pressure under restriction, whilst the fourth shatters the material by impact under free suspension. Rod mills have useful crushing applications in the finer range, particularly in conjunction with ore reduction. More detailed reference to these is made in the subsection 'Grinding Equipment'.

Selection of the type of crusher for a particular application is governed by the following conditions:

1. The largest size piece in the feed.
2. Characteristics of the material, *e.g.* hardness; abrasiveness; stickiness and moisture content.
3. Hourly tonnage to be crushed.
4. Required size of discharge product.

Multi-stage crushing is usually employed if the ratio of feed size to product is more than about 8:1 although impact types may be used for higher reduction ratios where the material is dry, friable and non-abrasive. High manganese steel castings are almost universally employed for the actual crushing members on account of the abrasion-resisting and work-hardening properties of this metal.

Since crushing machines are usually of very robust construction, maintenance is often neglected. Regular adjustment for wear, efficient lubrication, systematic inspection and tightening of bolts, *etc.*, will be reflected in lower maintenance costs and power consumption, also in maximum output of consistent product.

JAW CRUSHER

There are many variations in design of jaw crushers but the majority are either of the Blake type or the Dodge type.

The Blake is the design most widely employed and is arranged with the pivot at the top of the swing jaw so that the greatest movement of this jaw is at the bottom, thus giving free discharge of the crushed material and consequent high capacity. Oversize pieces pass out with the product due to the varying size of the discharge opening. The size of the jaw crusher is denoted by the dimensions of the top of the mouth opening and the size range is from about 12 in. × 6 in. to 84 in. × 60 in. The nominal setting is taken as the distance from the tip of the corrugations on one jaw face to

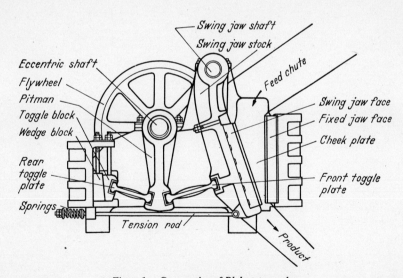

Figure 1. *Cross section of Blake type crusher*

the root of the corrugations on the opposite jaw face, measured at the bottom of the jaws when they are in the fully open position. The angle between the jaws is usually about 20°.

Originally these machines were of cast iron construction, but cast steel was introduced to meet the duty of increased sizes. Semi-steel and welded steel frames are used for some sizes, and the jaw and cheek plates are invariably of manganese steel.

Large crushers are sectionalized to facilitate manufacture, transportation and erection, the latter particularly in those cases where the crusher is required to pass down a mine shaft for installation underground.

Table 1. Sizes and Capacities of Blake Crusher.
(*Submitted by and printed by courtesy of* Fraser & Chalmers Ltd.)

Size of crusher in.	*Approx. capacity at settings stated**				*Speed of eccentric shaft r.p.m. No. of crushing strokes*	*h.p. of motor (approx.)*	*Approx. total weight* tons
	Min. setting in.	*Capacity* tons/h	*Average setting* in.	*Capacity* tons/h			
12 × 6	1	2	2	4½	260	10	2¼
12 × 9	1½	3½	2¼	5½	260	10	2¼
16 × 9	1½	5½	2¼	9	260	20	3½
20 × 10	1½	7½	2¼	12	260	25	5¼
20 × 12	2	10	2½	13	260	27½	5¼
24 × 14	2	14	2½	19	260	30	9½
24 × 18	2½	19	3	25	260	35	12¼
30 × 18	2½	24	3	30	260	55	14½
36 × 24	3	36	6	90	200	90	26
42 × 30	4	60	8	155	180	110	36
48 × 36	4½	90	8	180	180	150	60
60 × 48	5	120	9	240	170	200	108

* Approximate capacity figures are based on crushing material such as limestone weighing about 90 lb./ft.^{3} in the crushed state.

Figure 1 shows a typical cross section of a Blake type crusher, and an examination of it shows that the eccentric shaft imparts a vertical motion to the pitman and the toggles transform this into a reciprocating motion of the swing jaw. An adjustable bearing is provided at the end of the rear toggle so that the jaw opening may be varied to suit the size of product required and also to readjust the setting to compensate for wear. One of the toggles is usually provided with bolts which are designed to shear under excessive load such as that which would be caused by tramp iron or other uncrushable material becoming trapped between the jaws. This precaution helps to avoid damage to expensive parts of the machine and enables repairs to be readily effected.

The movable jaw is held back against the toggle by means of the spring and this spring should only be tightened sufficiently to prevent 'chatter' of the toggles. The jaw plates usually have corrugated crushing faces and the pitch is about the same as the size of product normally anticipated from the crusher. In certain cases jaw plates are made with staggered teeth to prevent long pieces of laminar material from sliding down the corrugations. There is a natural tendency for the moving parts to come to rest in the most inert position, coinciding with the lowest position of the pitman. Accordingly, the power required to put the crusher in motion is greater than that absorbed under running conditions and the following practice assists in reducing the starting load. It has been found that the best position for the pitman is about 15° behind the upper dead centre in the direction of rotation at the time of starting, and steps should be taken to move the pitman into this position before starting up. On some large crushers suitable gear is provided to facilitate rotation of the flywheel for positioning the pitman. Another notable point is that crushers with grease-lubricated bearings are frequently preferred for easier starting because grease is not displaced so readily as oil under stationary loading.

Table 1 gives particulars of a range of Blake crusher sizes and their anticipated performance.

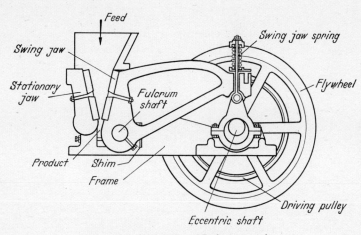

Figure 2. *Cross section of Dodge type crusher*

A Dodge crusher is illustrated in *Figure 2*. It will be noted that the fulcrum shaft is located near the bottom of the swing jaw and accordingly maximum movement takes place at the inlet, with little variation in the size of the discharge opening when the crusher is in operation. This prevents the discharge of oversize pieces, but the restriction causes packing at the outlet with consequent low capacity and production of a large amount of fines by attrition in the packing zone. The machine operates satisfactorily on hard, clean-breaking minerals which flow freely, but is unsuitable on soft materials which tend to pack.

The Dodge crusher is capable of a large reduction ratio and is recommended for small-scale plant where two-stage crushing would not be justified. Owing to its simple construction, this machine is relatively cheap and easy to adjust and maintain, but output is limited.

GYRATORY CRUSHER

This type of machine comes into the primary crusher range, but works on a somewhat different principle to the linear motion of jaw crushers previously described. Various companies manufacture this type of machine with detail modifications, although the basic principle of operation is common to all. A section through one of these machines is shown in *Figure 3*.

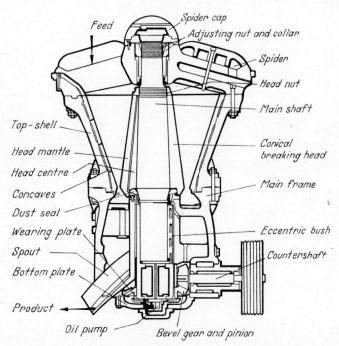

Figure 3. *Cross section of characteristic gyratory crusher.* (*By courtesy of* Allis-Chalmers Manufacturing Co., U.S.A.)

The vertical shaft, on which is mounted the cone or conical head, is carried eccentrically in the large lower bush and is free to rotate. This bush is mounted concentrically in a housing and is rotated through the bevel gears by means of the horizontal driving shaft.

At each revolution of the large lower bush the conical head gyrates around the conical bowl or ring, the gap between the two crushing surfaces first opening and then closing as the eccentric movement proceeds. The vertical shaft is free to rotate around its own axis and rotates slowly in the direction of travel of the shaft when the machine is empty, but in the reverse direction when crushing is taking place.

A range of sizes is available from laboratory scale up to giant machines weighing some 300 tons and capable of handling 1,000 tons/h of rock. *Table 2* (gyratory crusher sizes and approximate capacities) is based on information provided by Allis-Chalmers Manufacturing Co., U.S.A.

Table 2. Sizes and Capacities of Gyratory Crushers.
(*Submitted by and printed by courtesy of* Allis-Chalmers Manufacturing Co., U.S.A.)

Approx. size feed opening in.	Approx. capacity at settings stated* short tons (2,000 lb.)/h							Gyrations per min	Eccentric throw in.	Max. h.p.
	Primary crushers									
	Open side setting of discharge opening									
	$2\frac{1}{2}$ in.	3 in.	$4\frac{1}{2}$ in.	5 in.	$7\frac{1}{2}$ in.	9 in.	12 in.			
30 × 78	150	205	390					175	$\frac{5}{8}$	150
36 × 96	370	385	475					150	$\frac{3}{4}$	200
36 × 96			630	720				150	1	265
36 × 96				1,000	1,320			150	$1\frac{1}{2}$	400
48 × 120					2,360	2,530		135	$1\frac{5}{8}$	500
60 × 150					2,750	3,500	5,000	100	$1\frac{1}{2}$	1,000
	Secondary crushers									
	Open side setting of discharge opening									
	$1\frac{1}{2}$ in.	2 in.	$2\frac{1}{2}$ in.	3 in.	$3\frac{1}{2}$ in.	4 in.	$4\frac{1}{2}$ in.			
24 × 66	195							175	$\frac{3}{4}$	175
24 × 66		325						175	1	250
24 × 66			415	475	525	575	620	175	$1\frac{1}{4}$	300
30 × 84			450	500				150	1	265
30 × 84					650	750		150	$1\frac{1}{4}$	330
30 × 84						850	965	150	$1\frac{1}{2}$	400

* Capacities are based on crushing dry friable material equivalent to limestone; allowing 100 lb./ft.³ for the crushed material. Capacities vary approximately in proportion to specific gravity of ore or rock crushed. Adverse factors which reduce normal capacity are toughness, high moisture content and clay content.

CONE CRUSHER

The cone crusher, a section of which is shown in *Figure 4*, is a modified type of gyratory crusher. The principle of operation is the same as that described for gyratory crushers, but the outer conical crushing surface is arranged almost parallel to the conical head. The spider and top bearing fitted in gyratory crushers are omitted in the majority of cone crushers giving a shorter centre shaft and better feeding facilities. The initial gap between these two surfaces is adjustable, the method depending on the type of crusher, *i.e.* by raising or lowering either the conical bowl or the vertical centre shaft carrying the conical head.

The shape of the conical bowl can be varied to suit the material and the size reduction being carried out, and as in jaw crushers, the wearing surfaces are invariably of manganese steel.

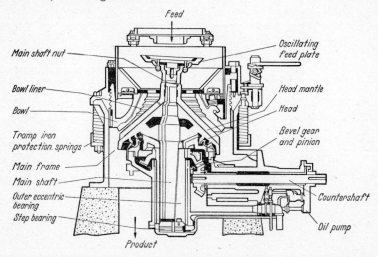

Figure 4. Cross section of cone crusher. (By courtesy of Nordberg Manufacturing Co.)

Provision is made in the design to minimize damage to the crusher when tramp iron enters with the feed material. For this purpose arrangements are made to allow the crushing members to move apart when excessive

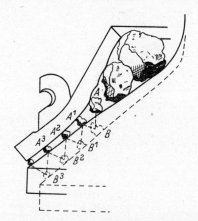

Figure 5. Diagram showing path of material through a cone crusher.
(*By courtesy of* International Combustion Products Limited)

resistance is encountered. In some cases this is achieved by spring-loading the bowl, and in other designs the head is supported by a predetermined hydraulic pressure. When the tramp iron has been ejected the crushing members automatically return to their previous setting.

These machines are manufactured in a range of sizes from approximately 2 ft. diameter up to 7 ft. diameter and their duty is usually more in the nature of intermediate than primary crushing.

Table 3 gives particulars of a range of cone crushers and anticipated performance.

Table 3. Sizes and Capacities of Cone Crushers.
(*Submitted by and printed by courtesy of* Nordberg Manufacturing Co.)

Size of crusher ft. in.	Approx. capacity at settings stated* tons/h				r.p.m.	h.p. of motor (approx.)	Approx. total weight tons
	Min. setting in.	Capacity	Average setting in.	Capacity			
2 0	$\frac{3}{16}$	15	$\frac{3}{8}$	25	650	20–25	$3\frac{1}{2}$
2 0	$\frac{1}{4}$	25	$\frac{3}{8}$	35	575	25–30	$4\frac{1}{2}$
3 0	$\frac{3}{8}$	55	$\frac{1}{2}$	75	580	50–60	$9\frac{1}{2}$
4 0	$\frac{3}{8}$	100	$\frac{3}{4}$	150	485	75–100	$15\frac{1}{2}$
4 6	$\frac{1}{2}$	125	$\frac{3}{4}$	170	485	125–150	20
5 6	$\frac{5}{8}$	220	1	350	485	150–200	37
7 0	$\frac{3}{4}$	450	$1\frac{1}{4}$	700	435	250–300	62

* Approximate capacity figures only, based on crushing material such as limestone weighing about 90 lb./ft.^{3} in the crushed state.

Table 4. Product Analysis for Cone Crushers

Passed by in.	Retained on in.	Per cent of total	Cumulative per cent
$\frac{3}{4}$	$\frac{5}{8}$	5·5	100
$\frac{5}{8}$	$\frac{1}{2}$	13·5	84·5
$\frac{1}{2}$	$\frac{3}{8}$	23·5	81·0
$\frac{3}{8}$	$\frac{1}{4}$	19·0	57·5
$\frac{1}{4}$	$\frac{1}{8}$	15·0	38·5
$\frac{1}{8}$	$\frac{1}{16}$	11·0	23·5
$\frac{1}{16}$	Pan	12·5	12·5

Cone crushers, having more or less parallel faces for the lower portion of the crushing members, give a closer range of sizes in the finished product than that produced by a gyratory; approximately 80 per cent of the product being to size. *Figure 5* illustrates the crushing action and the path of material moving through the crusher. A typical size analysis of granite passed through a cone crusher is given in *Table 4*. (Feed size $1\frac{1}{2}$ in. and down.)

ROLL CRUSHER

Both single and double roll crushers are in current use on a variety of applications and are manufactured in a wide range of sizes. Although they are not used as extensively as other types of crushers they are very effective within their range of usefulness.

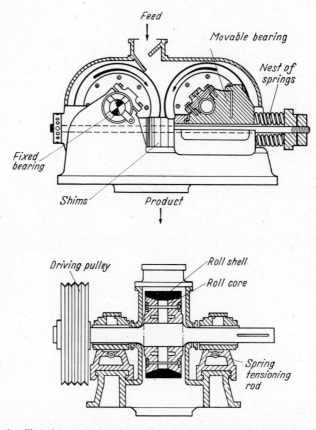

Figure 6. Typical example of crushing rolls. (By courtesy of J. Harrison Carter Ltd.)

Where the mineral is required to be divided into a number of different sized grades, with a minimum production of fines, double rolls are employed in conjunction with screens. Smooth-faced rolls can be applied for fine crushing down to gradings approaching 100 mesh, whilst large diameter 'toothed' rolls are used for the initial breaking of soft ores, such as Northamptonshire ironstone, down to a size of about 6 in. or 8 in. Single roll crushers are used for minerals not normally handled in gyratory crushers, e.g. manganese ore and soft stickier types of minerals.

The effective range of reduction of double roll crushers is about $2\frac{1}{2}$ or 3 to 1, but the range may be greater or less than this according to the setting of the rolls and the method of feeding. A section through a double roll unit is shown in *Figure 6*. One shaft carrying a roll revolves in fixed bearings whilst the second is carried in movable bearings which can slide in a horizontal plane. The distance between the rolls is fixed by the shims and this is maintained by a nest of strong springs. Should a piece of tramp iron or hard material be fed to the crusher the rolls will part against the spring pressure thus allowing the material to pass without causing damage.

The roll shells are, again, usually made of manganese or chromium steel and are renewable. Modern practice is to drive both rolls, often by independent motors; in older machines one roll only was driven, the other being driven by friction of the material being crushed.

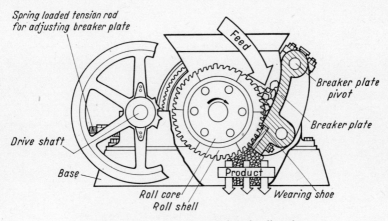

Figure 7. Cross section of a typical single roll crusher.
(*By courtesy of* British Jeffrey-Diamond Limited)

The capacity of crushing rolls increases with both the face width and the diameter of the roll. The angle of nip is the angle formed by the tangents to the roll faces at the points of contact with a particle to be crushed. This angle varies from 0° to 45°, but is generally something like 20° and is given by the formula:

$$\cos \frac{N}{2} = \frac{r+a}{r+b}$$

N = angle of nip, r = radius of rolls, a = one-half of space between rolls, and b = radius of the particle.

For average conditions, assuming an angle of nip of 25° and the size of the finished product being $2a$, the appropriate diameter of rolls required can be calculated. Failure to nip can be overcome by corrugated or toothed rolls but these tend to cause much more vibration and for the finer products smooth rolls are essential.

The single roll crusher, a section of which is shown in *Figure 7*, breaks partly by pressure and partly by impact. It is suitable for fairly soft friable materials such as coal, gypsum, *etc.*, and produces less fines than a swing hammer crusher. The roll, which is normally provided with a combination of long and short teeth, the actual shape and number of these depending on the material to be crushed, revolves at a relatively high speed and drives the material against the breaker plate. The breaker plates are usually fitted with renewable wearing shoes.

SWING HAMMER CRUSHER

This type of machine is manufactured by a large number of firms and although the details vary considerably, the principle of crushing, mainly by impact, is common to all.

A swing hammer crusher consists essentially of a rotor revolving at speed, the rotor being built up with hammers supported by suspension bars carried in hammer discs, these discs being keyed to the rotor shaft. The whole assembly revolves in suitable bearings usually in a horizontal plane although some machines are in existence with the shaft held vertically. A section through a characteristic machine is shown in *Figure 8*.

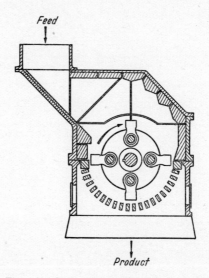

Figure 8. Cross section of swing hammer crusher

The speed of the shaft may vary between 500 and 1,000 r.p.m. for coarse crushing, depending upon the size of machine and the duty. With finer products and lighter duties the speed range is higher. The greater part of the inside periphery of the body is lined with manganese steel breaker plates, whilst the lower portion of the periphery may be fitted with grids, or on some machines, just left open. The hammers are normally held in a radial position by centrifugal force, but the mounting pins act as pivots enabling them to fly back when tramp iron, *etc.*, accidentally enters the machine.

The size, shape and number of hammers vary widely in different machines as well as the design of the discharge grids. Most forms of hammers are reversible so that a new face can be used after a certain amount of wear has taken place. Some machines are also designed to allow reversed direction of rotation to balance the wear on hammers and the discharge grids. A 'stirrup' type of hammer which imparts greater impact than simple bar types is shown in *Figure 9*. A greater wearing surface per hammer disc is presented to the material with a corresponding saving in maintenance.

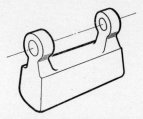

Figure 9. Stirrup type hammer as used in swing hammer crushers. (*By courtesy of* The Patent Lightning Crusher Co. Ltd.)

Another machine in this group is the ring-hammer crusher, which, as the name implies, employs rings instead of hammers. In use, the rings move gradually about their retaining pins, thus tending to equalize wear. They are also free to move radially inwards when encountering tramp iron or other uncrushable material.

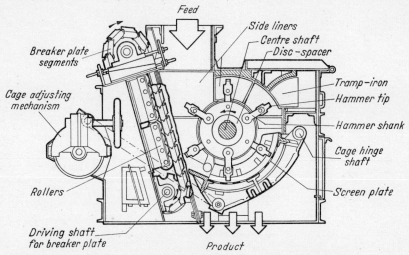

Figure 10. Cross section of swing hammer crusher with moving breaker plate. (*By courtesy of* Fraser & Chalmers Engineering Works)

A modified type of hammer mill for dealing with damp and sticky material such as bauxite, gypsum and chalk is shown in *Figure 10*. This machine is equipped with a moving breaker plate which travels at a speed of approximately 20 to 25 ft./min, so preventing the discharge of the machine from choking.

Table 5. Swing Hammer Crushers—Operating Data

Type of machine	Material	Feed in.	Moisture per cent	Product in.	Approx. kWh/ton
Pennsylvania Crusher*	Limestone	12 and less	Up to 7	$-\frac{3}{4}$ screen	2 to 2·7
Christy and Norris Disintegrator	Limestone	2	$2\frac{1}{2}$	$-\frac{1}{16}$	7 to 12
Christy and Norris Disintegrator	Limestone	2	$2\frac{1}{2}$	$-\frac{1}{8}$ 70%—100 mesh on soft stone 50%—100 mesh on hard stone	3·75 to 7·5
Lightning Crusher†	Bauxite	4	2	$-\frac{1}{2}$	1·5
Lightning Crusher .	Graphite (Ceylon)	$2\frac{1}{2}$	1	$-\frac{1}{8}$	7·0
Lightning Crusher .	Gypsum	$4\frac{1}{2}$	2	$-\frac{3}{4}$	1·4
Lightning Crusher .	Slate	$2\frac{1}{2}$	1	$-\frac{1}{4}$	3·4

* Pennsylvania Crusher by Fraser & Chalmers Engineering Works.
† Lightning Crusher by Patent Lightning Crusher Co. Ltd.

Performance data on various types of swing hammer crushers handling different grades of material of varying sizes are given in Table 5.

ROD MILL

A description and illustration of the rod mill appear in the subsection headed 'Grinding Equipment'.

The application of rod mills for tertiary crushing is increasing. In large-scale crushing plant, where jaw or gyratory crushers are used for primary breaking, followed by gyratory or cone for secondary and tertiary crushing, the effective limit of size of crushed product is approximately $\frac{5}{16}$ in.

If the mineral to be crushed is damp or sticky, output would be reduced, or discharge opening would be increased, entailing a coarser product than desired. Under such conditions rod milling may be preferred for the final crushing stage.

The range of rod mill reduction is roughly from $1\frac{1}{2}$ in. and less to approximately 10 mesh. The advantages claimed for the wet grinding rod mills are:
1. Dust is eliminated from the tertiary crushing department and the subsequent grinding units receive a finer feed than that produced from crushers.

2. The range of reduction in the regrinding mills is lessened, with decreased consumption of steel and power absorbed when grinding to the same final product.

NOTES ON SMALL-SCALE CRUSHING PLANT

Certain difficulties arise when handling a few tons of raw materials containing large pieces which require subsequently to be ground to a very fine state of division. The harder the mineral, the finer it should be crushed, and the selection and design of the crushing plant depends upon several factors, *e.g.* capital cost versus mechanical efficiency; the value of the finished product; maintenance and labour required. There are so many variables to be considered that no hard and fast line can be drawn as to the plant involved. The following are examples of processes which can be used.

With a high priced finished product the following scheme has many advocates. A jaw crusher of ample dimensions to take the rock 12 in. or so, followed by a granulator of the Blake type, in which the receiving opening is longer and narrower than the standard design of jaw crusher. These crushers are run at a relatively higher speed. The jaw plates are closely serrated and the throw of the eccentric is proportionately less than the standard types. Such a crusher will deliver a product roughly $\frac{3}{4}$ in. and less, while following with a pair of rolls and a screen for determining maximum size (say $\frac{1}{4}$ in.) provides a consistent feed to the receiving bins of a grinding plant. Other plant would eliminate the roll section and use a larger size mill for the final reduction.

When the original raw material is delivered at a maximum size of 4 in. and less, then a single granulator would be used in closed circuit with a screen to produce minus 1 in. or $\frac{3}{4}$ in. Tertiary gyratory crushers have not yet been developed for such small-scale work.

With softer types of minerals and materials the problem is comparatively simple as high speed crushers can be used, preferably in closed circuit with screens, to control the size of product delivered to the mill bins.

POWER REQUIRED FOR CRUSHING AND GRINDING

As the crushing operation becomes more expensive and complicated when producing finer sizes of crushed product, and the grinding operation is less efficient when dealing with coarse feed, the question naturally arises as to the most economical size at which the process should change from crushing to grinding.

Figure 11 illustrates the variation in power consumption when crushing to different sizes, and the increase in grinding power for dealing with coarser feed sizes. By combining these two curves it is possible to determine the optimum size. The curves should only be regarded as being indicative, as the actual power for crushing would depend upon the initial size of feed to the crusher, whilst the actual power consumption of the mill would largely depend upon the fineness of product. However, it can be taken that the ½ in. optimum size shown on the curve agrees with general experience on a wide

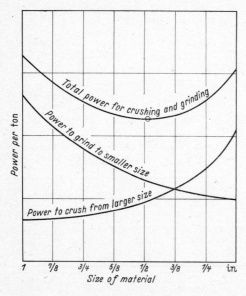

Figure 11. Curves showing effect on power consumption when crushing to, and grinding from, various sizes. (*By courtesy of* International Combustion Products Limited)

variety of materials and applications. In the case of soft materials, such as gypsum, it is possible to accept a coarser feed to the mill without detriment, whilst on very hard materials, such as quartz or carborundum, a finer mill feed has definite advantages.

GRINDING EQUIPMENT

The amount of energy required to reduce a material to a given particle size depends on a number of factors, the most important of which is the relative hardness of the material. A simple and commonly used method of assessing the hardness is known as Mohr's scale and consists of a series of ten standard minerals numbered from 1 to 10, the softest being talc and having the number 1; the hardest being diamond and having the number 10. A material of unknown hardness can thus be classified by its ability to scratch or be scratched by the standard minerals. For example, a material that scratches 6 and is scratched by 7 would be classified as having a hardness of 6·5 on the Mohr scale.

The full scale of minerals is given below.

1. Talc
2. Gypsum
3. Calcite
4. Fluorspar
5. Apatite
6. Feldspar
7. Quartz
8. Topaz
9. Sapphire
10. Diamond

Another factor which has an important bearing on the 'grindability' is the physical structure of the material. A foliated or flaky material, such as graphite or some types of talc, will have very different grinding characteristics from those of a crystalline material such as quartz or feldspar. Some materials have a tendency to be tough and rubbery in nature; this makes them difficult to grind since deformation of the particle occurs with little actual reduction in size. Cereals such as soya beans and maize come into this category together with natural gums, seaweeds, *etc*. Other factors which affect the grinding characteristics are moisture content of the material, sensitivity to temperature changes, and tendency of the material to be either deliquescent or hygroscopic.

The energy which is put into a grinding mill may be divided into (*a*) useful energy which is actually used to effect reduction in the particle size, and (*b*) that which is wasted in overcoming friction, inertia of the mill, and deformation of the particle without actual reduction in size.

Crushing and grinding mathematics are based very largely upon empirical statements; this is inevitably the case since so many variables exist in size reduction processes, and a completely homogeneous material rarely occurs. Even the same class of mineral varies widely in grinding characteristics. As an example, it can be stated that when conducting grinding tests under parallel conditions in a Raymond roller milling plant, the output obtained on a soft Northumberland limestone was twice that obtained on a hard Derbyshire limestone. An extreme case is graphite. Various grades of this material received from different countries and tested in the same mill have yielded output ratios up to 5:1, grinding to the same fineness of product. The ultimate criteria in this respect, therefore, remain either reliable field data or full-scale tests on a representative bulk sample of the particular material. Reputable manufacturers of grinding plant maintain a fully equipped test house for this purpose.

CLASSIFICATION OF GRINDING MILLS

The reduction or grinding of various minerals and other substances covers a wide and diversified field, and in order to simplify the application of the many forms of grinding machines, the main types have been classified under the following groups.

Group 1: Slow Speed

Listed under this heading are ball and pebble mills, tube mills and rod mills, which are universally used for grinding abrasive materials, the grinding media employed being simple in shape and comparatively cheap. The ball or pebble charge can be easily replenished whilst the mill is in motion, thus maintaining optimum output, and all these machines are capable of being run for 24 hours a day over long periods. Grinding is effected by a combination of impact and attrition.

Group 2: Medium Speed

Consisting of various types of roller mills, this group is generally used for the grinding of materials with a hardness not exceeding 4 in the Mohr scale, and covers the reduction of various minerals and substances mostly in the medium and fine grinding range. Less floor space is required and less power consumed than the mills in Group 1 when grinding to a definite residue on a given mesh, *e.g.* 1 per cent residue on 300 mesh.

Group 3: High Speed

This group of machines includes various hammer mills and disintegrators, and is relatively inexpensive. It is generally applied for coarse and intermediate grinding of non-abrasive materials, and usually the machines only require a small amount of floor space. The indicative graph shown in *Figure 12* illustrates the maintenance characteristics of the above series of mills.

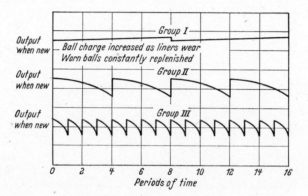

Figure 12. *Curves showing relative life of grinding parts in different types of mills and effect on output.* (*By courtesy of* International Combustion Products Limited)

Group 4: Fluid Jet Pulverizers

This covers various designs of machines in which reduction is effected by bombarding particles against each other, usually in a stream of compressed air or superheated steam. Generally for use with materials not exceeding a hardness figure of 4, they give a superfine product well down in the micron range. The forerunner of this type of grinding unit was introduced over 20 years ago.

Exceptions

There are always exceptions to the general rule, as the following will show. Ball mills and tube mills have been used for grinding barytes, gypsum, lime and other soft minerals. The selections, in these cases, are based on the particular service requirements which entail 24 hours a day, seven days a week operation, for long periods without a stop.

In the case of Group 2, roller mills have been selected for grinding coke for the manufacture of electrodes since the particle shape of the finished product for certain purposes is preferable to the shape obtained from Groups 1 and 3. The maintenance is high but is justified in such exceptional cases. Swing hammer mills have also been used for breaking up raw flint for concrete aggregates which require sharp edges.

A brief reference should be made to another type of crusher-cum-mill known as the 'Hadsel', and a later type for dry grinding known as the 'Cascade' mill, in which lumps of the raw material are used as grinding media. Six installations of the 'Hadsel' mill have been made for wet crushing and grinding minerals in one stage, but further attempts were abandoned some years ago in favour of the traditional ball and rod mill grinding.

It may be stated that this form of reduction was efficient under certain conditions, but has not shown sufficient savings for general application. Opinions differ as to the cause of this partial failure although certain explanations can be suggested. Minerals drawn from underground and quarried rock, vary in size and in distribution of sizes. If one particular grading is most suitable, other gradings may not be so. A high proportion of fines would not constitute a good feed as the fines would not fracture readily by their own falling weight and there would be insufficient crushing effect by the remainder of the material. In breaking down a hard and tough coarse feed it may form a large proportion of particles of intermediate size which do not possess sufficient energy to shatter themselves when falling on to the breaker bars. This method of grinding is, however, currently employed in dry grinding with air sweeping, for example, in the reduction of asbestos rock.

BALL AND PEBBLE MILLS

The mills in Group 1 are more widely used than any other form of grinding apparatus, since they are of very robust construction which enables them to handle a wide range of materials, and furthermore, the same design of mill (with detail modifications) can be used for wet or dry grinding.

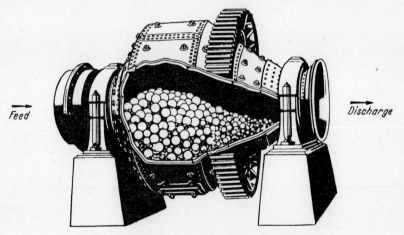

Figure 13. Hardinge conical ball mill. (*By courtesy of* International Combustion Products Limited)

This type of mill consists essentially of a cylindrical or conical shell revolving in a horizontal plane and containing a charge of grinding media such as steel balls, porcelain balls, or flint pebbles. The inside of the mill shell is lined with manganese steel, chromium steel or special hard cast iron liners when steel balls are used, and Belgian Silex, porcelain, or other non-metallic wear-resisting material when pebbles or porcelain balls are used. Pebble mills are only used when it is essential to avoid metal contamination of the product and it should be noted that the output of a given size of mill varies in almost direct proportion to the specific gravity of the grinding media employed. Therefore, wherever conditions permit, it is more economical to use a metal ball charge.

The speed at which these 'tumbling' mills revolve should be sufficient to promote the correct cascading action of the grinding charge and this speed varies with the diameter of the particular mill and the duty it is performing. The common denominator, however, is the 'critical speed', being the speed at which a particle on the inside periphery of the shell liners will just centrifuge. Critical speed is given by the formula:

$$N = \frac{54 \cdot 19}{\sqrt{R}}$$

where N = critical speed r.p.m. and R = inside radius of mill liners in feet. In practice, most mills run between 50 and 80 per cent of the critical speed. Sections through various types of ball mills are shown in *Figures 13–17*.

Figure 13 shows the Hardinge conical ball mill which is used for wet and dry grinding in both open and closed circuit systems. During rotation the conical shape of the mill causes a natural automatic segregation both of the grinding media and of the material being ground. The illustration shows the incoming feed crushed in the largest diameter of the mill by the largest

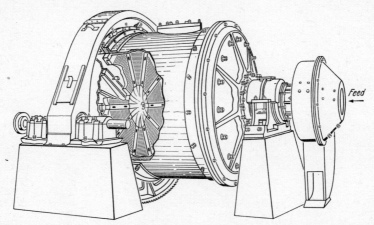

Figure 14. Marcy ball mill. (*By courtesy of* The Mine and Smelter Supply Company, U.S.A.)

balls, with the greatest superincumbent weight, the greatest height of fall, and with the highest peripheral speed. As the material travels towards the discharge the crushing force is gradually diminished because the grinding media are smaller and drop from a lesser height, thus reducing power consumption. The reduction by attrition, however, is increased on account of the greater surface exposed by the smaller balls. The Hardinge conical pebble mill is of similar construction and shape but the mill is lined with Silex blocks, porcelain, or similar material and flint pebbles or porcelain balls are used for grinding media.

Figure 14 shows the Marcy ball mill which has a cylindrical shell and is of the grate discharge type, that is, the discharge end of the mill is fitted with a grate through which the ground material must pass. This particular mill has a series of lifters incorporated with the grate which lift the material up and finally discharge it through the trunnion in the normal manner.

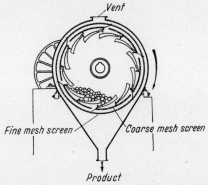

Figure 15. Stag ball mill. (*By courtesy of* Edgar Allen & Co., Limited)

The 'Stag' ball mill, *Figure 15*, is arranged with a screen mesh around the exterior of the mill which sizes the finished product. The liners are arranged in such a way that ground material passes between them on to a coarse screen which serves to protect the fine screen from large oversize material and from excessive wear. Oversize particles on the screens pass back into the mill as rotation proceeds. The feed material enters the mill through one trunnion.

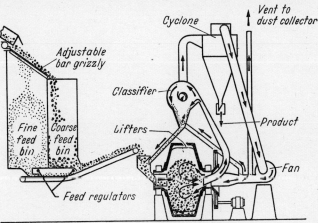

Figure 16. Hardinge cascade mill and air classifier with sized and recombined feed control. (*By courtesy of* Hardinge Co. Inc., U.S.A.)

The Hardinge cascade mill, *Figure 16*, has a diameter two to three times its cylindrical length and is provided with conical ends; it is designed to operate using the crude ore itself as a grinding medium. The mill is air swept to remove the finished product and simultaneous hot air drying can be incorporated if desired. The crude ore is sized on a grizzly screen and fed to the mill in suitable proportions to maintain adequate lump material in the mill and thus perform the necessary grinding.

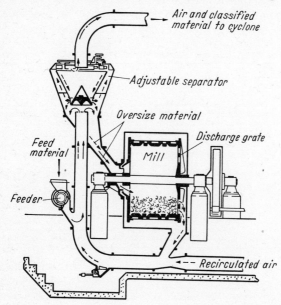

Figure 17. British Rema ball mill and air separating plant.
(*By courtesy of* Edgar Allen & Co., Limited)

Figure 17 shows the British 'Rema' ball mill which is also of the grate discharge pattern. The air system incorporated in this mill is arranged to pass all the raw feed material through a separator to extract the fines before it enters the mill, the product from the mill being returned back through the separator for extraction of fines, the oversize being returned to the mill for further grinding

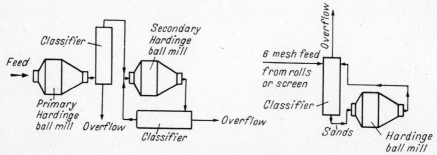

Figure 18. Typical flow diagrams for 'closed circuit' wet grinding. (*By courtesy of* International Combustion Products Limited)

Ball mills are pre-eminent in the wet grinding field, notably in the sphere of metallurgical ore reduction in which millions of tons of minerals are ground annually. Whilst it is possible to operate these mills in open circuit, *i.e.* with a single pass of material through the mill, this method results in over-grinding of the fine particles produced in the early stages and, conversely, there is always the possibility of really coarse particles escaping from the mill. The presence of excess 'superfines' and/or coarse pieces can cause serious difficulties in many subsequent processes. Apart from this aspect it is more efficient from a grinding standpoint to arrange a classifier in closed circuit with the mill, especially when fine products are required with controlled upper particle size limit. Flow diagrams of closed circuit wet grinding systems (single and two-stage milling) are shown in *Figure 18*.

Table 6. Marcy Ball Mills Wet Grinding Gold Ore in Closed Circuit with Classifiers

Ref.	Mill size ft.	Feed size in.	Product fineness	Approx. kWh/ton
Marcy (701)	7 × 6	$\frac{5}{8}$	82%—200 mesh	21
Per	9 × 7	$\frac{3}{8}$	80%—200 mesh	16·5
Marcy Rod Mill and Ball Mill	$9\frac{1}{2}$ × 12 Two 10 × 10	$1\frac{1}{2}$	55%—200 mesh	12

Tables 6, 7 and *8* give operating data on the wet grinding of gold ore in various types of ball mills, whilst *Tables 9, 10* and *11* give grinding data on lead–zinc ore, and similar information on copper ore appears in *Tables 12, 13* and *14*. Although this data relates to ore reduction, it will be found helpful, by interpolation, to serve as a guide on other grinding applications.

Table 7. Hardinge Ball Mills Wet Grinding Gold Ore in Closed Circuit with Classifiers

Ref.	Mill size ft. in.	Feed size in.	Product fineness	Approx. kWh/ton
705	7 × 36	$1\frac{1}{2}$	67·5%—200 mesh B.S.	15
693	10 × 72	$-\frac{1}{2}$	62·7%—200 mesh B.S.	19
883	8 × 60	$\frac{1}{2}$	84%—200 mesh B.S.	19
K.S.	5 × 36	$\frac{1}{2}$	97%—100 mesh B.S.	20
B.P.	8 × 72	$\frac{3}{8}$	82%—200 mesh B.S.	20
685	7 × 48	11% + $\frac{1}{4}$	74·9%—200 mesh B.S.	19
669	7 × 48	7% + $\frac{1}{4}$	76·6%—200 mesh B.S.	24 hard ore
681	8 × 36	$\frac{1}{2}$	85·0%—200 mesh B.S.	13 soft ore
St. J.	8 × 60	$\frac{5}{8}$	98%—200 mesh B.S.	23, 2-stage

Table 8. *Cylindrical Ball Mills Wet Grinding Gold Ore*

Ref.	Mill size ft.	Feed size in.	Product fineness	Approx. kWh/ton
Ref. 884 (Ash Peak)	6 × 10	$\frac{1}{2}$	68%—200 mesh	22
Ref. 884 (Ash Peak)	8½ × 9	$\frac{3}{4}$	56%—200 mesh	18·8
674	9 × 10	$\frac{3}{8}$	65%—200 mesh	17
689	8 × 8	$\frac{5}{8}$	60%—200 mesh	13·5
721A	5 × 8	$\frac{3}{8} \times 1\frac{1}{2}$	80%—200 mesh	19·5
797	6 × 10	$\frac{1}{2}$	68%—200 mesh	18
753	8 × 10	$\frac{7}{16} \times 2\frac{1}{2}$	28%—200 mesh	6 primary grind
675	9 × 10 5 × 14 tube mill	$-\frac{1}{4}$ 65%—200 mesh	57%—200 mesh 99·7%—200 mesh	8 22·5 regrinding flotation concentrates
670	5 × 16	$\frac{1}{2}$	84%—200 mesh	21

Table 9. *Hardinge Ball Mills Wet Grinding Lead–Zinc Ore in Closed Circuit with Classifiers*

Ref.	Mill size ft. in.	Feed size in.	Product fineness	Approx. kWh/ton
828 NT	7 × 36	$-\frac{1}{4}$	50%—200 mesh B.S.	7·5
511	6 × 36	$1\frac{1}{2}$	65%—200 mesh B.S.	10·3
786	9 × 48	$\frac{3}{8}$	75%—200 mesh B.S.	10·5
469	8 × 36	$\frac{3}{8}$	60%—200 mesh B.S.	9·8
694	10 × 48	$\frac{3}{8}$	61·4%—200 mesh B.S.	8·3
468	8 × 36	$\frac{3}{4}$	62·1%—200 mesh B.S.	7·6
587	8 × 60	$\frac{1}{2}$	39·0%—200 mesh B.S.	5·25
698	10 × 48	$\frac{3}{8}$	55·2%—200 mesh B.S.	4·5
684	8 × 60	$\frac{3}{4}$	91·5%—200 mesh B.S.	12·75
525 524	8 × 48 10 × 48	$\frac{5}{8}$ $\frac{5}{8}$	42·4%—200 mesh B.S. 42·1%—200 mesh B.S.	4·6 4·2
MT1	8 × 60	$\frac{3}{4}$	75%—200 mesh B.S.	11

Table 10. Marcy Ball Mills Wet Grinding Lead–Zinc Ore

Ref.	Mill size ft.	Feed size in.	Product fineness	Approx. kWh/ton
744	6 × 5	$\frac{1}{2}$	94%—150 mesh B.S.	9·25
	6 × 4$\frac{1}{2}$	$\frac{1}{5}$	40%—150 mesh B.S.	3·4

Table 11. Cylindrical Ball Mills Wet Grinding Lead–Zinc Ore

Ref.	Mill size ft.	Feed size in.	Product fineness	Approx. kWh/ton
563	6 × 6	$\frac{1}{4}$	95%—200 mesh B.S.	11·6
CC/2	7 × 6	$\frac{3}{4}$	54·7%—200 mesh B.S.	8·2
902	10 × 8	−2	80%—200 mesh B.S.	15

Table 12. Hardinge Ball Mills Wet Grinding Copper Ore in Closed Circuit with Classifiers

Ref.	Mill size ft. in.	Feed size in.	Product fineness	Approx. kWh/ton
MA	10 × 72	$\frac{1}{2}$	52%—200 mesh	9·1
RO	10 × 60	$\frac{5}{8}$	68%—200 mesh	10
574	8 × 60 8 × 72	$\frac{3}{8}$	40·2%—200 mesh	8·2
702	10 × 66	$\frac{1}{2}$	61·6%—200 mesh	8·6
633	10 × 66 10 × 66	$\frac{1}{2}$ $\frac{1}{2}$	84·8%—200 mesh 68·5%—200 mesh	14 8
583	10 × 66	$\frac{1}{2}$	90%—200 mesh	11·75
533 Cy	8 × 36	$\frac{3}{8}$	53%—200 mesh	7·14
531 Cy	8 × 30	4 mesh	53%—200 mesh	3·50
589	5 × 36	—48 mesh 74%—200 mesh	97%—200 mesh	6 regrinding concentrates
533	8 × 22	2·6% + 100 41·4%—325 mesh	0·56% + 100 51·64%—325 mesh	5 regrind

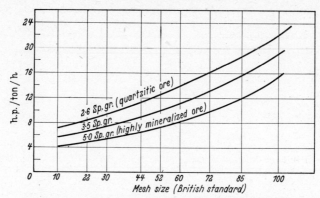

Figure 19. Curves showing approximate power required for wet grinding in ball mills from ½ in. feed to 90 per cent below mesh size stated. (By courtesy of International Combustion Products Limited)

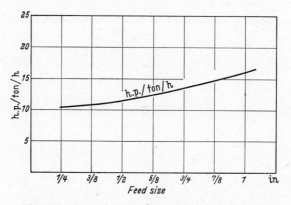

Figure 20. Curve showing effect of feed size when wet grinding pyritic ore to 70 per cent minus 200 mesh in a ball mill. (By courtesy of International Combustion Products Limited)

Table 13. Marcy Ball Mills Wet Grinding Copper Ore in Closed Circuit with Classifiers

Ref.	Mill size ft.	Feed size in.	Product fineness	Approx. kWh/ton
651 PE	8 × 15	24·4%—4 mesh	42·5%—200 mesh	8·25 2-stage grind
551 IN	8 × 6	−2	56·4%—200 mesh	7·54
681 Cy	7 × 5	$\frac{3}{16}$	—35 mesh	6·63 primary grind
PE	6 × 12	$1\frac{3}{4}$	54·75%—200 mesh	8 2-stage
Ca	8 × 6	$\frac{3}{4}$	58%—200 mesh	8·2
MA	9 × 8	$\frac{1}{2}$	51%—200 mesh	9·2
228	6 × 4½	2	50%—200 mesh	10·0

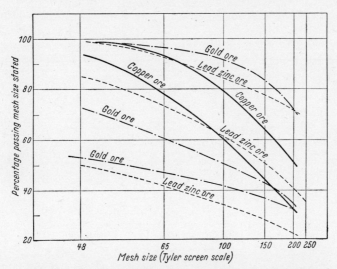

Figure 21. Particle size distribution curves of various types of metallurgical ores wet ground in ball mills. (By courtesy of International Combustion Products Limited)

The curves in *Figure 19* are included to supplement the data in the foregoing tables. These curves indicate the power consumption per ton per hour of product when grinding different grades of ore to various degrees of fineness, from $\frac{1}{2}$ in. feed size. Naturally, the size of feed material also affects the output of a mill, and the curve in *Figure 20* shows the effect of varying sizes of feed on power consumption per ton per hour of product, when wet grinding pyritic ore in a ball mill (with classifier) to a fineness of 70 per cent passing 200 mesh.

Table 14. Cylindrical Ball Mills Wet Grinding Copper Ore

Ref.	Mill size ft.	Feed size in.	Product fineness	Approx. kWh/ton
829	7 × 6	$\frac{1}{2}$	82%—100 mesh	18
556	7 × 5	$\frac{1}{2}$	98%—200 mesh	18 (4 sp. gr.) 2-stage Copper, lead–zinc ore. Sulphide
520	8 × 12	3 mesh	74·2%—200 mesh	10·3
552	8 × 12 6½ × 12	$\frac{3}{8}$	62%—200 mesh	7·35 2-stage
541	7 × 12	1	80%—200 mesh	11·2
532	7 × 10 rod mill 3–7 × 10 ball mills	1¼	47%—200 mesh	5·8 2-stage

It is found when grinding a particular material to different degrees of fineness, that the size gradings of the resultant products follow a certain pattern. Particle size grading curves for gold ore, lead–zinc ore and copper ore, each wet ground in ball mills (with classifiers) to several different degrees of fineness are shown in *Figure 21*.

Table 15. Hardinge Ball and Pebble Mills Dry Grinding

Mineral	Feed size in.	Moisture	Product	Approx. kWh/ton	Type of mill
Anhydrite	1 down	0·5%	82%—200 mesh B.S.	15	Air swept
Anthracite	¼ down	5%	90·5%—200 mesh B.S.	43·8	Air swept with hot air
Anthracite	¼	5%	87·3%—200 mesh B.S.	36·9	Air swept with hot air
Anthracite	¼	5%	70%—200 mesh B.S.	20	Air swept with hot air
Carborundum	1	Dry	93%—100 mesh B.S.	48	Air swept with cold air
Coal (hard)	1¼	3·2%	99%—100 mesh B.S.	26	Air swept with cold air
Coke	1	3·2%	89%—120 mesh B.S.	20·5	Air swept with cold air
Frit	−¾ down	Dry	All—40 mesh B.S.	60·0	Hardinge pebble mill (porcelain balls)
Frit	⅛ down	Dry	All—60 mesh B.S.	45	Hardinge pebble mill (porcelain balls)
Feldspar	¼ down	Dry	98%—200 mesh B.S.	57	Porcelain balls
Graphite	¼ down	0·5%	90%—240 mesh B.S.	75	Air swept ball mill
Sillimanite	20 mesh	1%	99·99%—200 mesh B.S.	125	Air swept ball mill
Sillimanite	20 mesh	1%	82%—200 mesh B.S.	75	Air swept ball mill
Slate (hard)	⅜	0·5%	99%—200 mesh B.S.	68	Air swept ball mill
Zircon sands	−60 mesh 1%—200	Dry	99%—300 mesh B.S.	120	Air swept ball mill

The extension of ball and pebble mill wet grinding practice has been widely applied to various industrial and chemical projects. Grinding calcined flint for pottery slip manufacture is an example of this principle except that, owing to the extreme fineness of the finished product, hydraulic

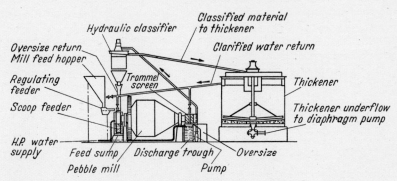

Figure 22. Diagrammatic arrangement of pebble mill wet grinding in closed circuit with hydraulic classifier and thickener for producing fine pottery materials. (*By courtesy of* International Combustion Products Limited)

classification is used in preference to 'mechanical classifiers', as it is important to avoid any disturbance caused by eddying in the final overflow from the classifier. The ground product ranges from 50–65 per cent less than 10 μ. A flow diagram for this type of grinding circuit is shown in *Figure 22*.

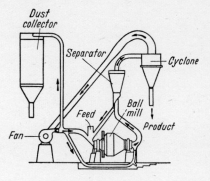

Fig. 23. Diagrammatic arrangement of Hardinge ball mill dry grinding in closed circuit with air classifier. (*By courtesy of* International Combustion Products Limited)

As previously stated, these mills are equally suitable for dry grinding operations and a list of abrasive materials thus treated is appended to indicate the range.

- Anthracite
- Carborundum
- Coke
- Chemical products
- Chromium ore
- Calcined flint
- Feldspar
- Iron borings
- Limestone (siliceous)
- Nickel matte
- Pyrites
- Quartz
- Slate
- Zircon sand

The mills operate in closed circuit with suitable classifiers and the finished products range, according to requirements, from −10 mesh to 0·5 per cent residue on 325 mesh and finer. A typical closed circuit arrangement for a dry grinding ball mill is shown in *Figure 23*.

The feed material should preferably be quite dry (when dry grinding) unless hot gas is introduced into the mill for simultaneous drying during the grinding operation, as the presence of moisture causes general stickiness of the material which in this condition retards grinding, and in severe cases, results in caking on the mill liners and balls and on the internal surfaces of the air classifier, *etc.* The curve in *Figure 24* shows the effect on the output

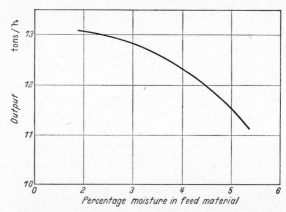

Figure 24. Curve showing reduction in output with increased moisture content when grinding coal in a conical ball mill. (Fineness of product 90 per cent minus 100 mesh B.S.). (By courtesy of International Combustion Products Limited)

of a ball mill when grinding coal containing various amounts of moisture, without provision for drying. Grinding data on various materials, dry ground to varying degrees of fineness in ball and pebble mills, is given in *Table 15*.

Tricone Mill

The Hardinge Tricone ball mill is a modification of the conical mill and has a tapered barrel portion between the two ends which causes ball

Table 16. *Improvements in Mill Performance Due to Decreased Ball Load*

	Now	Before
Ball load, tons	35	55
Ball load, % volumes	29%	45%
Tonnage	2,250	2,130
Horsepower requirements	370	490
Ball consumption lb./ton	0·84	1·02
Grind: sulphides—200 mesh	65%	62%
Grind: gangue—200 mesh	35%	40%
Tails: copper	0·068%	0·08%
Tails: sulphur	2·5%	4·0%
Sod: ethyl xanthate lb./ton	0·28	0·34

segregation. It is essentially a slow speed mill for fine and intermediate grinding. One of its primary features is the fact that, with this shape, a greater unit capacity is obtained per unit of diameter.

Large-scale experiments have been carried out at the Tennessee Copper Company. Lewis, superintendent at Copperhill, who carried out these experiments, reduced the normal ball charge with beneficial results to capacity and extraction of values. See *Table 16*, reprinted from *Engineering and Mining Journal*, September 1953.

This tricone mill was run at a much lower speed than normal, whereby the action of the balls was one of rolling rather than cascading. Under these conditions a large quiescent pool is formed which induces classification within the mill.

It should, however, be noted that in this particular experiment there was a considerable difference in specific gravity between the ore-carrying values and the barren siliceous gangue, which naturally aided classification of the pulp in the mill.

Tube Mills

The tube mill is the oldest form of mill in which tumbling media are used to effect reduction. It consists of a cylindrical shell, the length of which is several times its diameter. The interior is often divided into compartments to enable the reduction of the material to be carried out in stages as it passes through the mill, particularly when it is used for grinding cement clinker. The first compartment, where the coarsest grinding is done, is charged with the largest balls, and the last compartment with the smallest. Grates and lifters are fitted between compartments, the operation of these being clearly shown in *Figure 25*. Due to the considerable length of these mills, grinding is usually completed in one pass through the mill, *i.e.* open circuit operation.

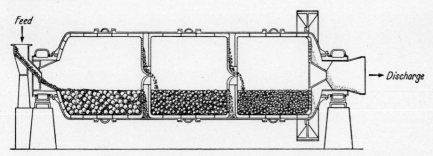

Figure 25. Typical cross section of compartmented tube mill showing ball-retaining grates and lifters. (By courtesy of Hardinge Co. Inc., U.S.A.)

In the larger cement plants, open circuit compound (compartmented) mills are largely used. The requirements of the finished grind differ from most other industrial undertakings in that a maximum number of different size particles are required. This has a relation to the packing density of the finished cement and to the setting qualities.

There are many installations of closed circuit grinding of cement clinker, but the product is rarely used for rapid hardening cement. In such a product there is less oversize than with the open circuit system on, say, 200 mesh, but there is also less minus 5 μ powder. On the other hand, products ground to one-half of 1 per cent on 180 mesh in open circuit, as in the cement field, would be unsuitable for many industrial purposes where specifications may require less than one-tenth of 1 per cent on 300 mesh. The power absorbed in grinding to the latter fineness by the open circuit system would be uneconomical compared with the closed circuit method.

Rod Mills

A rod mill consists of a cylinder, with vertical or slightly tapered ends, revolving in a horizontal plane and containing a charge of steel rods. The length is normally about twice the diameter and the inside of the mill shell is usually lined with manganese or chromium steel liners. The rods are

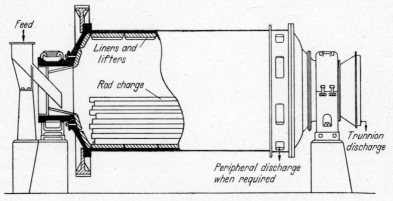

Figure 26. *Typical cross section of rod mill.* (*By courtesy of* Hardinge Co. Inc., U.S.A.)

about 3 in. shorter than the length of the mill. Its role is mainly that of a fine crusher or primary grinder, and a typical section through the rod mill is shown in *Figure 26*. These mills normally run at a lower speed than ball mills of the same diameter and can be used with either trunnion or peripheral discharge.

Grinding in a rod mill is by line contact, and as the material at the feed end is coarser than that near the discharge end it is obvious that the rods are spaced wider apart at the feed end than at the discharge end. This tapering of the rod charge has a screening effect and reduces the possibility of discharging very coarse pieces—an important factor when preparing a fine feed for subsequent grinding and also in those cases where the product is required for direct use in processing.

Since the rods are spaced by the coarsest pieces of material lying between them it follows that the smaller pieces are not subjected to the direct grinding action of the rods until the coarser particles have been reduced in size. This feature tends to produce more uniform sizing, and it is an established fact that the product from the rod mill is more granular than that from a ball mill in which point contact of the grinding media occurs. The granular property of a powder is a highly desirable feature in many processes. On the other hand, in the case of ore reduction, a granular product would not be sufficiently fine to release the metallic values for subsequent extraction.

Table 17. Rod Mills—Wet Grinding (Open Circuit Unless Noted)

Ref.	Mill size ft.	Material	Feed size	Product fineness	Approx. output tons/h	Approx. kWh/ton
S.H.	6 × 12 closed circuit with classifier	Gold ore quartz and shale	15% + 1 in.	40%—200 mesh	10	10
St.J.	6 × 12	Lead–zinc quartzitic	$-\frac{1}{4}$ in.	14 mesh	18	5·2
F.Z.	4 × 10	Zinc quartzitic	75% + $\frac{1}{2}$ in.	$-\frac{3}{8}$ in. 44%—20 mesh	16	2·3
P.D.	6 × 12	Copper porphyry	$-1\frac{3}{4}$ in.	10 mesh	25	4·5
O.D.	4 × 8	Copper ore	14% + 4 mesh	99·1%—20 mesh	6	4·5
I.C.	5 × 10 closed circuit with classifier	Copper ore	$-\frac{3}{4}$ in.	38%—200 mesh	10	6
Qu.	11½ × 12	Lead–zinc ore	21% + 1 in.	5·8% + 4 mesh 15%—200 mesh	200 to 250	3
G.E.	5 × 10	Jasperized granite	$-1\frac{1}{2}$ in.	20 mesh	4	16

Table 18. Rod Mills—Dry Grinding

Ref.	Mill size ft.	Material	Feed size	Size of product	Approx. output tons/h	Approx. kWh/ton
H.A.	5 × 10	Coke breeze	$\frac{3}{4}$ in.	3% + 8 mesh	5	12
A.S.	3 × 8	Sodium sulphate crystals	$\frac{1}{2}$ in. + $\frac{3}{32}$ in.	$-\frac{3}{32}$ in.	3	5·5
St.J.	5 × 10	Coke 12% H₂O	$-1\frac{1}{2}$ in.	45% + 10 30% − 20 mesh	9	7·5
Al.	5 × 12	Coke	1 in. 2·7% + 4, 69%—14 mesh	10% + 20 mesh	4·5	16·5
Am.S.	5 × 10	Coke 10–12% H₂O	37·2% + ½ in. 45% + 8 mesh	3·1% + ¼ in. 67% + 30 mesh	5·5	9

It is interesting to note that about twenty years ago many regrinding rod mills, in closed circuit with classifiers, were installed on copper ores in the U.S.A., but after a few months of experimenting the rods were taken out and replaced by a charge of steel balls.

Rod mills are mainly used in open circuit, but, in those cases where it is essential to control the upper particle size of the product, advantages are gained by operating these mills in closed circuit with screens. Apart from the application in conjunction with ore reduction in ball mills mentioned in the section 'Crushing Equipment', rod mills are used for the dry grinding of coke, sand, lime, *etc.*, where granular products are of prime importance. These mills are used for the preparation of sand-lime bricks, grinding from 1 in. to approximately 10 mesh in one stage. The rod mill, by reason of its line contact and the granular grind, has advantages over the ball mill in that coke with a moisture content of approximately 10 per cent can be effectively dealt with in a rod mill grinding from approximately $1\frac{1}{2}$ in. to 10 mesh. Performance data of rod mills on the wet grinding of metallurgical ores are given in *Table 17* and dry grinding results appear in *Table 18*.

ROLLER MILLS

Roller mills have a wide application in the industrial and chemical processing of non-abrasive raw materials. A partial list of materials to show the range is given below:

>Barytes
>Calcined magnesite*
>Chalk
>Chemical products
>Clay
>Coal
>Copal gum
>Fullers earth
>Gypsum
>Hematite
>Limestone
>Phosphates
>Titanium oxide

* N.B. Ball mills are normally used for dead burnt magnesite.

Raymond Roller Mill

As a point of historical interest, the well-known Raymond mill is an adaption, for dry grinding, of the original wet grinding Huntington mill used in ore reduction at the latter end of the last century. The maintenance cost, using abrasive ores, was extremely high, but it was most satisfactory as an amalgamator when grinding gold ores.

The Raymond roller mill, a cross section of which is shown in *Figure 27*, combines dry grinding and air separating simultaneously. The main base casting is supported on the two foundation piers. Underneath this casting is carried a pair of bevel gears, in a housing, which transmit the drive from the horizontal shaft to the vertical shaft. This vertical centre shaft carries a spider casting at the upper end from which hang the pendulums or roller arms which may be two to five in number, depending on the size of the machine, and at the lower end of each is mounted a grinding roll, which is free to rotate on the pendulum shaft and mates with the bull-ring seated in the main base casting. Each pendulum is mounted on the spider by a small cross shaft, which allows the pendulum to swing radially from the centre shaft. In the stationary position the rolls only lightly touch the bull-ring, but when the centre shaft is rotating centrifugal force holds the rolls to the ring. Grinding is carried out by the rolling action of the rolls as they run around the bull-ring.

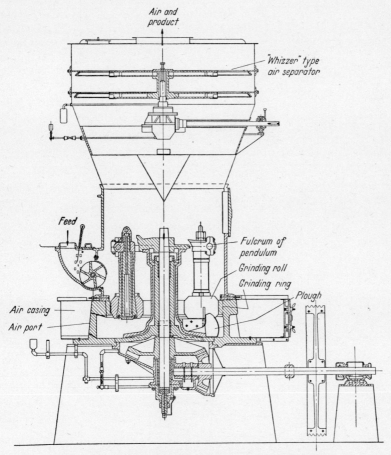

Figure 27. *Section through Raymond roller mill.* (*By courtesy of International Combustion Products Limited*)

Around the side of the base casting and situated beneath the bull-ring are a series of tangential air ports through which the air enters from the surrounding air casing. Also attached to the centre shaft in line with the base casting is a second spider which carries a series of ploughs, one preceding each pendulum. These ploughs scoop up the material in the base and project it between the rolls and bull-ring where grinding is effected.

Raw material is fed into the mill through a rotary feeder mounted on the side of the mill body. This method is employed to prevent ingress of air into the grinding system. Mounted directly on top of the mill is the

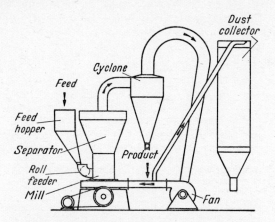

Figure 28. Diagrammatic arrangement of Raymond roller mill and combined air separating plant. (By courtesy of International Combustion Products Limited)

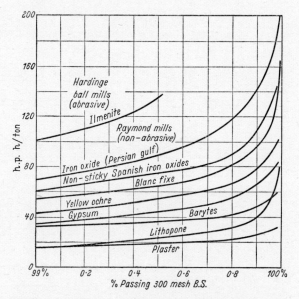

Figure 29. Graph (with comparison curve) showing relative power figures for fine grinding various materials in Raymond roller mills. (By courtesy of International Combustion Products Limited)

separator, the function of which is to return all oversize to the mill for regrinding, whilst the material ground to a finished size is carried by the air stream on to the cyclone and so out of the mill system. A diagrammatic arrangement of the mill and air plant is shown in *Figure 28*. A similar circuit is employed with other mills shown in *Figures 31, 32* and *33*. (The action of the separator is described more fully in a later subsection.)

This type of mill is very widely used for grinding coal, limestone, gypsum, barytes, phosphate rock, chalk, oxides, ochres, precipitated

Table 19. *Raymond Roller Mills—Dry Grinding*

Material	Feed size in.	Moisture per cent	Product fineness	Approx. kWh/ton
Barytes	½ down	0.5	99%—300 mesh B.S.	50
Bone charcoal	30 mesh	3.5	96%—300 mesh B.S.	62
Chalk	1 down	0.2	95%—150 mesh B.S.	25
Chalk	1 down	0.2	95%—300 mesh B.S.	45
Diatomaceous earth	1 down	Dry	96%—60 mesh B.S.	10
Fullers earth	¾ down	—	99.8%—200 mesh B.S.	40
Gypsum	¾	1	90%—120 mesh B.S.	15
Iron oxide	½	0.5	99%—300 mesh B.S.	100
Slate	⅜	0.5	99%—240 mesh B.S.	28
Sulphur	¾	0.5	99%—300 mesh B.S.	58
Synthetic resin	½	Dry	99%—200 mesh B.S.	140
Talc	½	0.5	99%—300 mesh B.S.	65
Titanium oxide	½	Dry	99.9%—300 mesh B.S.	100
Umber	½	Dry	99.9%—300 mesh B.S.	50

Table 20. *Test Data on Raymond Mill Grinding Five Different Grades of Oxides to Approximately 99.8 per cent—240 Mesh B.S. and 99.98 per cent—240 Mesh B.S.*

Material	Yellow ochre	Red ochre	Oxide B	Oxide A	Persian Gulf
Fineness—240 mesh B.S.	99.80%	99.80%	99.8%	99.8%	99.75%
Capacity lb./h	925	2,000	570	960	670
Net power BHP	23.8	23.8	29.0	24.2	26.7
Feed size	1½ in. to dust	1½ in. to dust	1 in. to dust	1 in. to dust	1 in. to dust
Moisture	1.70%	3.45%	0.75%	0.04%	1.70%
Fineness—240 mesh B.S.	99.98%	99.98%	99.96%	99.98%	99.96%
Capacity lb./h	625	1,200	385	450	400
Net power BHP	21.0	20.70	26.7	23.8	23.8
Feed size	1½ in. to dust	1½ in. to dust	1 in. to dust	1 in. to dust	1 in. to dust
Moisture	1.70%	3.45%	0.75%	0.04%	1.70%

chemicals, pigments, *etc.*, and generally for minerals not exceeding a hardness figure of 4 on Mohr's scale. The finished product from this type of mill can range from approximately 80 per cent passing 100 mesh to 0·1 per cent residue on 325 mesh. The form of separator fitted to the mill depends upon the type of material and the fineness to which it is ground. In this connection, attention is drawn to the fact that when fine grinding, it is most

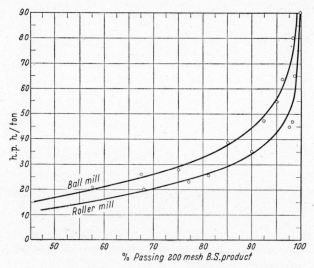

Figure 30. Comparison of power per ton expended by ball and roller mills grinding limestone—⅜ in. feed—1 per cent moisture. (By courtesy of International Combustion Products Limited)

important to give accurate specifications of the fineness required. The curves in *Figure 29* show how the power absorbed increases rapidly in the finer stages. The specification 'all through 200 mesh' means little; if such a product is required the specification should state the permissible residue on a finer mesh—for example, 99·9 per cent passing 300 mesh—thus providing a degree of measurement.

When grinding materials within their scope through a given range of reduction, roller mills are more economical than ball mills and the relative power consumption of these two types of mill when grinding limestone is shown by the curves in *Figure 30*. Operating information on a range of materials ground in Raymond roller mills is given in *Table 19* and test data on five different grades of oxides appears in *Table 20*.

The Bradley-Poitte mill *(Figure 31)* grinds on the same principle as the Raymond roller mill previously described. The design of this machine differs in that the drive is from above the grinding chamber, thus allowing the machine to rest directly on the floor. The type of separator fitted can again be varied to suit the fineness of product required as in the Raymond mill.

The Babcock and Wilcox type 'E' mill illustrated in *Figure 32* consists essentially of two horizontal grinding rings between which is arranged a row of steel balls. The lower ring, on which is also mounted the rotating separator, is driven through gears from a horizontal shaft. Material is fed into the middle of the top ring, pulverizing action taking place as it finds its

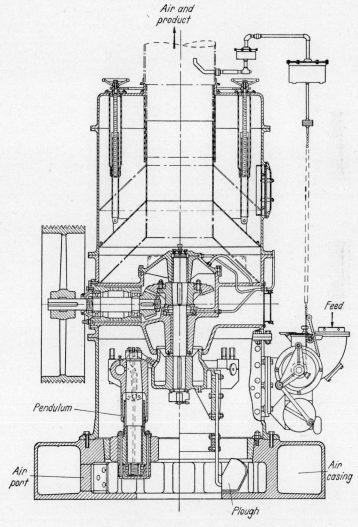

Figure 31. Section through Bradley-Poitte mill. (By courtesy of Bradley Pulverizer Company)

way, aided by centrifugal force, between the rotating balls from whence it drops over the edge of the bottom ring into an upward current of air. Separation is effected by the rotating classifier, the fines passing out at the top with the air, whilst the oversize particles are returned by gravity for further grinding. This type of mill is widely used for grinding coal in connection with the pulverized fuel firing of boilers.

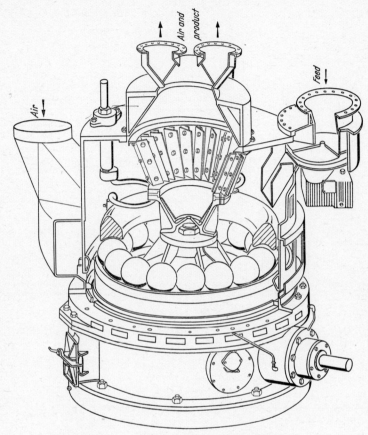

Figure 32. *Section through Babcock & Wilcox 'E' type mill.*
(*By courtesy of* Babcock & Wilcox Limited)

The Lopulco mill (*Figure 33*) is a medium speed machine grinding between heavy inclined conical rolls and a rotating table. The table is driven through a totally enclosed gear box at a speed suitable for the material being ground. The two diametrically opposed inclined rolls have conical grinding faces and rotate freely on their spindles, being driven through contact with the material which is being ground on the revolving table. Adjustable stops on the mill sides prevent the rolls coming into contact with the table and this eliminates metal-to-metal wear. The rolls are spring-loaded by coupling the ends of the trunnion arms with spring gear which can be adjusted externally to give the required degree of load. A notable feature of this system is that any uneven distribution of material on the table is automatically compensated by the spring gear, thus ensuring a

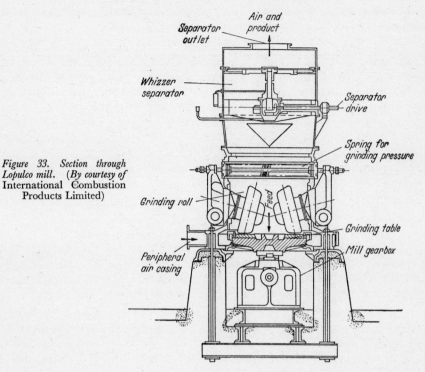

Figure 33. Section through Lopulco mill. (*By courtesy of* International Combustion Products Limited)

balanced load on the table under all conditions. A separator of the revolving blade type is mounted in a circular casing immediately above the mill and driven by a separate motor. A variable speed drive provides

Table 21. *Lopulco Mills*

Mineral or Material	Feed size in.	Moisture per cent	Product	Approx. kWh/ton
Coal	½	8	75%—200 mesh B.S.	16
Limestone . . .	¾	1	55%—100 mesh B.S.	10
Limestone . . .	¾	1	56%—200 mesh B.S.	12
Limestone . . .	1	—	85%—200 mesh B.S.	17
Limestone . . .	−¾	1	80%—200 mesh B.S.	23
Gramophone records . .	¼	Dry	98%—200 mesh B.S.	62
Phosphate (Morocco) .	1	2	75% 100 mesh B.S.	10·5
Clay	−3/16	10	99%—30 mesh B.S.	10·5
Bog ore (for gas purification)	¾	1	70%—100 mesh B.S.	12·8
Congo copal gum . .	½	—	85%—200 mesh B.S.	112

adjustment for obtaining the required product fineness, and oversize is returned by gravity to the mill for regrinding.

The Lopulco mill has a very extensive field of application in the pulverizing of coal for combustion and is also used for grinding industrial materials such as limestone, gypsum, phosphate rock, ranging from a product size of minus 10 mesh to 99 per cent passing 200 mesh. *Table 21* gives performance data on a list of materials ground in this mill.

HIGH SPEED MILLS

For coarse and intermediate grinding the following list of materials indicates some of the applications of high speed mills:

Asphalt	Fullers earth
Antimony sulphide	Lead arsenic
Borax	Magnesium carbonate
Copper oxide	Oil cake
Calcium phosphate	Pitch
Chalk	Resins
China clay	Soya beans
Fish meal	Synthetic resin

The many designs of high speed mills, including the peg and serrated disc mills, have certain applications which are unsuitable for mills in Groups 1 and 2 owing to the difficult nature, as distinct from hardness, of the material, *e.g.* asphalt, pitch and other compounds.

On the other hand, high speed hammer mills have been used to grind coal for tin smelting, *etc.*, the fineness being 87 per cent passing 200 mesh. Each high speed mill feeds one of several tin smelting furnaces, occupies very little room, and forms an ideal unit for this purpose; the hammers being reversed after two or three weeks' use. With such mills, it is a significant factor that, when grinding some of the harder varieties of raw materials, the capacity and/or fineness drops rapidly as the hammers or grinding elements wear. The comparative rate of wear is shown in *Figure 12*, to which reference has previously been made in this section.

The Atritor pulverizer, *Figure 34*, is a self-contained unit which pulverizes entirely by attrition—hence its name. The principal components are a regulating feeder, pulverizer, and fan, and the machine functions as follows:

Material is delivered to the pulverizer chamber from the hopper by means of the table feeder. The quantity of material fed being regulated by an adjustable knife controlled by an external handwheel. As the material passes into the first pulverizing compartment it is subjected to the disintegrating effect of a number of hammer segments on the rotor. After this initial reduction in size the material is carried by the air stream over the periphery of the rotor into the second, or main pulverizing zone. In this zone the movement of the particles is complex due to the turbulence set up by the rotor segments and the fixed interrupter pegs. The material is then drawn by the fan through the eye of the diaphragm, past the rejecters (which return oversize particles for further grinding), and out of the mill into a suitable collecting cyclone.

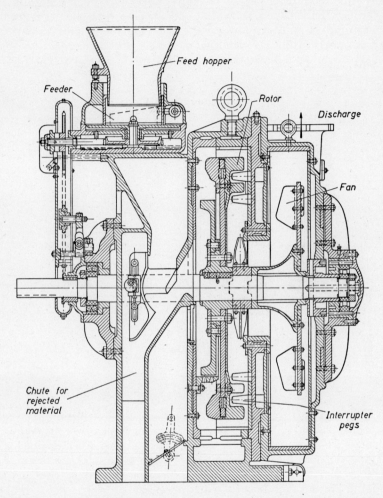

Figure 34. Section through Atritor pulverizer. (By courtesy of Alfred Herbert Ltd.)

The Atritor is used for grinding such materials as clays, chalk, dyestuffs, gypsum and limestone, and also for grinding fuel for combustion purposes. As with most of the air swept mills already described, the Atritor can be used for combined drying and grinding by the introduction of hot air into the air system.

Pin Disc Mill

This class of pulverizer is suitable for the grinding of friable, fairly soft substances such as foodstuffs, drugs, chemicals in crystal form, and some types of fibrous roots. A typical fineness of product being 70–80 per cent passing 100 mesh.

A section through a Kek mill, which is representative of this type, is shown in *Figure 35*. The material to be ground is fed at a controlled rate down a tube on to the centre of the lower disc. Circular rows of pins are fitted to the lower revolving disc between rows of similar pins mounted in the upper stationary disc. As the material is propelled by centrifugal action towards

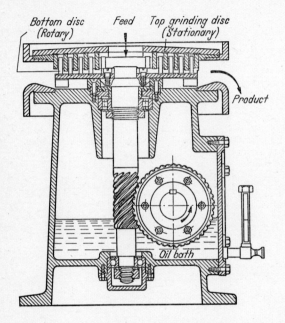

Figure 35. Section through Kek pin disc pulverizer. (By courtesy of Kek Limited)

the periphery it has to pass between these pins, thus, reduction is effected by impact and attrition. The ground material then spills over the periphery of the machine and is collected in an annular container which is fitted with a bagging spout at the bottom and air release filter stockings at the top. These machines are easily damaged by tramp iron or hard material and it is essential to prevent such material entering the mill.

Bar Type Disintegrator with Revolving Cages

This type of disintegrator consists of two or more concentric squirrel cages, the adjacent cages revolving in opposite directions in a vertical plane and each being driven independently. It is made in cage sizes up to 48 in. diam. requiring in the order of 70 h.p. and is used for the breaking of clays, marls, chemical manures, spent oxide, *etc*. Material is fed into the centre of the cages and receives a series of rapid blows as it works its way out to the periphery where it then drops to the bottom and is collected in a hopper.

Swing Hammer Pulverizer with Air Plant

Swing hammer pulverizers are also manufactured with air swept grinding chambers, the ground material being removed by an air stream which is led to a cyclone collector where the material is separated; the relatively clean air returning to the pulverizer for recirculation. An example of a machine of this type is the Raymond Impax Pulverizer which is shown in *Figure 36*.

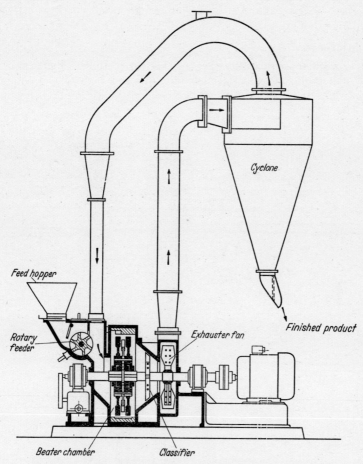

Figure 36. Raymond Impax Pulverizer. (By courtesy of International Combustion Products Limited*)*

The material is delivered to the grinding chamber by the rotary feeder; as grinding proceeds the material is caught up in the air stream and after passing through the fineness regulator and fan is conveyed to the cyclone. The ground material is separated from the air stream by means of the cyclone and is discharged through an airtight non-return valve at the bottom of the cyclone, whilst the relatively clean air passes out at the top and returns to the pulverizer inlet for recirculation.

This kind of machine can be used very effectively for grinding materials with a hardness of 4 or less on Mohr's scale, and the size of product can be varied from a fairly coarse material to one containing approximately 90 per cent passing 200 mesh. Where finer products are desired an air separator can be incorporated in the system, with the oversize from the separator automatically returning to the mill for regrinding. On the coarser products the fineness is adjusted by a simple movement of the regulator inside the pulverizer body. Simultaneous grinding and drying can be achieved by introducing hot air or gas into the circuit.

The Micronizer

This machine is probably the best known of those which come under the fourth group of pulverizers, namely, those which effect reduction by bombarding the particles of material against each other. The general

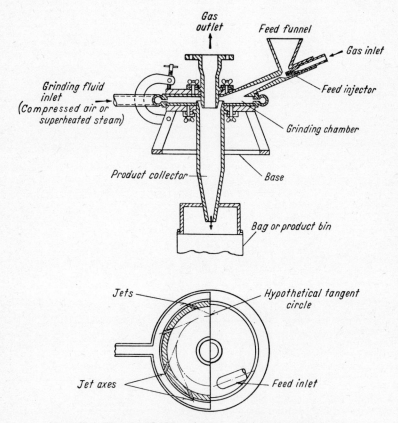

Figure 37. *Micronizer fluid jet pulverizer.* (*By courtesy of* Micronizer Processing Company, Inc., U.S.A.)

arrangement of the Micronizer is shown in *Figure 37* and essentially the machine consists of a shallow circular grinding chamber wherein a circulating charge of material is acted upon by a number of gaseous fluid jets issuing through orifices spaced around the periphery of the chamber. The gaseous fluid is usually either compressed air at approximately 100 lb./in.^{2} or superheated steam at pressures from 100 to 200 lb./in.^{2} and temperatures ranging from 400° F to 1,000° F. The grinding chambers vary from 2 to 30 in. in diameter and from $\frac{3}{8}$ to 2 in. axial height at the periphery.

Table 22. *Micronizer-Grinding Data for Various Materials.* (*By courtesy of* Micronizer Processing Company, Inc., U.S.A.)

Feed size approximately 40 to 60 mesh

Material	Mill diameter in.	Grinding medium Type	Grinding medium Flow	Feed rate material lb./h	Approx. avg. particle size μ
DDT (50%)	24	Air	1,000 c.f.m.	900	5–6
Sulphur	24	Air	1,000 c.f.m.	1,300	3–4
Calomel	24	Air	1,000 c.f.m.	800	2–3 12 top
Bakelite	24	Air	1,000 c.f.m.	300	5 40 top
Titanium dioxide	30	Steam	4,000 lb./h	2,250	Less than 1
Cryolite	30	Steam	4,000 lb./h	1,000	3
Barytes	30	Steam	4,000 lb./h	1,800	3–4
Talc (varies)	30	Steam	4,000 lb./h	2,000	2
Iron oxide pigment	30	Steam	4,000 lb./h	1,000	2–3
Anthracite coal	20	Air	1,000 c.f.m.	1,000	5–6
Fullers earth	20	Steam	1,200 lb./h	600	3–4 (15 top size)
Procaine-penicillin	8	Air	100 c.f.m.	10	5 (20 top size)

Air–100–105 lb. pressure 60° F.   Steam–150 lb. pressure–550° F.

The jet orifices vary from 3 to 16 in number, equally spaced, and are arranged tangentially to a common circle, so promoting rotation of the contents of the grinding chamber. As the pressure inside the grinding chamber is substantially atmospheric, the expansion of the compressed gaseous fluid leaving the jets releases energy and causes high speed rotation of the contents with turbulence and bombardment of the particles against each other. Radial acceleration causes the material to concentrate adjacent to the periphery, and local to each jet violent agitation occurs and it is this which is responsible for most of the reduction.

There is an intense centrifugal classifying action within the grinding chamber and the fluid eventually leaves the machine through the central opening at the top whilst the fine product is separated by cyclonic action and drops into a container underneath. Practically all applications of the Micronizer are for products in the sub-sieve range, say 20 μ and below. The information in *Table 22* listing performance data of different size machines, handling various materials, has been provided by the Micronizer Processing Company, Inc., U.S.A.

Wheeler Fluid Energy Mill

Another pulverizer in the fourth group is the Wheeler Fluid Energy Mill, a cross section of which is shown in *Figure 38*. Raw material (minus 4 mesh or smaller) is delivered from the feed hopper into the reduction chamber by means of an injector. Pressurized fluid enters the chamber through the nozzles and bombards against each other particles entrained in the circulating fluid. As the stream of fluid rounds the top bend,

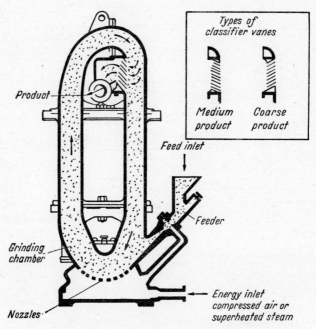

Figure 38. *Wheeler fluid energy reduction mill.* (By courtesy of C. H. Wheeler Manufacturing Co., U.S.A.)

centrifugal force classifies the particles and, by adjustment of the separator vanes, particles below a predetermined size are carried out with the excess fluid leaving the pulverizer. Oversize particles continue with the stream and re-enter the reduction chamber for further grinding. Pressure in the system is usually slightly above atmospheric.

NOTES ON COAL PULVERIZING FOR BOILER FIRING PURPOSES

With the original 'central' system (or storage bin and feeder system) the pulverized coal preparation plant consisted of raw coal storage bunkers feeding to drying equipment from which the dried coal passed to mills for grinding to the required degree of fineness, and the product was transferred to the boiler feed storage bunkers by compressed air pumps. This system resulted in high capital, maintenance and operating costs which were appreciably reduced when hot gas taken from the boiler passes was introduced into the milling circuit to effect simultaneous drying with the grinding process.

The most economical method has been achieved by the 'unit' firing system which provides for grinding, drying and delivering the product direct from the mill to the burners in one operation. During the past 30 years of development of pulverized fuel firing in this country, the 'unit' firing system has almost completely superseded the 'central' system, the latter now being installed only where the most refractory type of coal, such as South Wales anthracite, is to be used. On the 'unit' firing system it is essential that the mill should have flexibility in output to meet variations in boiler fuel requirements and it is this feature of the Lopulco mill (together with satisfactory maintenance and power consumption characteristics) which has been responsible for its extensive use in this field.

The hot air used for drying in the milling circuit is used as the primary air for combustion and it is possible to deal with bituminous coals having a total moisture as high as 20–25 per cent. In order to obtain good furnace conditions with an average bituminous coal it is necessary for the pulverized coal delivered to the burners to be of such fineness that approximately 95 per cent will pass through a 100 mesh B.S. sieve or approximately 77 per cent will pass through a 200 mesh B.S. sieve. This degree of fineness, derived from practice, represents the optimum to give good furnace conditions with economic milling plant power consumption and minimum wear of grinding parts, *etc.*

A typical modern pulverized fuel fired steam generating unit having an evaporation of 550,000 lb./h and producing steam at 950 lb./in.^{2} and 925° F would burn approximately 30 tons/h of coal. To provide this quantity of pulverized coal the boiler would be equipped with four sets of milling plant arranged on the 'unit' firing system, each set comprising a raw coal feeder, a pulverizing mill with classifier and an exhauster or primary air fan. Each of the four mills would be capable of producing approximately 10 tons/h of pulverized coal, but only three of the mills would normally be used to fire the boiler. The fourth mill is provided to ensure that full boiler output can be maintained when inferior coals have to be burnt or when one of the other three mills is taken out of service for routine maintenance. With six steam generating units of the above capacity installed in a modern base load power station, the weekly coal consumption would be in the order of 25,000 tons.

By considering the large number of modern power stations now in operation, it is possible to obtain a rough idea of the enormous tonnage of coal and anthracite milled in this phase of industry alone.

4

THE MECHANICS OF PULVERIZERS

HAROLD HEYWOOD

CLASSIFICATION

DETAILED descriptions of industrial pulverizers have been given in Section 3. Pulverizers may be classified into the following three groups according to the mechanical principles of operation:

1. Slow speed: ball and tube mills, including rod mills.
2. Medium speed: ring roll pulverizers, including ball and ring mills.
3. High speed: impact mills, beater mills.

The majority of mills used for the pulverization of coal are air swept, that is, the finely ground material is removed by an air stream. The three variables concerned in the operation of a pulverizer having a given coal feed are power, throughput and fineness of the product. Under normal conditions of operation the fineness of the product is maintained constant within fairly close limits, and consideration will first be given to the relationship between power and throughput.

Relationship between Power and Throughput

Pulverizers may be classified into two ideal groups; those in which the power is constant, independent of the rate of throughput, and those in which power is proportional to the rate of throughput. These ideal relationships are not fulfilled by industrial mills, which have intermediate performance characteristics and according to the design tend towards one or other of the above ideals.

The ball and tube mill approaches closely to the constant power group, since the useful energy is absorbed by lifting the ball charge, and the rate of throughput of feed material has relatively little effect. A high speed impact mill approaches fairly closely to the proportional power group if the fan power is deducted from the total power input, since the friction losses are relatively small and consequently the grinding power is more nearly proportional to the rate of throughput. Ring-roll mills, and others operating on similar principles, have power characteristics intermediate between these two groups. The relationship between power and throughput for these three types of mill is shown diagrammatically in *Figure 1*.

The general case of a variable power mill is next considered in detail. *Figure 2* (*a*) shows the relationship between power in kilowatts and throughput in tons per hour. The energy in kilowatt-hours per ton at any given rate of pulverizing is obtained by dividing the power in kilowatts by the corresponding throughput. Thus at an output represented by the point B, the energy in kilowatt-hours per ton is AB/OB, which is also the slope of a line drawn from the point A on the curve to the origin O. The energy demand per ton of coal will, therefore, be a minimum when such a line drawn from the origin has the minimum slope, *i.e.* becomes tangential to the curve, as at the point *C*. At greater rates of throughput the mill becomes overloaded and the energy per ton increases. The relationship between

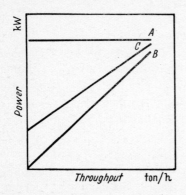

Figure 1. Ideal types of pulverizers
A = constant power;
B = proportional power;
C = variable power

energy per ton and throughput for this type of mill is shown graphically in *Figure 2 (b)*, and may be stated as follows:

The optimum conditions as regards power and rate of throughput are defined by the point at which a line drawn from the origin of the graph is tangential to the curve relating these two variables.

In the ideal case of the constant power mill this tangent point will not occur until the rate of throughput is infinite, and hence the energy per ton of coal will continue to decrease as the throughput increases until the limiting capacity is attained. In the case of the ideal proportional power

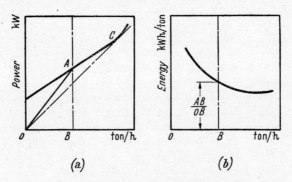

Figure 2. Characteristics of variable power pulverizers: (a) power and throughput; (b) energy and throughput

mill, the power curve coincides with the tangent passing through the origin and hence the energy per ton is constant at all throughputs. These relationships may be derived mathematically for the three types of mills on the assumption that power and throughput are related over the normal working range by a straight line.

Let P_t = Total power at the optimum throughput of L tons per hour.
P_e = Power to drive mill when empty.
a = Proportion of optimum capacity at which the pulverizer is working.

Power demand at capacity aL tons per hour = $P_e + a(P_t - P_e)$.

Taking the total power P_t at the optimum capacity L as the basis for reference:

$$\text{Relative power at capacity } aL = \frac{P_e}{P_t} + \frac{a(P_t - P_e)}{P_t}$$

$$= \frac{P_e}{P_t}(1-a) + a$$

$$\text{and relative energy per ton} = \frac{P_e}{P_t}\frac{(1-a)}{a} + 1$$

The ratio P_e/P_t would be 1 for the constant power mill, though tests on a ball mill have shown the actual ratio to be about 0·9. Typical values of the ratio P_e/P_t are 0·4 for a ring roll mill and 0·25 for an impact mill. The relative energies at reduced rates of pulverization are given in *Table 1*, which shows the advantage of the high speed impact mill when there is a wide range of throughput.

Table 1

Type of mill	Ratio P_e/P_t	Relative energy per ton at reduced throughput		
		Normal	Half	Quarter
Ball and tube	0·9	1	1·9	3·7
Ring-roll	0·4	1	1·4	2·2
Impact	0·25	1	1·25	1·75

Relationship between Power, Throughput and Fineness of Product

The effect of fineness of grinding of the mill products has been ignored in the preceding subsection, but must be included for a complete analysis of the energy characteristics of a pulverizer. In order to incorporate the fineness of grinding as a factor with the other variables, this must be expressed in the same system of units, and for this purpose is measured by the specific surface, *i.e.* the total surface area of all the particles in square feet per pound of material. It is also convenient to transform the energy units from kilowatt-hours per ton to foot-pounds per pound and to express the throughput in pounds per hour. (1 kWh/ton = 1,185 ft.Lb./lb.)

These variables may be plotted along three axes that are mutually perpendicular as in *Figure 3*, and the complete characteristics of the pulverizer are then shown by a curve in space as illustrated in *Figure 4* and *Figure 5*. The projections of this curve in space on to the three reference planes give three individual graphs which represent respectively the relationship between energy and throughput; energy and specific surface; specific surface and throughput. *Figure 4* shows an isometric diagram of such a system and *Figure 5* shows the reference planes opened out into one plane. The curve in *Plane 1* corresponds to the curve shown in *Figure 2 (b)*.

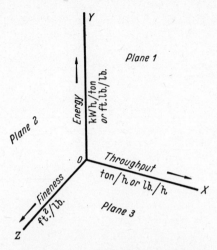

Figure 3. Axes and reference planes for relating energy, throughput and fineness

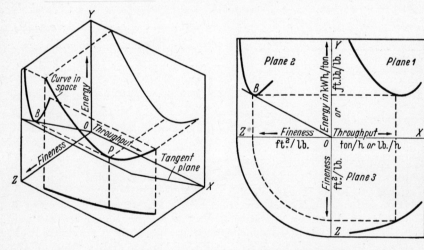

Figure 4. Curve in three dimensions relating energy, throughput and fineness

Figure 5. Projections of three-dimensional curve relating energy, throughput and fineness

The effective output of the pulverizer is measured by the surface area of the product, and therefore the optimum conditions of operation will be attained when the energy required to produce 1 ft.^{2} of surface is a minimum. The two factors of energy and specific surface are related by the curve in *Plane 2*, and the optimum conditions correspond to the point at which a line drawn from the origin is tangential to this curve, *i.e.* at point B in *Figure 5*. Hence by analogy with the previous subsection it can be stated that:

> The optimum conditions as regards energy, throughput and fineness are defined by the point at which a plane passing through the axis of throughput is tangential to the curve in space relating these three variables.

This plane and the tangent point P are shown in *Figure 4*, and *Figure 5* shows that this point does not coincide with the point of minimum power per ton of throughput.

Analysis of Energy Losses

Commencing with the electrical energy supplied to the motor driving the pulverizer, losses occur in the following sequence.

Electrical losses in the driving motor.
Frictional losses in the transmission and pulverizer bearings.
Air resistance of rotating parts.
Exhauster fan power (unless separately driven).

The difference between the input power and these losses gives the power available inside the mill for the purpose of grinding the product, but further losses occur in the application of this energy to the actual process of fracturing the material. These additional losses are caused by:

Contacting of the grinding media without the interposition of feed material.
Friction between the particles which does not create new surface.
Small impacts that are insufficient to shatter the particles.
The energy absorbed as elastic strain energy to stress the particles to the point of fracture.

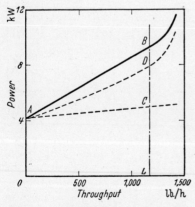

Figure 6. *Analysis of power characteristics of variable power mill*

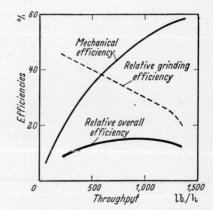

Figure 7. *Efficiencies of a variable power mill*

This analysis of the energy characteristics of a variable power mill is shown in detail by *Figure 6*. The line *AB* represents the total power consumption with varying throughput. At a given output, *OL*, the total power supply is therefore represented by the distance *BL*. The electrical and mechanical losses are represented by the distance *CL*, leaving the difference *BC* as the power available for grinding. The combined electrical and mechanical efficiency of the pulverizer is given by the ratio *BC* to *BL*. In the section 'Principles of Crushing and Grinding' it has been explained that laboratory tests may be used to assess the actual work needed for the fracture of various materials, this being expressed in foot-pounds per square foot of surface increase. Hence, if the specific surface of the product at the various throughputs is known, it is possible to calculate the minimum energy needed for pulverization. The minimum power for fracturing is represented by the distance *BD* plotted downwards from the total power curve, and hence the relative grinding efficiency is the ratio of *BD* to *BC*. The overall efficiency is represented by the ratio *BD* to *BL* and is the product of the mechanical and relative grinding efficiencies. *Figure 7* shows these efficiencies as estimated for a high speed impact mill, and indicates that such a machine has about 15 per cent of the efficiency of an ideal laboratory crushing process, operating on the same material.

Dynamics of Ball Mills

Special consideration is given below to a study of the essential characteristics of the ball mill, this being of particular importance in the chemical engineering and mining industries[1, 2, 3, 6].

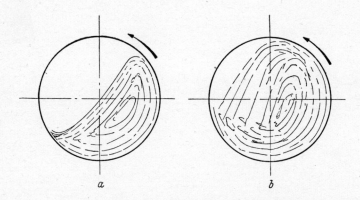

Figure 8. (a) *Cascading (low speeds)*; (b) *cataracting (high speeds)*. (*By courtesy of* British Coal Utilization Research Association)

A ball mill consists essentially of a cylindrical casing containing a charge of balls usually occupying one-third to one-half of the mill volume and rotating so that the ball charge has a circulating motion within the casing.

The ball charge is lifted by friction against the casing (assisted by the fitting of lifter bars in some cases) and the three states of motion are as follows:

1. Cascading, which occurs at relatively low speeds whereby the balls are only carried a short distance by the motion of the casing, and the action is mainly one of rolling.
2. Cataracting, which occurs at higher speeds whereby the balls are carried nearly to the top of the casing, leaving in a parabolic trajectory to impact on the other balls and feed.
3. Centrifuging, which occurs at a speed higher than a certain critical speed of rotation sufficient to maintain the balls in contact with the casing by centrifugal force during a complete revolution.

The cascading and cataracting motions are shown by diagrams (a) and (b) in *Figure 8*. The pattern of ball motion is thus controlled by the mill speed expressed as a fraction of the critical speed. This critical speed, N_c, is calculated so that at the topmost position of the ball the centrifugal force upwards is equal to the gravitational force downwards, and is

$$N_c = \frac{265}{\sqrt{D}}$$

where D is the internal diameter of the casing in inches.

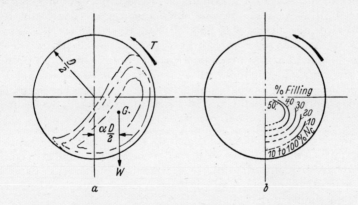

Figure 9. (a) *Forces acting on charge in ball mill;* (b) *locus of centre of gravity of ball charge.* (By courtesy of British Coal Utilization Research Association)

The above expression is termed the nominal critical speed since the ball diameter is neglected in comparison with the diameter of the casing. This assumption may not apply to small mills and the effective critical speed is calculated using the radius of rotation of the ball centre, *i.e.* the effective critical speed becomes:

$$N_c = \frac{265}{\sqrt{D-d}}$$

where d is the ball diameter in inches.

In practice, the effective critical speed may be considerably higher than the nominal if the ball charge is sufficiently large to produce more than a single layer covering the interior surface of the mill, and also the balls may not attain the same peripheral velocity as the casing because of slip. For general comparison, however, it is usual to express the speed of rotation in terms of the nominal critical speed, and therefore complete centrifuging may not occur until a rotational speed of, say, 120 per cent of the nominal critical speed is attained.

The forces acting in a mill are shown by *Figure 9 (a)*. The rotating ball charge may be considered to be concentrated at the centre of gravity, G, which is displaced horizontally from the axis of the mill by a distance $\alpha D/2$, α being the eccentricity of the ball charge. The weight of the ball charge acts vertically downwards from G so that the torque T required to turn the casing is $W\alpha D/2$, W being the weight of the ball charge. The power required to drive the mill, neglecting friction of bearings, etc., is $2\pi T.N/60$, where N is the speed of rotation in revolutions per minute, and this power is available for grinding the material which is interspersed with the ball charge.

When the mill is at rest the centre of gravity of the ball charge will be in a vertical line through the centre of rotation of the mill. As the casing revolves the eccentricity of the ball charge increases and this displacement continues to increase to a maximum at a speed of rotation about 60 or 70 per cent of the critical. Beyond this speed a proportion of the balls is flung towards the descending part of the casing, as will be seen from *Figure 8 (b)*, so that the centre of gravity reapproaches the vertical line through the mill axis. At the critical speed the centre of gravity of the ball charge coincides with the mill axis, no power will be absorbed (apart from friction) and therefore no grinding effect will occur. The locus of the centre of gravity of the ball charge is shown approximately by the diagram in *Figure 9 (b)*, and has been drawn for various proportions of ball filling.

The action of a ball mill is controlled by gravity, and dynamic similarity between mills of different sizes is attained if the Froude number is constant. This number is equal to the ratio v^2/lg, where v is a characteristic velocity, l a characteristic dimension, and g is the gravitational acceleration. Hence for dynamic similarity:

$$\frac{v_1^2}{l_1 g} = \frac{v_2^2}{l_2 g} \quad \text{or} \quad \frac{v_1^2}{v_2^2} = \frac{l_1}{l_2}$$

Let l be the casing diameter D, and v the peripheral velocity of the casing; thus $v = \pi D N$. Dynamic similarity for the different sizes of mill will be attained when:

$$\frac{D_1^2 N_1^2}{D_2^2 N_2^2} = \frac{D_1}{D_2} \quad \text{or} \quad \frac{N_1}{N_2} = \sqrt{\frac{D_2}{D_1}}$$

The critical speed N_c has already been shown to be proportional to $1/\sqrt{D}$, and the conditions for dynamic similarity are that mills operate at the same proportion of the critical speed and with the same relative volume of ball charge. The ball diameter should also be a constant fraction of the casing diameter for strict similarity, but if the ball diameter is small relative to that of the casing, this has only a secondary effect. The presence of feed material and its characteristics will also affect the pattern of ball motion, but if all these factors are constant then the eccentricity should be constant for different mills. *Figure 10* from a research by CONNOR[5], shows how the eccentricity α varies with mill speed and filling. These experiments were made with a smooth mill casing and show that at low mill fillings there was insufficient friction to produce centrifuging of the ball charge. The presence of dust increased the grip of the casing on the balls, and lifters fitted to the inside of the mill casing would have had a similar effect.

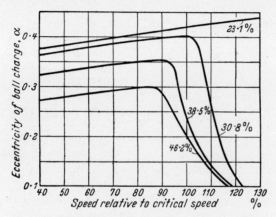

Figure 10. *Effect of variation in mill speed and percentage filling on eccentricity. 20 in. dia. smooth casing; ½ in. dia. steel balls with dust.* (From J. M. CONNOR, Univ. London Thesis)

Considering mills with length equal to diameter, operating at the same proportion of the critical speed N_c and with the same relative filling of ball charge, so that W is proportional to D^3, then torque is proportional to $\alpha . D/2 . D^3$, and if α is constant, torque proportional to D^4. Since N_c is proportional to $1/\sqrt{D}$, power will be proportional to $D^{3.5}$.

A series of tests were made with ball mills varying from 8–26 in. in diameter grinding coal[4], and the index of D when operating under conditions of dynamic similarity was found to be 3·55. This is in good agreement with the theoretical value of 3·5, when allowance is made for the difficulties of making exact measurements of the net power. The mills used in these comparative tests were run under the following conditions:

(i) Smooth casing.
(ii) Speed 68 per cent of the nominal critical speed.
(iii) Ball charge one-third the volume of mill having a bulk density 279 lb./ft.³. Equal proportions by weight of balls ¾ in., 1 in., 1¼ in., 1½ in. diameter.
(iv) Coal feed graded between 5 and 25 mesh B.S. sieves and 8 per cent of ball charge weight.

Various coals were ground in the four mills to a fineness of 95 per cent through 100 mesh B.S. sieve and the total number of revolutions, N_t, recorded at this stage. Since the mill torque is proportional to D^4, the work done in grinding is proportional to $N_t.D^4$; but the weight of coal ground is proportional to D^3, and hence the work done per pound of coal should be proportional to $N_t.D$. This product was found to be approximately constant and it may be deduced that mills working under conditions of dynamic similarity have similar grinding characteristics for a given material, though considerably more research is needed on this subject.

Selective Grinding Effects

Small-scale laboratory mills operate on a batch grinding process, whereas large industrial mills operate with a continuous feed and output. If the feed is a mixture of minerals having different grinding characteristics, selective grinding effects may occur. This situation frequently exists in industrial practice; ores consist of the metallic concentrates and associated rocks, even clean coal has different petrological constituents and almost invariably has some proportion of shale and pyrites. When a non-homogeneous material is subjected to grinding, the softer component is first reduced in size and only later is the harder component broken down. Taking as an example an industrial mill grinding a mixture of coal and hard shale, the initial output immediately after starting would have a higher proportion of coal than the feed to the mill, and there would thus be a concentration of shale inside the mill grinding chamber. As grinding proceeds equilibrium must eventually be attained between the composition of the output and the feed, but the material inside the mill will always contain a larger proportion of the harder constituent. As an example, in a test on an impact-type mill, the feed and output had an ash content of 15 per cent, but the ash content of material located within the pulverizing zone was 23 per cent[4].

Table 2. Ash Content of Mill Products

Type of coal	Sieve fraction mesh	Mill diameter in.		
		18	$13\frac{1}{2}$	8
D.C.S. cleaned coal	+ 100	3·8	4·0	7·5
	100/200	2·6	2·7	2·9
	− 200	2·7	2·7	2·4
R.S. slack coal	+ 100	17·4	22·0	41·6
	100/200	14·8	15·3	18·5
	− 200	11·5	12·7	12·2

When a mixture of materials is batch ground in a small mill, the softer component is concentrated in the fine material and the harder component in the coarse. If, therefore, the amount of fine material produced is taken as a measure of the grindability of the mixture, this may be predominantly influenced by the characteristics of the softer component, and may not represent the characteristics of the sample as a whole. Some extracts from the tests described are given in *Table 2*, showing the ash content of the sieved fractions of ground coal[4].

These figures show that selective grinding is more prevalent with the higher ash content coal, and also that the effect is greater with the smallest ball mill in which the intensity of grinding is least. This selective grinding effect must be eliminated as far as possible if grindability tests using small mills are to be used as a guide to predict the output of large industrial mills.

Circulating Load in Ball Mills

The object of ball mill grinding in mineral dressing processes is to separate or release the valuable ore constituent from the associated minerals. To obtain this effect and yet obviate excessively fine grinding of the materials, it is usual to pass the feed through the mill at a relatively high rate and to separate by classifiers that fraction of the output which has been ground sufficiently fine, the oversize material being returned to the mill for further

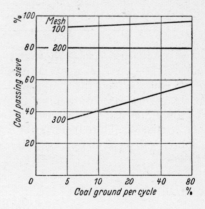

Figure 11. Effect of dust removal on size analysis of product

grinding. This circulating load may be as much as five times the actual weight of the product leaving the classifiers as undersize material. The size characteristics of materials ground with a circulating load differ considerably from similar products ground by a batch process or without the classifier system. The circulating load in coal grinding mills is not as great as in mineral dressing processes, but there is usually a classifier at the mill exit which returns oversize material to the mill feed. Experiments were made by the writer, on the 8 in. ball mill used for grindability tests, in order to examine the effect of circulating loads on the particle size distribution of the products, and also to study the relationship between energy applied to grinding and surface increase. Samples of Barnsley soft coal weighing 500 g were ground to a fineness of 80 per cent passing through 200 mesh B.S. sieve in a number of cycles, after each of which all the material that would pass the 200 mesh sieve was removed and the oversize returned to the mill for further grinding.

In the normal method of operation as a grindability test, each cycle consists of the number of revolutions required to grind 10 per cent of the coal to pass 200 mesh sieve; these cycles being repeated until 80 per cent of the sample passes 200 mesh sieve. For the purpose of these experiments additional tests were made with 5, 20, 40 and 80 per cent cycles, the latter corresponding to batch grinding in a single stage of operation. The complete sieving analyses of the products expressed graphically in *Figure 11* show a considerable difference between the size distribution for high circulating rates and batch grinding, the latter product having a much greater specific surface.

Table 3. 8 in. Ball Mill Tests on 500 g of Barnsley Soft Coal

Per cent through 200 mesh/c	Revs. for 80 per cent through 200 mesh	Grammes through 200 mesh/rev.	Energy/surface ft.Lb./ft.^{2}
5	921	0·434	11·2
10	1,020	0·392	12·8
20	1,095	0·365	13·8
40	1,470	0·272	—
80	2,610	0·153	14·7

Test results in *Table 3* show that the effect of a circulating load is to increase the weight of material ground, but that this has a relatively low specific surface, and there appears to be comparatively little difference in the energy to surface ratio in foot-pounds per square foot of surface increase. The slight increase with batch grinding is due to the cushioning effect of the finer particles. Thus batch grinding is not necessarily less efficient than grinding with a circulating load, though the products have different characteristics.

REFERENCES

[1] Rosin, P. et al., *Ber. Reichskohlenrates*, Nos. 2, 3, 9, 12. See also *Arch. Wärmew.*, 6 (1925) 289; 7 (1926) 54, 81; 8 (1927) 69, 109; 8 (1927) 239
[2] White, H. A., *J. chem. Soc. S. Afr.*, 5 (1905) 290; 15 (1915) 176; 30 (1929) 1; 31 (1930) 1
[3] Gow, A. M., Campbell, A. B., and Coghill, W. H., *Trans. Amer. Inst. min. (metall.) Engrs*, 87 (1930) 51
[4] Heywood, H., 'Pulverized Fuel', *Inst. Fuel Conf.* (1947) 594
[5] Connor, J. M., *Ph.D. Thesis*, Faculty Science, London University, 1936
[6] Rose, H. E., and Evans, D. E., *Proc. Inst. Mech. Engrs.*, 170 (1956) No. 23

5

SPECIAL APPLICATIONS OF GRINDING MACHINES

R. A. SCOTT

In the wide range of grinding equipment described in the preceding section it has not been required to consider the influence of comminution on the constitution of treated material. Many products handled in the foodstuffs, chemical and other industries demand special attention to the particle form, heat sensitivity, *etc*. This has resulted in the development of special machines and this short contribution deals with equipment for grinding cereals and other natural substances.

The design of Roller mills and the earlier Attrition mills (stone grinders) is described, followed by more particular reference to the practice of Grain Milling.

ROLLER MILLS

Certain sorts of soft friable material are commonly ground on roller mills. These machines consist essentially of long cylindrical rolls carefully machined with smooth or with longitudinally fluted surfaces and geared together so as to run in closely spaced, counter-rotating pairs. The material to be ground is spread evenly along the length of the rolls by means of a suitably designed feed gate mechanism and falls in an even curtain on to the surface and into the nip of the rolls. Where mild action is required, for example, where few fine particles are to be produced or the development of too much heat may damage the ground material, the roller mill can be made to give a regular and controlled action and a high capacity. It is economical in power, especially where thin uniform feeds of carefully sized material are ground in a machine with accurately set clearance between rolls.

The grinding action of smooth rolls is partly one of crushing and partly one of shear. Some control of the severity of the shearing action may be obtained by driving one roll at a faster speed than the other; in fine grinding on smooth rolls it is common to gear them so that one revolves at about one-and-a-quarter times the speed of the other. The differential speeds of the rolls are of particular use when the machine is employed for grinding material of organic origin or material which is 'sticky'. When such material is crushed the small particles tend to be squeezed together to form flakes. The shearing action which arises from the differing peripheral speeds of the rolls may be made great enough to discourage the formation of flakes.

Roller mills are commonly designed for special purposes and accordingly vary considerably in construction and arrangement. The principal constructional differences lie in the disposition of the rolls and *Figure 1* shows the so-called vertical, horizontal and diagonal arrangements. The vertical arrangement is now not often found. The horizontal is the simplest arrangement; the diagonal roller mill is usually the most elaborate and precise.

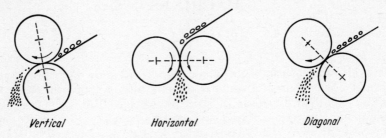

Figure 1. Roll arrangements.

Horizontal roller mills are used where mild but not specially precise grinding is required in provender, chemical and other industries. Where the pressure accompanying grinding is particularly large the individual rolls

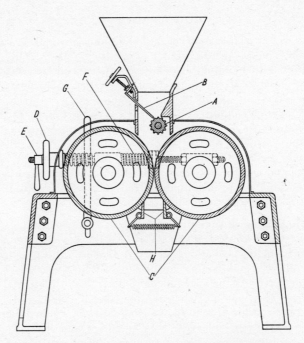

Figure 2. Horizontal roller mill. (From J. F. Lockwood, by courtesy of Northern Publishing Co., Ltd.)

are made large in diameter and small in length. Thus, for heavy work the rolls may be 18 in. diameter and 32 in. long; for lighter work they may be 9 or 10 in. diameter and up to 60 in. long. The simplest designs of horizontal roller mill provide for the driving of only one of the two rolls; the other is indirectly driven at the common peripheral speed by friction through the stock itself. Where the material to be ground is sticky or is organic in origin, mills of this design tend to produce flakes rather than small discrete particles. For this reason the more elaborate designs of horizontal mill provide for separate drive to each roll through gearing so that the rolls run at different peripheral speed. *Figure 2* illustrates a simple form of horizontal roller mill.

Material to be ground is fed to the top hopper and is carried by the corrugated roller A, under the edge of the feed gate B, into the nip of the rolls C. The handwheel D allows the gap between rolls to be set, and is locked in position by the clamp E. A spring, F, thrusts the rolls towards each other but contracts to allow large hard particles to pass. The throw-out lever G allows the gap to be opened manually in emergency. Scrapers, H, are provided to clean the surfaces of the rolls.

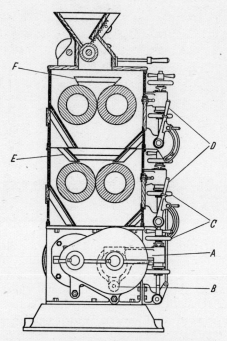

Figure 3. Three-high roller mill. (From J. F. Lockwood, *by courtesy of* Northern Publishing Co., Ltd.)

Horizontal roller mills are sometimes built with three pairs of rollers arranged vertically one above the other in the same machine; three successive mild grinding operations may then be completed in one pass through a single machine. An illustration of such a machine is shown in *Figure 3.*

The rolls are separately driven. One roll of each pair is carried on arms such as *A*, pivoted at *B*, and capable of alignment by means of the handwheels *C*. Levers, *D*, throw the rolls out of action by opening the roll gap. Intermediate sieves may be fitted at *E* and *F*. The diagonal arrangement of rolls is used particularly where very precise control of grinding is essential.

A typical example of such a double-sided machine of this type is shown in *Figure 4*. The upper and lower rolls *A* and *B* are driven through separate gearing so that the upper runs at the higher speed. The feed to be ground falls in an even layer on to the top of the lower grinding roll *B*, and is carried by rotation of the roll into the nip. In this manner no great accumulation of material occurs at the nip and each individual particle enters with least hindrance from its neighbours. The ground material is

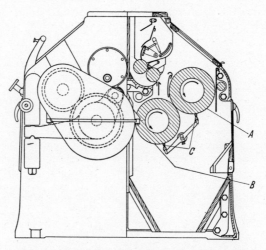

Figure 4. Diagonal roller mill, double sided. (From J. F. LOCKWOOD, by courtesy of Henry Simon Ltd.)

delivered at the underside of the rolls and falls to a collecting hopper from which it is carried away by gravity or by air suction. The stream of stock is conveniently accessible to the operator for inspection and sampling and this is of importance where frequent and careful check of the quality of the ground material is required. Brushes *C*, or knives, may be fitted to detach any particles of ground material which tend to stick to the surfaces of the rolls.

For uniform grinding action the individual particles must engage with the surfaces of the roll in closely similar fashion throughout the length of the rolls. To this purpose, the roller mill is usually built as a robust, accurately constructed machine. The rolls are mounted between closely located adjustable bearings and their surfaces are ground to a smooth and regular form. The pressure between the rolls due to the crushing and shearing stresses is considerable and it is, therefore, usual to taper the last 6 in. of the surface of the upper roll, as illustrated in *Figure 5*, so that when the rolls are deformed under grinding pressure, the spacing is uniform. The extent of the necessary taper varies according to the length of the roll and the severity of the grinding process and is usually determined empirically. For smooth 'reduction' rolls, such as those used in the cereal milling industries, a roll 60 in. long and of 10 in. diameter at the centre of its length is ground so that the diameter at its ends is about two-thousandths of an inch less than at the centre.

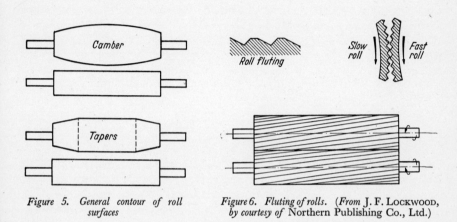

Figure 5. General contour of roll surfaces

Figure 6. Fluting of rolls. (From J. F. LOCKWOOD, by courtesy of Northern Publishing Co., Ltd.)

For a range of special grinding operations, such as the opening-up of wheat grains in the flour milling 'break process', the roller mill is required to release the interior fragments of the grain without extensive cutting-up of the outer bran skin. For such applications the rolls are fluted along their length with grooves, generally of saw-tooth shape (see *Figure 6*). The flutes are of such depth that, for example, a whole grain may rest in and

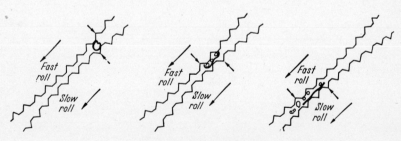

Figure 7. Action of fluted rolls

be gripped by the walls of a groove in the lower roll. As the grain is conveyed into the nip between the rolls, the faster moving flutes of the top roll open the grain coat and release the friable 'endosperm' of the interior (see *Figure 7*). The flutes are cut so as to run in a mildly helical course along the length of the roll, as illustrated in *Figure 6*.

The rolls are usually made from cast iron, chilled on the outside, or are centrifugally cast in steel so that the working surface remains hard and durable even after repeated machining of worn rolls. The roll spindles are separately machined and are pressed under considerable force into holes left for that purpose in the roll shell before the final shaping of the roll surface.

Roller mills are fitted with robust main bearings in the form of either long, precisely manufactured plain bearings of phosphor bronze, or roller bearings held in self-aligning housings. The bearing system is generally designed so that the spacing of the rolls, rather than the grinding pressure, is held constant. For this purpose the bearings of the main roll may, for example, be mounted on pivoted arms provided with positional adjustments so that the roll axes may readily be set both for parallelism and for spacing. In addition, provision is generally made to allow the gap to open when occasional hard, foreign particles, which might otherwise damage the grinding surfaces, enter the nip. *Figure 8* illustrates a typical bearing arrangement.

The arm A is carried at one end by an adjustable eccentric pivot B and supports the bearing for the lower of the two main rolls C. The wheel D adjusts the height of the stop at the outer end of the arm and the spring E thrusts the arm against the stop. When unusually large forces act between the rolls, the springs deflect and allow the gap between rolls to open. Independent adjusting wheels, D, exist for each lower bearing; when these are set the clearance between rolls may be altered by the simultaneous raising or lowering of the stops through the action of the control knob F.

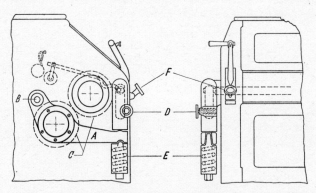

Figure 8. Typical adjustments for bearing alignment: diagonal roller mill. (From J. F. LOCKWOOD, *by courtesy of* Henry Simon Ltd.)

Diagonal roller mills, particularly when fitted with fluted rolls, are sometimes provided with additional automatic means for moving the rolls apart when the feed of material stops. The appropriate signals are obtained from the movement of the 'self-balancing' feed gate mechanism. Without such provision, finely set, fluted rolls driven at different speeds may become seriously damaged.

Whether the main purpose of grinding is to effect a gentle reduction in size or to release by differential action some selected component of the primary material, regular action and smooth working are only to be obtained when the feed is closely graded in size and uniformly spread along the length of the rolls. In the more precise forms of roller mill the feed is spread along the length of the rolls by means of a feed gate mechanism

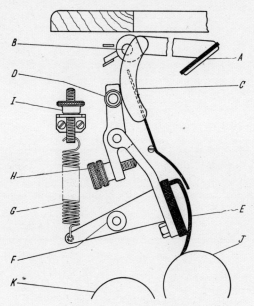

Figure 9. *Automatic feed gate arrangement for diagonal roller mill.* (From J. F. LOCKWOOD, *by courtesy of* Henry Simon Ltd.)

of special construction. A typical form of the mechanism is shown in *Figure 9*. A feed roll, usually fluted to grip the stock, runs along the length of the machine at a position just above the main rolls. The feed roll, driven through gearing at an appropriate speed, tends to draw material through the gap between its upper surface and the lower edge of the feed gate. An additional feed roll taking the form of two outwardly propelling half-lengths of worm is sometimes fitted to assist the spreading of materials to the ends of the gate.

Sufficient material is allowed to accumulate in the space behind the gate to form a pile extending to the ends of the gate. For as long as the pile is maintained, all positions along the length of the gate are fully supplied with feed, and a curtain of stock, of depth equal to the gap between feed roll and feed gate, is shed from the mechanism. An additional feature of the mechanism serves to keep the pile from building to such a size that entry to the machine is choked with the excess stock. For this purpose a sensing member, such as a pivoted plate or a framework of tree-like form, is placed so as to intercept the upper region of the pile of stock. Any small increase in the height of the pile, *i.e.* that which comes from an increase in rate of supply of material to the machine, imposes an extra downward pressure on the sensing member. This pressure acts through a series of linkages to lift the supply gate sufficiently to allow the material to run through the gate at an increased rate. The greater the feed to the machine the greater the gap between the gate and the feed roll and the faster the material runs through the gap. In this manner the pile is not allowed to vary greatly in size as the supply rate changes and yet always extends sufficiently for the feed to be spread uniformly along the roll. The plate A of *Figure 9* is pivoted at B and an attached lever C thrusts against a roller, D, mounted on the composite lever and feed gate E, pivoted at F. A spring, G, restrains the upward movement of the gate so that the gate opening varies in appropriate manner with the force on the plate A. The gate mechanism is set by adjustment of the screw H and the nut I. The main feed roll J carries the feed of particles forward either to the main rolls or on to an intermediate, faster feed roll K which serves to disperse the particles of stock.

The capacity of a roller mill depends on the length and speed of the rolls and on the particle size of the feed and the product. *Table 1* shows typical figures for the capacity of fluted rolls used in the initial stages of grinding wheat, and of smooth rolls used in later stages of reduction in size of wheat

Table 1. Capacity of Roller Mills on Wheat Granules

Feed	Product	Capacity (lb./h/in.)	Nature of roll surface
Wheat grain 100% over 2,500μ	30% through 1,000μ	33–66	fluted
95% over 1,000μ	35% through 710μ	16–32	fluted
100% over 530μ	20% through 130μ	13–26	smooth

granules. For the fluted roll operations the roller mill may be taken as having 10 in. diameter rolls running with the faster roll at 350 r.p.m. For the smooth roll operations the roller mill may again be taken as having 10 in. diameter rolls running at 215 r.p.m.

Higher speeds and substantially higher feed rates are physically possible but are attended by appreciable increase in the severity of action, and, in consequence, by some deterioration in quality.

ATTRITION MILLS

The rubbing of hard, rough stones against softer material is the basis of the primitive corn-grinding quern and of the pestle and mortar. The well-known millstone grinder embodies the same features and is perhaps the earliest form of continuous grinding machine. The common principle which underlies the grinding action of millstones and of other attrition grinders is the shearing of the softer material trapped in the space between moving hard surfaces. In most attrition grinders the hard surfaces are constructed to have a rough or pitted texture. The asperities of the surface tear through the softer parts of adjacent particles, whilst the hollows of the surface grip other particles so that as they move they distort and fracture the particles of the remaining mass. The principal forms of attrition grinder are the horizontal millstone, the vertical stone grinder and the vertical serrated disc mill.

MILLSTONES

Early types of millstone comprise a lower stationary stone and a driven upper stone supported by the lower. The severity of grind is not easily controllable, for the pressure depends markedly on the quantity and distribution of material between the faces of the stones. In elaborations of the primitive mill the lower stone is rotated and the upper stone is secured at a controlled distance above the lower one. Feed enters through a hole in the centre of the upper stone and the ground material is delivered at the rim. The upper stone may also be driven in counter-rotation to the lower one. *Figure 10* illustrates what is perhaps the best known form of the mill.

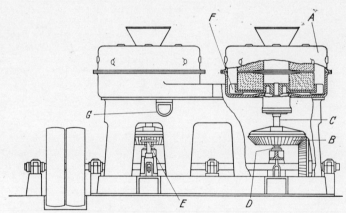

Figure 10. Under-running hursting mill. (From J. F. LOCKWOOD, *by courtesy of* Northern Publishing Co., Ltd.)

The upper part A of the casing carries the stationary upper stone, the gears B drive the shaft C, which rotates the lower stone. A bearing face D takes the grinding thrust and the face may be adjusted in position by the wheel E to control the gap between the stones. Material to be ground enters centrally and after grinding is swept out of the casing by scrapers F to the outlet port G.

The millstones are of carefully selected stone and where necessary are compositely constructed by the cementing together of smaller pieces. The surfaces of the stones are specially furrowed; the nature of the grooves affects the rate at which the material moves out towards the circumference of the gap as well as the distribution of material and the severity of grinding.

Stone grinders are now more commonly of the vertical faced type and a typical machine of this construction is shown in *Figure 11*. The casing A carries a fixed vertical stone B. The moving stone C is rotated by the shaft D, driven by the pulley E. A handwheel F, locked by a clamp G, sets the clearance between the faces of the stones. A feed gate H and feed wheel I control the feed rate to the mill. The feed roll obtains its motion from the main shaft *via* a pulley J. A clutch at K may be operated to stop the feeding. A magnet, L, collects foreign steel particles which might otherwise damage the grinding surfaces. Vertical stone grinders may have

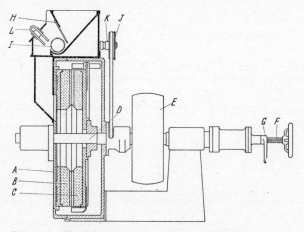

Figure 11. *Vertical stone grinder.* (*From* J. F. Lockwood, *by courtesy of* Northern Publishing Co., Ltd.)

one or both stones directly rotated. Generally, they allow better control of spacing of the stones and a higher grinding capacity. A common form of attrition mill has a similar construction but uses discs of synthetic abrasive material or of metal with serrated faces.

Further details of the nature, geometry and grinding characteristics of millstones and of vertical disc grinders of stone or metal are to be found under the heading 'Provender Milling'.

GRAIN MILLING

Grain milling industries offer examples of a wide range of typical grinding processes for soft materials; the two most important sections are concerned respectively with flour milling and with provender milling. Grinding processes carried out in these industries are described below at some length.

FLOUR MILLING

The unique and culinarily important elastic properties of wheat-flour dough derive from the special composition of the wheat proteins. The special characteristics of the dough are, however, adversely affected by the presence of quite small quantities of bran and by any extensive damage to the individual starch cells. The purposes of grinding in flour milling are therefore: (1) separation of bran coat from the central starchy material of the grain, and (2) reduction of central material to a size suited to the subsequent baking process. The first of these purposes calls for a grinding process

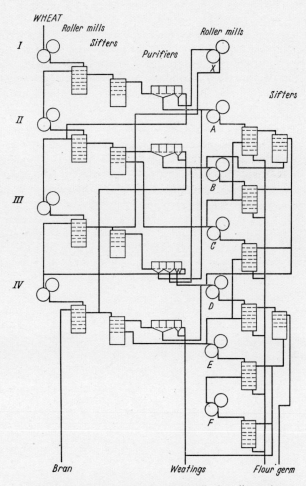

Figure 12. *Flow diagram for simple mill of small capacity*

which acts more severely on the interior starchy material (endosperm) than on the bran. The second calls for a mild grinding process which is capable of breaking down the endosperm into ultimate small groups of cells without fracture of many individual cells.

The main part of the bran is separated from the endosperm by treatment in a succession of roller mills fitted with progressively shallower and closer flutes. At each stage much of the endosperm breaks up readily while the fibrous bran coat remains attached to the larger particles. The ground product is sifted to remove the larger particles and these are then reground on the succeeding fluted roller mill. The succession of fluted roller mills—usually five in number—are known as the 'break system'.

The finer particles at each stage of the break process are rich in endosperm but, nevertheless, contain some bran particles, much of which is removed by subsidiary processes. The purified product is separated into narrow ranges of particle size and each range is fed to the smooth-surfaced roller mills of the reduction system which grind the endosperm, in gentle stages, to flour. A much simplified flow diagram of the whole process is given in *Figure 12*. The combination of roller mills, sieving and separating machinery is integrated to form a continuous and largely automatic plant.

A considerable amount of the equipment of the flour mill is devoted to cleaning and preparation of the raw material. Separation itself is made possible through the different mechanical properties of the fibrous bran coat and of the cellular mass of the endosperm. A part of the preparation process is therefore directed at bringing the content and distribution of moisture in the grain to an optimal level. The moister the bran the more tenacious it is, and for best effect it must be as moist as is consistent with retaining sufficient free sifting character in the ground endosperm. The best mean moisture content varies according to the hardness of the wheat but is usually in the range 15–17 per cent. By delaying the addition of water until late in the process of preparation of the wheat it is possible in practice to keep the bran a good deal moister than the endosperm.

In *Figure 7* the action of the first break roll is illustrated in somewhat idealized form. The flutes of the upper roll cut the bran coat and open out the grain so that the bran lies substantially in one piece over the flutes of the lower roll. The central endosperm breaks up and is carried forward by the flutes of the upper rolls. The larger particles which overtail the sieves that follow the first break are subjected to a somewhat similar shearing and scraping action while held in the finer flutes of the lower roll of the second break roller mill. The flutes of the first break roll are of saw-toothed profile as shown in *Figure 6*. The rolls are usually mounted in the machine so that the flutes run 'sharp to sharp'. When freshly fluted they are mounted 'dull to dull', since otherwise the flutes tend to cut through rather than to scrape over the bran coat.

First break rolls are usually fluted with 10–12 flutes to each inch of circumference. Later break rolls are fluted with correspondingly closer flutes. The flutes are cut to run in a mildly helical course along the roll (helix angle about one in seven). The rolls are normally 10 in. in diameter and 20–60 in. long according to the feed rate. The upper roll usually runs at 350 r.p.m., and the lower roll speed is geared down in a ratio of two-and-a-half to one.

It has already been remarked that at each stage of the break process the smaller particles which pass through the sieves, although predominantly endosperm, comprise a certain proportion of bran particles. If these are ground to flour in the reduction by rolls they cause a substantial deterioration in quality of the product. The streams of material at these points of the process are, therefore, passed to machines called 'purifiers' whose chief purpose is to remove the main portion of small bran particles.

The nature and construction of this machine are described in Section 6 and a typical form of it is illustrated in *Figure 24* of that section. The machine consists essentially of a substantially horizontal oscillating sieve supported on inclined hangers and fitted with an exhaust hood so that a uniform current of air may be drawn up continuously through the travelling layer of stock. The sieves increase progressively in aperture size from head to tail of the machine. The upward air currents tend to lift the flatter branny particles so that these remain near the top of the layer of stock and suffer little risk of passing through the finer covers. At the same time, the combined action of the oscillating motion and the upward lift of the air makes the layer of stock mobile so that, in particular, the finer particles of endosperm migrate down towards the sieve surface. The products of the head end of the sieve where the mesh is small are therefore fine, clean endosperm. The stock which passes through the succeeding sieves is of progressively increasing particle size; in this manner the 'throughs' of the sieves form a succession of accurately sized fractions, while the 'overtails' contain the flat bran particles and the coarse particles of mixed character (weatings).

The sized products from the purifiers each pass to their appropriate reduction roller mill, and further sifting machines follow each reduction roller mill. In this grinding process, as in the break process, large particles in the ground product are the most likely to contain bran. The finest products of the early reduction rolls generally yield the best flour. The coarser products may pass to further reduction rolls to yield more flour. Material which persistently proves too large to pass through the sieve covers is finally rejected from the process.

Separate streams of flour are produced from the sifters which follow each of the numerous members of the long succession of reduction rolls. These flour streams vary in quality according to the general degree of purity of the feed to the individual rolls. The streams are blended to form one or more final products according to the requirements of the consumer.

The reduction rolls, like the fluted break rolls, are usually 10 in. diameter and between 20 and 60 in. long. The upper roll usually runs at 215 r.p.m. and the lower roll speed is geared down in a ratio of about one-and-a-quarter to one. Capacities vary according to the particle size of the feed. Early members of the series of reduction rolls (see *Figure 12*), which grind particles of about 700 μ in size, take a feed of 13–26 lb./h per inch length of the roll; later members take about half that quantity.

The excellence of product of the final flour depends very markedly on the uniform action of both break and reduction roller mills. Much attention has therefore been devoted to refinement of roller mill design; the diagonal roller mill, illustrated in *Figure 4*, is typical of currently used machines. The rolls must be of accurate form and capable of precise adjustment, and the feed must be uniformly spread along the length of the rolls. The machines are fully enclosed and the casings are connected with an exhaust system which draws a controlled quantity of air through gaps left for that purpose in the casing. Dust and free moisture produced during the grinding operation thus tend to be drawn away from the machine. In the absence of an exhaust current the ground stock runs less freely and moisture condenses on the cooler parts of the casing with consequent formation of moist deposits of stock.

PROVENDER MILLING

The preparation of properly compounded animal feeding stuffs, of constitution and texture suited to the diverse requirements of the trade, is known as compound or provender milling. Several separate purposes are accomplished in provender grinding. Hard grain and compressed oil cake must be reduced to a size and texture acceptable to the class of animal that is to consume them. Where several materials contribute to a compound, some reduction in size may be needed to facilitate homogeneous mixing. An increasingly large proportion of provender is ultimately produced in the form of compressed cubes or pellets of carefully controlled composition; in this case, the components are ground partly to facilitate the proper blending of components and partly to give correct texture for pelleting.

The materials of the industry vary widely in character but usually contain both tough, fibrous elements and friable, cellular elements. The end products of grinding may be required as 'sharp', relatively coarse particles, or as smooth, fine ones. The grinding process varies, therefore, according to the nature of the raw material and to the special characteristics required of the grist. Typical of the raw materials are cereal grains, leguminous seeds, grasses and the numerous by-products such as bran, oil cake and bone left over from other industries. These materials are reduced in size by means of impact mills, roller mills, attrition mills and by specially devised cutting machines.

The choice of mill depends partly on convenience and partly on the special characteristics required in the ground product. Attrition mills produce a smooth, soft, fine product. Impact mills are versatile and have high capacity. Roller mills produce a uniformly sized product, particularly from hard cereal grains, but are generally unsuitable for grinding fibrous or branny materials. Corn cutters are well adapted to produce coarse angular grits. A typical provender mill, therefore, contains a combination of all these machines.

Attrition Mills

Attrition mills in which material is ground between millstones or between serrated metal discs, are widely used in provender milling. Simple horizontally arranged millstones or hursting mills (*Figure 10*) were traditionally the main means of grinding provender materials. Such millstones, with upper stone fixed and lower one driven, are now chiefly found in the smaller mills. Larger capacities and better control of the spacing of the stones can be obtained by vertical positioning of the faces of the stones.

Figure 11 illustrates a form of vertical stone grinder in which one stone is stationary and the other is driven from a shaft. This shaft which carries the moving stone is free to move axially in its bearings and the pressure between the grinding faces can therefore be adjusted by means of a screw

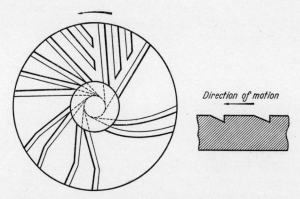

Figure 13. *Various arrangements of mill stone furrows.* (From J. F. LOCKWOOD, *by courtesy of* Northern Publishing Co., Ltd.)

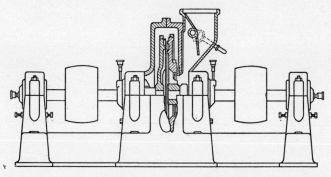

Figure 14. *Attrition grinder with separately driven metal discs.* (From J. F. LOCKWOOD, *by courtesy of* Northern Publishing Co., Ltd.)

which applies a thrust to the shaft through the medium of a heavy spring. Feed enters through a central aperture in the fixed stone and passes outwardly to the periphery of the stones, the faces of which are furrowed with a series of grooves. A cross section of the furrows and typical dispositions over the surface is illustrated in *Figure 13*. The shape, depth, pattern and inclination to the radius of the individual furrows are critically important in affecting the uniformity of the grinding and the capacity of the mill. The fineness of the product, especially where the feed contains tough fibrous components, is affected by the nature of the stone. For example, soft natural stones made from Derbyshire Peak stone, produce a particularly smooth meal from the softer central parts of oats and barley. Composition stones are more regular in structure and require less frequent dressing; they are generally made up of sharper cutting fragments and are used particularly where husk is required to be ground as fine as the interior of the grain.

Higher speeds and high capacities are attained in attrition mills in which the grinding members are serrated metal discs—usually of chilled cast iron. The grooves may have similar shapes to those illustrated in *Figure 13* and both faces of the discs are serrated so that they can be reversed when one face is worn down. Metal disc attrition mills have either one rotating disc, or two rotating in opposite directions, as shown in a machine of the latter type illustrated in *Figure 14*. The feed enters through a central hole, A, in one disc; both shafts, B and C, are adjustable in axial position to give a controllable spacing between the discs, D and E. In the case illustrated the discs may be of 15 in. diameter and each run at a speed of up to 2,500 r.p.m.

Impact Grinders

Impact mills or hammer mills of various design may be used for the production of fine granular material in one operation from oil cake, cereals, bran and other raw materials of the industry. The machines are relatively simple in construction and do not call for particularly skilled maintenance since hammers and screens are easily replaceable.

The severity of the grinding action on any one material depends especially on the clearance between fixed and moving members, the speed of the impact elements and the duration of the stay of particles in the grinding compartment. These factors may each be readily controlled and the machine is thus conveniently adaptable to a wide range of duties. The commonly used varieties of impact mill are fitted internally with a perforated screen or grid which contains the particles in the grinding compartment until the individual fragments are sufficiently small to pass through the apertures. The screens are easily removable and a change in screen aperture therefore forms a convenient way of changing the mean particle size of the product.

Most of the established forms of impact mill find some application in the grinding of the generally fibrous provender materials. Thus, large pieces of oil-cake, deriving from the extraction of oil from cotton seed, copra and palm kernel, may be put through a preliminary process of breaking up in hammer mills, such as that shown in *Plate I*, fitted with heavy, coarse projecting surfaces to catch the initial impact of the incoming cake. In the machine illustrated the cake enters at the top of the casing and is thrown by the pivoted hammers A against the breaker plate B. The hammers are normally held against pins by the centrifugal force accompanying the rotation, but can retract if necessary to prevent the creation of unduly large impact forces. Further reduction in size takes place at the impact of the fragments of cakes against the hammers and against the bars, C, of the

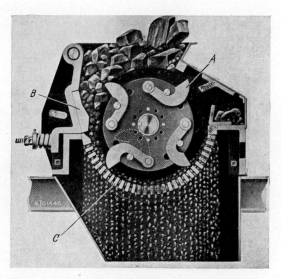

Plate I. Cake breaker with impact hammers. (Reproduced from 'Compound Milling', by N. O. SIMMONS, *by permission from* Leonard Hill (Books) Ltd.)

screen or grid which surrounds the lower part of the grinding chamber. The product finally passes out through the apertures of the grid.

Such a machine will accept slabs of two or three square feet in area and reduce them in size sufficiently to pass through a 1 in. mesh screen. For purposes such as cake breaking the rotor is commonly driven at speeds of about 5,000 ft./min, a speed somewhat lower than that used for fine grinding on impact mills.

Impact mills suitable for grinding cereals to a meal or flour consistency usually run at peripheral speeds upwards of 10,000 ft./min. Such mills do not necessarily need the provision of a breaker plate and, consequently, an outer screen usually extends through the greater part of the circumference of the grinding chamber. The screens are made to be readily removable to accord with the varying requirements in fineness of the product.

In the larger models of mill a single machine may consume up to a 100 h.p. Power consumption per unit weight of feed for grinding to a given fineness depends critically on the specific nature of the material to be ground and on its moisture content. For cereal grain such as barley, containing 15 per cent of moisture, the power consumed is of the order of 30 h.p.h/ton for a product which all just passes through a screen of 850 μ in aperture.

Roller Mills

Several forms of roller mill are used in provender milling. Toothed roller mills may be used instead of hammer mills for the initial stages of reduction in size of oil-cake. A cake breaker of this variety is illustrated in *Figure 15*. Cake is guided into the nip of the upper pair of rolls by the plates A. The broken fragments fall into the nip of a second pair of rolls mounted below the first and formed with rather smaller teeth. Where a fine product is required, an adjustable, toothed, concave plate, B, is fitted below one of the lower rolls and such large fragments that remain are further broken up against the teeth of this plate. One roll of each pair is adjustable for alignment by the setting of an eccentric pivot, C, on which its supporting arm is carried. A catch, D, and a notched wheel, E, secure the pivot in angular position.

Diagonal roller mills, similar to those which have been described for use in the grinding of wheat, are also used for the grinding of other cereal grains such as maize and barley. Such mills produce a meal of good uniformity in particle size and are generally more economical in power. They are not, however, particularly suitable for the grinding of the bran or husk and such material is preferably removed at an early stage of grinding by means of an air elutriator or 'aspirator' and ground on separate plant. Apart from their use for reduction in size, roller mills are used for the flattening or flaking of cooked grain in the production of maize flakes. For this purpose heavily constructed mills with horizontal rolls are used. The rolls may be 18 in. diameter and run at 150 r.p.m. with no speed differential between rolls.

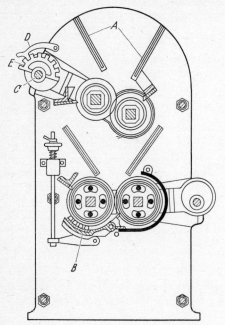

Figure 15. Cake breaker with toothed rolls. (From J. F. Lockwood, *by courtesy of* Northern Publishing Co., Ltd.)

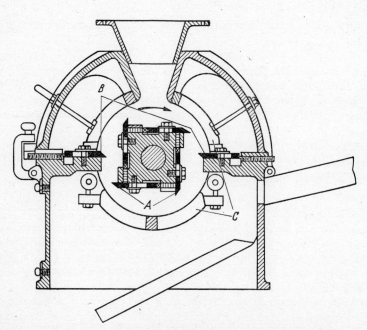

Figure 16. Corn cutter. (From J. F. Lockwood, *by courtesy of* Northern Publishing Co., Ltd.)

Corn Cutters

For certain purposes, for instance, the production of some types of poultry food, hard grain such as maize must be cut into discrete pieces to form a 'gritty' product. *Figure 16* shows a cross section of a Barron corn cutter in which a heavy cast iron chamber encloses a rotor to which are attached four knives, A, set so that they just clear the ends of two stationary knives, B, attached to the casing. Material enters an aperture at the top of the casing and when reduced sufficiently in size, leaves by way of the perforated screen, C, which forms part of the wall of the chamber. The moving knives trace out a path of about 6 in. diameter and rotate at speeds up to 750 r.p.m. A corn cutter with, say, six rotating and two fixed knives, may in a typical case handle 2 tons/h of maize, cutting it so that all passes through a screen of $\frac{9}{32}$ in. diameter perforations, with a power consumption of 5–8 h.p.h/ton.

BIBLIOGRAPHY

LOCKWOOD, J. F., *Flour Milling*, Northern Publishing Co., Ltd., Liverpool, 1946

SMITH, L., *Flour Milling Technology*, Northern Publishing Co., Ltd., Liverpool, 1945

LOCKWOOD, J. F., *Provender Milling*, Northern Publishing Co., Ltd., Liverpool, 1947

SIMMONS, N. Owen, *Compound Milling and Associated Subjects*, Leonard Hill Ltd., London, 1955

NATTRASS, J., *Grinders*, Pamphl. No. 13, Technical Education Series. The National Joint Industrial Council, 52 Grosvenor Place, London, June, 1936

MOORE, H., and MOORE, A. S., *Seed Crushing, Compound and Provender Milling*, Vols. I and II, Northern Publishing Co., Ltd., Liverpool, 1947 and 1948

6

SCREENING, GRADING AND CLASSIFYING

R. A. SCOTT

The wide range of operations of separating granular material into fractions of characteristic size, shape, density, electric or magnetic property, for example, involves many fundamentally different processes. In each process, however, the appropriate separation can only take place if an adequate freedom of motion of the individual particle exists. In this sense a distinction must be made between those processes in which particles are treated in an aggregated mass, as in the sieving or dry jigging of a bed of particles, and those processes in which particles are sufficiently dispersed during treatment to be out of direct contact with their neighbours—as where particles are elutriated or sedimented in air or water.

In the first part of this section those processes are considered which involve the treatment of beds or dense streams of particles. This is introduced by a discussion of the principles which govern mobility and stagnation in beds of solid particles. The second part of the section is concerned with those processes which involve the treatment of relatively dispersed particles, and is introduced by a discussion of the laws of settling of particles in fluid media.

MOBILITY OF PARTICLES IN A GRANULAR BED

The ease with which free movement may be induced in the individual particles which form a granular bed is closely associated with the character of the packing of the particles. A bed of individual particles packed at random contains many large gaps or voids between particles; the voids are associated with the fixed characteristic shapes of the particles and with the arbitrary manner in which each particle lies in relation to its neighbour.

Various special influences act to determine the nature of the voids found in practice. The particles which form the bed thrust downwards with their weight in a manner which tends to make the depth of the bed, and therefore the volume of the voidage, as small as possible. Where, however, the bed is not subject to other disturbing forces the friction between particles acts to prevent the compacting. In consequence the voidage usually forms a much larger proportion of the total volume than corresponds to ideal 'close packing' of the particles.

The character of close packing and its difference from the more open packing found in practice is illustrated in *Figure 1* (*a*) and (*b*) for regular spheres arranged in a vertical sheet. *Figure 1* (*a*) shows the regular close-packed structure with its minimum of voidage; *Figure 1* (*b*) shows the greater voidage of a typical irregular structure as stabilized by the friction between particles. The same type of departure from close packing exists for irregular as well as for spherical particles, see *Figure 2* (*a*) and (*b*), and for three dimensional as well as for two dimensional arrays of particles.

When a means can be found for reducing the effect of friction, an increase in closeness of packing can sometimes be obtained. In industrial compacting operations the means sometimes takes the form of pounding the mass of particles. Such a process is effective partly because the pounding causes general agitation among the particles and partly because the direct compressing forces act like gravity to favour close packing.

Agitation, however introduced, frequently leads to an increase in compaction. The friction between particles is reduced because of their

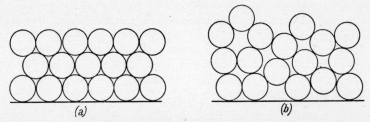

Figure 1. (a) Bed of spherical particles in close packing; (b) bed of spherical particles in open packing

motion and, in particular, at least some of the small displacements prove favourable to the movement of small particles into the more stable position of a lower neighbouring void.

The extent of compaction of a bed of material is significantly related to freedom of flow in the mass. The more a bed is compacted the more difficult it becomes to initiate a general flow in the bed. The reason for

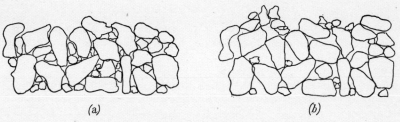

Figure 2. (a) Bed of irregular particles in close packing; (b) bed of irregular particles in open packing

the connection is that flow cannot, in general, take place without numerous small changes in mutual position of the individual particles. In the compacted state, voids are already filled which might otherwise be large enough to provide a new position for a disturbed particle. The compacted state of a bed of particles is therefore a relatively immobile state.

The special properties and conditions which lead to freedom of flow in granular material may be summarized as the existence of numerous voids not much smaller in size than the average particle and the capacity of individual particles to slip past one another. These will be considered in turn.

Where particles are smooth, regular and do not range too widely in size, numerous voids nearly equal in size to the average particle exist naturally. For this reason particles with a restricted range in size flow more easily than particles of wider range in size. Where particles are widely distributed in size and are in a moderately compact state, particles in the centre of the mass can change position only in company with the co-ordinated movement of very many particles and it is difficult to initiate flow except by some special means. Material in this condition cakes in the corners of containers and arches in storage bins. Material which otherwise does not flow easily may often be made to do so by 'expanding' the compact bed of particles by injection of air through, for example, diffusing pads in a lower part of the bed. This operation is the basis of 'fluidization' and will be referred to in more detail later in this section.

A bed of particles may often be made to flow more freely by transmitting vibration to the particles. If a confined bed is agitated in this manner the bed itself tends ultimately to become more compact and in this sense the vibration has an opposite effect to that desired. If, however, the bed is free, by virtue of the absence of some retaining wall, to fall to some lower level, the vibration may initiate and maintain the flow. The action of the vibration in this case is both to reduce the particle friction and to increase from time to time the spacing between particles.

Free flowing properties of a granular mass may also be maintained by the flow itself. Once flow has commenced in the mass the consequent low particle friction and relatively large voidage tend to maintain it indefinitely. The friction between the moving particles is less when in motion than when stationary; also the dynamic state of the moving mass is one in which the voids change continuously in size as neighbours adjust in minor degree their relative positions.

In the processes which are dealt with later in this section one or other of the means referred to above is generally used to assist in the rearrangement and flow of the material. Where it is not convenient to maintain a sufficient mass flow to ensure that stagnation does not exist at one place or another, flow is facilitated by one of the particular forms of vibration discussed below.

Horizontal Rotary Motion

If a horizontal deck is supported so that it can move freely in its own plane and is then driven by a crank so that each point traces out a horizontal circle of common radius, then a bed of material lying on the surface tends to move wholly with the surface until the severity of the motion exceeds a certain limit. At this more or less definite stage the mass of material slips on the surface so as to take up a generally steady motion of rather smaller radius. In this special condition where the material and the deck no longer move together, the individual particles likewise do not retain their earlier relative positions; the mass exhibits those small local motions which favour the free flow of the material as a whole.

The nature of the general motion is illustrated in *Figure 3*, which shows the traces of typical particles as heavily lined circles. The dots represent particles at a particular instant and the lighter circles represent the traces of the points of the deck lying immediately under the particles at the same instant; in general, the deck moves not only with a greater motion than the particle, but also in advance of it.

For a particle moving steadily on the deck the ultimate circular motion of the particle is such that the centrifugal force accompanying the circular movement is exactly equal and opposite to the friction force between particle

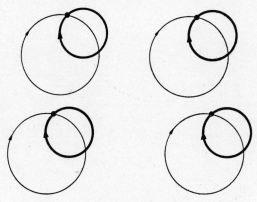

Figure 3. *Nature of particle movement on a horizontal rotary deck*

and deck. The friction force at any instant lies in the direction of relative movement of the particle and the deck. For this direction to be the same as that of the centrifugal force, the relative motion at any instant must be along the line joining the particle to the centre of the circular track.

Figure 4 illustrates the course of the particle (heavy circle) and the course of that point of the deck which lies for the moment immediately below the particle (light circle, centred at 0'). The arrows drawn tangentially to the circles represent respectively the velocities of the particle and of the point of the deck; the remaining arrow represents, therefore, the relative velocity of the particle with respect to the deck. For equilibrium, this arrow must be directed through the centre 0 of the particle trajectory and the friction force $32 \cdot 2\mu$ and centrifugal force $4\pi^2 f^2 r$ on a particle of unit mass must be equal in size. Thus, the radius r of the particle trajectory must satisfy the relation:

$$r = 32 \cdot 2\mu / 4\pi^2 f^2$$

where μ is the 'coefficient of sliding friction' of a particle (or of a representative small portion of the bed of particles), f is the frequency of rotation in revolutions per second, r is the radius of the course of the particle in feet, and the acceleration due to gravity is taken as 32·2 ft./sec².

Once slipping has commenced r depends, for a fixed μ, solely on rotational speed; the higher the speed the shorter the radius. *Figure 5* shows the relationship between radius of particle motion and rotational

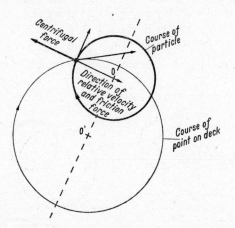

Figure 4. Forces on a particle on a horizontal rotary deck

speed: the solid line shows the relation calculated on the assumption of μ equal to 0·55; the circled points show observations made on an actual deck moving with a circular motion of 1·50 in. radius.

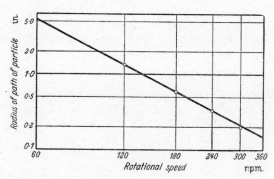

Figure 5. Radius of path of particles on a deck in horizontal rotary motion

While the radius of the course taken by the particle is independent of the radius of the motion of the deck once slipping has commenced, the relative velocity between the particle and deck is dependent on the radius of the deck motion.

It may readily be shown that according to the simple theory outlined above, relative velocity is given by:

$$2\pi f(a^2 - r^2)^{\frac{1}{2}}$$

where a is the radius of the deck motion in feet. For a given value of a the relative velocity increases steadily as the speed is increased above the critical slipping speed given by

$$f_c = (32 \cdot 2\mu/4\pi^2 a)^{\frac{1}{2}}$$

So much for the general motion of the mass; the individual particles have a more complicated motion. The force which keeps the main mass in motion is supplied by the deck through friction with the lowest layer of particles. These particles in turn transfer the necessary force to particles in the next layer. Since, however, the force needed to maintain the circular motion of each particle is comparable with the weight of each, such particles that rest insecurely on those below may rock or slip and thereby initiate similar local motions throughout the mass.

Horizontal rotary motion at various combinations of speed and stroke is used, for example, in screening machines to help the movement of fine particles through the apertures and coarse particles over the surface of the screen.

Horizontal Oscillation

A horizontal deck which is driven with a horizontal oscillatory motion carries with it a superimposed layer of particles, as in the case of rotary motion, only for as long as the accelerations of the deck remain small enough for the friction forces between the deck and the particles to maintain an equal acceleration in the particles.

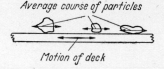

Figure 6. Nature of particle movement on a horizontal oscillating deck

Once again, if the coefficient of sliding friction is μ, slipping occurs at all speeds of oscillation above a value f_c given by equating the friction force $32 \cdot 2\mu$ to the acceleration force $4\pi^2 f_c^2 a$ which would act on unit mass of particles moving with the deck. Hence the critical speed is given by:

$$f_c = (32 \cdot 2\mu/4\pi^2 a)^{\frac{1}{2}}$$

where $2a$ is the total excursion of the deck.

For higher speeds, or for larger strokes at the same speed, some slippage occurs. The material oscillates with an amplitude generally rather less than that of the deck, for the acceleration—never greater than $32 \cdot 2\mu$—is too small to allow a mass of particles to move further than a fixed distance $64 \cdot 4\mu/f^2$ in any one half-cycle no matter what the amplitude of the deck (see *Figure 6*).

Once again the particles of the mass receive forces from their neighbours comparable with their individual weights; some particles rock or slip and these movements initiate those general small motions which aid in the flow of the whole mass.

Vertical Oscillatory Motion

When a horizontal deck is made to move with a vertical oscillating movement particles cease to move wholly with the deck when the downward acceleration of the deck, during any part of the motion, exceeds the acceleration due to gravity. At this point the tray tends to move away from the particle at a faster rate than the material can fall. A similar result follows if the deck is made to move with a circular motion in the vertical plane or with a motion inclined to the horizontal; at some point in the cycle the particle loses contact with the deck if the vertical component of the deck acceleration exceeds the acceleration due to gravity.

Upward Flow of Air Through the Bed of Particles

A steady flow of air or gas through a bed of particles exerts a viscous drag on the individual particles. The force depends on the shape and spacing of the particles as well as on the viscosity and velocity of the air. If an upward directed air flow of sufficient velocity exists the upward forces become comparable with the weights of the particles. Further increase in flow rate which might be expected to lift the bed as a whole, lifts, in fact, groups of particles away from their neighbours since the bed is never homogeneous in structure. The result of the upward air current is therefore to give a general mobility to the particles in the bed; this aeration process, generally known as 'fluidization', is demonstrated in *Figure 7 (a)* and *(b)*.

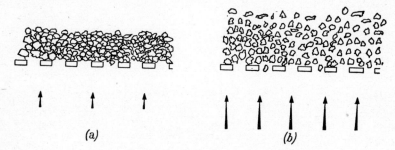

Figure 7. (a) *Effect of air flow on bed of particles: low upward velocity;* (b) *effect of air flow on bed of particles: fluidization*

Further consideration of the nature of flow characteristics of a bed of particles illustrates the process and gives a guide to the magnitude of upward air velocity at which fluidization occurs. Flow through the bed, as through a porous mass generally, is accompanied by a steady fall in air pressure across the bed. Suppose air is drawn downwards through a bed at a steady rate; where the particles are sufficiently small and close packed, the drag is of a simple viscous nature and the pressure drop in unit depth is directly proportional to air velocity (Darcy's law). By virtue of the pressure drop the bed of particles is thrust down on to each unit area of a supporting membrane with a force equal in magnitude to the pressure drop. If the air is now considered to flow in an upward instead of a downward direction the bed will be supported by the air pressure alone when the pressure drop per unit depth (pressure gradient) of bed is large enough to equal the weight of material on unit area of the support per unit depth of material (density of material). This is the condition corresponding to fluidization.

Fluidization velocities for air passing through beds of material of density of, say, 1·5 vary according to size, shape and closeness of packing; for material in the size range 30–60 μ a bed begins to become 'fluid' at velocities of about 3 ft./min and takes on a clearly visible shimmering motion at velocities of the order of 5 ft./min. Particles of similar density but of $\frac{1}{8}$ in. diameter require an upward velocity of about 175 ft./min.

So far the treatment has been concerned with materials which are homogeneous in density. Where particles differ widely in density, a loosening of the bed may, under certain circumstances, cause large as well as small particles of the denser component to fall to the lower strata. Certain forms of machine have been devised which use this principle for the dry separation of denser from coarse material.

SCREENING MEDIA

The separation of particles according to their ability to pass through apertures of fixed dimensions is known variously by the terms 'grading', 'screening' and 'sieving'. Industrial screening machines involve the use of a variety of different mechanical members to form the screening surface. The basic requirement of a screening medium is regularity of aperture; the many forms of grid, woven mesh and perforated sheet represent the different means for obtaining regularity, together, according to circumstance, with robustness, rigidity, an adequately large number of perforations in each unit area, and economy of cost.

Reference has been made earlier in this volume to the manner in which the size and shape of the particle determine whether it may pass an aperture of given size in the screen surface; similar general considerations govern the choice of best size and shape of the aperture. Where first importance is attached to high capacity for passing the particles substantially smaller than the aperture, screens should be chosen with a high proportion of screen surface area available as aperture. A bar or slot structure provides a large percentage of open area and for coarse screening an adequate strength may be maintained in the screen members for them to hold their positions. The slot apertures, however, tend to clog less with particles of near aperture size than do circular holes.

For close grading in size the choice of screening medium depends on the character of the separation desired and on the nature of the particles; circular apertures separate according to the medial dimension of the particle, while long slots separate by smallest dimension, at least in cases where sufficient time is allowed for particles to present every aspect to the aperture. Slots are used for distinguishing between elongated seeds such as wheat and oats which differ principally in narrowest width dimension. Circular holes are used for the rapid and accurate separation of roughly equidimensional granular particles. The square and polygonal apertures of the woven screening media come of necessity rather than choice. As distinct from circular apertures, such apertures pass a proportion of particles of larger flake-like character since such particles are capable of passing when oriented with medial dimension along a diagonal of the aperture. For such square or polygonal apertures the criterion of passage is not quite as clear-cut as for circular apertures and a substantially increased sieving time is required if all particles able to pass are to await a favourable opportunity.

The various types of screening media, together with their characteristics, are detailed below.

Bars, Grids and Wedge Wires

Where the scale of size of the screens is large enough, apertures may be formed by a grid of spaced bars. Such screens are particularly robust and therefore withstand heavy mechanical stresses. The sets of bars are commonly kinked at intervals so that intervening slots present a staggered course along their length. Particles of moderate size, which might otherwise slide smoothly on some broad face along the bars, continually change their aspect in relation to the screen and have less chance of overtailing falsely.

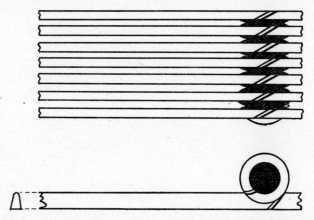

Figure 8. *Wedge wire screen.* (*By courtesy of* Lockers (Wedge Wire) Ltd.)

Wedge wire screens consist of a grid of spaced bars of wedge shaped cross section. The bars are assembled so that the gaps between members diverge away from the upper face of the screen (see *Figure 8*). Particles very nearly of aperture size are less prone to become caught between bars and choke the screen. Wedge wire screens range upwards in slot width from about $\frac{1}{8}$ mm.

Mechanically Perforated Plate

Plates or sheets of, for example, mild steel punched with regular apertures are commonly used for sieving particles varying in size from $\frac{1}{32}$–3 in. Such screens are convenient where high rigidity is needed, for instance, where the superimposed load is large or where the screen is vibrated. Perforated screens have good abrasion resistance and the apertures are uniform in size. The shape of the aperture can be chosen to suit the nature of the sizing operation: regular rounded particles may best be separated from extraneous larger material by screening through round holes; particles with one long axis may best be graded through elongated apertures or slots. Mechanically perforated screens are rarely formed with apertures of gross open area much greater than 50 per cent of the sheet. Substantial solid portions must be left between holes so that the sheet may withstand the action of the punches. Coarsely perforated metal screens may be produced economically; the smaller and more numerous the holes the more costly the screen. Apertures of such sheets are rarely of smaller size than 0·6 mm (600 μ).

Photomechanically Prepared Metal Screens

Perforated screens with round (and sometimes square) holes much smaller than 600 μ may be produced by photomechanical means. Nickel screens formed in this fashion are available in pieces as large as one metre square with apertures of 40–750 μ and open areas ranging from 1–36 per cent. These screens are generally costly to produce.

Woven Wire Screens

In a high proportion of industrial sieving operations, use is made of screens of woven wire, the wires being woven in a regular square net (plain weave). Regularity is produced by careful control of the warp and weft

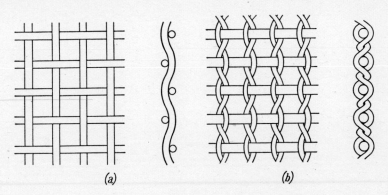

Figure 9. (a) Plain weave in screening cloth; (b) Leno weave in bolting cloth

spacings, and stability of position of the individual wires is maintained by the crimp, that is to say, by the permanent crest-and-trough form of the intertwining wires [see Figure 9 (a)]. Woven wires are produced with apertures ranging from about 1 in. down to 200 mesh (76 μ) for ordinary industrial processes. Woven wire test sieves, both regular in aperture and durable, are available with apertures as small as 400 mesh (37 μ).

Table 1. Woven Wire Cloth

B.S. No.	Size of aperture side of square in.	Number of meshes per linear inch	Size of wire S.W.G.	Screening area per cent
300 × 49	0·0021	300	49	40·5
240 × 48	0·0026	240	48	38·3
200 × 47	0·0030	200	47	36
180 × 46½	0·0033	180	46½	36
160 × 45½	0·0036	160	45½	33·7
140 × 44½	0·0041	140	44½	33·3
120 × 43	0·0047	120	43	32·1
100 × 41	0·0056	100	41	31·4
90 × 40	0·0063	90	40	32·2
80 × 39	0·0073	80	39	34·1
70 × 38	0·0083	70	38	33·7
60 × 37	0·0097	60	37	34·6
50 × 35½	0·0120	50	35½	36
40 × 34	0·0158	40	34	39·9
36 × 33	0·0178	36	33	41
30 × 31	0·0217	30	31	42·5
24 × 29	0·0281	24	29	45·4
20 × 28	0·0352	20	28	49·6
16 × 27	0·0461	16	27	54·4
14 × 26	0·053	14	26	55·7
12 × 25	0·063	12	25	57·6
12 × 22	0·055		22	43·9
10 × 24	0·078	10	24	60·8
10 × 21	0·068		21	46·2
8 × 22	0·097	8	22	60·2
8 × 19	0·085		19	46·2
6 × 20	0·131	6	20	61·5
6 × 18	0·119		18	50·8
5 × 19	0·160	5	19	64
5 × 17	0·144		17	51·8
4 × 18	0·202	4	18	65·3
4 × 16	0·186		16	55·4

Characteristics of Perforated Plate Screens for Industrial Use. *From* B.S. 481 : 1933

Accurately woven wire cloth of small aperture size (100 mesh and smaller) is, however, expensive to produce.

The stiffness of coarse wire cloths makes them suitable for supporting moderate loads of stock. The fine wire cloth must be stretched taut on the

Table 2. Clear Mesh Woven Wire

B.S. No.	Size of aperture side of square in.	Size of rod or wire S.W.G.	Screening area per cent
$\frac{1}{8} \times 12$	$\frac{1}{8}$	12	29·8
$\frac{1}{8} \times 14$		14	37·2
$\frac{3}{16} \times 10$	$\frac{3}{16}$	10	35·3
$\frac{3}{16} \times 12$		12	41·4
$\frac{3}{16} \times 14$		14	49·1
$\frac{1}{4} \times 12$	$\frac{1}{4}$	12	49·9
$\frac{1}{4} \times 14$		14	57·4
$\frac{1}{4} \times 16$		16	63·4
$\frac{3}{8} \times 10$	$\frac{3}{8}$	10	55·6
$\frac{3}{8} \times 12$		12	61·3
$\frac{3}{8} \times 14$		14	67·9
$\frac{1}{2} \times 6$	$\frac{1}{2}$	6	52·2
$\frac{1}{2} \times 8$		8	57·4
$\frac{1}{2} \times 10$		10	63·4
$\frac{5}{8} \times 6$	$\frac{5}{8}$	6	58·5
$\frac{5}{8} \times 8$		8	63·4
$\frac{5}{8} \times 10$		10	68·9
$\frac{3}{4} \times 3$	$\frac{3}{4}$	3	56·0
$\frac{3}{4} \times 6$		6	63·4
$\frac{3}{4} \times 8$		8	67·9
$\frac{3}{4} \times 10$		10	73·0
$\frac{7}{8} \times 3$	$\frac{7}{8}$	3	60·3
$\frac{7}{8} \times 6$		6	67·2
$\frac{7}{8} \times 8$		8	71·5
1×3	1	3	63·8
1×6		6	70·4
1×8		8	74·3
$1\frac{1}{4} \times 1$	$1\frac{1}{4}$	1	65·0
$1\frac{1}{4} \times 3$		3	69·3
$1\frac{1}{4} \times 6$		6	75·1
$1\frac{1}{2} \times 1$	$1\frac{1}{2}$	1	69·4
$1\frac{1}{2} \times 3$		3	73·3
$1\frac{1}{2} \times 6$		6	78·6
$1\frac{3}{4} \times 1$	$1\frac{3}{4}$	1	72·9
$1\frac{3}{4} \times 3$		3	76·4
$1\frac{3}{4} \times 6$		6	81·2
$2 \times \frac{3}{8}$	2		70·9
2×1		1	75·6
2×3		3	78·9
$2\frac{1}{2} \times \frac{3}{8}$	$2\frac{1}{2}$		75·6
$2\frac{1}{2} \times \frac{5}{16}$			79·0
$3 \times \frac{3}{8}$	3		79·0
$3 \times \frac{5}{16}$			82·0
$3\frac{1}{2} \times \frac{1}{2}$	$3\frac{1}{2}$		76·6
$3\frac{1}{2} \times \frac{3}{8}$			81·6
$4 \times \frac{1}{2}$	4		79·0
$4 \times \frac{3}{8}$			83·6

From B.S. 481 : 1933.

supporting frame when the load of material is heavy and particularly if large unsupported areas of screen are to be kept in high speed oscillatory motion.

Extensive ranges of woven wire screens are manufactured in series of changing numbers of wires to the inch. This number is generally known as the 'mesh number' and is used to characterize the screen. For a given mesh number, screens may be obtained in any of a number of standard

Table 3. *Wire Cloth for Fine Mesh Test Sieves*

Nominal mesh No.	Nominal width of aperture		Nominal diameter of wire in.	Screening area (approx.) per cent
	in.	μ		
5	0·1320	3,353	0·0680	44
6	0·1107	2,812	0·0560	44
7	0·0949	2,411	0·0480	44
8	0·0810	2,057	0·0440	42
10	0·0660	1,676	0·0340	44
12	0·0553	1,405	0·0280	44
14	0·0474	1,204	0·0240	44
16	0·0395	1,003	0·0230	40
18	0·0336	853	0·0220	36
22	0·0275	699	0·0180	36
25	0·0236	599	0·0164	35
30	0·0197	500	0·0136	35
36	0·0166	422	0·0112	36
44	0·0139	353	0·0088	38
52	0·0116	295	0·0076	37
60	0·0099	251	0·0068	35
72	0·0083	211	0·0056	36
85	0·0070	178	0·0048	35
100	0·0060	152	0·0040	36
120	0·0049	124	0·0034	35
150	0·0041	104	0·0026	37
170	0·0035	89	0·0024	35
200	0·0030	76	0·0020	36
240	0·0026	66	0·0016	38
300	0·0021	53	0·0012	41

From B.S. 410 : 1943

sizes of wire; in consequence the apertures do not constitute a simple regular series. B.S. 481:1933 embodies certain standard series of wire screens for industrial use and the principal dimensional characteristics of these are set out in *Tables 1* and *2*. Many variations from these series are available to order from established manufacturers.

Apart from screens available for ordinary industrial use, several separate series of test sieve are standardized. Thus, B.S. 410:1943 defines two series of test sieves with in each case a rather larger percentage of open aperture area than for the industrial screens of *Tables 1* and *2*. These test sieve specifications are given in *Table 3* but with tolerances omitted. A substantially different series of test-sieve apertures is incorporated in the American Standard Specification Z 23.1–1939; the consecutive apertures are in geometric progression with a ratio between members of $\sqrt{2}:1$.

Woven Textile Fabrics

Woven textile fabrics are probably among the earliest forms of screen. In the earlier history of grain milling woollen cloth was used for sieving or bolting the flour after stone milling the grain. In more recent times silk

Table 4. Coarse Aperture and Medium Aperture Silk Bolting Cloth
(Grit Gauze)

No.	Meshes per linear inch	Apertures μ	Apertures in.	Sieving area per cent
14	13½	1,585	0·0624	70
16	15½	1,352	0·0532	68
18	17½	1,169	0·0460	64
20	19	1,046	0·0411	63
22	21	939	0·0369	61
24	23	850	0·0334	60
26	25	778	0·0306	59
28	27	715	0·0281	58
30	29	660	0·0259	57
32	31	611	0·0240	55
34	33	563	0·0221	52
36	35	520	0·0204	50
38	37	494	0·0194	51
40	39	466	0·0183	50
42	40½	439	0·0172	49
44	42½	414	0·0162	48
46	44½	388	0·0152	46
48	46½	369	0·0145	45
50	48½	354	0·0139	45
52	50½	339	0·0133	45
54	52½	325	0·0127	44
56	54½	307	0·0120	43
58	56½	291	0·0114	41
60	58	282	0·0111	41
62	60	275	0·0108	42
64	62	262	0·0103	40
66	64	253	0·0099	40
68	66	242	0·0095	39
70	68	239	0·0094	40
72	72	220	0·0086	39

has been the traditional material for the construction of bolting cloth and a very wide range of accurately woven open-mesh silk screens has been available for many years. Increasing use is now being made of nylon yarn for textile screening cloths; its special advantage lies in its high resistance to abrasion and corrosion and in the evenness of the yarn which makes it possible to weave cloth with extremely high regularity of aperture.

Where the weave is fine or the cloth is woven from single thick threads of monofilament nylon, for example, a simple plain weave is used—as in the case of wire cloths. These cloths retain their initial regularity because of the rigidity of the threads and the set or crimp which they develop as each

Table 5. *Fine Aperture Silk Bolting Cloth* (XX Series)

No.	Meshes per linear inch	Apertures μ	Apertures in.	Sieving area per cent
1	48	368	0·0144	48
2	54	319	0·0125	46
3	58	296	0·0116	46
4	62	278	0·0109	46
5	66	255	0·0100	44
6	74	219	0·0086	41
7	82	193	0·0075	39
8	86	183	0·0072	38
9	97	155	0·0061	35
10	109	129	0·0050	31
11	116	117	0·0046	29
12	125	112	0·0044	29
13	129	106	0·0041	28
14	139	92	0·0036	25
15	150	85	0·0033	23
16	157	81	0·0031	22
17	163	77	0·0030	22

thread passes in turn over and under successive cross threads. Most silk and many nylon textile bolting cloths are made in full leno or gauze weave [see *Figure 9 (b)*] or in half leno weave.

In full leno weave the threads of the warp are twisted in pairs first in one direction and then in the other so as to give a structure which locks each weft thread in position. In half leno weave twisted warp threads alternate with single threads.

Textile screens are woven with great consistency from cloth to cloth and with very regular aperture, in a long series of standard aperture sizes.

The traditional silk screens for general industrial use are usually characterized for coarse and medium cloths (grit gauze) by the number of threads per 'Vienna inch' of the weave; the dimensional characteristics of the best known cloths of this sort are given in *Table 4*. The finer silk screens form a separate numbered series of cloths of increasing smallness of aperture; aperture size and percentage open aperture area for the commonly used 'XX series' of bolting cloths are given in *Table 5*. Nylon screens of generally similar aperture sizes are also available.

Table 6. *Standard Aperture Dimensions for Textile Screening Cloth*

Nominal aperture width μ	Nominal aperture width μ	Nominal aperture width μ
1,600	475	118
1,500	450	112
1,400	425	106
1,320	400	100
1,250	375	95
1,180	355	90
1,120	335	85
1,060	315	80
1,000	300	75
950	280	71
900	265	67
850	250	63
800	236	60
750	224	56
710	212	53
		50
670	200	
630	190	
600	180	
560	170	
530	160	
500	150	
	140	
	132	
	125	

From B.S. 1812 : 1951

B.S. 1812:1951 (*Table 6*) embodies a series of textile screening cloths with apertures forming a regular geometric series.

Bolting cloth of silk and of nylon is produced in fairly large quantities and at least for the finer cloths (apertures smaller than 350 μ) is cheaper than wire cloth. The taut mounting of textile screening cloth is particularly simple since the elastic nature of the textile threads makes it easy to stretch the cloth uniformly over the sieve frame.

RATE OF SCREENING—SCREENING EFFICIENCY

The effectiveness of a screening process is determined by the extent to which the small particles capable of passing through the apertures are separated from the feed. For fixed applications, where the feed is always identical in character, the efficiency is sometimes expressed numerically by the percentage ratio of the weights of material passing and of material potentially able to pass through the screen. Such figures may serve as useful indices of performance in a process, but have little general application.

The performance of the screen as a whole depends on the individual rates of passage of material through each local unit area of screen. The rate of sieving in a screening process is controlled by several factors of which the most important are:

(a) the size of the material especially in relation to the aperture,
(b) the free running properties of the material,
(c) the motion of the screen,
(d) the height of the bed of material supported by the screen surface,
(e) the proportion of small to large particles in the feed.

Some of these factors are affected by such matters as shape and surface character of the particles and are not readily expressible by simple parameters. Detailed formulae, which have from time to time been set up to predict rates of screening, are necessarily empirical and limited in application; certain broad conclusions are, however, helpful to a better understanding of the screening process and these are considered below.

Some consideration has already been given to the shape and size of particle and screen aperture as criteria of whether a particle may pass through the aperture. Particles of size comparable with that of the average aperture offer the greatest resistance to passage. This is partly because such particles may need to present a proper aspect to the aperture before they may pass, and partly because such particles continually tend to block the aperture and therefore restrict the rate of passage of smaller particles. Particles substantially smaller in size than the screen aperture, sieve at relatively high rates and fine material may be observed passing in profusion through the head end of a continuous screening machine provided that care is taken to keep the screen apertures moderately free. The screening of coarse rather than fine material gives generally higher rates of screening (expressed in pounds per square feet per hour), provided that in each case the particles considered are substantially smaller than the appropriate screen apertures.

Devices for freeing the apertures of clogging particles materially improve the sieving rates of most machines. Thus, a screen aperture formed so that the opening increases in width between upper and lower surfaces—as for wedge wire screens and tapered hole screens—improves the sieving rate. In many industrial screening machines wedged particles are dislodged by direct physical means such as making the screen motion sufficiently severe for this purpose or by fitting loose brushes, pads or rubber balls immediately below the screen surface so that under the violence of the motion they bounce against the surface and act directly to counter the wedging. The density of aperturing of the screen surface has an obvious effect on screening rate. The density is often expressed in the form of the percentage of open area of the screen.

Free-running properties of the feed have very marked effects on rate of sieving. In this sense moisture content has a large influence on the rate of sieving, for instance, dry sand which is capable of running freely through a stationary screen may be sieved only with great difficulty when damp. The shape and distribution in size of the particles affect free running character and rate of screening, and material which contains rounded particles of limited range in size gives the greatest rates of screening.

A considerable degree of mobility must be maintained in the particles of the mass of feed if substantial rates of screening are to be achieved. The necessary mobility is ordinarily induced in one of the ways described earlier in this section and, indeed, each type of mechanism finds some place in screening processes. This mobility of the particle leads to two important consequences:
(1) The smaller particles tend to work down through the stock so as to concentrate in the layers of material more closely in proximity with the screen.
(2) The particles at the screen face are in a state of continual agitation which makes them less subject to the restraints of friction so that they move more frequently into the immediate neighbourhood of the screen apertures.

Where the mobility is obtained by giving an oscillation or a circular motion to the screen surface, the nature and severity of the motion are important variables. A more or less definite minimum of agitation is usually necessary before even the most free running material will run continuously through the screen apertures. This 'threshold' agitation is generally such that the average particle accelerations are not far short of the value of the acceleration due to gravity. It corresponds roughly, therefore, to the condition already considered on p. 133 in which the mass of particles no longer moves wholly with the screen motion.

Increasing the severity of the agitation beyond the threshold of mobility generally increases the rate of sieving, particularly with damp or cohering particles, or with a mass of particles having a wide range of particle size and consequent tendency to flow sluggishly. A point in severity is reached, however, beyond which further increase, for example, in speed of horizontal oscillation, causes a falling off in the rate of sieving. An example is given in *Figure 10* for wheat passing through a screen perforated with holes 5·1 mm in diameter and therefore potentially capable of passing the aperture. For the case cited the screen was driven with a horizontal longitudinal oscillation, over a range of speeds, at each five separated lengths of stroke.

A similar rule relating to severity of agitation applies to horizontal, inclined and vertical oscillating motion, and also to rotary and combinations of rotary and oscillating motion. Screening machines are not, however, always to be found running at speeds which give the maximal rate of

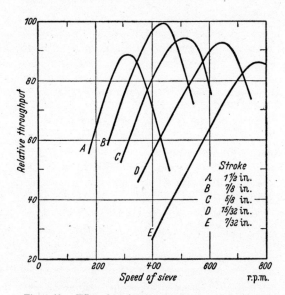

Figure 10. Effect of speed and stroke of horizontal oscillation on rate of screening of wheat. (From E. D. SIMON, by courtesy of Northern Publishing Co., Ltd.)

sieving. This may be either because constructional characteristics of the machine set an upper limit to speed or, alternatively, because a somewhat milder action allows the screen to discriminate to better effect between particles of differing shape.

Many combinations of speed and travel of the screen lead to moderate mobility in granular material (see *Figure 10*). The optimal combination depends, in any particular case, on such variables as moisture content, particle size and aperture size. In general, however, screening of large particles is most efficiently carried out with motions involving comparably large excursions of the screen.

The depth of the bed of material supported by the screen affects the rate of screening. Full effect of the screen is, of course, not obtained until the feed is sufficient to cover the screen completely. Increase in depth of the bed increases the rate of screening until a depth is reached at which no substantial further increase is found. This effect is illustrated for cereal granules on a screen deck oscillated with a stroke of $\frac{7}{16}$ in. at 650 r.p.m. for a nylon screen cloth of aperture size 710 μ (see *Figure 11*). The rate of passage of particles through the screen increases steadily until a depth of about 0·2 in. is reached; thereafter, it falls off a little.

The dependence of rate of screening on depth of bed has an important bearing on the operation of inclined screens. If the inclination is very steep the material travels at a high speed over the screen surface; the depth

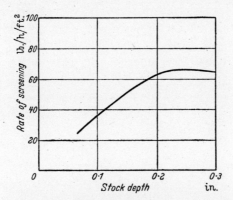

Figure 11. Effect of depth of stock on rate of sieving in oscillating screen

of bed may then be too small for efficient screening and relatively low rates of screening may be obtained. A less steep inclination of the screen then improves the sieving efficiency since it leads to an increased depth of bed.

SCREENING AND GRADING MACHINES

The many types of grading and screening machine that exist differ essentially in the means which they employ

(*a*) to induce particles to pass through the screen apertures,

(*b*) to maintain a steady transport of oversize particles across the surface of the screen, and

(*c*) to prevent clogging of the apertures with oversize particles.

In simple screening machines used generally for coarse grading, the screen is stationary and the material falls on the screen in a continuous stream. Blinding of the screen is largely prevented by the scavenging action caused by the impact of the heavy particles in the stream. Where the particles are smaller or where many do not differ much in size from the screen aperture, stock lies dormant on a stationary screen since the particles wedge in or arch over the apertures. Screens constructed for the accurate grading of all but the largest size of particle are therefore driven with some sort of special mechanical motion partly designed to dislodge the wedged particles and partly to maintain a continual flow of fresh material both to and along the screen surface.

In the subsection which follows screening machines will be divided, for purposes of description, into classes according to the nature of the means used to ensure freedom of movement of the particles.

Stationary Screens and Grizzlies

The simplest forms of screen have stationary screening surfaces, usually inclined steeply downwards so that the feed falls in a continuous and moderately fast stream on to the surface. Stationary screens are principally used for rough grading either where the bulk of the feed easily passes through the broad meshes, as in the well-known sand screen, or where heavy particles in the feed fall on the screen bars and excite sufficient vibration by impact to clear the apertures of wedged particles.

The grizzly is an example of the steeply inclined stationary screen and is chiefly characterized by a robustness of construction which allows it to withstand the impact of large pieces of falling material. The screen surface

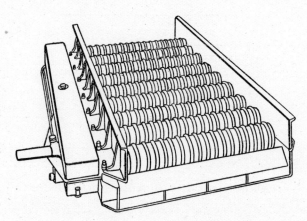

Figure 12. Roll grizzly. (By permission, from 'Elements of Ore Dressing', *by* A. S. TAGGART. *Copyright* 1951. John Wiley & Sons, Inc.)

is usually formed by bars of wedge-shaped section which are spaced so that the apertures between them increase steadily from head to foot of the screen. The diverging gaps and the impacts of the particles help to keep the apertures free.

Screens may have decks composed of two or more parallel rolls each formed with circumferential corrugations. The gaps between rolls allow smaller particles to pass through while large pieces travel over the deck. A roll grizzly in which the rolls are rotated by mechanical drive in order to aid the transport of material over the deck is shown in *Figure 12*.

Reels and Trommels

The reel consists of a generally horizontal rotating frame of cylindrical or polygonal cross section, over which the screen is fastened. Material to be sifted is fed to the inside of the cylinder and falls towards the screen in a generally tumbling action induced by the rotation. Fine material is collected after passing through the screen, while oversize material migrates away from the inlet end and is discharged at the far end of the cylinder

Plate I. Rotor of compound trommel. (By courtesy of G. A. Harvey Ltd.*)*

Plate II. High speed gyratory sieve. (From J. Hurst, *by courtesy of* Russell Constructions, Ltd.*)*

The covers are frequently graded in fineness along the length of the reel with the finer covers towards the inlet port; in this manner several sizes of product may be produced with a single screen. Double-ended reels are sometimes used with inlet ports at extreme ends and with a central discharge for oversize particles. This arrangement conveniently condenses two screens into a single machine and proves useful where provision must be made for many separate operations. Single-ended reels usually have their

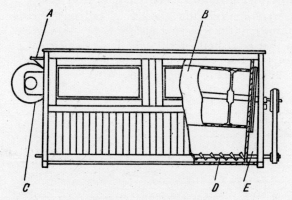

Figure 13. Single-ended reel. (By courtesy of Henry Simon Ltd.)

axes inclined downwards towards the discharge end so as to augment the forward flow of material, see *Figure 13*. Stock enters through the inlet A to the screen B which is rotated by the pulley C. Throughgoing material is collected by the worm D; overtails leave at the port E.

The trommel is closely akin to the reel. In its more elaborate forms it may comprise a series of several concentric screens which progress from coarse aperture towards the axis to fine aperture for the outermost screen. Material is fed to the inside of the inner screen, the finest material passing through all the screens in turn; overtails from each screen leave the machine by separate ports. *Plate I* illustrates the rotor of a compound trommel.

The screening surface itself may be of perforated plate or of woven wire or cloth; in the latter case the screening surface is commonly cleaned continuously by the action of a stationary brush which bears on the outside of the moving screen and reduces the tendency of the slightly oversize particles to blind the apertures of the screen.

Reels are driven at speeds somewhat below the 'critical' speed at which particles, carried round the inside of the screen in a circular motion, are acted on by a centrifugal force greater than their weight. Under such conditions particles would form a stagnant layer over the inside of the screen. Critical speed in this sense is given by the expression:

$$f_c = \left(\frac{32 \cdot 2}{4\pi^2 a}\right)^{\frac{1}{2}}$$

where a is the radius of the reel in feet, f_c the reel speed in revolutions per second and the acceleration due to gravity is taken as $32 \cdot 2$ ft./sec^2.

For the continuous separation of coarse, free-flowing material (that is, material greater than $\frac{1}{8}$ in. mesh) the trommel or reel has the merit of robustness and simplicity, but for finer screening it is now seldom used.

Reels with Throwing Action ('Centrifugals')

A form of reel in which the stock is picked up by internal revolving blades and thrown at the screening surface gives a throughput several times greater than that of a simple reel. The higher efficiency is partly due to better dispersion of the particles and partly to the greater area of screening cloth effectively in use at any one moment.

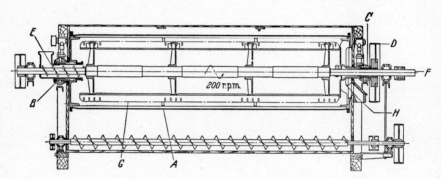

Figure 14. Reel (centrifugal). (By courtesy of Henry Simon Ltd.)

Figure 14 shows a diagram of a centrifugal screening machine. A cylindrical rotor carries the screening cloth A and is supported on end bearings B and C and driven by the pulley D. The end bearing B is mounted on a tube E, through which stock is fed to one end of the reel. A central shaft F, which is separately rotated, passes through the screen rotor and carries the beater assembly G. Material which passes through the screen is collected in a hopper below the rotor and delivered to an outlet port. Larger particles pass along the length of the screen and emerge through a discharge port H.

In a typical machine the screen may have a diameter of 2 ft. 6 in. and a length of 9 ft. and be driven at a speed of 17 r.p.m. The internal rotor of blades may then be of 2 ft. $2\frac{1}{2}$ in. diameter and revolve at a speed of about 200 r.p.m.

Simple Rotary Sifters

Several forms of screening machine contain a horizontal screen surface which is supported on links or canes and which is driven so that each point of the surface traces out a circular path of common radius. The drive may consist of some form of crank or eccentric; the canes prevent the screen from rotating bodily. *Figure 15* illustrates a two-decked form of such a machine. The box *A* is supported from the main frame *B* by the canes *C*, and is fitted with two screens *D*. An intermediate inclined tray *E* delivers throughgoing material of the upper screen to the head of the lower screen.

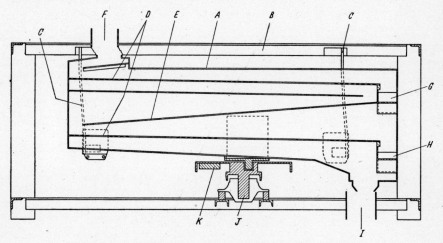

Figure 15. Rotary screen. (*By courtesy of* Henry Simon Ltd.)

Feed enters at *F* and overtails emerge at *G* and *H*. Material passing through the lower sieve leaves at *I*. The box is driven from below by a central, eccentric bearing arrangement *J*, which carries a balance mass *K* to counteract the centrifugal force arising from the circular motion of the box.

Plansifters

Plansifter is the name usually given to sifters which move with the simple rotary motion described above but which are constructed of a multi-tiered arrangement of sieves. Such screening machines are used where a considerable number of successive sizing operations are to be made on the primary material. Plansifters are arranged so that the primary material is fed into the top of the machine and is divided and directed so that the treated materials pass by gravity through the height of the machine and are diverted according to their size, to be retreated on lower screening surfaces.

A single supporting frame may carry as many as 96 sieves. Such machines are widely used in the flour milling industry and a typical example is illustrated in *Figure 16*. This machine is of the type known as the 'free-swinging' plansifter and is characterized by the hanging of the machine from above and by the special nature of the drive. The weight of the machine is carried by four groups of long canes *A*, so that the main frame *B* may swing as a conical pendulum. The rotary motion is imparted from above through a centrally placed vertical shaft *C*. The lower end of the shaft drives the main body of the machine through a crank *D*. As the shaft rotates the body moves in a circular path of radius equal to the eccentricity of the crank, whilst a balance weight E, attached to the shaft and crank assembly in a direction diametrically opposite to that of the crank, moves round the axis of the shaft. The mass and disposition of the balance weight are such that when the pendular motion has become stable the centrifugal forces accompanying the rotation of the weight are equal and

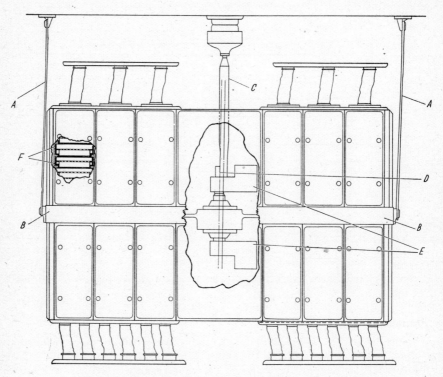

Figure 16. *Large plansifter.* (*By courtesy of* Henry Simon Ltd.)

opposite to those accompanying the circular motion of the main body of the machine. In this condition, the shaft hangs vertically and rotates on its own axis; its function is to transmit a simple torque to the driven mechanism, just sufficient to maintain the motion against windage and similar losses of energy.

In the machine shown in *Figure 16* the sieves, F, are relatively long and narrow, *viz*: 29 in. long by $12\frac{1}{2}$ in. wide, and are mounted in trays which serve to collect and redirect the throughgoing material. The material falls on to one end of the sieve and as fresh material arrives the mass migrates down the length of the sieve under the propulsive action of the head of accumulated material. The finer particles which pass through the sieve find their way in a similar manner to a separate outlet port. In the design illustrated in *Figure 16* the ports for 'throughs' and for 'overtails' cover only half the width of the sieves, and this arrangement is convenient where elaborate flow schemes need to be accommodated. *Figure 17* shows one of the many ways in which various sieved fractions may be obtained from a single stock.

A smaller version of multi-decked plansifter is shown in *Figure 18*. In this machine a box A contains two sections of six sieves, B, each. The box is suspended in the machine frame C from short canes D and is driven by a crankpin E attached to a shaft F, which runs in a bearing system G, located in the lower part of the main frame. A balance mass H, attached to the shaft eccentrically, neutralizes the centrifugal thrust caused by the circular motion of the sieve box. Stock may be directed to flow over more than one cover in succession, in the manner illustrated in *Figure 17*.

Table 7. *Relative Rates of Sieving of Reels and Plansifters when used on Flour Mill Stocks*

Percentage of feed smaller than 110 μ	Rate of sieving through bolting cloth lb./h/ft.^{2}		
	25	50	75
Reels (centrifugal)	2–4	4–8	5–12
Plansifters	6–12	12–24	16–33

Plansifters are designed to run at definite combinations of speed and stroke. The machine illustrated in *Figure 16* is normally driven so as to make 205 r.p.m. with a circular motion of $3\frac{1}{2}$ in. diameter; that shown in *Figure 18* is normally driven at 260 r.p.m. with a circular motion of 2 in. diameter. The speed is in each case substantially above the lower limit referred to on p. 133, at which the stock just does not move bodily with the screen. In complex plansifters of the type described above, the speed and stroke play an important part in determining the rate at which the streams of stock move along the horizontal surfaces in the machine, and in turn, the depth of stock at various points along the sieves.

The screening surfaces of the plansifter, as with other types of sifting machine, tend to clog with particles which are slightly larger in size than the apertures of the screen. Some form of cleaning device is therefore fitted below each sieve. The cleaner may take the form of a brush or of a canvas pad which is contained between the screen and a coarse wire retaining net below the screen; the cleaners are agitated by the motion of the machine and brush against the under surface of the screen with a rapid irregular motion.

The rate of sieving of like materials through like screening media is generally rather greater for plansifters than for reels; the improved efficiency is, in part, due to the better utilization of the surface of the screen in the

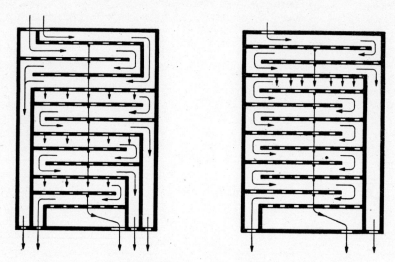

Figure 17. Flow scheme in plansifter. (*By courtesy of* Henry Simon Ltd.)

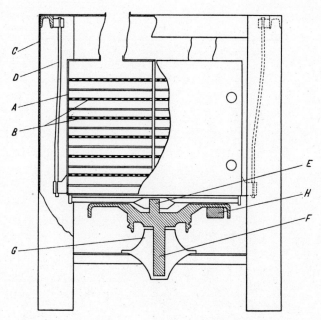

Figure 18. Small plansifter. (*By courtesy of* Henry Simon Ltd.)

case of the plansifter. Good performance of the plansifter is only obtained, however, if the sieve is covered with a uniform layer of stock of adequate depth and if effective means are provided for continuous cleaning of the screen surface. *Table 7* gives an indication of the relative rates of sieving of reels and plansifters when used on flour mill stocks.

Screens with Combined Rotary and Oscillatory Motion

A common form of sieving machine is provided with a special composite motion made up of a horizontal rotation at the head end and a horizontal linear oscillation towards the tail. The sieve frame is usually supported on canes and driven by a crank at a point near the head end; towards the tail end a lateral restraining member allows that end to move only longitudinally with respect to the length of the sieve. The rotary action at the head end

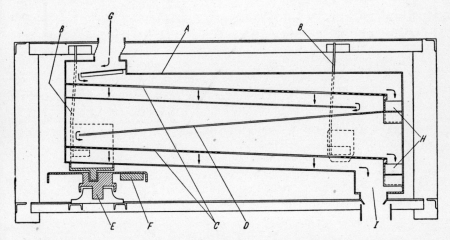

Figure 19. Screen with compound horizontal motion. (*By courtesy of* Henry Simon Ltd.)

tends to give a mild agitation under which action the screen cover is less prone to 'blind' with particles of near-mesh size; the small particles in the feed, therefore, pass through the head end of the screen without undue hindrance while the oscillating motion at the tail end serves to complete the separation with no loss of sieving rate due to the 'cushioning' action of the finest particles.

Figure 19 shows a two-decked sieving machine of this construction. The main box A is supported on canes B and is fitted with screening decks C connected with an intermediate inclined deck D, which carries the throughgoing material from the upper sieve to the head end of the lower sieve. The head end of the box is driven in rotary motion by the eccentric arrangement E and a mass F counterbalances the centrifugal force caused by the rotating component of the motion of the box. Feed enters at G and overtails of the top and bottom sieves leave by separate ports H; throughgoing material of the lower screen leaves by the port I.

High Speed Rotary Screens

Screens with horizontal rotary motion may be designed to run at comparatively high speed and with correspondingly small movement; *Plate II* shows a high speed gyratory or rotary sieve which runs at about 1,500 r.p.m. with a circular motion of about $\frac{5}{32}$ in. diameter. The main moving part of the machine consists of a series of circular sieves held in a frame which supports at the same time a rotating horizontal flywheel, to which is attached an out-of-balance mass. As in the free-swinging plansifter, the main sieve frame tends to move with such a circular motion that the centrifugal force associated with the motion just counterbalances the centrifugal force due to the moving out-of-balance mass.

Each screen is stretched tightly on a circular frame formed generally with a circular hole at its centre. The stock is fed to the centre of the screen and moves out across the surface. Throughgoing material passes on to a shallow hopper-shaped guide lying immediately under the screen and leaves the machine by way of the central circular hole. Material which overtails the rim of the first screen is guided by a further shallow hopper to the central region of the second screen and there undergoes further sieving. No special means are used for the continuous cleaning of the screen surfaces since the high speed of the motion apparently gives sufficient agitation to minimize clogging of the screen.

Screens with Horizontal Oscillatory Motion

Oscillating screens of low and medium speeds are used widely for diverse screening operations. Single-tier and two-tier arrangements of screens of very robust construction are used for screening coal, ore, chemicals and other hard and abrasive materials. Where the particle size is large, the speed is usually low and the stroke long; it is a general rule that length of the stroke is kept at least to the same order of size as the width of the screen aperture.

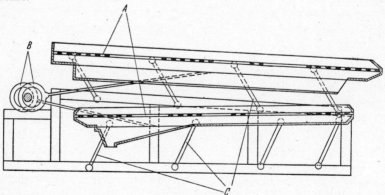

Figure 20. Screen with horizontal oscillating motion. (*By courtesy of* Automatic Coal Cleaning Co., Ltd.)

A horizontal oscillating machine of this robust type is shown in *Figure 20*. The oscillating motion is provided by eccentrics and connecting rods. In the machine illustrated the two screening decks A, supported in the main frame, are driven in counter phase by separate eccentrics B mounted on a common shaft. In this manner the main part of the acceleration force imposed on the driving shaft and frame by the motion of one of the massive decks is counterbalanced by the force due to motion of the other deck. In a typical application the speed of the machine may be 400 r.p.m., say, and the stroke 1 in. The decks are supported on inclined links C, set at an angle to the vertical so as to provide a vertical component in addition to the horizontal component of acceleration. The inclination serves to free the stock and to assist in propelling it along the screens. Less heavily constructed machines using a horizontal oscillating motion are made, usually in totally enclosed form, for the sieving of non-abrasive materials such as foodstuffs and certain chemicals.

Screens with Substantial Vertical Rotary or Oscillatory Motion

Special problems arise in the construction of screens which are to run with a substantial component of vertical motion. If the screen itself is light or is mounted so that it flexes easily, the screen may not be stiff enough in the vertical direction to transmit to the material at the centre the force required to keep it moving with the full motion of the frame. The material then tends to lie still, while the screen wires stretch to accommodate the motion of the frame. Screens with large vertical components of motion are therefore ruggedly built and provided with screen surfaces of high rigidity in the form of perforated plate or of tensioned, heavy gauge wire. *Figure 21* illustrates such a machine. The cross members A support the screening deck B and are secured to well reinforced side members C. The screen surface is tensioned by means of the bolts D at the rear end. An armature E spans the side members and is driven by an electromagnet at F, housed in a heavy saddle G. The frame is supported at its rear end by the suspension rods H and at its front end, *via* the springs I connecting the frame to the saddle, by the suspension rods J. The magnet coils are driven with rectified alternating current and the frame and deck take up a motion largely at right angles to the deck.

Figure 22 illustrates a form of screening machine in which all points of the deck move with roughly circular motion in the vertical plane, under the action of a centrally mounted eccentric shaft. The screen frame A, to which the screening deck B is secured, is resiliently mounted on rubber supporting units C which allow the frame freedom of motion in the vertical plane. A central driving shaft D runs in bearings E mounted in the side members of the screen frame. Eccentric spindles F extend from each end of the shaft and run in bearings G mounted in the centre of the supporting yokes H. An eccentric spindle carries a pulley I driven from the motor J mounted on the main frame K. The shaft system imparts a circular motion to the screen frame and screen. A flywheel carries an offset mass M which serves to balance the centrifugal forces arising from the motion of the screen frame. The main frame may be supported by slings N. Transmission of residual forces to the main frame is minimized by providing further rubber supporting units O between the yokes and the main frame.

The problem of keeping a substantial motion over the main area of a screen which moves with a vertical motion is greatest when the driving speed is high and the screening medium is fine and therefore lacking in rigidity. A successful approach to the problem is to apply the oscillatory

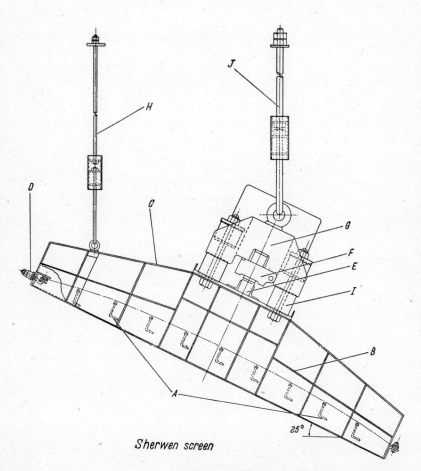

Figure 21. Screen with motion in vertical plane: electromagnetically driven. (By courtesy of Fraser & Chalmers Ltd.)

movement direct to the central areas of the screen itself. The Hummer screen is designed to make use of this principle, and an arrangement of it is shown in *Figure 23*.

The driving unit is electromagnetic and is mounted centrally over the width of the screen on a member rigidly attached to the stationary main frame. The armature is supported below the electromagnet by a leaf spring, the ends of which are secured to the sides of the screen frame. The centre of the leaf spring is in turn connected to strips which run centrally along the screen and are firmly bolted to the screen cloth. The position of the armature in relation to the electromagnet is adjustable by means of a hand screw which thrusts a helical spring against the centre of the leaf

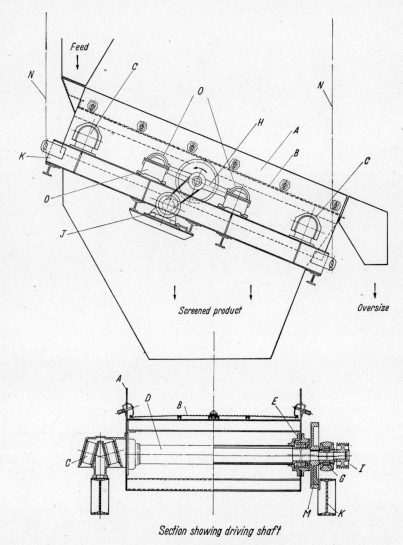

Figure 22. *Screen with motion in vertical plane: mechanically driven.* (*By courtesy of International Combustion Products Ltd.*)

spring. The hand screw is set so that the ends of the armature strike, with controlled severity, the blocks mounted on the electromagnet support. Recurrent vertical impacts are imparted effectively to a large portion of the screen cloth.

The electromagnet may be driven from a 50 c/s supply through a rectifier to give 3,000 vibrations a minute, or may be driven at lower speed from a motor generator. The machine may be single or multi-decked; the larger screens are driven from a pair of electromagnets spaced along the centre-line.

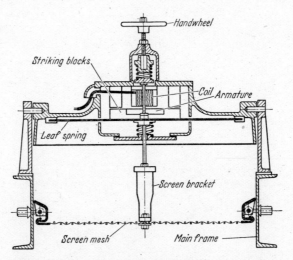

Figure 23. Screen with vertical motion applied directly to centre of screen cloth. (By courtesy of International Combustion Products Ltd.*)*

Screens with Combined Oscillatory Motion and Fluidization

In certain operations in which importance is attached to separation by shape as well as by size, it is customary to use a form of screening machine in which air is drawn upwards through the sieve while the sieve is oscillated with a substantially horizontal motion. Action of the upward lift of the air and of the direct vibration causes the granular particles, particularly the moderately small ones, to form a layer near the surface of the screen and at the same time concentrates the flaky or fibrous material towards the top layers of the bed of stock. Throughgoing material is therefore rich in granular particles, whilst the flaky or fibrous particles are to be found in the overtails.

Such screening methods are used in asbestos separation and very widely in the flour milling industry where the process of separating flat bran particles from granular endosperm particles is of prime importance.

Figure 24 shows an illustration of a flour mill 'purifier' in which a long sieve frame 7 ft. long and 18 in. wide is driven at one end by an eccentric, A, and pitman, and is supported on hangers, B, which are usually inclined to the vertical at an angle of about 15° so that the bed of stock is thrown forward at each oscillation and tends to move steadily along the sieve towards the tail. An air current of carefully controlled flow rate is drawn in an upward direction through the bed of material by way of an exhaust hood C, which covers the sieve. Air valves situated between the hood and the central exhaust trunk D control the distribution of the air flow along the length of the sieve.

The purifier is capable of making precise separations in size since the stratification of the stock keeps large particles away from the sieve itself. The feed to the machine is graded in size in a preliminary screening machine of conventional design. The total range in size of the fraction passed to

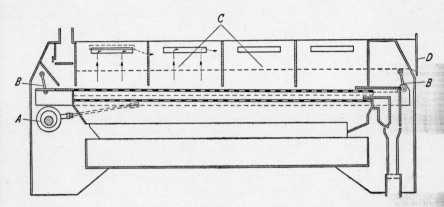

Figure 24.   Oscillating screen with air fluidization.   (By courtesy of Henry Simon Ltd.)

the purifier is, apart from the minority of small particles which adhere to the larger ones, not usually much greater than two to one. The sieve covers are chosen so that the sieve aperture increases progressively in size from head to tail of the machine and provision is made for the throughs to be directed according to size to form separate streams of characteristically different size. In the machine illustrated, the screening surface is shown in the form of a double-decked sieve with provision for removing a further fraction which overtails the lower sieve. Single, double and treble arrangements of sieve are all used in purifier construction; the extra sieves do not affect the basic nature of the separations but on occasion make for easier setting of the machine. The upward drag of the air currents tends to slow down the passage of particles through the apertures of the sieve. This, together with the fact that little of the feed is much smaller in size than the sieve aperture, results in moderately small capacity for the machine.

LAWS OF SETTLING

The screening process of separation is based on size and shape; complementary processes are based on differences in buoyancy and in resistance to fluid flow of particles immersed in a gaseous or liquid medium. Density separation by flotation, terminal velocity separation by elutriation or centrifuging and settling-rate separation in concentrated dispersions are typical of the latter processes. The principles of separation are considered below in terms of the general laws of settling. The early part of the discussion relates to the laws governing the drag force on isolated particles moving through a fluid. This is followed by observations on the manner in which the laws are modified when many particles in close proximity move through the fluid.

The upthrust of a fluid on a submerged body is equal to the weight of fluid displaced; for a fluid of density ϱ the resultant downward force due to gravity on a particle of mass m and density ϱ_0 is therefore given by

$$m \frac{\varrho_0 - \varrho}{\varrho_0} g \qquad \ldots \ldots (1)$$

A particle tends to float or sink according to whether its density is less or greater than the density of the fluid.

Where particles are suspended in a medium of lower mean density the resultant force due to gravity causes each particle to commence to fall with an accelerating motion. The movement of the particle through the neighbouring medium brings into action restraining forces which tend to resist the movement. The forces increase with the velocity of relative movement so that ultimately each particle moves with a steady speed for which the retarding force due to fluid resistance is balanced by the force due to gravity. This velocity is known as the 'critical' or 'terminal' velocity of the particle, and many sorts of industrial separator make use of the characteristic terminal velocities of the constituents.

The terminal velocity of a particle falling freely in a fluid depends on the effective weight of the particle and on the extent of the fluid drag which accompanies the motion of the particle through the fluid. The nature and the magnitude of the drag forces which arise from the relative motion of particle and fluid are outlined below.

The forces associated with the flow of liquids spring from two sources: viscosity and inertia. In each case, the forces are manifest in the changes of pressure throughout the fluid. The viscous forces reflect the steep gradients of fluid velocity such as exist in the shearing of the fluid adjacent to a solid surface. The inertial forces reflect the changing fluid velocities such as accompany a local constriction or curvature of the fluid flow.

When a single particle or a group of particles is present in a flowing fluid, drag forces, arising from each of the above causes, act on the surfaces and tend to make the particle move with the flow. The most general laws which cover the simultaneous action of both types of force have been known for over a century and are embodied in the Stokes-Navier equations[1]. Useful explicit solutions are, however, only obtainable in very limited circumstances and, for present purposes, the treatment will be divided for the most part into two separate sections, according to whether the viscous or the inertial forces predominate.

Where a viscous fluid flows at moderate velocity past a surface, the drag force is proportional to the mean velocity of flow. Stokes showed that under these circumstances the force R on a spherical particle of diameter a in a fluid of viscosity η and velocity v relative to the particle is given by

$$R = 3\pi a \eta v$$

This expression is known as Stokes' law. It applies with moderate accuracy, for example, to fine particles of water falling freely in air or to small steel balls falling freely through oil. For particles which are not spherical, a similar law applies in which, however, the diameter may be replaced by a characteristic length dimension l of the particle and a new numerical fact then replaces 3π

$$R = k\eta l v \qquad \ldots (2)$$

The flow of viscous liquids past particles or past fixed boundaries is an example of streamline flow. The special characteristic of streamline flow is that the motions of the individual elements are smooth and regular: each element is followed by a succeeding element which traces out a similar path.

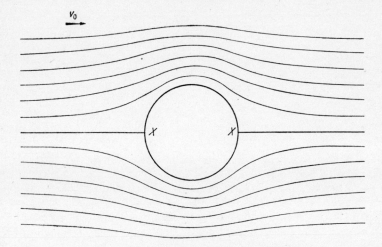

Figure 25. Streamline flow past a spherical body

For flow of liquids of low viscosity and especially in those regions away from solid surfaces, the streamline flow follows a pattern determined by the inertia of the moving elements of fluid. Such flow approximates to 'potential flow' characteristic of the ideal (that is, frictionless) fluid. The laws governing potential flow may be deduced simply from a study of two basic requirements of the flow: first, that no fluid is destroyed and, secondly, that the local pressures and momentum changes are consistent with Newton's laws of motion. A full account of this type of motion is to be found in text books on hydrodynamics.

Figure 25 illustrates streamline flow past a spherical body, and *Figure 26* illustrates streamline flow past an oblique disc. A special characteristic of streamline flow is that the flow is exactly of similar form on the upstream and downstream sides of a symmetrical body. Bernoulli's law for fluid flow derives from consideration of potential flow and may be expressed as follows. If, at a point distance h vertically above an agreed datum, the static pressure of a fluid of density ϱ is p and its velocity is v, then the expression $p + \frac{1}{2}\varrho v^2 + \varrho g h$ remains constant along a streamline through the point. Thus, if the flow lines are constricted so that the velocity increases, the static pressure falls locally by the extent of the increase in $\frac{1}{2}\varrho v^2$.

While it is common to find that the general pattern of flow in the regions between particles approximates to simple potential flow, the flow near a surface is not of this form because of the influence of viscosity. For low fluid speeds, the flow near the surface is of laminar form but in a state of

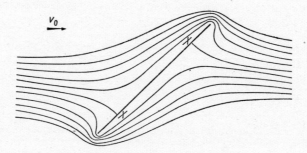

Figure 26. Streamline flow past an oblique disc

shear. At higher speeds the motion of fluid near the surface becomes more complicated. If the viscosity is small and the velocity and general dimension of a body are large, the flow near the body ceases altogether to be steady. The general direction of the flow remains definite and is depictable by flow lines as in *Figure 25*; the detailed pattern is unsteady. This form of motion is called turbulent and where it exists, a second type of restraining force between the particle and the moving fluid is produced. *Figures 27* and *28* illustrate the forms of turbulent flow past the sphere and the oblique disc. The pattern of the flow is no longer identical at upstream and downstream sides of the body. The high inertia of the flow coupled with the growing instability of the fluid near the surface tends to cause the stream to move away from immediate contact with the body after the broadest section of the body has been reached. The flow lines fail to close in completely immediately behind the body. The region of stagnant fluid behind the body is known as the 'wake' and is filled with irregular eddies which form from the rolling up of fast moving filaments of fluid at the surface of the body and their subsequent detachment from the forward moving stream.

The special drag forces which are associated with fully turbulent motion are proportional to the density of the fluid, to the area of section of the body normal to the flow, and to the square of the velocity of the mean flow. The forces may be considered as arising for the most part from the destruction of forward momentum of the stream as it meets the upstream face of the obstacle. The mechanism of the drag may be illustrated in more detail by reference to the case of the spherical particle of *Figures 25* and *27*.

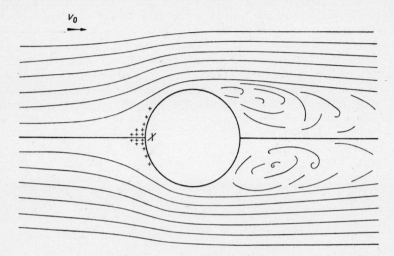

Figure 27. Turbulent flow past a sphere

Consider first the case where the flow is streamline as distinct from turbulent. On the upstream side of the sphere the stream is retarded as the flow lines diverge to pass round the obstacle. The velocity of the central filament which meets the surface at X is reduced from its initial value v_0 in the

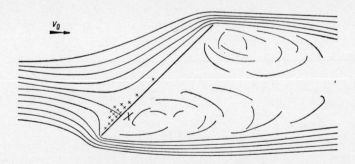

Figure 28. Turbulent flow past an oblique disc

undisturbed stream, almost to rest. In consequence, the static pressure rises above the value p in the surrounding fluid by an amount $\frac{1}{2}\varrho v_0^2$ (Bernoulli's theorem). The kinetic energy which the fluid would otherwise have possessed is converted to the fluid potential or pressure energy. In general, at other points of the upstream surface the static pressure rises above that in the body of the liquid by an amount equal to the local reduction in $\tfrac{1}{2}\varrho v^2$.

The forward thrust on the upstream face of the body is the sum of the excess of static pressure forces acting on each unit area and is substantially proportional to the product of the area of cross section of the sphere and the 'dynamic pressure', $\tfrac{1}{2}\varrho v^2$ of the impinging stream. For a spherical particle in streamline flow the flow closes in behind the body to give a field of flow near the downstream surface (see *Figure 25*). The second system of increased static pressure thrusts the sphere in a direction counter to the flow. For any symmetrical body in streamline flow the resultant thrust is zero. Where the flow is fully turbulent, the thrust caused by the slowing down of the stream acts only on the upstream side of the particle (see *Figure 27*). This forward thrust, augmented in some degree by viscous traction at the surface and by forces resulting from detachment of eddies to form the wake, constitutes the turbulent drag on the particle.

For a particle with a cross sectional area A perpendicular to a turbulent stream of fluid of mean velocity v and density ϱ, the drag force may be written:

$$F = k' \cdot \tfrac{1}{2}\varrho v^2 A \qquad \ldots\ (3)$$

where k' depends on the shape of the body, but does not change much with velocity. This is Newton's law of fluid resistance, and it applies, for example, with moderate accuracy to stones falling freely through air.

As the velocity of flow past a body increases steadily from a low value, the drag force is at first due to viscous drag at the surface and obeys a law generally of Stokesian form. When the velocity has grown to very high values, the drag force has the Newtonian form characteristic of turbulent conditions. No satisfactory general theoretical treatment covers the intermediate field and it is therefore convenient to resort to dimensional theory for collation of the full range of drag data.

If the expression for drag force F for a body of given shape is regarded as compounded of terms involving powers of the essential variables ϱ, η, v and l, where l is a typical dimension and ϱ, η and v have the meanings already assigned to them, it may be shown by equating powers of 'dimensions', that F must take the form

$$F = \varrho v^2 l^2 \psi\!\left(\frac{vl\varrho}{\eta}\right) \qquad \ldots\ (4)$$

where ψ is some function of the parameter $vl\varrho/\eta$. The parameter is known as Reynolds number R_e and is 'non-dimensional', that is, its value does not depend on which one of the series of self-consistent systems of units is used.

Equation (4) represents the general form of the drag equation. For the purposes of application, it is usually written

$$F = C_D \cdot \tfrac{1}{2}\varrho v^2 A \qquad \ldots\ (5)$$

where A is the cross sectional area perpendicular to the flow and where C_D

has a value simply proportional to $\psi(R_e)$ and which alters characteristically therefore, with Reynolds number.

The function C_D is not in any way restricted in form. However, for the two special regions in which inertia and viscosity respectively, have most influence on the flow near the particle, C_D takes moderately simple forms. Thus, for example, formulae (3) and (5) are equivalent if $k' = C_D$. For ranges of ϱ and v, for which k' is observed to be roughly constant, C_D and therefore $\psi(R_e)$ must be roughly constant. Equally, formulae (2) and (5) are equivalent if

$$3\pi a \eta v = C_D \cdot \tfrac{1}{2}\varrho v^2 \cdot \tfrac{1}{4}\pi a^2$$

or

$$C_D = 24\frac{\eta}{\varrho a v} = 24/R_e$$

Thus, for this region C_D must vary inversely with R_e. In general, the viscous region, with drag force given by formula (2) corresponds with low values, not merely of viscosity, but more generally of Reynolds number. The turbulent region, with drag force given by formula (3) corresponds with high values of Reynolds number. The general nature of the variation of C_D with R_e is illustrated in *Figure 29* for flow past a sphere.

The forces so far discussed are those on isolated particles. Free falling particles which are well separated from each other are retarded in their fall by forces of this magnitude; the terminal velocity is given by equating the drag force given by equation (5) with the gravitational force given by equation (1). Thus, the terminal velocity u is given by

$$mg\frac{\varrho_0 - \varrho}{\varrho_0} = C_D \cdot \tfrac{1}{2}\varrho u^2 A$$

or

$$u^2 = \frac{mg(\varrho_0 - \varrho)}{\tfrac{1}{2}C_D \varrho \varrho_0 A} \quad \ldots \ (6)$$

where the value of C_D must be taken to correspond with the appropriate value of Reynolds number.

For turbulent flow C_D is approximately constant at the value of 0·4 for a sphere of diameter a. The terminal velocity u is then given by

$$u^2 = \frac{ag(\varrho_0 - \varrho)}{0 \cdot 3\varrho}$$

For viscous flow the expression for terminal velocity reduces to

$$u = \frac{a^2 g(\varrho_0 - \varrho)}{18\eta}$$

It will become clear in the course of later discussion that the general nature of many industrial separating processes can be described in terms of the steady equilibrium of opposing forces of fluid drag and of inertial origin. A full description of detailed behaviour of particles involves, however, consideration of their accelerated motion. Where inertial effects are involved the heavier particles may not reach a steady velocity in the limited period during which the particle moves through the separating chamber. In such cases the shape of the trajectory of the particle during the period of acceleration may have important bearing on the mechanism of separation.

Many problems involving the motion of isolated particles are soluble, at least in principle, if the drag forces can be represented by a known function of the velocity of the particle relative to the fluid (see for example, *Figure 29*).

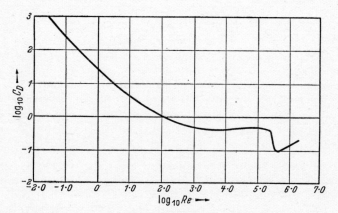

Figure 29. *Variation of* C_D *with* R_e *for spherical particles.* (*After* MUTTRAY, *from* S. GOLDSTEIN, *by courtesy of the* Clarendon Press, Oxford)

The necessary computing technique is, nevertheless, elaborate, and often the results are not presentable in a form easy to interpret. The following treatment will be limited to the simplest examples of particles moving in a straight line; these examples suggest, however, important generalizations of wider utility. LAPPLE and SHEPHERD[2] have formulated methods for calculating simple trajectories in two dimensions. LANGMUIR and BLODGETT[3] and also DAVIES and AYLWARD[4] have shown how to obtain solutions to more complex problems.

If a particle is projected with substantial velocity into a fluid, the drag slows the particle down so that it eventually comes to rest. Consider a particle of mass m, diameter a and density ϱ_o moving with velocity v relative to a fluid of density ϱ and viscosity η. The equation of motion is

$$m'\frac{dv}{dt} + \tfrac{1}{2}\varrho C_D v^2 = 0 \qquad \ldots\ldots (7)$$

m' is the effective mass of the particle and consists of the mass m augmented by a term contributed by the acceleration of fluid which accompanies the motion of the particle (the 'accession to inertia'). For purposes of the present example m' will be assumed equal to m.

The equation (7) reduces to

$$m\frac{dv}{dt} = -ka\eta v$$

for viscous motion, and this has the solution

$$v = v_0 e^{-ka\eta t/m}$$

where v_0 is the particle velocity when t is zero. Thus, the velocity of the particle falls to $1/e$ of its initial value after an interval of τ seconds, where

$$\tau = \frac{m}{ka\eta}$$

τ may, therefore, be termed the time constant or relaxation time of the motion.

The distance s travelled during the interval t may be deduced by integration as $s = v_0\tau(1 - e^{-t/\tau})$; for $t = \tau$, this is of the order 0.63τ. For a spherical particle of density 2 and of 100 μ diameter projected with a velocity of 1 cm/sec (about twice the terminal velocity) into water, the distance s travelled before the initial velocity has fallen to $1/e$ is about 1/1400 cm. For turbulent flow equation (7) becomes

$$m\frac{dv}{dt} = -\tfrac{1}{2}\varrho C_D v^2$$

where C_D is equal to about 0.40 for spherical particles.

This solves to give

$$\frac{1}{v} - \frac{1}{v_0} = \frac{\varrho C_D A}{2m} t$$

and by integration

$$s = \frac{2m}{\varrho c_0 A} \log_e \left(1 + \frac{\varrho C_D A}{2m} v_0 t\right)$$

For a spherical particle of density 2 and diameter 1 mm projected in air with a velocity of 500 cm/sec (about 1,000 ft./min) the distance travelled before the velocity falls to one half of its initial value is about 350 cm (11.5 ft.).

The drag force on a particle is modified by the proximity of fixed walls or by the presence of neighbouring particles. The drag force on individual particles as they settle in a vessel is increased by the presence of the walls and the particles therefore settle more slowly. For flow at low Reynolds number and where the walls remain at a large distance l, the drag force on a spherical particle of diameter a is increased beyond that calculated from Stokes' law by the factor $1/(1 - qa/e)$ where q has a value between 1 and 5, according to the nature and disposition of the wall. (See HAWKSLEY[5].) When the wall lies relatively close to the particle the drag force increases more rapidly than the above expression indicates. At high Reynolds numbers walls situated at moderate distance (20 particle diameters) appear to have little influence on drag.

The presence of neighbouring particles has a marked effect on drag force[6], but calculation of the size of the modified drag presents obvious difficulties. However, estimates have been made by several investigators[7, 8] of the modified drag on an assemblage of equal spheres arranged in some specified manner. Thus, for low concentration BURGERS[7] suggests an increased drag force in the ratio of $1 + \lambda c$ where c is the volume concentration of the spheres and where $\lambda = 55/8$ for randomly arranged spheres.

For high concentrations of suspended particles several investigators[9, 10] have devised expressions for the ratio in which the settling rate is increased. The calculations are based on assumed configurations of the particles during settling and the values of the ratio vary accordingly. RICHARDSON and ZAKI[10] have shown experimentally that the ratio may be expressed in the form $(1-C)^{4.65}$ which is in moderate agreement with the expression $(1-C)^{9/2}$ developed by Hawksley on the assumption that particles take up an equilibrium configuration for which the settling rate is a maximum. In highly concentrated suspensions, as for example, where particles are spaced only a few diameters apart, the drag forces are considerably modified by the presence of neighbouring particles. For particles whose behaviour is not principally determined by physico-chemical interaction and which exhibit, therefore, no marked characteristics of a gel, the mass of particles tends to settle at a common rate, even though the individual particles may vary considerably in size. Each particle must, in this case, be regarded as contributing a share of the total drag force of the cloud.

The fundamental reasons which cause particles in dense suspensions to move as a single cloud are not well understood. The interactions which come into play between particles as they fall through the fluid must tend to make the particles take up an equilibrium configuration in which further relative movement is largely inhibited. However, the fact that little relative movement takes place between particles is of some assistance in the explaining of the rate of settling of dense suspensions. The cloud of steadily falling particles is restrained in its fall by the forces due to the relative upward movement of the fluid through the interstices or 'voids' between particles. In this sense the friction forces acting in total on the particles are of the same form as the forces acting on a stationary bed of particles when fluid is forced through the bed.

The slow rate of fall of the cloud of particles as a whole comes from two main causes. On the one hand, large viscous forces arise from the flow of liquid in the narrow capillary passages between particles. On the other, the relative upward velocity of the fluid with respect to the particles is considerably increased by the upward flow of the fluid which is displaced by the downward movement of the particles. As the cloud of particles settles at a steady rate the effective weight of the immersed particles must be supported by the fall in fluid pressure across the bed caused by the relative upward flow of the fluid. For a fully compacted cloud of particles, the flow of fluid relative to the individual particles would be expected to accord with the upward flow of fluid required to fluidize a similarly compacted stationary bed of the particles (see 'Fluidization and Fluidized Beds', Vol. 6). It should, however, be noted that since the upward flow is compounded of the speed of fall of the cloud and the speed of the upward flow of displaced fluid, free settling rates of concentrated suspensions may be lower than, although of the same order as, the fluidization velocity referred to above.

It has already been remarked that the coherent cloud of particles, typically to be found in a 'hindered' settling of concentrated suspensions, may retain many small particles in the voids between the larger particles. The settling column no longer behaves, therefore, as a freely sedimenting column in which the coarse particles may be concentrated by early removal of fully settled particles. When bodies substantially larger in size than the bulk of particles are immersed in the hindered settling column, the body settles at a rate completely different from that of the suspended material. The buoyancy forces which provide the upthrust on the body are in this condition determined by the average density of the suspension. A large particle of density lower than the suspension but higher than the fluid medium may therefore float rather than sink, and separations may thus be made in hindered settling columns in terms of particle characteristics not directly related to terminal velocity. These special separating properties form the basis of action of several forms of liquid classifying plant.

Under certain conditions of particle size and shape and of physico-chemical interaction between particles, the uniform cloud so far described does not represent the stable form of the suspended material. Individual particles which are by accident at suitably situated points with respect to their near neighbours, group together to form aggregates. These aggregates are separated one from another by clearly defined gaps or channels and in this condition the separated aggregates or flocs settle as if each aggregate were a discrete particle. Flocculation provides wide channels for the return flow of fluid in a settling column and therefore tends to reduce the average fluid drag on the mass and, at least relatively to hindered settling of the uniform cloud, to increase the mean settling rate.

BIBLIOGRAPHY AND REFERENCES

FAHRENWALD, A. W., and STOCKDALE, S. W., 'Effect of Sieve Motion on Screening Efficiency', *Bur. Min. Rep.* No. 2933

HURST, J., *Times Rev. Ind.*, Feb. (1954) 29

1 GOLDSTEIN, S., *Modern Developments in Fluid Dynamics*, Oxford University Press, 1938
2 LAPPLE, C. E., and SHEPHERD, C. B., *Industr. Engng Chem. (Industr.)*, 32 (1940) 605
3 LANGMUIR, I., and BLODGETT, K., *Gen. Elect. Res. Lab. Rep.* RL–225 (1949)
4 DAVIES, C. N., and AYLWARD, M., *Proc. Phys. Soc.*, B 64 (1951) 889
5 HAWKSLEY, P. G. W., *Mon. Bull. Brit. Coal Util. Res. Ass.*, XV (1951) 105
6 STEINOUR, H., *Industr. Engng Chem. (Industr.)*, 36 (1944) 618, 840, 991
7 BURGERS, J. M., *Prod. Koninklijke Ned. Akad. Wet.*, B 44 (1941) 1045
8 DEBYE, P., and BUECHE, A. M., *J. Chem. Phys.*, 16 (1948) 573
9 HAWKSLEY, P. G. W., *Physics of Particle Size Analysis*, Institute of Physics, 1954
10 RICHARDSON, J. F., and ZAKI, W. N., *Chem. Engng Sci.*, 3 (1954) 65

7

TABLING AND JIGGING

G. H. HIGGINBOTHAM

TABLING and jigging are both well established methods of separating materials according to their specific gravities. Jigging is one of the oldest processes for the concentration of ores and for the washing of coal, and is still widely used. Tabling was first employed on a large scale about 1896, when the Wilfley table was developed, and reached its peak around 1910. For a period the introduction of froth flotation led to a decline in the use of tables, but the need to separate minerals not amenable to flotation has stimulated their use.

While the apparatus and methods used in these two processes differ widely, the basic principles involved are identical, namely, hindered settling combined with flowing stream separation. These basic principles will therefore be discussed prior to detailed descriptions of the two operations.

THEORY

In tabling and jigging, and in other processes of gravity concentration, the operative forces are dependent upon the specific gravity, size and shape of the particles to be separated. Under the action of these forces each particle moves along a path according to its own characteristics until two or more groups of particles are formed sufficiently separated to be removed from the process. In moving, however, each particle influences the motion of the surrounding fluid which in turn produces an effect upon the neighbouring particles. These effects are so complicated as to defy rigorous analysis, but it is possible to establish a number of basic principles.

The three types of force to which particles being separated are subjected are as follows:

1. Force of gravity $= mg$
2. Force of buoyancy $= V\varrho_2 g$
3. Forces of resistance $= F$

where $m =$ mass of the particle
        $g =$ acceleration due to gravity
        $V =$ volume of the particle
        $\varrho_2 =$ density of the surrounding fluid.

A particle moving under the action of these forces will accelerate to a constant or terminal velocity, when:

$$mg = V\varrho_2 g + F \qquad \ldots\ldots (1)$$

Of the three types of force involved, the forces of resistance F are the most complex, arising from the motion of the particle through the surrounding fluid and including forces resulting from interaction with neighbouring particles. Even for spherical particles moving in a viscous fluid the values of F can be defined only within specific ranges. For example, under conditions of laminar flow the force of resistance experienced by a smooth sphere is given by:

$$F = 3\pi \eta d v \quad \dots \quad (2)$$

where η = coefficient of viscosity
   d = diameter of the sphere
   v = velocity of the sphere
If $v = v_t$ = terminal velocity of the particle
and ϱ_1 = density of the particle

then inserting the value of F in equation (1) and substituting for the volume and mass of the particle, the equation becomes:

$$\tfrac{1}{6}\pi d^3 \varrho_1 \cdot g = \tfrac{1}{6}\pi d^3 \varrho_2 g + 3\pi \eta d v_t \quad \dots \quad (3)$$

$$V_t = \frac{d^2}{18} \frac{(\varrho_1 - \varrho_2)}{\eta} g \quad \dots \quad (4)$$

This equation is Stokes' law and is applicable under the conditions for which F was defined. It is also possible to derive Newton's and Rittinger's equations in this manner when:

$$F = \frac{\pi}{8} \varrho_2 d^2 v^2 \quad \text{(Newton's)} \quad \dots \quad (5)$$

or

$$F = Q \frac{\pi}{8} \varrho_2 d^2 v^2 \quad \text{(Rittinger's)} \quad \dots \quad (6)$$

where Q = an experimental coefficient of resistance.

It is significant to note that of the three forces acting upon a particle, that due to buoyancy, namely $V\varrho_2 g$, is large and is probably the most important, and it is dependent upon ϱ_2, the mean suspension density defined by the second general conclusion which will be referred to later.

The motion of spherical particles in a viscous medium is the present limit for a rigorous analysis of the problem of gravity concentration. Nevertheless it is possible to draw general qualitative conclusions as to the behaviour of particles under hindered settling conditions. Three such conclusions form the basis of the separation of minerals by tabling and jigging, namely:

1. If two equi-settling particles of different density, size and shape are allowed to settle in a restricted area, they will settle in such a way that the denser particle will fall faster.

2. A suspension of solids in a fluid behaves as though it were a liquid having a density equal to the mass of the suspension divided by its volume. A particle immersed in the suspension will, in fact, be buoyed up in accordance with Archimedes' law.

3. In the operation of a gravity separation process the density of the bed of material should be as high as possible.

To supplement the last two conclusions it is necessary to note from the laws of settling that each includes a term of the difference in density between the particles and the fluid in which they are immersed. For two particles the ratio of these differences to some power is referred to as the hindered settling ratio; given by

$$H_R = \left(\frac{\varrho_1 - \varrho_2}{\varrho_3 - \varrho_2}\right)^m$$

ϱ_1 = density of first particle
ϱ_3 = density of second particle
m = index = 1 for Stokes' law
= $\frac{1}{2}$ for Newton's law

The hindered settling ratio is, in fact, a measure of relative settling rates, and it follows that the higher the density of the surrounding fluid the higher will be the hindered settling ratio for two given particles. Hence the easier will the separation of the particles become.

The last conclusion (3) noted above must not be taken to its ultimate conclusion or the particles will be so interlocked that no separation will be possible. A compromise must be found between maximum density of particles and sufficient mobility to effect a separation. In tabling and jigging, this compromise is effected by inducing oscillation in the bed of material. The particles are moved from the rest and allowed to settle under conditions of high bed density at each oscillation, the oscillation being sufficient to achieve mobility and yet not disturb the separation being effected, by throwing the particles into random distribution.

In addition to oscillation, particles being separated by tabling are also subjected to a flowing current to a greater extent than particles in a jig. This flowing current produces a marked effect, since the flow rate varies with depth, thus inducing rotation of the particles[2]. The net result of this current is twofold, firstly material of a low density is carried faster than material of high density. Secondly, in mixtures of particles, the coarsest are carried along the bottom of the stream faster than fine ones.

Finally, it should be noted that when jig-washing coal, it is possible to separate a wide size range of material in one unit. A possible explanation of this has been presented by CHAPMAN[3], and his calculations show that the distances travelled by particles during the acceleratory period of fall are primarily dependent upon density rather than size, the effect being most marked during the first incidence of fall. *Figure 1*, due to Chapman, compares the distances fallen by 0·25 and 1·0 cm diameter particles of shale with coal particles 1·0 cm and 10·0 cm diameter. A position occurs on this graph at approximately 0·12 sec when the finest shale particle has fallen further than the coarsest coal particle. From this it is concluded that provided the time of fall is sufficiently short and that the process is repeated often enough it is possible to effect a gravity separation independent of size.

The following subsections deal with the practical aspects of tabling and jigging of materials.

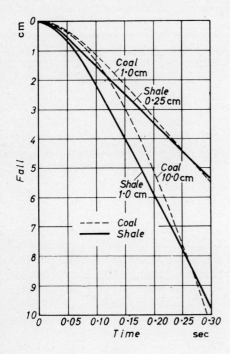

Figure 1. Distance of fall of coal and shale particles in water. (From W. R. CHAPMAN[3])

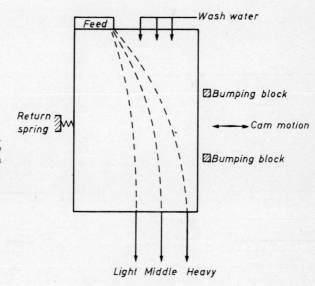

Figure 2. Bumping table. (From S. J. TRUSCOTT, by courtesy of Macmillan & Co.)

TABLING

General Considerations

The wet concentrating table as it is known today was developed from the bumping or percussion table[4], a diagrammatic representation of which is shown in *Figure 2*. This table was inclined and was fed in one upper corner, the solids being flushed down the table with wash water. As the solids moved down the table they were subjected to a jerking motion at right angles to their direction of travel and as a result moved across the flowing stream. The denser particles moved further than the lighter particles, and the larger ones moved further than the smaller. Consequently, with a narrow size range feed, a separation according to specific gravity was achieved.

Two major improvements in design resulted in the modern table. First, the plain deck was replaced by one having riffles along the direction of motion, a diagrammatic view of the Wilfley table incorporating these riffles being shown in *Figure 3*. The introduction of this riffled deck led to increased capacity coupled with more accurate separation. Stratification occurs behind the riffles, the heavier particles sinking to the bottom where they escape the streaming motion of the wash water, being carried to the discharge point by the table motion only. The lighter particles are brought up into the flowing stream and are carried down the table to their distinct discharge.

The riffles are generally of wood, soft pine or oak, and are nailed to the table surface which is normally of linoleum. Sand concrete has been suggested as a material for table tops, but it requires wooden nailing strips for the riffles and is also a heavy structure to support and to oscillate.

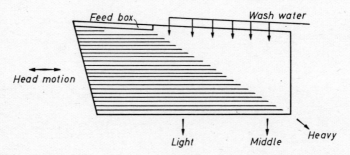

Figure 3. Wilfley table. (From S. J. TRUSCOTT, by courtesy of Macmillan & Co.)

Wired glass tops with impressed riffles have also been considered but nothing seems to be as satisfactory as linoleum. The depth of the riffles will vary according to the size being treated but they are normally ½ in. deep at the feed end tapering to a feather edge at the discharge point.

The second improvement in table design was the replacement of the camshaft and bumping blocks by a differential motion. The principle of this motion is to make the table approach its reversing point and recede from it at a greater speed at one end of its travel than at the other. A conveying action may thus be achieved, equally as effective as bumping, but with greater smoothness and permitting more accurate control. While a number of mechanisms have been devised for imparting this differential motion or 'head motion' to the table, they are all based upon the same principle of eccentric, pitman and toggles. A simple illustration is given in

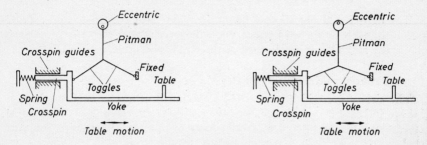

Figure 4. Toggle motion

Figure 4 showing the two extreme positions of the mechanism. As the end of the pitman descends the horizontal speed of travel of the crosspin is gradually reduced. Conversely, as the pitman rises, the speed of the crosspin progressively increases. By attaching the crosspin through a yoke to the table, the reversal of direction of the table is made sharply at one end of its travel while at the other end it is made relatively slowly. It is common practice to incorporate into such mechanisms a spring which is compressed on one stroke and which extends under the control of the toggles on the return stroke. The smoothness of the motion is improved by the addition of this spring which overcomes any 'bottom dead centre effects'. The use of this motion and the mounting of the tables results in low power requirements, most tables being driven by a motor of less than 2 h.p.

Principles of Separation

The separation of material by tabling is effected by a combination of hindered settling and flowing stream selection, and this process of separation may be considered in stages as follows. The feed is distributed over the first section of the table from the feed box, and coming under the action of the table motion, it is stratified in such a manner that the larger particles come to the surface. The larger light particles are carried away rapidly by the wash water across the riffles, while the larger heavier particles, however, are carried across the table by the table motion far more rapidly than they are moved by the wash water and are therefore discharged with the heavy product. With the removal of the larger light particles, and as the material moves diagonally across the table, the medium sized light particles are brought to the surface of the bed. They lose the protection of the riffles as these decrease in depth and are consequently carried across the table by the wash water.

As the material nears the end of the table it consists of the finest light particles together with heavier particles of a larger average size. It is considered that hindered settling now plays a more significant part in the separation. Eddying is taking place behind each riffle and the size and specific gravity of the particles are such that the finest lightest particles are thrown to the surface, where they are carried away by the wash water.

The lighter solids of all sizes have, therefore, been separated and carried across the table by the wash water, while the heavier particles have progressed along the riffles being discharged at the end of the table.

Sizes Treated and Table Capacity

The size of ore which is treated by tabling is generally between 5 mesh B.S. ($\frac{1}{8}$ in.) and 300 mesh B.S. In the case of coal preparation a size range of $\frac{1}{4}$ in.–0 in. is considered most amenable to tabling, although in America coal as coarse as $1\frac{1}{2}$ in. has been cleaned by this process.

It is considered, however, that separation by tabling is only fully effective down to sizes of the order of 52 mesh B.S. and that on sizes finer than 100 mesh B.S., little or no cleaning takes place. One reason for this size limitation which has been proposed is that between 52 and 72 mesh a transition occurs from the laws of eddying resistance to viscous resistance, and that this adversely affects the separation.

One further feature should be noted in relation to the sizes subjected to tabling, the separation will be more difficult as the size range of materials being treated becomes wider. Conversely, however, too narrow a size range can result in a loss of mobility on the table. A compromise must therefore be achieved, and having once fixed the size range and set the table for this condition, then it must be maintained as closely as possible.

The capacity of a table will naturally vary according to the size and nature of the material being treated. Considering a table between 10 and 15 ft. in length, on a coarse ore a capacity of 2–5 ton/h can be expected, reducing to a $\frac{1}{2}$–1 ton/h when treating finer material. When treating, say, $\frac{1}{4}$ in.–0 in. coal a capacity of 6–10 ton/h is obtained on an average, although figures of 15–20 ton/h have been quoted; with coarse coal feeds, capacities of 8–12 ton/h are obtained.

Types of Table [5, 6]

Wilfley—The Wilfley table, developed in 1896, incorporated for the first time the differential head motion and the riffled deck. This table, illustrated in *Plate I*, is generally 15–16 ft. long reducing in width from 6 ft. at the feed end to 5 ft. at the discharge end. Riffles of wood, nailed to a linoleum surface, cover more than half the deck surface ending in a diagonal line from the discharge corner. The differential motion is supplied by an eccentric, pitman and toggles as already described. The distribution box,

Plate I. Wilfley table. (*From* S. J. Truscott, *by courtesy of* Macmillan & Co.)

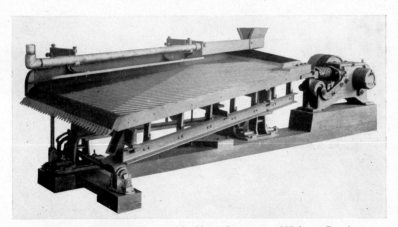

Plate II. Holman half-size sand table. (*By courtesy of* Holman Bros.)

[*To face p. 178*

Plate III. Butchart table. (From S. J. Truscott, *by courtesy of* Macmillan & Co.*)*

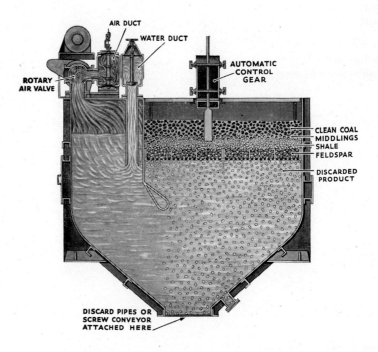

Plate IV. Feldspar jig. (By courtesy of Automatic Coal Cleaning Co.*)*

[*To face p.* 179

extending between 4 and 5 ft. down the table, is attached to and moves with the table. Typical performances of Wilfley tables are shown in *Table 1*.

Mention should be made here of the Garfield table which is essentially a Wilfley table but with the riffles carried right across the full surface of the table.

Table 1. *Typical Performances of Wilfley Tables*

Mineral	Assays—per cent				Capacity tons/day
	Feed	Concentrate	Middles	Refuse	
Zinc	5·5	59·8	8·0	0·5	14·0
Zinc	3·8	58·2	7·7	2·1	21·5
Lead	6·0	55·0	4·0	1·0	10–20
Lead	5·0	72·0	5·0	0·4	30
Tungsten	2·0	62·0	—	0·2	30
Coal	25·0 ash	9·03 ash	16·7 ash	70·0 ash	—

Ferraris or Buss—The head motion of the Ferraris table is a simple eccentric with the table mounted on laths fitted at an angle of 75° to the vertical.

An active travel of material results and the shape and positioning of the riffles has accordingly been modified, the riffles running the full length of the table and converging towards the discharge end of the table.

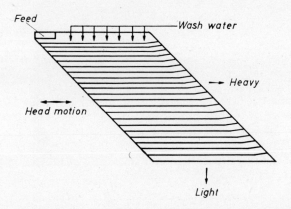

Figure 5. *Deister sand table.* (*From* A. F. TAGGART, *by courtesy of* J. Wiley & Sons)

Holman—This machine, originally the James table, combines the head motion of the Wilfley table with the lath mounting of the Buss. It differs from these two latter tables, however, in that the motion is not along the length of the table but at an angle of about 30° to the length, the riffles being parallel to the motion (*Plate II*). The riffles are of wood fastened to linoleum along the greater part of their length, but are extended by thin brass strips. The speed of the table is generally of the order of 300 r.p.m.

with a short stroke around ½ in. The positioning of the drive leads to a reduction in space occupied by the table without loss in efficiency of separation.

Deister Sand—The Deister sand table is illustrated diagrammatically in *Figure 5*, the table motion is parallel to the riffles, and the deck rises towards the heavy product discharge causing this product to climb against the flow of liquid. The head motion of the table is notable in that it consists of a combination of a quick return slide-crank mechanism and a bell lever.[7]

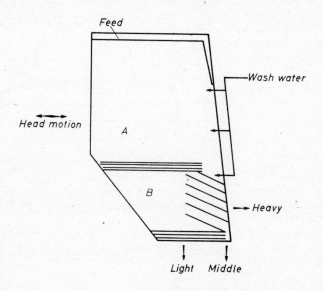

Figure 6. Deister slime table. (*From* A. F. TAGGART, *by courtesy of* J. Wiley & Sons)

Deister Slime—Used in the concentration of very fine material, the Deister slime table employs settling in two pools to afford the required separation. Referring to *Figure 6*, the feed to the table travels across a smooth surface to the pool A, which is about ½ in. deep. The heavy particles settling in this pool climb along the riffle barrier towards their discharge to the right

Table 2

Test No.	Feed	Assays—per cent					
		Feed Cu	Concentrate		Middles Cu	Tailings Cu	Slime Cu
			Cu	Insol.			
1	Finest slime	1·70	9·3	67·7	1·70	1·48	1·10
2	Slime-vanner feed	1·02	7·6	67·8	0·93	0·62	0·70
3	Smooth sand-vanner feed	0·89	10·5	58·3	2·03	0·46	0·65
4	Wilfley middles	1·13	13·5	54·3	2·15	0·85	0·63

of the table, and material carried over the riffles undergoes a second separation in pool *B*. *Table 2* gives the results of tests carried out with a Deister slime table by the Nevada Consolidated Copper Co.

Deister–Overstrom—The Deister–Overstrom table is widely used for the treatment of minerals and in coal washing, particularly in the United States. The table shape is roughly rhomboidal with the riffles diagonal to the long edge as in the Holman table, *Figure 7*. The riffles are, however, at right angles to the flow of the wash water and the head motion operates parallel to them. The rhomboidal shape has been chosen to increase the number of riffles per table, resulting in a greater number of retreatments of

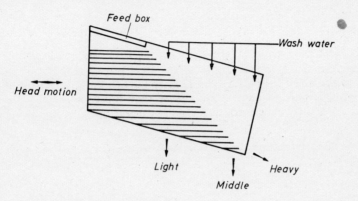

Figure 7. Deister–Overstrom table. (From A. F. TAGGART, *by courtesy of* J. Wiley & Sons)

the material behind each riffle and hence an increase in the accuracy of separation. Results of the operation of this type of table are given in *Table 3*.

Table 3

Material	Assays—per cent			
	Feed	Concentrate	Middles	Tailings
Copper	1·15	11·0	0·8	0·57
Lead	—	73·0	—	—
Tungsten	0·3	58·0	2·0	0·14

When applied to coal treatment the Deister–Overstrom table has an extra large deck, with slight modifications to the height of the riffles. Results when treating anthracite are given in *Table 4*.

Plat–O—Illustrated diagrammatically in *Figure 8*, the Plat–O table is substantially rectangular in shape and has the distinctive feature of an area raised above the general table level at the diagonal end of the riffles. On

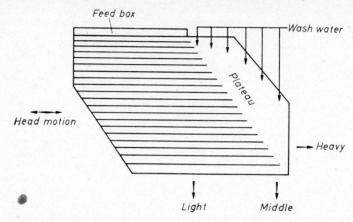

Figure 8. Plat–O table. (*From* S. J. Truscott, *by courtesy of* Macmillan & Co.)

reaching the end of the riffles the heavier particles are forced to climb on to this raised portion or plateau. Here they must move across the area against the action of the wash water, each particle being free to move almost

Table 4

Size in.	Feed ton/h	Clean coal ton/h	Ash content—per cent		
			Feed	Clean coal	Refuse
$\frac{9}{16} - \frac{5}{16}$	19·3	11·2	43·3	15·4	80·6
$\frac{5}{16} - \frac{3}{16}$	18·0	12·1	35·8	15·0	78·2
$\frac{3}{16} - \frac{3}{32}$	14·4	10·5	33·5	17·5	74·7

individually. Any misplaced light particles therefore lose the protection of the heavier ones and are washed to the light product discharge. This table has been applied both to mineral and to coal separation, results with copper ore are shown in *Table 5*.

Table 5

Test No.	Assays—per cent				Recovery per cent
	Feed Cu	Concentrate		Tailings	
		Cu	Insol.		
1	2·62	6·03	38·0	1·56	54·6
2	2·64	6·65	31·3	1·22	64·9
3	2·38	6·16	33·4	1·09	65·9
4	2·17	6·71	26·5	1·02	62·5
5	2·38	7·00	23·4	1·00	67·6
6	2·49	6·98	26·4	1·10	66·3
7	2·41	7·17	18·4	1·13	63·0
8	2·23	6·50	24·9	1·10	63·4

Butchart—The essential feature of the Butchart table is the bend in the riffles as they cross the deck. The arrangement of the riffles is shown in *Figure 9*, while a view of the table is given in *Plate III*. The riffles are laid parallel to the longer side of the table for a length comparable to the Wilfley table riffles. They then bend diagonally across to the flowing

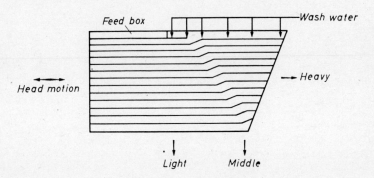

Figure 9. Butchart table

stream before continuing to the end of the table again parallel to the longer table edge. The advantage of the bend is that particles travelling along behind the riffles are suddenly forced to travel against the water streaming across the table. The lighter particles are therefore washed out and a more accurate separation achieved. At the time of its introduction the Butchart table was the first to have its head mechanism fully enclosed and running in oil, which was rightly considered a distinct advantage. Results obtained with this machine are given in *Table 6*.

Table 6

Material	Assays—per cent				Capacity ton/day
	Feed	Concentrate	Middles	Tailings	
Copper	1·8	11·5	—	1·05	175
Copper	1·15	10·5	0·75	0·58	45
Copper	2·80	17·8	—	2·16	214
Lead	8	77	8	0·45	55
Lead	5	70	4	0·4	30
Lead-zinc	—	42–51	—	—	45

Operating Variables

Given the solids to be treated, of a fixed size, size range, shape, *etc.*, and having established the type and size of table suitable for carrying out the required separation, then consideration must be given to the operating variables[8]. The two primary ones are:

1. Rate of feed.
2. Solids content of the feed pulp.

Considering these two variables together it will be noted that a peculiarity of tabling is the extreme sensitiveness to changes in the feed. Under given feed conditions the clean product and refuse discharge along adjacent sides of the table, the table corner generally being the dividing line. If the rate of feed is increased without alteration to the table settings, then the thickness of the bed on the table will increase. Consequently, the accuracy of the separation will be reduced and a certain amount of heavy product will report to the light product discharge. Conversely, a reduction in the feed rate would lead to a decrease in bed thickness with a reduction in the degree of separation, and this would result in a loss of light product in the heavy product. As distinct from the rate of feed of solids to the table, it is also important that the water associated with the feed should be maintained constant. From foregoing remarks on the principle of separation it is apparent that the water fed with the solids, and as wash water, has a considerable bearing upon the transport of the light product to its discharge point and upon the separation within the bed. Maintenance of a fixed volume rate of feed at a fixed solids concentration is therefore of prime importance in table operations.

Having established the rate of feed the optimum operating conditions may be obtained by correct adjustment of the following variables.

Head Motion

The shape of the differential motion applied to the table will depend upon the actual mechanism employed. The speed of the table is adjustable over a range of 150–375 r.p.m. but it is normally within the range 240–300 r.p.m. Similarly, the length of the stroke can be varied from a fraction of an inch to over one inch, by adjustment of the head mechanism. In the case of tables oscillated by means of the Sherwin electromagnetic vibrator, the speed is much greater than the figures given above but with a very much reduced stroke.

Slope

Adjustment of a table slope can be made in both the longitudinal and transverse directions. The object of these two adjustments is twofold; first, to achieve a good distribution of material across the table, and then to ensure the correct discharge of the products. To obtain these requirements the heavy product end of the table is generally raised above the feed end, with the transverse inclination as small as possible, consistent with correct material distribution.

Wash Water

As already mentioned, the quantity of water used is of considerable importance in tabling. Having fixed the water fed to the table with the raw material to be treated, then the wash water can be adjusted and fixed. A minimum of water should generally be used, but this must be sufficient to keep the products discharging uniformly along the edge of the table. Alterations in the separation being made can be obtained by directing additional spray water on certain areas of the table.

Product Launders

In certain separations it may be desirable to make a cut between the clean product and the refuse at some point other than at the table corner. To this end it is desirable that the launder system should have an adjustable splitter which may be set to give the required division between the two products.

The following are very general conclusions on the adjustments applied to a table:

(a) Fine feed—less water; less feed; shorter, faster stroke.
(b) Coarse feed—more water; more feed; longer, slower stroke.
(c) Rough separation—more water; more feed; longer stroke; more tilt.
(d) Accurate separation—less water; less feed; shorter stroke; less tilt.

Dry Tabling

The foregoing discussions have been centred on wet tabling of materials but mention must now be made of dry tabling. The principles employed are much the same as in wet tabling except that the mobility and stratification of the particles is carried out under the action of upward air currents.

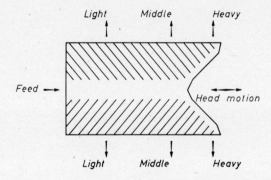

Figure 10. Single deck pneumatic table. (*From* D. R. MITCHELL, *by courtesy of the* American Institute of Mining and Metallurgy)

HIRST[9] investigated the pneumatic separation of coal in size fractions from $1\frac{1}{4}$ in. top size to zero. He showed that neither the mechanical oscillation nor the upward air current caused any appreciable stratification according to size. Hirst also found that the fluidity of the bed increased from a minimum at zero air pressure to a maximum when the pressure only just permitted the material to stay on the deck. The bed of material is most fluid when actual rupture or boiling is occurring, but in this condition any stratification is being destroyed. The pressure at which maximum fluidity occurs without boiling is referred to as the 'critical pressure'. To achieve the best results there must be a minimum passage of air through the bed, and this, coupled with maximum fluidity, is best achieved when the material has a wide size range.

It is in the field of coal preparation on sizes below ¾ in. that the pneumatic table has been largely employed, and in this field most widely for treating American bituminous coals[10]. Examples of a pneumatic table can be seen diagrammatically in *Figures 10* and *11*, which show the American oscillatory table with the rectangular deck and the twin deck table respectively. In the latter unit the two decks are mounted in such a manner that they counterbalance each other.

The use of pneumatic tables has a limited application due to the necessity for de-dusting the material to be treated and the need for a consistent moisture content in the feed. Further, when air is the suspending

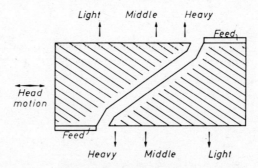

Figure 11. Twin deck pneumatic table. (*From* D. R. Mitchell, *by courtesy of the* American Institute of Mining and Metallurgy)

medium instead of water, then the hindered settling ratios referred to earlier are considerably reduced, making the separation of two given materials more difficult. These disadvantages must be offset against the fact that dry products are obtained and by the avoidance of slurry difficulties.

Conclusions

The table is an ore and coal preparation unit designed specifically to treat the finer sizes of material. In capital cost it is not cheap but it is simple in construction with low running and maintenance costs. The floor space requirements are high due to the comparatively low capacity of each machine; however, one operator can control a number of units. To some extent tabling has been replaced by froth flotation but it maintains a distinct advantage in that it can treat effectively the coarser material which is not amenable to froth flotation. Tables can also be employed to separate two minerals which are obtained together in a froth flotation concentrate. In its operation it has the advantage that the separation is taking place in full view of the operator, who, when skilled, can tell immediately whether this separation is accurate. Against this, as stressed previously, the table is very sensitive to changes in feed conditions, particularly when it is set for an accurate separation.

JIGGING

General Considerations

Jigging is a process for the stratification, according to specific gravity, of a mass of particles in water under the action of alternate upward and downward pulsations. Commercially, the process is continuous, with the removal of the light and heavy fractions as distinct products.

As an ore preparation process, jigging is probably the most widely applicable and the most widely used. It is capable of application to materials of size ranges 8 in.–0 down to ¼ in.–0 or smaller. Specific gravities of material treated range from 1·4 for coal, to as high as 19 for gold, while the percentage of the denser product can be as high as 90 per cent when treating an ore, to as low as 10 per cent when treating coal.

The application of jig washing in Great Britain is largely confined to the coal preparation field, approximately 70 per cent of all coal mechanically cleaned in Britain being treated in Baum jig washers. In 1953 this percentage amounted to a total capacity of Baum jigs of 38,500 ton/h.

Despite its wide use, the actual mechanics of jigging is still a controversial subject and presents several problems. The principles of the separation will therefore be presented in general terms, and in the following sections a classification of jig washers will be given, together with details of a number of typical units.

Principles of Separation

The separation of material in a jig washer, according to specific gravity is carried out under the laws of hindered settling, the principles of which have been stated previously[11]. The practical application of these laws is best understood by considering what happens during a complete jigging cycle. The material being treated rests on a perforated plate and is

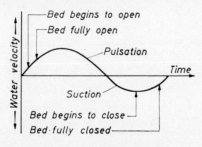

Figure 12. Sine wave jig cycle

subjected to upward and downward water pulsations. *Figure 12* shows a simplified curve of the water velocity through the bed on a time axis. It will be noted that a sine curve has been used (although this may not be the best possible) and since, generally, a steady input of water is delivered to the jig, the curve is displaced in relation to the horizontal axis.

During the upward or pulsation stroke, the top of the bed opens and comes into a teeter condition followed by the middle and lower layers progressively. The upward acceleration is such that the whole bed is lifted and therefore no separation, according to size or shape, occurs. The water velocity increases to its maximum and with the bed fully mobile reduces to zero prior to the suction stroke. The pulsation stroke has thus prepared the bed by bringing it into a mobile condition, but it has effected little separation and that only on the largest sizes. A secondary effect of the pulsation stroke is that it frees any light particles which may have been trapped on the sieve plate by heavier material.

At the end of the pulsation stroke free falling conditions apply momentarily for the only time during the separation process. The downward or suction stroke now begins, during which period the major separation takes place, the material being subjected to hindered settling under conditions of increasing bed density. At the end of the suction stroke the bottom, middle and top of the bed close in that order, the instant of closing being the condition most conducive to good separation. The next pulsation stroke immediately commences and the cycle is repeated.

It is apparent that there is a critical period for the jig cycle to produce maximum efficiency at a given capacity. If too long a cycle period is used, capacity must be reduced to produce the same efficiency of separation. If, however, too short a jig cycle is employed, the effectiveness of the cycle is lost. The optimum cycle is such that:

(*a*) the whole bed is mobile on the pulsation stroke;
(*b*) the whole bed closes completely on the suction stroke;
(*c*) at the end of the suction stroke the next pulsation commences immediately.

The ideal jig cycle is at present not established but it has been proposed[12] that a cycle coming closest to the ideal consists of a short rapid acceleration followed by a period of constant water velocity. This is an excellent arrangement for opening the jig bed, and ideally it is followed by a suction stroke during which the water falls, as near as possible, solely under the influence of gravity. Experimentally, the nearer such a cycle is followed, the greater are both accuracy of separation and the throughput of the operation.

Classification of Jigs

As one would expect with such a widely used process, there are a number of distinct types of jig and numerous variations of each type to be found[13-15]. The earliest jig separators were hand jigs, a typical example of these machines being illustrated in *Figure 13*. The mechanical jigs which were developed from this simple machine can be divided primarily into two groups, the first in which the material being treated is moved through the water (movable sieve), and the second in which the water moves through the material (fixed sieve). This second group can then be subdivided according to the method of producing the movement of the water, for example, plunger, diaphragm, *etc.* Finally, both the major groups can again be subdivided into two classes depending upon the method of removal of the heavy product. The first class is that in which the major portion of the heavy product is retained on the sieve, being removed continuously under some control mechanism. In the second class, however, the heavy product is

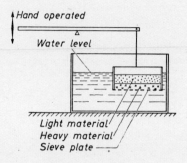

Figure 13. Hand jig

jigged through the sieve on which an artificial bed of heavy material larger than the sieve apertures has been placed.

In *Table 7* the classification described above is presented together with some typical examples of separators used for the treatment of minerals and of coal, and in the following subsection a number of these washers are described.

Table 7. Jig Classification

Type	Pulsation	Ore		Coal	
		Jigging over sieve	*Jigging through sieve*	*Jigging over sieve*	*Jigging through sieve*
Movable sieve	—	Halkyn	Hancock Franz James	Wilmot Pan Stewart James	
	Plunger	Harz Shields and Thielmann	Cooley	Elmore Montgomery Reading Lehigh Delaware Luhrig	Foust Luhrig
Fixed sieve	Diaphragm	Bendelari	Denver	Jeffrey	
	Water	Richards	Pan-American	Vissac	
	Air			Baum	Feldspar
	Vanes		Neill		

Description of Typical Jig Washers

The mechanical process of jig washing can be divided into three distinct considerations:

1. The means of feeding the raw material to the separator.
2. The process of effecting the separation according to specific gravity.
3. The means of removing the light and heavy material from the washer as two distinct products.

In feeding the separator it is important to ensure that the material is distributed evenly across the full width of the unit, and that it is delivered at a constant rate. This may be achieved by introducing a balance bunker into the circuit prior to the jig washer. The raw material is delivered from this bunker at a constant rate, for example, by a vibrating feeder, going from the bunker directly to the wash box. Alternately, the feeder may deliver into a fixed trough down which the raw material is flushed by water into the washer. This latter practice is commonly employed in coal preparation.

Methods of effecting the separation and of removing the products are peculiar to the particular form of washer. The various methods of effecting the separation are illustrated under jig classification, *Table 7*, while the method of removing the light product is generally to allow it to be carried across to and over the sill of the jig.

The heavy product may be removed by jigging through the sieve or it can be drawn off through a gate and dam discharge under hand control. Alternatively, if automatic control for heavy product discharge is employed it is based upon one of three interrelated variables:

1. The weight of heavy product on the gate.
2. Height of the heavy product layer on the sieve plate.
3. Water pressure below the sieve plate.

The application of these different methods is illustrated below by describing typical examples of the various jig washers.

Hancock Jig

The Hancock jig[16] is typical of the movable sieve jig used for the treatment of ores and is illustrated in *Figure 14*. It consists essentially of a water-filled rectangular tank (1) in which a movable sieve (2) is suspended and which is oscillated by the cam and levers below the machine, through a link mechanism to the supporting crossbars (3). This latter actuating mechanism is so arranged that the sieve tray moves forward and upward simultaneously, moving backward as it falls under gravity. The stroke is generally $\frac{3}{4}$ in. both horizontally and vertically at a rate between 180 and 195 per minute.

Raw material fed on to the screen progresses along it by virtue of the backward and forward motion, while it is separated according to specific gravity by the up and down component of the motion through the water. The heavier product is removed progressively, passing through the sieve into the four or five compartments below. The first two or three compartments contain the true heavy product, while the last compartment receives the final light product, and the middling product discharges into the intermediate compartments. A grid on the sieve plates retains the lower

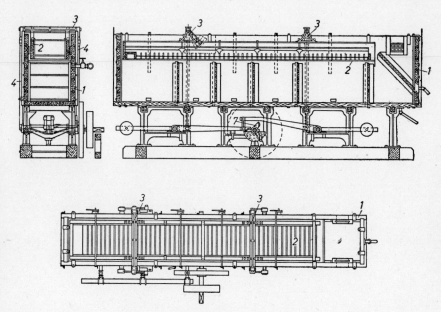

Figure 14. Hancock jig. (*After* RICHARDS *and* LOCKE. *From* GAUDIN's 'Principles of Mineral Dressing'. 1939. McGraw-Hill Book Co. Inc.)

portions of the heavy product to form a bed of material, but in certain cases an artificial bed of steel balls, for example, may be used through which the heavy product must pass.

The feed to the washer is generally dry, and water is fed in through the first three compartments, rising through the sieve and overflowing the tank at such a level that there is some 3 in.–4 in. cover over the material being treated on the sieve.

Table 8

Compartment	1st	2nd	3rd	4th	5th	Tailings
Assay % Lead	79·7	71·8	58·5	24·7	2·0	0·66

The Hancock jig has its principal application in the treatment of low grade ores, recovering small quantities of concentrate and rejecting the bulk of the material as middlings or tailings. The feed is normally a deslimed minus ⅜ in. material, when the capacity of the largest machine, a 20 ft. ×2 ft. 8 in. sieve, is between 300 and 600 tons/day. The performance of a Hancock jig treating 500 tons/day of galena–dolomite ore 9 mm–2 mm in size is given in *Table 8*.

Halkyn Jig

The Halkyn movable sieve jig differs from the Hancock jig in that the heavy product is jigged over the sieve rather than through it. This jig is illustrated in *Figure 15* from which it can be seen that the sieve is divided into three sections, each of which ends in an adjustable gate through which the heavy product is discharged into the different compartments below the

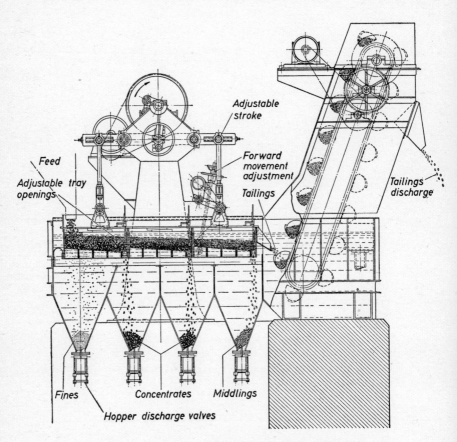

Figure 15. Halkyn jig. (*By courtesy of* R. O. Stokes & Co.)

sieve. The first of these compartments takes any undersize material which has entered with the feed, this material passing through this sieve. The second and third compartments receive the true heavy product, while the last compartment receives a middlings product. The light product is discharged from the end of the sieve and is removed by bucket elevator.

The jig mechanism is situated over the jig box and is adjusted to produce a slow upward movement of the sieve followed by a rapid downward return. This is equivalent to a rapid pulsation stroke and a slow suction stroke.

The capacity of the jig is of the order of 1 ton/h per square foot of sieve area, but it is, of course, dependent upon the material being treated. Power requirements are about 1 h.p./3 ft.² of sieve area; while typical sieve sizes are $1\frac{1}{2}$ ft. × 5 ft. or 2 ft. × 8 ft.

Wilmot Pan Jig

Figure 16 is a diagrammatic arrangement of the Wilmot pan jig which consists essentially of a sieve box suspended on a water-filled, V-shaped tank. The sieve box is moved by two eccentrics, placed one on either side of the box and which impart a forward as well as vertical motion. The

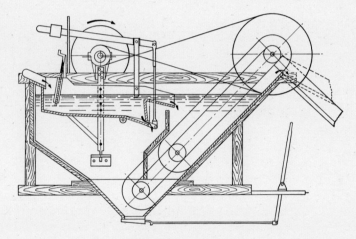

Figure 16. Wilmot jig. (*From* D. R. MITCHELL, *by courtesy of the* American Institute of Mining and Metallurgy)

separation takes place on the sieve plate, the light product discharging from the end of the sieve box into a bucket elevator, while the heavy product is drawn off through a drop gate. This heavy product discharge is worthy of note since it represents an elementary form of automatic control. The heavy product moves forward along the sieve to the drop gate, and the gate opens against the action of an adjustable weight on a lever arm. A reduction in the amount of heavy product to be discharged causes the gate to close, and an increase causes it to open further. Alternatively, the drop gate mechanism may be connected to a float resting on the layer of heavy product in the sieve box. In this case, an increase in the thickness of the heavy product layer causes the gate to open further and *vice versa*.

The Wilmot jig has been used principally for the treatment of closely sized anthracite in the United States, when the speed of operation is usually 130–150 strokes/min, each of $1\frac{1}{2}$ in.–$1\frac{3}{4}$ in.

Harz Jig

Of the fixed sieve, plunger type jigs for ore treatment, the Harz jig is outstanding in respect of its simplicity and its extensive use. There are many variants of this unit, but *Figure 17* illustrates a single compartment of the jig arranged for treatment of coarser ores by jigging over the sieve.

This single jig compartment consists of a rectangular, hopper-shaped vessel, divided longitudinally by a centre board. On one side of this division is the fixed sieve plate while on the other a reciprocating plunger is loosely fitted. The movement of the plunger under the action of the

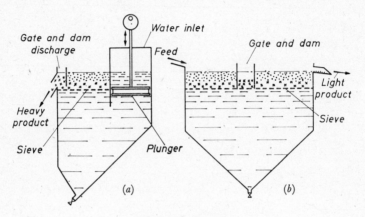

Figure 17. Harz jig

overhead eccentric causes the water to flow between the two sections and through the sieve plate. The feed is delivered on to the sieve plate at one end of the compartment and is moved along the compartment by the pulsations, being stratified progressively. The light product is finally discharged over the end of the compartment while the layer of heavy product which has been formed on the sieve plate is removed continuously through a gate and dam discharge at the side of the compartment. Having set the gate and dam for a specific separation, any increase in the quantity of heavy product produced will increase the amount removed over the dam and *vice versa*. Undersize material passing through the sieve plate of the Harz jig is removed periodically by hand.

The number of compartments depends upon the size of the material being treated and upon the difficulty of the separation to be made. For a simple separation of a coarse material two compartments may suffice. The first compartment would make a true heavy product, the light product passing forward to the second one, where a second separation into middles and true light product is made. For a difficult separation of fine material, as many as five compartments may be used to form the complete jig.

Whatever the number of compartments, the jig as a whole has a fixed number of strokes per minute, but the stroke of each individual plunger may be varied. Water is usually added to the plunger compartment to replace that removed with the products, and each compartment has its independent adjustable supply. This water has a marked influence upon the pulsation and suction strokes, reinforcing or opposing the strokes according to the quantity added, and to the amount removed with the products and from the bottom of the compartments. It is by setting the number of strokes, the length of each individual stroke, and the quantity of back water, that the correct separation is achieved in the jig with any particular feed.

The material treated in a Harz jig must be comparatively closely sized with a top size of $1\frac{1}{2}$ in. and a minimum of about 50 mesh. The capacity can range from 1 to 4 tons/ft.^{2} of sieve area per 24 h. The coarser the feed the higher the capacity. Power requirements are between 0·1 and 0·15 h.p. per square foot of area.

Cooley Jig

The Cooley jig, a variant of the Harz jig, deserves mention because it is used with a coarse sieve through which the heavy concentrate is removed. An artificial bed of suitable size and specific gravity may be placed on the sieve bed to control the removal of the heavy product. The jig has been used largely for treatment of zinc ores and may have from seven to nine compartments, with capacities from 3 to 12 tons/ft.^{2} per 24 h according to the material being treated and upon the degree of separation required.

All Cooley jigs, working as they do through the sieve, operate with a strong suction. This suction is achieved easily by the continuous removal of material from the bottom of the compartment with its consequent steady downflow of water.

Elmore Jig

This jig, shown diagrammatically in *Figure 18*, is typical of the plunger type jig applied to coal preparation. It also illustrates the development and application of the star wheel to removal of the heavy shale under automatic control.

A single compartment is illustrated, but two or more compartments may be used. The clean coal overflows the jig while a star wheel extending across the full width of the jig removes the discard. This star wheel operates under the action of a float which is so weighted that it rests upon the layer of heavy material at the end of the jig. As the quantity of heavy product in the jig bed increases, the float rises higher at each pulsation stroke until it reaches a certain level when the star wheel is brought into operation by a linkage mechanism. The heavy discard is then removed until the float drops back to its predetermined level.

Any material passing through the sieve plate settles to the foot of the compartment from which it is removed by a screw conveyor into a bucket elevator. This bucket elevator also receives the heavy discard from the star wheel discharge.

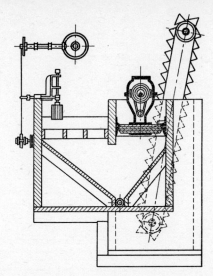

Figure 18. Elmore jig. (From D. R. Mitchell, *by courtesy of the* American Institute of Mining and Metallurgy)

The Elmore jig has been used in the United States for the treatment of bituminous coals at capacities of up to 50 tons/h.

Foust Jig

Figure 19 shows one compartment of the Foust plunger jig, notable because of its synchronized twin plungers which result in an improvement in the water distribution through the sieve. The area of the two plungers is arranged to be slightly greater than the sieve area.

When treating coal the jig is so designed that the discard is removed through the sieve plate, being drawn off periodically by hand from the

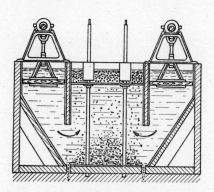

Figure 19. Foust jig. (From D. R. Mitchell, *by courtesy of the* American Institute of Mining and Metallurgy)

foot of the compartments. Minus 1 in. coal is normally treated in this type of jig which may have two or more compartments, with capacities as high as 45 tons/h.

Bendelari Jig

In the Bendelari jig (*Figure 20*) the plunger is arranged below the sieve plate and is sealed by a rubber diaphragm; in this way the leakage of water around the plunger, a frequent trouble of the Harz jig, is eliminated.

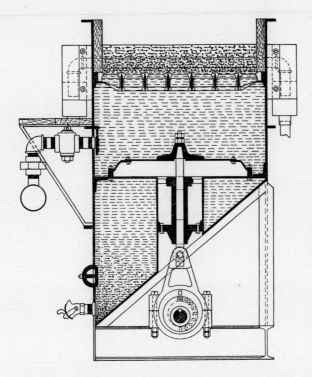

Figure 20. Bendelari jig. After F. N. BENDELARI. *From* GAUDIN's 'Principles of Mineral Dressing'. 1939. McGraw-Hill Book Co. Inc.)

This arrangement is also advantageous in the precise control of the jigging action by adjustment of the water added above the diaphragm on each suction stroke. The quantity of water used can be regulated to give the required degree of pulsation or suction.

Denver Jig

The use of a diaphragm-sealed plunger in the treatment of ores by jigging through the bed is illustrated by the Denver jig. This jig, constructed in six sizes from 4×6 in. to 3×4 ft., with one or two compartments, is used to treat the fine sizes of mineral with a maximum capacity in the largest unit of the order of 500 tons/day.

A feature of the machine is the mechanism of a rotating water valve synchronized with the plunger. This valve is closed during the pulsation stroke but on the suction stroke it opens to admit a predetermined quantity of water below the diaphragm, *Figure 21*. In this way the suction stroke

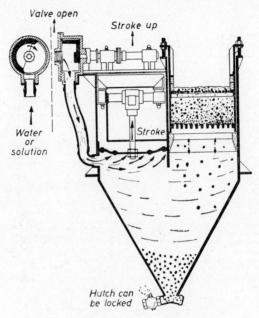

Figure 21. Denver jig. (*By courtesy of* Denver Equipment Co. Ltd.)

may be partially or completely opposed. In the latter case this means that the solids and water will fall on to the sieve under gravity alone, a condition which has been referred to previously as approaching the ideal.

The length and speed of the stroke, together with the material forming the artificial bed on the sieve plate are other variables used to control the jig action. The light product is jigged over the sill of the wash box, while the heavy concentrate passing through the sieve plate may be drawn off continuously under control or batchwise by hand.

Jeffrey Diaphragm Jig

The Jeffrey diaphragm jig has been used to treat bituminous coal in the United States and may be constructed either with a single compartment or with a number of them. Referring to *Figure 22*, a rubber diaphragm is sealed below the sieve plate of each compartment and is actuated by a plunger through a system of levers from a cam. Various shaped cams are available giving fierce or mild suction strokes according to their design. The heavy discard travels across the sieve plate and is removed by a star wheel running across the full width of the box at the end of each compartment. The operation of the star wheel is controlled by a float resting on the discard layer, which results in automatic discard removal. The sieve plates slope towards the discard removal point and are adjustable in

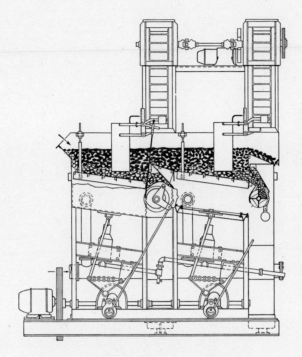

Figure 22. *Jeffrey diaphragm jig.* (*By courtesy of Jeffrey Manufacturing Company*)

inclination. The top of each diaphragm also slopes towards the elevator boot at the end of each compartment; any discard passing through the sieve plate falls on to the diaphragm and moves across its surface into the refuse elevator.

The jig normally operates at about 33 strokes/min, each stroke being up to 8 in. in length, while the maximum size of coal treated is of the order of 6 in.

Richards Pulsator Jig

Figure 23, shows a section through one compartment of the Richards jig, the pulsations in which are produced by the inlet of water from a main under the action of a valve rotating at 150–200 r.p.m. Water shocks are avoided by the use of an air vessel above the water inlet point. The heavy product is normally removed by a gate and dam discharge at the side of each compartment, while the light product passes across each compartment and over the discharge lip. In order to obtain an even distribution of the water, and hence of the pulsation, across the jig, the area of the sieve must be small and generally does not exceed 6 in. square. The capacity of the machine is correspondingly low, being of the order of 100 tons/ft.²/day.

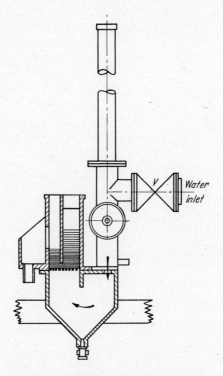

Figure 23. Richards pulsator jig. (After RICHARDS and LOCKE. From GAUDIN's 'Principles of Mineral Dressing'. 1939. McGraw-Hill Book Co. Inc.)

The feed must also be closely sized, and while a clean heavy product is obtained, the light product tends to require retreatment. Water consumption with a jig of this type is naturally high but other power requirements are low.

Pan-American Jig

The Pan-American jig was primarily designed for the recovery of fine gold, treating a material minus ½ in. or ¼ in. size to produce a high grade heavy concentrate. *Figure 24 (a)*, shows the jig, which has a fixed sieve supporting a layer of steel shot. The heavy product is jigged through this artificial bed, falling into the compartment below from which it is removed continuously or intermittently according to its quantity and the type of separation being made. The pulsations are achieved by the admission of water under pressure through the valve shown in *Figure 24 (b)*. The design of the valve is worthy of note in that it is fully automatic and produces 400–600 pulsations/min. The water from the main, acting on the diaphragm, raises the valve from its seating and causes a pulsation in the box. This, in turn, reduces the pressure in the line and the valve reseats

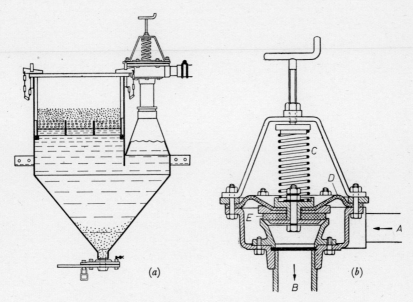

Figure 24. Pan-American jig. (By courtesy of Pan-American Engineering Co.)

under the action of the spring. The pressure rapidly builds up again and the process is repeated at a high frequency.

This type of jig, which is made with one or two compartments from 12 to 24 in. square and with capacities of the order of 200–250 tons/ft.²/day, is normally used in ball mill classifier circuits or in 'placer' operations where it may be used as a roughing jig or as a cleaner.

Vissac Jig

The Vissac jig, illustrated in *Figure 25*, really combines the action of air and water to produce the pulsations. Water entering A compresses the air in this chamber which is sealed by the upper rotary valve. As the valve opens compressed air forces the water into chamber B causing the curved vane to move to the right thus displacing water into the compartment D and causing the pulsation stroke. As the upper water valve closes the lower one opens, the vane moves to the left, and the water it displaces passes out of chamber B by the passage F.

The heavy discard removal is by means of a star wheel not shown in the figure. The operation of this wheel is carried out automatically according to pressure changes in the vessel D, any increase in pressure causing the wheel to operate. This jig has been used to treat sized bituminous coal principally in Canada and the United States.

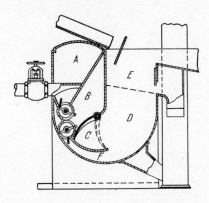

Figure 25. Vissac jig. (From D. R. Mitchell, by courtesy of the American Institute of Mining and Metallurgy)

Baum Jig

Introduced into Britain by Simon-Carves in 1900, the Baum jig washer[17] is the principal unit used in the mechanical cleaning of British coals, treating approximately 70 per cent of the coal washed. The wash box consists essentially of a steel tank of riveted or welded and bolted construction, with

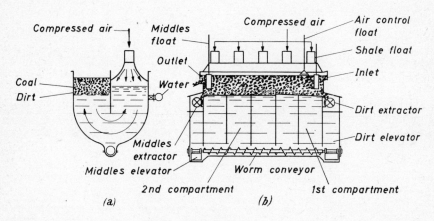

Figure 26. Baum jig: (a) cross-section; (b) longitudinal section. (By courtesy of Simon-Carves Ltd.)

a cylindrical base and divided longitudinally into two compartments. This is illustrated diagrammatically in *Figure 26 (a)*, while *Figure 26 (b)* is a longitudinal section showing the division of the wash box into compartments, which are in turn subdivided into cells. The pulsations are obtained by the admission of compressed air at 1–3 lb/in.² to one side of the wash box through an air valve, each compartment having its own air valve which may be a piston type operated from an eccentric, or a rotary valve. The latter type of valve is becoming widely used because it is more readily adjustable, permitting the jig cycle to be more carefully controlled. The air valves on a wash box will all be operating at the same speed, but the compartments, normally two or three, may be out of phase with each other. Each cell also has its own water supply with independent valve control.

In the illustration, two compartments are shown each having a dirt extractor and a dirt elevator, and it is to be noted that the sieve plates slope towards the dirt extraction points. The positioning of the dirt elevators at each end of the box is generally regarded as British practice. In such an arrangement, the heavy dirt being rapidly stratified does not have to be transported along the wash box, but is removed directly below the feed point. This is a distinct advantage when treating coal of 6–7 in. top size to zero, the large dirt in which would require an active pulsation to move it along the sieve plate with a possible reduction in the accuracy of separation in the finer sizes. An alternative arrangement of dirt extraction points is to place them at the end of each compartment furthest away from the feed point. This method is normally considered to be German practice, and is generally employed with a smaller size of coal or with a narrower size range coal than is treated in the British Baum.

With a two compartment, two elevator wash box, the first compartment is set to separate the heavy discard which is removed in the first elevator. The second compartment makes a finer separation and the second elevator product may be sent to waste as a discard or it may be retreated as a middles product. In a three compartment box the centre elevator can also make a discard or a middles, according to the coal being treated and the setting of the box, but the final compartment normally makes a middles product which is crushed and retreated. A worm conveyor at the base of the wash box is normally fitted, conveying to the dirt elevator any material passing through the sieve plates.

The removal of the discard may be effected by an application of the gate and dam discharge[18], while referring to *Figure 26 (b)* floats are shown which operate the dirt extractors. The float is of course rising and falling with each jig stroke but the maximum height to which it rises is related to the amount of discard present on the sieve plate. An increase in the height to which the float rises above a fixed level, therefore, brings the dirt rotor into operation, since the optimum operating condition is a fixed thickness of dirt bed on the sieve plate. This control may be effected by linking the float electrically or mechanically to an air operated clutch, the movement of which causes the dirt extractor to rotate.

Alternatively, the dirt rotor may be eliminated and the last section of the sieve plate shrouded and connected to atmosphere through a valve[19]. Pulsations will only be experienced in this last section when the valve is open, and these can be used to remove the dirt over a sill. The valve opening is therefore controlled by the float, which responds to the amount of shale present. An arrangement of this method of control is shown in *Figures 27 (a)* and *(b)* which illustrate the *A.C.C.O.* dirt extraction gear. The float in this equipment is in a cylinder extending below the sieve plate, and *Figure 27 (a)* shows the condition when the thickness of the dirt bed is too thin to require dirt removal. When the quantity of dirt increases the float rises higher [*Figure 27 (b)*] and the striker *h* raises the stop *g* to open the air valve *d*. The pulsation then moves the dirt over the sill into the dirt elevator compartment. Valve *d* is closed by a fractional horse power motor, through a slipping clutch, as the float drops to its required operating level.

The Baum jig can be designed to treat a raw coal in size ranges from 8 in.–0 at capacities of up to 250 tons/h, two or more jigs being used in parallel for higher tonnages.

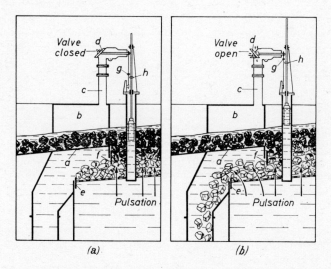

Figure 27. *Dirt extraction gear.* (*By courtesy of* Automatic Coal Cleaning Co.)

It should be noted here that when treating a mineral in a jig washer the specific gravity of separation is often set readily by the distinct and different specific gravities of the two or more minerals. The specific gravity difference between a clean coal and a discard is in most cases by no means as distinct, and the specific gravity of separation required may vary according to the analysis of the coal to be treated. The Baum system is applicable to easy or normal coals with relatively high specific gravities of separation in the range 1·5 to 1·75. The system is thus applicable to the majority of British coals.

Other features to be noted are that the finer sizes are generally separated at a higher specific gravity than the coarser sizes when a given coal is treated in a jig washer. Also the larger the size of coal the more accurate is the separation generally achieved.

Feldspar Jig

The Feldspar jig is, in fact, a modified version of the Baum jig designed specifically for the treatment of small coal generally minus $\frac{1}{2}$ in. in size. It derives its name from the fact that feldspar, by virtue of its specific gravity and shape, is used to form an artificial bed through which the heavy discard is removed. The wash box may be composed of three to six compartments each fitted with its own air and water valves and with its own control mechanism. *Plate IV* shows a cross sectional view through one compartment of the *A.C.C.O.* Feldspar jig which incorporates these features and which clearly illustrates the feldspar held in a grid on the perforated plate[20]. The clean coal moves along the wash box and is flushed over the sill as in the Baum box. Discard is removed through the feldspar under automatic control which in the example illustrated operates on the wash box pulsation. A float in the jig bed controls an air valve connected to the air chamber of the wash box. As the quantity of discard decreases, the valve is opened, releasing air from the chamber and thereby reducing the pulsation and the the rate of discard removal. Conversely, as the quantity of discard in the feed increases so the pulsation is increased. The discard from each compartment may be combined and removed by bucket elevator or a third product, a middles, can be prepared by keeping the material from the last one or two compartments distinct from the other refuse. By the use of automatic control and careful setting accurate cleaning on sizes as fine as $\frac{1}{50}$ in. can be obtained with a degree of cleaning in even finer sizes. Capacities of from 30 tons/h to 150 tons/h can be washed in the Feldspar jig.

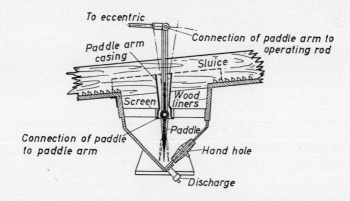

Figure 28. Neill jig. (From S. J. TRUSCOTT, *by courtesy of* Macmillan & Co.)

Neill Jig

This jig was designed for the specific purpose of treating fine material in a sluice to remove any valuable heavy concentrate which would otherwise escape.

Figure 28 shows a section through the jig. The paddle is oscillated about a horizontal axis by means of an eccentric placed above the sluice. The pulsations which result are effective across the full area of the jig since there is no pulsation compartment. The screen generally supports an artificial bed of steel shot through which the heavy product is jigged into the compartment.

Operating Variables

Because of the wide variety of jig washers in use, employing different methods of producing the pulsations and different methods of discard removal, *etc.*, it is almost impossible to specify the relative influence of the operating variables. In very general terms the more important operating variables are as follows.

Rate and Size Range of Feed

The jig washer is generally less sensitive to changes in feed rate than the table but, nevertheless, for the optimum efficiency the feed rate should be kept constant. Furthermore, the more constant the size range of material being fed to the jig the more consistent will the washing results be. A balance bunker placed before the wash box contributes much to consistent feed conditions.

Jig Cycle

The adjustment of the jig cycle is of considerable importance in correct operation of a jig washer[21]. The rate of jigging is related to the material being treated, its size range, the depth of bed employed and the tonnage to be treated. The relative times of pulsation and suction strokes are in many cases adjustable to suit the conditions of the separation being undertaken.

Air Pressure and Plunger Stroke

Interrelated to the jig cycle is the air pressure setting in those jigs employing this means of producing the pulsation. The higher air pressures are normally employed in order to obtain a complete mobility of the bed on the pulsation stroke when the larger sizes of material are treated. In the plunger type of jig, the length of the stroke is the counterpart of air pressure as an operating variable.

Water Feed

The water feed is of importance in all wash boxes and its even distribution must be of first consideration. In cases where water is used to produce the pulsation it is of prime importance since its pressure and quantity control the strength of the pulsation stroke. Its use to flush the raw feed into a wash box is generally kept to a minimum since an excess of flush water can result in the finer sizes of material being swept along the wash box with insufficient cleaning. In other jig boxes the bulk of the water supplied is used to oppose and to control the suction stroke. It is to be noted that in cases where the material is being jigged across the box too fierce a suction stroke can result in fine light particles being drawn through the sieve plate. Such material is directed to the heavy product either contaminating it in the cases of mineral treatment or resulting in a loss of clean product when washing coal.

Sieve Plate and Artificial Beds

The size of the apertures in the sieve plate can be considered as an operating variable since firstly they control the size of material which passes into the jig hutch, and secondly, they form a fixed resistance to both the pulsations, but more especially to the suction stroke. Very generally, if the maximum suction is inadequately cleaning the finer sizes, then a coarser perforation will effect an improvement.

When jigging through the sieve the aperture sizes must always be larger than the maximum size of material treated. Conversely, the size of any material forming an artificial bed must be larger than the aperture size. The choice of material to form an artificial bed is based first upon specific gravity, which must be less than that of the heavy product, and then on other considerations such as size, shape, stability and availability. Normally steel shot is used in mineral dressing, while feldspar predominates in coal preparation.

Control of Product Removal

The light product is normally jigged across the wash box and over a sill of a preset height. The heavy product, however, forms a layer next to the sieve plate or above any artificial bed and must be removed from this position. Its removal consists essentially of maintaining a fixed thickness of bed of heavy material in the wash box and withdrawing this material at the same rate as that which it enters the box with the feed. The thickness of a bed required is the variable, being related to the pulsation stroke and to the capacity of the washer, and influencing the specific gravity of separation. Having decided upon the thickness required, the control mechanism must be set to maintain this thickness.

CONCLUSIONS

The wide application of jig washers and the variety of types in use is testimony to the versatility of this method of mineral treatment. Originally a process applied to the coarser sized minerals of the order of 1 in. in size, it has been developed for the treatment of coal from 8 in. to $\frac{1}{50}$ in. size and for the separation of the finer sizes of minerals. Its high relative capacity makes it a comparatively cheap process to operate, but an adequately large supply of water is required. While in many cases requiring careful setting, the jig washer is stable in operation, the application of automatic removal of the heavy product being a major contribution to this stability.

Reference should finally be made to dense medium jigging which has been applied to the treatment of coal in the United States[22]. This process consists of maintaining in the jig circuit a quantity of fine heavy particles. These effectively increase the specific gravity of the fluid which has already been shown to increase the hindered settling ratio, resulting in an easier separation.

BIBLIOGRAPHY

[1] THOMAS, B. D., 'Coal Preparation', *Amer. Inst. min. (metall.) Engrs* (1950) 249–73
[2] GAUDIN, A. M., *Principles of Mineral Dressing*, McGraw-Hill Book Co. Inc., London (1939) 280–317
[3] CHAPMAN, W. R., 2nd *Int. Coal Prep. Congr., Essen* (1954) Pap. A115
[4] GANDRUD, B. W., 'Coal Preparation', *Amer. Inst. min. (metall.) Engrs* (1950) 435–69
[5] TAGGART, A. F., *Handbook of Ore Dressing*, Chapman and Hall, London (1927) 717–62
[6] TRUSCOTT, S. J., *Textbook of Ore Dressing*, Macmillan & Co., London (1923) 360–77
[7] TAGGART, A. F., *Handbook of Ore Dressing*, Chapman and Hall, London (1927) 736
[8] YANCEY, H. F., and BLACK, C. G., *U.S. Bur. Min. Rep.* No. RI 3111 (1931)
[9] HIRST, A. A., *Trans. Instn Min. Engrs, Lond.*, 79 (1929) 463
[10] CARRIS, E. C., and MITCHELL, D. R., 'Coal Preparation', *Amer. Inst. min. (metall.) Engrs* (1950) 542–52
[11] HIRST, A. A., *Coal Preparation Symposium*, Leeds University Mining Dept. (1952) 68–82
[12] BIRD, B. M., and MITCHELL, D. R., 'Coal Preparation', *Amer. Inst. min. (metall.) Engrs* (1950) 391–434
[13] TAGGART, A. F., *Handbook of Ore Dressing*, Chapman and Hall, London (1927) 666–716
[14] GAUDIN, A. M., *Principles of Mineral Dressing*, McGraw-Hill Book Co. Inc., London (1939) 250–79
[15] TRUSCOTT, S. G., *Textbook of Ore Dressing*, Macmillan & Co., Ltd., London (1923) 298–326
[16] RABLING, H., *Trans. Amer. Inst. min. (metall.) Engrs*, 57 (1917) 309–21
[17] WALLACE, W. M., *Colliery Engng*, 30 (1953) 31–3, 153–7
[18] TAYLOR, F., *Birmingham University Mining Magazine*, Whitehouse & Co., Ltd., Birmingham (1950) 17–26
[19] MICHELL, F. B., 'Recent Developments in Mineral Dressing', *Inst. Min. & Metall.* (1953) 263–6
[20] MICHELL, F. B., *Iron Coal Tr. Rev.*, 160 (1950) 487–90
[21] HIRST, A. A., and WALLACE, W. M., 2nd *Int. Coal Prep. Congr., Essen* (1954) Pap. A116
[22] BIRD, B. M., *Proc. Ill. Min. Inst.* (1947) 107

8

FLOTATION

J. E. FELSTEAD

A SEPARATING operation is sometimes required which is independent of the difference in density between the mineral and its associated impurity and which owes its efficiency to a more pronounced difference in physical character. Such an operation is flotation, more correctly called froth flotation. Flotation is the general term for describing an operation where fine particles of one or more minerals or chemical compounds, in the form of a pulp or slurry, can individually or selectively be caused to adhere to bubbles of air agitated within the pulp, and finally to rise to the surface of the separating vessel or flotation cell, the remaining constituents of the pulp being separately collected or subjected if necessary to further flotation treatment.

In a general sense, the term 'flotation' appears to indicate the act of floating upon the surface of a liquid and signifies that the floating body has less density than the liquid on which it lies. In concentration by flotation the solid particles are, in nearly every case, of much greater density than the liquid on which they float. The reason for their assumed buoyancy being that the floating froth consists of a flocculated mass of mineral and air bubbles which is consistently lighter than an equivalent volume of water.

The mechanism of flotation is brought about by the addition of one or more chemical compounds or reagents, some of which wholly or partially modify the surface characteristics of the respective constituents of the pulp, and others which when dissolved in or dispersed through the liquid of the pulp enable the liquid to form a froth which will remain stable for a relatively short period.

Flotation can therefore be resolved into three distinct operations. First, the preparation of the surfaces of the particles which constitute the mixture of constituents it is required to separate, so that one or more of the constituents can adhere to the air bubbles rising through the pulp, whilst the remaining minerals are wetted with water and are rejected by the bubbles. Second, the formation of a semi-stable froth by the reaction of the frothing agent with the aqueous medium in the pulp. Finally, the separation of the water repellent particles from the water-attracted particles in the pulp and the removal of the former from the surface of the flotation cell.

By the variable nature of the physical and chemical properties involved in the flotation process and by the careful control of the flotation reagents, the scope and application of flotation has increased to cover a very wide field.

Flotation treatment is applied extensively to the concentration of metalliferous minerals both sulphides and oxides and it is mainly in this field that its use as a means of selection is utilized. The main use of flotation in Great Britain is in the cleaning of coal, whereby the ash-forming constituents can be eliminated to a considerable degree. Other applications are in the concentration of a variety of materials such as sulphur, graphite, apatite, barytes, limestone, *etc.*, all of which can be floated to free them from siliceous impurities.

The limitations of flotation are, however, concerned with the size range over which it is effective. It is neither effective on material coarser than about 16 mesh nor satisfactory for material approaching colloidal size. However, with a fresh material to separate, there is rarely sufficient of this ultra-fine portion to make itself felt and flotation is entirely successful, but with a decomposed or very clayey material difficulties are often encountered.

PRINCIPLES OF FLOTATION

To explain the action of flotation it is necessary to appreciate what is meant by the phenomenon of wetting by means of water or other liquids. Essentially, wetting is the spread of a liquid over a solid to the displacement of air or of another liquid.

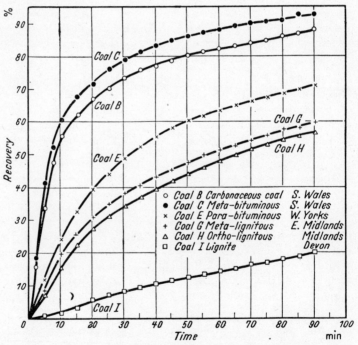

Figure 1. *Flotation of coals of different rank after preparation under water.* (From R. M. HORSLEY and H. G. SMITH, *by courtesy of* Fuel, London)

According to their affinity for water, it is convenient to classify solid substances into the three following groups:

(1) Those which are described as being 'hydrophobic' or, in chemical terms, non-polar in nature. These are substances which strongly resist wetting by water and are said to possess natural floatability, *e.g.* paraffin wax, which is a saturated hydrocarbon.

(2) Those compounds which are 'hydrophilic', *i.e.* they tend to be associated with water molecules and are easily wetted. Chemically speaking, they are called 'polar' compounds. Most minerals which contain oxygen and tend to become hydrated, such as clays, fall into this group and all these compounds are non-floatable.

(3) There is a large number of substances which have intermediate properties between these two groups, the molecules being non-polar in one part and polar in another. Such compounds are called hetero-polar and are only partially floatable. Coal substance falls into this group because of its hydrocarbon nature, and since it contains only relatively few polar groupings, it possesses natural floatability. This floatability can, however, vary with the rank of the coal because low rank coal contains more oxygenated, polar or hydrophilic groupings than high rank coal and for that reason is less readily floatable. *Figure 1* illustrates the variation of floatability with coal rank.

Contact Angle

A contact angle is the angle of contact of a bubble of a liquid with a surface of a solid and is a convenient measure of the force of adhesion between the bubble and the solid surface and its floatability. The contact angle marks the position of equilibrium between the solid–liquid and liquid–air surfaces on a wetted surface, *i.e.* it is the position of equilibrium between three forces; the surface tension of the water, that of the solid and that of

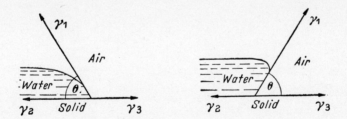

Figure 2. Contact angle

the interface between the water and the solid. Each of these three forces tends to contract the particular surface from which it is derived, but the only contraction that can actually take place is the displacement of the point of contact along the solid surface. The extent of wetting is therefore determined by the relationship between the solid surface tension and the interfacial tension. If the solid surface tension γ_3 is the greater, the water is pulled over the solid till an acute angle θ is reached (*Figure 2*), when the component of the water tension γ_1 together with the interfacial tension γ_2 is sufficient to bring about equilibrium. Under these conditions, it is said that the solid shows a preference for water. If the interfacial tension be the greater, the water will be drawn back and an obtuse angle will be formed, it is then said that the solid has a preference for air.

Reference to *Figure 2* will show that when a solid shows affinity for water,

$$\gamma_3 = \gamma_2 + \gamma_1 \cos \theta$$

and when a solid shows affinity for air

$$\gamma_3 + \gamma_2 \cos \theta = \gamma_2$$

$\gamma_1 \cos \theta$ therefore is a measure of the degree of wetting. When the contact angle is $0°$, $\cos \theta = 1$ and the degree of wetting is at a maximum, *Figure 3(a)*. When the contact angle is $90°$, $\cos \theta = 0$ and there is no tendency for the water to spread or contract, the degree of wetting may

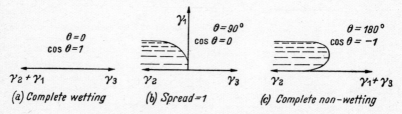

Figure 3. Variation of wetting with increase of contact angle.
(a) Complete wetting; (b) Spread = 1; (c) Complete non-wetting

therefore be regarded as unity, *Figure 3(b)*. Finally, when the contact angle is $180°$, $\cos \theta = -1$, the water will contract its extent and the degree of wetting is at a minimum, *Figure 3(c)*.

Since there is always some adhesion between solids and liquids in contact, there is no such thing as complete non-wettability: *i.e.* a contact angle of $180°$. The contact angle may vary between the two extremes depending upon whether the liquid is tending to advance over the dry solid surface or recede from a previously wetted one. The difference between the advancing and receding angles is known as the hysteresis of the contact angle and is explained by the fact that the amount of adhesion between liquid and solid for a dry surface is different from that for a previously wetted one. The cleaner the surface, the smaller is the hysteresis of the contact angle. In the case of a large advancing contact angle (small adhesion between liquid and solid) it appears that the presence of some film prevents the liquid adhering closely to the solid; after contact with the liquid, this film is wholly or partially removed, so that the contact between liquid and solid becomes more complete and the adhesion increases giving a smaller receding angle. The film may be air which is displaced by the liquid being absorbed into the surface of the liquid and therefore removed, or it may be a film of greasy material which can only be removed by solution or displacement at the solid surface by molecules of a liquid.

If the surfaces of the solid are rough, the observed contact angle will not be the true angle. The effect of roughness is to increase the apparent angle, if the advancing angle is greater than 90°, and to decrease it if the true angle is less than 90°.

Where a sufficiently large surface area of the mineral to be studied is available, it is possible to measure the contact angle by bringing an air bubble in contact with a polished coal surface immersed in water or in a solution of the flotation reagent, and the angles of contact are measured on a screen, on to which is projected an image of the bubble and its reflection in the polished surface, *Figure 4*.

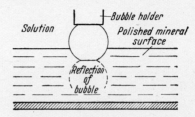

Figure 4. Measurement of contact angle

The apparatus used for contact angle measurements is based on the design of TAGGART, TAYLOR and INCE[1] and later modifications by DEL GUIDICE[2] and WARK and COX[3].

Floatability

Contact angle measurements are a measure of the wetting characteristics of a particular mineral. A solid is said to be completely wetted if the contact angle is zero; and incompletely wetted if there is a finite contact angle. Minerals which float more readily are non-wettable minerals, because non-wetting indicates a relatively high interfacial tension and the greater the interfacial tension in relation to the water tension, the more stable the attachment of the air–water surface and the greater the floatability.

To distinguish between wetting and floatability it may be said that wetting is the power of water to spread over the mineral, displacing air, and floatability is the power of mineral particles to attach themselves to an air–water surface and particularly to air bubbles.

Application of Wetting and Floatability to Flotation

The same conditions suffice for a mineral particle meeting an air bubble below the surface of a liquid as exist with a free surface of standing water, *Figure 3*.

A contact angle of 0° indicates a minimum attachment of an air bubble to a mineral surface. At 90° the attachment would be unity and at 180° the attachment would be at a maximum, *Figure 5*. For attachment of a particle to a bubble, therefore, it is necessary for the contact angle to exceed zero and be as large as possible and for the volume of the bubble to be such, that together with the mineral particle it forms a system with a density less than water.

The above discussion has been concerned with the static contact angle, *i.e.* the contact angle measured under static conditions. However, in a flotation cell, centrifugal forces are generated which tend to increase the effective weight of a particle. Therefore contact angle must exceed the static contact angle for the particular particle bubble system in order to allow equilibrium to be maintained between bubble and particle.

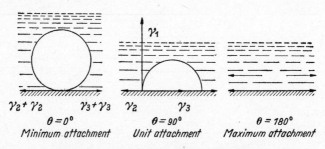

Figure 5. *Contact angles with air bubbles in water*

In practice, mineral particles form more or less a complete lining to a bubble large enough to float them all, and particles weakly attached receive support from particles more strongly attached. Some of the particles in the course of agitation are projected to the surface and once on the surface find bubbles large enough to support them.

FLOTATION REAGENTS

The interfacial tensions normally existing between minerals and water are generally insufficient in the magnitude and range to be used effectively in flotation, and it is the function of the flotation reagents to intensify the characteristics of the interfaces in each of the required directions in order to separate the mineral components from the associated constituents.

The reagents can be divided into five classes:
1. Collectors or selectors.
2. Frothing agents.
3. Depressant and activating agents.
4. Regulators.
5. Sulphidizers.

Collectors

There are often differences between the contact angles of valuable and unwanted minerals in the natural state, but these are not, as a rule, sufficient to achieve a good separation by flotation. Advantage is therefore taken of the property of various reagents to be selectively adsorbed at, or to combine with, the surface of the mineral it is desired to float. Such reagents are called collectors and their function is to increase the contact angle of the valuable mineral. Collectors contain a polar group to attract water and a non-polar group to repel water. The polar group is adsorbed on the surface of the mineral to be floated, while the non-polar groups are oriented outwards from the surface.

Collectors, all of which must be adsorbed to be effective, generally fall into three main classes:
1. Oils.
2. Acids containing a hydrocarbon group and the sodium and potassium salts of these acids.
3. Bases containing a hydrocarbon group and the salts of these bases.

Oils can be frothers in addition to collectors and contain -OH, -COOH or -OCH$_3$ groups. In the second group the -COOH group of the fatty acids has a great affinity for water, whilst derivatives of the amines are the most important members of the third group.

Oxidized oils and waxes of a paraffin or naphthalene base and fatty acids and their soaps have been used as collectors and frothers[4]. The most common collector in this range is oleic acid, which is the cheapest of all the reagents that give satisfactory results, and has the advantage of being a liquid at ordinary temperatures. Oleic acid has been used with good results on oxidized copper and lead ores, calcite, and a number of non-sulphide minerals. Stearic and palmitic acids are also used as collectors, the latter being used in the concentration of African oxidized copper ores because it is more readily available there than oleic acid. Most of the fatty acids and their soaps have marked frothing properties, but nevertheless are usually considered as collectors and pine oil is commonly used in conjunction with them.

For flotation of metalliferous sulphide particles, oil has now been replaced almost entirely by chemical collectors. These are certain sulphur derivatives of organic composition in which the sulphur is loosely combined by an unsaturated bond and the formation of the non-polar coating is brought about by chemical metathesis and the production of definite compounds on the surface of the mineral. Most of the reagents belong to one or the other of two chemical groups, the alkyl dithiocarbonates or the dithiophosphates. The former are more commonly known as xanthates and the latter as aerofloats.

The parent substance of the xanthates is carbon bisulphide CS_2 and that of the aerofloats is phosphorus pentasulphide P_2S_5. The xanthates, therefore, have a carbon nucleus and the aerofloats a phosphorus nucleus. Besides the difference in nucleus, the xanthates and the dithiophosphates differ in certain other aspects. The valency of phosphorus requires two alcohol groups, but carbon only one. The dithiophosphate series is the more numerous because it may have either alcoholic or phenolic groups attached to the nucleus, whereas xanthates are limited to the aliphatic series.

The main points of similarity, however, are that both series consist of three essential parts; a nucleus, which is carbon in one series and phosphorus in the other; a double bonded sulphur atom and a monovalent SX group which is common to both; and one or more alcohol groups (which are non-polar) attached to the nucleus. As a general rule, it can be stated that the activity of a collector from a homologous series of compounds increases as the non-polar part of the reagent molecule increases. This is illustrated below in *Table 1* which applies to xanthates, although similar results could be expected from the aerofloats.

In general, xanthates of the higher alcohols are more potent, but their solubility is less and their price higher. For practical purposes, ethyl xanthate is satisfactory, for while butyl and amyl salts are more powerful, their high cost is only justified with certain ores.

Table 1. Copper Recovered Per Cent

Pounds of reagent per ton of ore	0·01	0·02	0·04	0·08
Ethyl xanthate	69·8	82·6	90·9	93·6
Isopropyl xanthate	76·4	89·8	94·9	96·8
Secondary butyl xanthate	82·4	91·3	95·3	96·8
Amyl xanthate	86·9	93·5	95·0	96·0

For coal flotation, many different types of oil are suitable as collecting agents and the choice is not critical. The properties of such oils are that they should have good spreading properties with low viscosity and be insoluble in water and easily emulsifiable. Reagents generally used as collectors are gas, diesel, paraffin and fuel oils, creosotes and other refined carbonization oils. In practice, when the oil is agitated with the slurry a conditioning period is necessary to allow time for the coal particles to encounter the oil droplets.

Coal can be floated by the use of a 'frother' only, but better results are obtained by the use of a 'frother' and 'collector' mixed in suitable proportions. It is important to keep the mixture at a temperature exceeding 70° F when in use, otherwise small particles of wax settle out from the oils and render the mixture a little less efficient and also result in operating difficulties.

The oily collectors invariably used in froth flotation of coal are less selective in their action than the chemical collectors utilized extensively in the froth flotation of minerals. Owing to the carbonaceous nature of the impurities associated with coal, the effectiveness with which flotation may be applied to the up-grading of slurries and the production of superclean coal is limited. Not only is the angle of contact at a coal surface increased to over 70° after treatment with an oily collector, but also the angle of contact at the shale surface is increased. For the efficient use of oily collectors, it is essential that the oil be finely dispersed. If this were not so, some particles may have a coating far in excess of their requirements, whereas other particles would be starved. If more oil is used to counteract this, then the excess oil would tend to float the shale particles. Maximum dispersion of the oil is therefore essential and no more than the minimum quantity necessary for effecting satisfactory flotation of the coal particles should be used.

It is found that all coals, with the exception of lignite, show contact angles exceeding 60° when treated with fuel oils. The low rank coals and anthracite give somewhat lower angles than those obtained with bituminous coals. Therefore, in order to obtain equal recoveries in flotation practice, a larger quantity of reagent will be required for anthracite and lignitous coals than for bituminous and carbonaceous coals. An approximate indication of the reagent requirements of coals of different rank is given in Table 2.

Table 2. Reagent Requirements for Coal of Different Rank

Type	Oil required lb./ton	F.R.S. rank
Anthracite	1–2·5	100
Low-volatile coals		200
Medium-volatile coals	0·25–2	300
High-volatile coals		400
High-volatile Medium rank	1·25–5	500, 600, 700
High-volatile Low rank	4·5–10	800, 900

Chemical collectors, such as xanthates have been suggested as possible collecting agents for coal flotation. It has been proved, however, that pure potassium ethyl xanthate alone is not a collector for coal, but that commercial xanthates or solutions of pure xanthates containing dissolved cupric or ferric salts possess collecting properties. This collecting action is brought about by oxidation, resulting in the formation of dixanthogen which is a powerful collector for coals of all ranks[5].

Frothing Agents

The requirements of a suitable froth in flotation are that it should have sufficient bouyancy to bear the mineral particles to the surface of the pulp. It must be stable enough to exist until the bubbles with their mineral load can pass over the lip of the flotation machine and not so stiff that it will be difficult to break up after passing from the cell. Finally, it should be fluid enough to flow over the lip of the cell without clogging.

In providing this froth, the functions of a frother are twofold: first, to produce within the pulp, a large air–water interphase to which the conditioned particles may migrate: and secondly, to enable the separated particles to be removed easily from the machine and to be readily recovered.

To perform these functions satisfactorily a frother should have the following characteristics:

(a) It must form in low concentrations a copious but not too persistent froth.
(b) It must froth equally well in alkaline or acid medium, *i.e.* it must be insensitive to changes in hydrogen ion concentration in the pulp.
(c) Ideally it should have no collecting properties, although several reagents exist which can serve both functions. These are only suitable, however, when selective flotation is not being undertaken.

(*d*) Its frothing properties must not be affected by collecting agents and salts even when both are present in high concentration.

(*e*) It must be readily dispersible in an aqueous medium.

One part of the frother molecule is polar and attracts water, while the other part is non-polar and repels water. Frother molecules concentrate at the interface between the fluid phases of flotation systems, the polar end adhering to the water phase and the non-polar part of the molecule oriented away from the water. The air bubbles formed under the liquid surface are more or less completely lined with a monomolecular layer of frother molecules which allows each bubble with its lining to come into contact with other bubbles without coalescing, thus forming a froth.

Commercial frothing agents are confined to that range of compounds which appreciably affect the surface tension of water, even when their concentration in the water is low. All good frothers contain one polar group, usually a group containing oxygen, in the form of -OH, -COOH or -CO, or nitrogen in the form of $-NH_2$ or -CN. Usually organic acids, alcohols and amines, are good frothers, whereas unsaturated hydrocarbons give little froth. All frothing agents contain an inactive alkyl or aryl group.

In practice, the most widely used frothers are pine oil and cresylic acid. Pine oils exist in several different varieties and include those obtained by the destructive distillation or steam distillation of pine wood. Steam distilled pine oil is more satisfactory as a frother since it contains organic compounds of lower molecular weight and has less collecting properties than destructively distilled oil.

Steam distilled pine oil is composed of terpene hydrocarbons, terpene ketones and terpene alcohols. Each of these substances has good frothing properties, but the alcohols (in which the OH is the polar group) are more satisfactory for this purpose than the hydrocarbons (in which double bonds act as the polar radical). The terpene alcohols of pine oils produce a more uniform froth and have smaller collecting properties.

Cresylic acid is composed of a number of higher homologues of phenol ($C_6H_5.OH$), *i.e.* cresols ($CH_3.C_6H_4.OH$) and xylenols ($C_2H_5.C_6H_4.OH$). In commercial cresylic acid, the xylenols are the dominant constituents. The use of cresylic acids is increasing at the expense of pine oil and it is found that they produce very satisfactory frothing agents.

Among the more locally used frothers is eucalyptus oil, which is obtained from the distillation of parts of eucalypti and is used extensively in Australia where the plentiful supply of eucalyptus trees makes it less expensive to use than other types of frothers. It does, however, tend to have more collecting properties than pine oil. Camphor oil is used in Japan in place of pine or eucalyptus oil.

Activating Agents

An activating agent is one which can modify the surface of an otherwise non-floatable or poorly floatable mineral so as to make it amenable to flotation with the usual collecting agents.

The reagents are adsorbed at the surface of the mineral and may often enter into chemical combination with the metallic radical of the mineral. An example of this is the use of copper sulphate in the flotation of sphalerite. The copper sulphate changes the surface of sphalerite to covellite, which is then readily floatable through the use of xanthates. Activating agents can also find a use in the treatment of non-metallic compounds.

As a general conclusion, it may be said that for an activating agent to work successfully, it must be capable of producing less soluble salts with the collecting agents than any of the metal compounds at the surface of the minerals to be floated.

The most effective activating agents are the salts of copper, lead and mercury, because their organic salts and oxides are the least soluble. The use of activating agents for reconditioning the surface of coal oxidized by weathering has been found to be of no value.

Depressing Agents

These are reagents which impart a coating to the mineral surface and prevent the adsorption of a collector, making it less floatable than the normal mineral surface. Like collectors, depressing agents must be selective in their action, affecting only one of the minerals to be separated and not the others.

Alkalis, particularly lime and sodium carbonate, are the depressors most commonly used. All alkalis at the same pH value are equally effective. Protective colloids such as glue, starch, casein and sodium silicate are the most important depressors for quartz and silicate minerals.

Alkali chromate is used in the depression of cerussite, the chromate forming a coating of lead chromate which is less soluble than lead carbonate and therefore reacts less actively with the collector than lead carbonate. Similarly, the action of potassium chromate on partly oxidized galena is one of forming a partial crust of lead chromate which resists reaction with the collector.

Cyanide compounds are important chemicals in this class of reagents, because of their action on iron sulphides when it is desired to separate these from the sulphides of the more valuable metals. Cyanide is also effective in depressing zinc when it is desired to separate galena from sphalerite as well as iron. Depressors are mainly used in coal flotation where pyrites are present in a finely divided condition and when it is necessary to produce a clean coal concentrate having a minimum sulphur content.

Regulators

The main function of a regulator is to modify the alkalinity or acidity of a flotation circuit, or in other words to control the pH of the solution. This modification of the pH has a pronounced effect on the action of flotation reagents and several difficult flotation operations are made possible as a result of it.

The action of a regulator may be considered as follows:
(1) To precipitate soluble salts out of solution.
(2) To clean the mineral surfaces and therefore facilitate the action of other reagents.
(3) To depress certain minerals.

The precipitation of soluble salts, whether present in the ore or in water, generally renders them inert to the action of flotation reagents, and prevents any combination with the promoters. This combination would probably result in the prevention of a certain quantity of reagent from taking part in any action with the mineral to be floated and consequently an increased quantity of reagent would be necessary to produce the required results.

The variation of the surfaces of some minerals, according to the pH of the associated solution, results in the use of regulators in 'cleansing' the surfaces of the mineral particles. This brings about more effective action on the part of promoters and other reagents and helps to increase selectivity. Controlling the pH by the action of regulators is very effective in depressing certain minerals. Examples of this are the use of lime in depressing pyrite, and of sodium silicate in dispersing and preventing quartz from floating.

For alkaline circuits the most common regulators used are lime, soda ash, and sodium silicate and for acid circuits sulphuric acid. From a practical standpoint, it is preferable to use a neutral or alkaline circuit. If, however, an acidic circuit is required, the use of special equipment will be necessary to withstand corrosion. For this reason, acids are not very widely used in flotation circuits, but one important application is in the flotation of oxidized pyrite for the production of a clean iron concentrate with a high recovery.

By regulating the quantity of collector used in flotation and by regulating the hydrogen ion concentration of the flotation circuit, a twofold adjustment of the promoter activity is provided and this regulation finds a wide use in selective work.

Sulphidizers

The function of this class of reagents is to precipitate a film of sulphide on the surface of oxidized minerals in order to make the surface more responsive to the action of promoters. The most widely used compound is sodium sulphide, which is commonly used in the flotation of lead carbonate ores, and slightly tarnished sulphides such as pyrite and galena.

USE OF REAGENTS

There are so many variables to be considered in the use of froth flotation reagents that each case has to be considered separately. However, there are some general considerations which could be applied to most cases and these concern mainly the question of addition of the reagent to the flotation circuit. The variables in reagent addition are:
(a) The amount of reagent.
(b) The place and method of addition.

Quantity of Reagent

If an excess or deficiency of reagent is added to the circuit, it will probably have a pronounced effect on the quantity and quality of the froth produced. In the case of a frothing agent, the effect produced is proportional to the concentration of the frother. As the concentration of frother increases, the frothing power increases; at first rapidly, but past a certain concentration the effect of further additions of frother is less than that of preceding additions. Finally, when the solution is saturated with the frothing agent, there is a total absence of frothing.

Increasing the concentration of the collector results in an increase in the contact angle, which reaches a maximum value and then remains substantially constant, irrespective of the further addition of collector. It is particularly effective in increasing the recovery of the coarsest particles. The coarsest particles are affected more than the fine ones because the contact angle required for static equilibrium with a given sized bubble is less for fine particles than for coarse particles[6]. There is a high consumption of reagents during the first stage of flotation, since reagents are used in lowering the surface tension of the pulp and in coating the mineral particles. Reagent consumption then varies linearly with the amount of mineral collected, except at the end of the flotation when it increases more rapidly during the elimination of the largest and heaviest particles.

Place of addition of Reagent

The usual place for the addition of reagents is in the cells themselves or just ahead of them. Some reagents require a variable period of conditioning for reaction with the mineral surfaces and this is usually carried out in a conditioning tank, although a flotation cell may be used for this purpose. Conditioning usually results in cutting the reagent cost down to a minimum and where chemical interaction between the reagent and the mineral surfaces takes place, conditioning before flotation is essential.

In the case of coal flotation using creosote oils as collectors, it appears that when frothing some types of coals a conditioner is not required. The reason for this is probably due to the fact that some coals possess a high natural floatability and can be recovered by the addition of a frother only. In these cases a separate tank for conditioning is not provided although the reagent consumption may be slightly higher than normal.

The addition of reagent into the cells as well as the conditioner provides a means of increasing the yield of frothed product, particularly in those cells at the end of the flotation circuit where the stage addition of reagents helps to maintain and amplify the reagent action. In some cases the reagent can be put into the circuit well ahead of the cells, although this is only suited to special circumstances where a variable time of conditioning is required for the reagent and the mineral.

The essential requirements for adding flotation reagents to the circuit are that the addition should be uniform and capable of accurate control. This can be achieved by a variety of different types of feeder, all of which provide a steady and uniform feed of reagent, capable of accurate adjustment, to the required points of addition in the plant.

FURTHER FACTORS AFFECTING FROTH FLOTATION

Apart from the various types and quantities of reagents used there are other factors to be taken into consideration in controlling the operation and performance of a flotation process. These will now be considered in turn.

Particle Size

There is an upper and lower limit of particle size for efficient recovery in froth flotation. The smaller the particles the smaller the recovery and this fact alone limits the field in which flotation can be applied. In addition, the coarsest particles are recovered first and the finest last, due to a gradual decrease in readiness to adhere to air bubbles from coarse to fine particles.

As regards the upper size limit for flotation, this depends upon the mineral particle itself, its density and the surface tension of the liquid associated with the pulp. An approximate comparison of the upper limit of particle size for the flotation of several minerals is given in *Table 3*.

Table 3. *Upper Limit of Particle Size for Flotation of Different Minerals*

Mineral	Density of mineral	Approximate upper size limit for effective flotation (mm)
Galena	7·5	0·2
Pyrite	5·0	0·3
Sphalerite	4·1	0·5
Calcite	2·7	0·5
Coal	1·35	1·0

For the recovery of coarser particles, within certain limits, a greater quantity of frothing agent is required than would normally be used in the treatment of finer pulps. This is due to the improvement of both the

Table 4. *Flotation Results on Different Raw Coal Samples.* (*From* J. L. LEWIS, *by courtesy of* The Institution of Mining Engineers)

Size $\frac{1}{16}$ in.–0

Mesh	Raw coal					
	Case 1		Case 2		Case 3	
	% Wt.	% Ash	% Wt.	% Ash	% Wt.	% Ash
Plus 30 mesh	39·5	12·5	48·0	41·5	47·6	28·0
30–60	32·5	10·8	18·1	40·5	23·3	29·2
60–90	9·6	12·4	8·0	39·8	10·1	32·2
90–200	9·6	9·9	5·1	43·4	6·7	28·8
Minus 200 mesh	8·8	24·8	20·8	59·4	12·3	68·8
Average % ash	—	12·8	—	45·0	—	33·2

contact angle of the particles and the number and quality of the air bubbles. If these limits are exceeded some of the finer particles of gangue material associated with the coarser particles may be recovered with the floatable material.

In coal flotation, particles up to 1 mm in size can be floated quite efficiently and some examples are given in *Table 4* and *Table 5*. Difficulties are encountered, however, where the minus 240 mesh fraction exceeds 40–50 per cent. In this case, the additional reagent required to float the coarse particles would probably result in some of the finer particles of shale in the minus 240 mesh range being floated in addition.

An apparatus which it is claimed can deal quite effectively with coarse coal has been patented recently[7]. The apparatus consists of a bank of cells of the usual sub-aeration pattern (see page 226) the main difference being that each of the cells is connected underneath the agitating compartments by a common channel. In this channel is fixed a screw conveyor for the removal of coarse settled tailings, which are subsequently collected by a

Table 5. Yields and Reagent Consumptions for Different Raw Coal Samples. (*From* J. L. LEWIS, *by courtesy of* The Institution of Mining Engineers)

	Froth flotation products					
	Case 1		Case 2		Case 3	
	% Wt.	% Ash	% Wt.	% Ash	% Wt.	% Ash
Yield of clean coal	88·7	4·7	51·4	13·3	64·3	5·4
Yield of tailings	11·3	75·7	48·6	78·5	35·7	83·3
Oil consumption lb./ton	2·5		4		3·2	

bucket elevator in the manner of a Baum jig washer. The floating of the fine coal is assisted by the introduction into the cells of fine suspensions of material having densities up to 0·3 below the specific gravity of the coal to be floated. It is claimed that the apparatus can treat coal 1–10 mm in size.

Temperature

It is important to control the temperature of a froth flotation circuit for a number of reasons some of which are as follows: Too low a temperature would necessitate prolonged mixing of some reagents and would increase the viscosity of water to such an extent that a certain amount of the material to be discarded might be entrained in the froth. In a circuit where the temperature is high, the end point of the operation is reached sooner, the viscosity of the water is decreased and the concentrate is therefore cleaner. At higher temperatures, the reagent is more mobile and is therefore more economical in use.

In spite of these advantages, however, it is not common practice to run a flotation circuit at temperatures much above normal, because the benefits obtained thereby can be obtained more economically by the careful choice of reagents and, where necessary, thinning the reagent with a suitable solvent.

Influence of Pulp Density

It is always advantageous in the feed to a flotation plant to use as high a concentration of solids as possible, as this saves water and increases the capacity of the flotation cells which, in turn, causes a saving in reagent consumption. This saving of reagents particularly concerns frothers and reagents that regulate the pH of the solution, *i.e.* reagents whose action depends upon the liquid phase or the gas–liquid phase.

However, a higher concentration of solids increases the amount of entrained solids in the froth and this leads to a froth carrying some of the undesirable material. As the concentration of solids in the froth feed decreases, the yield of concentrate also decreases but the quantity of reject material increases. As a general rule the pulp dilution should be less for primary flotation or roughing operations, but greater for final cleaning operations.

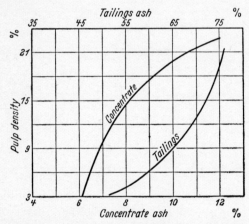

Figure 6. *Effect of varying pulp density on performance of coal flotation plants.* (*From* R. E. ZIMMERMAN, *by courtesy of the* American Institute of Mining Engineers)

In the case of coal flotation, tests have shown that the yield of clean coal drops and the ash content of the clean coal increases as the solids concentration in the feed to the cells increases above the optimum figure for a particular slurry. If the percentage of fines minus 200 mesh in the feed exceeds 30 per cent, it is advisable that the feed to the plant should not exceed a concentration of 10 per cent and in many cases 7·5 per cent. The higher the minus 200 mesh fraction in the feed, the lower should be the solids concentration in the feed to the plant. In addition, as the percentage of minus 200 mesh particles increases in the feed the capacities of the flotation cells are reduced. However, extremely dilute pulps require an excessive consumption of flotation reagents and water. The effect of pulp dilution on a feed consisting almost entirely of material finer than 200 mesh is shown in *Figure 6*.

MACHINES USED IN FLOTATION

Apart from the flotation cells, where the actual flotation process takes place, the machinery associated with froth flotation includes a number of items of ancillary equipment mainly consisting of conditioning tanks, thickeners and filters. Since the two latter items occupy an important place in chemical engineering practice and are adequately dealt with elsewhere, they will not be considered at this stage.

FROTH FLOTATION CELLS

A good froth flotation cell should fulfil the following requirements as far as its technical capabilities are concerned:

(a) It should be continuous in operation.
(b) It should be capable of generating fine air bubbles. This will partly depend on the reagents being used as frothers and also on the means of agitation within the cell.
(c) It should disperse these bubbles within the pulp as far as possible. In order to accomplish this, a certain amount of circulation within the cell is necessary.
(d) It should be able to bring about a separation of the mineral-laden froth from the pulp and this necessitates the formation of a quiescent zone in the machine prior to the removal of the froth.
(e) It should be capable of discharging the mineral-laden froth into a suitable container. In order to do this, some form of mechanical scraping or natural overflow from the cell is required.

Generally, modern flotation cells can be divided into three main groups depending on how the air is introduced within the pulp. These are as follows:

(1) Sub-aeration cells in which air is drawn in by suction or by direct introduction to the base of a rotating impeller.
(2) Cascade cells in which air is introduced by tumbling of the pulp.
(3) Pneumatic cells in which air is introduced directly by blowing through the pulp.

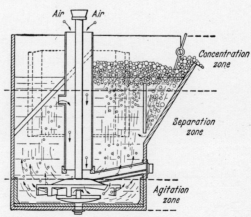

Figure 7. The three zones of a sub-aeration flotation cell. (By courtesy of the Denver Equipment Co., Ltd.)

A flotation plant usually consists of identical cells arranged in series in such a manner that each cell receives as feed the unfloated material from the preceding cell.

Sub-aeration Cell

This is one of the most widely used types of flotation cell. Its action depends upon the process of agitation and creation of the correct bubble structure at the very base of the cell where proper aeration is essential. The principle on which this cell works can be illustrated by reference to *Figure 7*, and it will be seen that there are three distinct zones with the cell which can be classified as Agitation Zone, Separation Zone and Concentration Zone.

Agitation Zone—The pulp, which is introduced into each cell by gravity flow, drops directly on top of the rotating impeller where it is thrown outwards by centrifugal action of the impeller. This action causes a positive

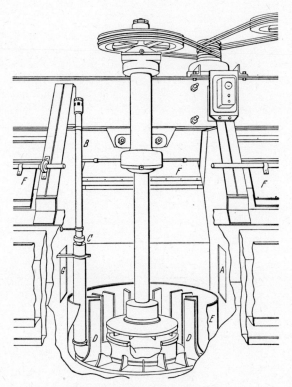

Figure 8. *Internal view of a sub-aeration flotation cell.* (By courtesy of Simon-Carves, Ltd.)

suction of air down the induction pipe or main standpipe, creating bubbles which are exceedingly small and therefore better able to support the largest number of mineral particles.

Separation Zone—This is a zone of quiescence, which is necessary to enable the bubbles with their mineral load to reach the surface of the liquid without the particles being shaken free by agitation and swirling within the cell. This quiescent zone is usually brought about by means of baffle vanes on the hood of the impeller or set round the periphery of the impeller.

Concentration Zone—As the mineral-laden bubbles move to the pulp level, they are carried forward to the overflow lip by the crowding action of succeeding bubbles. Quick removal of the froth is therefore desirable and this is usually brought about by froth paddles which assist the overflow. The overflow of the froth can either be from a single overflow lip or from a double overflow lip.

The working and construction of a sub-aeration flotation cell can be followed by reference to *Figure 8* which illustrates a 'Simcar-Geco' cell used in coal flotation.

The pulp, together with the correct amount of reagent or reagents, enters the cell through the inner cell, part A. The impeller at the base of the vertical spindle draws in air through the induced air pipe B, together with a quantity of pulp which enters the induced air pipe at point C. The volume of induced air is controlled by a special port-type air valve and the quantity of pulp is controlled by an adjustable cone piece at C. In some designs of cells, air is induced through hollow impeller spindles, but by having separate air pipes the amount can be controlled at will and the risk of blockage in the air system is eliminated.

The impeller flings the aerated pulp through the ring of baffles D, which destroy the vortex without obstructing the rising bubbles, thereby creating a quiescent zone above the agitation zone. The pulp then meets the wearing ring E, which is specially covered in rubber to resist abrasion. The froth bubbles, carrying the mineral particles with them, rise to the surface of the pulp and collect in a relatively thick mat which is driven over an adjustable froth lip by revolving paddles, F, on each side of the cell, and is finally collected in launders fixed alongside the row of cells. The rejects from the cell pass through the port G into the next cell where the process is repeated, the flow of pulp through the battery of cells being maintained by allowing a slight fall in the pulp level from cell to cell.

In the Denver type of cell (*Figure 9*), which is very commonly used for metallic minerals and coal, air induction through a hollow impeller spindle is used for aeration purposes and in some cases air or gas under pressure may be introduced. This controlled air or gas pressure is required in some cases to provide the correct bubble structure and also for improving the conditions for the chemical and physical reactions of the flotation reagents. Such a cell therefore finds a useful application in selective flotation. Another feature of the Denver type of machine is the positive pulp level control in each cell. This is accomplished by adjustable weir plates which can be operated by a hand wheel or, on some larger machines, by a gear mechanism.

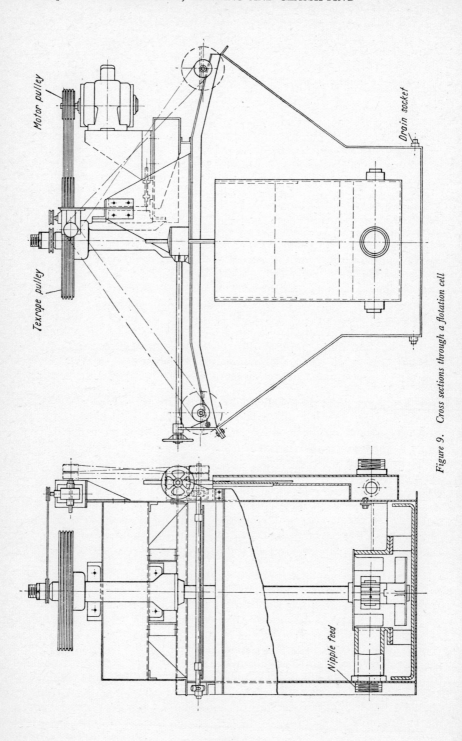

Figure 9. Cross sections through a flotation cell

The impellers on many types of sub-aeration cell are fitted with a stationary hood. The purpose of this hood is to support the stationary air stand pipe, if one is used, and the wearing plate or diffuser, and it also prevents the impeller from being buried by pulp if the machine is shut down for repairs or between shifts.

The advantages of sub-aeration cells can be summed up as follows:

(a) In most cases, the air is self-induced and consequently no air compressor is necessary.
(b) All the pulp must pass through each agitator as it flows through the system.
(c) Any desired depth of pulp can be maintained, irrespective of other cells, if the machine is fitted with adjustable overflow weirs.
(d) Any cell or cells can be used for conditioning purposes without disturbing the other cells. This is accomplished by lowering the pulp level below the point where the froth will overflow the froth lip.
(e) Reagents can be added at any point in the flow. This makes possible step reagent addition, which is sometimes desirable for improving selectivity and yield of concentrate.
(f) One or more of the middling products can be returned to the same machine if desired.
(g) Unit construction of the cells enables extra cells to be readily added to an existing battery if necessary to cater for changes in the raw feed.

Cascade Cells

This type of cell depends upon the principle that if a sheet of water issuing under the requisite head is passed over an overflow weir plate, it will take a certain amount of air with it and will introduce and distribute this air by virtue of its kinetic energy.

A plant comprising a number of cascade cells is so arranged that the cells are stepped one below the other in series. The pulp, after splashing into a head box feeding the first cell, takes a certain amount of air with it and after yielding a proportion of its mineral content to the froth bubbles produced by the influx of air, flows out through the bottom of the cell and into a second head box where the process is repeated.

Compared with the violent agitation of some of the sub-aeration machines and the abundant air supply of the pneumatic machines, the aeration effected by the cascade type of cell is mild and limited. As a result, the froth is not so copious and in many cases only the readily floatable particles are recovered so that only a partial concentration is made. A simple cascade flotation cell is illustrated in *Figure 10*.

The pulp, together with entrained air, enters the cell at A and into a quiescent zone C. The air bubbles rise to the surface carrying their mineral load and are directed to discharge launder E by an inclined baffle plate D. The liquid overflows at discharge F so that a constant bleed of liquid is maintained over the discharge nozzle B through which the bulk of the liquid flows. Air is drawn in through a series of air holes in the pipe walls adjacent to the nozzle and the whole process is repeated in a similar cell connected in series directly below.

Because of its simple construction and low operating cost, this type of machine is particularly suitable for insertion in the final stream of flotation discard. It can, however, be used ahead of a main flotation process as, for instance, in replacing a normal tabling process prior to flotation.

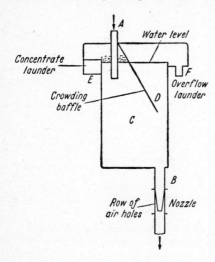

Figure 10. Cascade flotation cell

Another type of cell utilizing a similar principle to that of the cascade machine is the Kleinbentink cell used in the Dutch State Mines and in Great Britain for coal flotation. (See *Figure 11*). This type of cell consists of a conical vessel at the bottom of which is connected a U-shaped pipe feeding into a head box. An adjustable weir in the head box controls the feed into the succeeding cell which is positioned a short distance below its predecessor, all cells being connected in series and in steps.

Inside the conical vessel is an inner compartment, in which agitation of the pulp takes place by means of a twin-bladed stirrer. The feed, after being suitably treated with reagents, overflows an adjustable weir and is led together with entrained air to the inner agitation compartment by means of a supply pipe. In the lower part of the agitation compartment, the stirrer causes intimate contact between the pulp and the entrained air, the material being forced through perforations in the walls of the agitation chamber into a more quiescent zone. Here a separation takes place by the adhesion of mineral particles to air bubbles and a froth is formed which rises to the surface of the cell. The particles not adhering to the air bubbles sink into the lower part of the cell and are carried off by means of a pipe to the feed box of the following cell, which is situated at a lower level. Part of the liquid entering the quiescent zone flows back into the inner agitation compartment where it is further agitated. The froth layer on the surface of the liquid is skimmed off into gutters, arranged radially from the centre of the cell, by means of wipers freely suspended from a slowly revolving frame.

The wipers move through the froth layer, taking part of the froth towards the gutter. Approaching the gutter, the wipers are lifted by leading steps and turn about, their hinges moving the froth into the gutter. Aeration of the cell, in addition to being provided by means of entrained air with the feed, could also be introduced directly into the cell by a supply pipe.

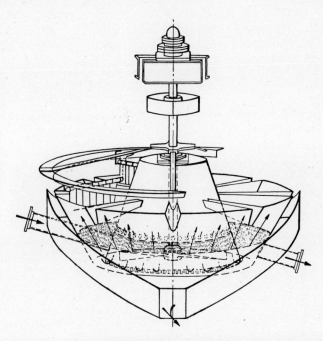

Figure 11. Kleinbentink flotation machine. (*From* H. LOUIS, *by courtesy of* Methuen & Co., Ltd.)

Pneumatic Cells

As the name implies, a pneumatic flotation cell does not rely on mechanical forms of self air induction but depends entirely upon the independent induction of air under pressure into the pulp. The most common machine in this series is the Callow Cell, *Figure 12*, which consists of a rectangular box 2 ft. wide by 8 ft. long with a sloping bottom, the depth being about 20 in. at one end and 45 in. at the other. This sloping bottom is a porous mat of closely stitched canvas held between wire gauze, which forms the cover to an air box into which a blower delivers air at a pressure of 5 lb./in.^{2}.

Because of the greater depth at one end, the hydrostatic head under which the air issues at this end is greater than that at the shallow end. To equalize these differences in hydrostatic head, the air box is divided into compartments each separately served from the air main and the amount of air passing into each compartment is regulated by a separate air valve.

The pulp is fed into the box at the shallow end and is directed downwards towards the porous mat by means of a baffle. On coming in contact with the air bubbles, the floatable particles attach themselves and rise to form a froth, whereas the tailings flow out from the deep end of the cell through an automatically controlled discharge.

The froth after building up from below overflows the side of the cell and is discharged into a launder.

The amount of air required in this machine is of the order of 10 ft.^{3}/min/ft.^{2} of porous area, but because of its low pressure the power consumed is less than that which would be required for an equivalent amount of mechanical aeration. Since this type of aeration does not assist in the mixing of reagents, this mixing must normally be done beforehand. Another objection to their use is the frequent blinding and corrosion of the porous mat by pyritic and limy accumulations.

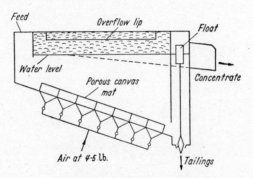

Figure 12. *Callow pneumatic cell*

Several other pneumatic type cells using a similar principle have been devised. Typical of these is the Inspiration Cell[8] and the MacIntosh Rotating Mat Pneumatic Machine[9]. The former is similar to the Callow machine except that the porous bottom of the cell is constructed of removable sections and the overall capacity is higher. To overcome the difficulties arising from the blinding of the porous mat in the case of Callow and similar machines, the MacIntosh machine was devised. Air under pressure is introduced from rotating porous socks into a truncated V-shaped trough, in which the rotor covered with the porous material is rotating slowly. The movement of the rotor prevents the permanent blinding of the porous mats.

Mats for all mat-pneumatic cells can be made of canvas or palma twill, perforated rubber, porous concrete or rubber impregnated canvas.

In all pneumatic machines, the volume of air used is relatively large, since it is issuing under pressure over a relatively big area. As a result, the froth produced is voluminous and at the same time frail, resulting in conditions which favour the flotation of fine material. The excess of air finds a ready outlet at the free surface of the froth and assists in the discharge.

Elmore-Vacuum Process

Mention must be made of a further process of froth flotation which does not depend on mechanical self-induction of air or the introduction of air under pressure. The principle used is the liberation of dissolved gases, or more commonly, air from a liquid when subjected to a vacuum, and the process is known as the Elmore-Vacuum process.

The feed, suitably treated with reagents, is introduced into the separating vessel, *Figure 13*, which consists of a cone with the apex upwards subjected to a vacuum of about 24 in. The material, once inside the cone, is agitated by means of revolving rakes, which at the same time assist in gradually ploughing the tailings to the periphery of the cone, where they are discharged down pipes extending about 30 ft. below. This discharge is controlled under water, and the discharge pipe, together with the feed pipe

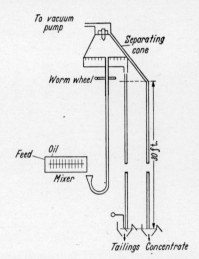

Figure 13. Elmore-Vacuum plant

into the separating vessel, form a natural syphon which produces a reduced pressure within the cone which is further assisted by the vacuum. During the progress of the material towards the periphery of the cone, the floatable mineral particles attach themselves to air bubbles released by the vacuum and rise to the apex of the cone where they overflow a circular weir into an outer hood from which the concentrate discharge pipe leads.

With coal, vacuum flotation is particularly suited for the treatment of material finer than $\frac{1}{8}$ in. and has the advantage over other froth flotation processes in that the cleaned coal is more easily dewatered. The process requires more reagent than for open cells and the amount of reagent used is related to the internal surface area of the coal. With vacuum cells, however, little or no frothing agent is required and the best results have been obtained when using petroleum oils. Wetter, dirtier froths are often produced if frothing agents are added during vacuum flotation.

CONDITIONING TANKS

The prime purpose of a conditioning tank in a froth flotation circuit is to provide an intimate mixing of the reagent or reagents with all the pulp

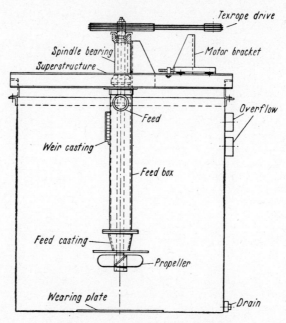

Figure 14. Cross section of a conditioner

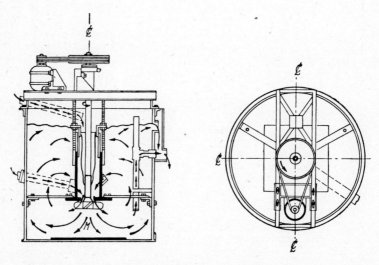

Figure 15. Denver (patented) super-agitator and conditioner. (By courtesy of the Denver Equipment Co., Ltd.)

particles and to provide sufficient reaction time for the reagents and the particle surfaces. By so doing, economies in the use of the reagents can be expected, in addition to more uniform frothing conditions. Where chemical reagents are being used, conditioning before frothing is a necessity, although for other types of reagents, particularly in the use of oils in coal flotation, the conditioning can take place in the first cell.

There are various different arrangements of conditioning tanks, each one mainly being concerned with its method of use and the duty it is required to perform. Basically, however, the general arrangement of a conditioning tank is as shown in *Figure 14*.

The feed enters at the top of the feed box around the propeller shaft and goes directly down to the propeller zone. In this way, positive agitation and circulation is assured and the possible short circuiting of the pulp eliminated. The problem of partially conditioned froth is prevented by this method, for as fast as the froth forms on the surface, it is drawn down the stand pipe and redistributed through the pulp. The propeller imparts a downward velocity to the pulp, thereby giving rise to intense agitation and circulation, although the speed of the propeller itself is only relatively slow.

This intense agitation and circulation prevents heavy solids from accumulating on the bottom of the tank during operation. The propeller is positioned at the proper height to produce optimum circulation and is also far enough above the bottom of the tank to prevent getting buried by the settling of heavy solids during shut down. In this way, it develops sufficient velocity to repulp settled solids, and the adjustable weir plate on the side of the feed box permits a control of the amount of recirculation within the conditioner.

In some makes of conditioner tank, notably that of the Denver type, provision is made for the recirculation of intermediate zones within the conditioner by the use of recirculation ports at various heights on the stand pipe. Part or all of these ports may be used, giving the operator positive control over conditions in the machine, *Figure 15*.

APPLICATIONS OF FLOTATION

Place in Mineral Dressing

Flotation can take any of the following places in mineral dressing procedure depending upon the conditions prevailing, the cost and the degree of concentration required:

(1) All-flotation.
(2) Primary flotation; secondary water concentration.
(3) Primary water concentration; secondary flotation.
(4) Primary water concentration; secondary flotation; final water concentration.

All-flotation is only practised where water concentration is impossible, *i.e.* where the gangue material has a high density or the mineral particles are flaky. Heavy gangue material may consist of heavy silicate, magnetite, specularite or undesirable sulphides, such as pyrite and pyrrhotite. All-flotation may be conducted as one operation or as 'straight flotation', or it may be used in stages with additional grinding in between.

Part flotation schemes are more common, however, and usually consist of a combination of water concentration with froth flotation. Usually, the water concentration process precedes the flotation so that the coarsest particles of the mineral can be removed soon after they have been liberated in the first stages of comminution. On the other hand, if the amount of removable mineral present is small, it may be more advantageous to treat it in the flotation process first, since a greater degree of comminution will be necessary before an adequate quantity of mineral is released.

In this state the material is in a better condition for flotation than if some of the mineral had been previously withdrawn by water concentration. Any granular mineral particles remaining can subsequently be recovered by water concentration. This secondary water concentration can at times be regarded as a means of correcting the irregular working of a preceding flotation process.

In treating complex ores flotation may come between initial and final water concentration or water concentration between two flotations. With initial and final water concentration, the former would probably be by jigs or roughing tables while the latter would be on slime tables. With initial and final flotation the latter would probably be by cascade cells.

In addition to the above applications of flotation to the main scheme of mineral dressing, flotation may also play a subsidiary part in treating concentrates obtained from the main processes of the system in use. These would involve the treatment of water concentrates by flotation, particularly where the concentrate is complex.

On the other hand, water concentration may be used to treat a flotation concentrate, particularly if it contains two sulphides and the mineral particles are granular.

The simpler equipment and moderate operating costs of flotation are points in favour of using it in place of water concentration in those cases where either process would be efficient. This is in spite of the fact that for its proper functioning further comminution may be necessary and its success may depend upon considerable skill in operation.

Flotation is applied to practically every non-ferrous mining field. From its successful application to zinc it has been applied to all the major base metals, being particularly useful in treating the large disseminated copper deposits. It has superseded water concentration in the fine treatment of these base metals, and although its working cost is much the same as water concentration, it brings an improved recovery and greater capacity.

References to some of these applications will be found at the end of this section, but one important application will be considered here, namely, the flotation of uranium.

Recovery of Uranium by Flotation

Flotation provides an alternative for the extraction of pure uranium in place of the expensive chemical methods which involve dissolving the ore in acids and precipitating the various impurities.

It can be achieved by conventional procedure using a reagent consisting of a material chosen from the class of anionic and non-ionic synthetic detergents together with a material chosen from the class of soluble salts of fatty acids and rosin acids[10]. By a suitable combination of the flotation reagents it has been found possible to recover about 90 per cent of the available uranium and to produce a suitable froth which can readily be broken by conventional techniques such as the use of water sprays and dilute acids.

The addition of other reagents such as activators, frothers and depressors, does not appear to improve the results obtained by the use of the combination flotation agent of synthetic detergents and soaps.

The best results are obtained by using a concentration of the flotation agent which is determined to be slightly above the minimum concentration necessary to produce a satisfactory froth. In order to obtain the best results it is necessary to determine the exact ratio of synthetic detergent to soap which gives the most satisfactory recovery and the easiest froth which can be broken down. A typical application of the process is cited below.

An eight-cell continuous flotation unit, consisting of cells having a capacity of one cubic foot, is employed in continuous operation to separate uranium oxide from gangue materials. The flotation agent, polyethylene oxide-phenol detergent is mixed with twice the quantity of refined tallow soap and is maintained at a concentration just slightly above that found necessary to maintain a suitable froth. No modifiers, activators or depressors are used. Uranium oxide-sand material is fed at the rate of 200 lb./h to the operating system, water is recirculated at the rate of about 4 gal./min around the entire system, and air is blown through each cell at an approximate rate of 1 ft.^{3}/min. Slightly more than 90 per cent of the available uranium is recovered in association with 1·2 per cent of undesirable material.

Flotation of Coal

The industrial application of froth flotation in the coal industry as a means of cleaning fine coals is rapidly increasing, although the percentage of coal preparation plants in Great Britain which have froth flotation units installed is quite small. The main use of coal flotation is in the upgrading of slurries, thus converting an almost waste material into a useful and saleable fuel. A slurry in coal preparation is usually defined as the fine coal settling in washery water and water-borne either to a fines treatment plant or to settling ponds. It is usually less than $\frac{1}{2}$ mm in size and in its wet condition generally has a calorific value of between 7,000 and 9,000 B.Th.U./lb. The loss of thermal value can be quite considerable, therefore, unless steps are taken to reclaim it.

The recovery of slurry has always been one of the main difficulties accompanying the wet cleaning of coal. Since the extensive use of underground mechanization in coal mining the problem has been further aggravated by an increase in the proportion of fines in the run-of-mine coal and also in its dirt content.

Until the introduction of flotation in the coal industry, the only method available for removing these fines from the circulating water in washeries was to run off the effluents to waste, either continuously or intermittently. The accumulation of this material in dumps or settling ponds often created a nuisance and caused pollution of rivers unless suitable measures were taken to trap the effluents in settling ponds and lagoons.

In view of the increasing ash content of slurries, and the need for maximum coal recovery, flotation methods are now being widely adopted.

The flotation of fines and slurries has been used on the Continent for many years before its introduction to Great Britain. The pioneers of this branch of coal treatment were the Dutch State Mines, who have operated the largest coal preparation plants in the world.

As far as coal slurry is concerned, the flotation process may be regarded as an additional step in the process of clarification of washery water by flocculation. The slurry, which was recovered in a raw state by flocculation methods, is separated into its clean coal and dirt constituents, the latter being still handled by the flocculation process. *Figure 16* shows a typical

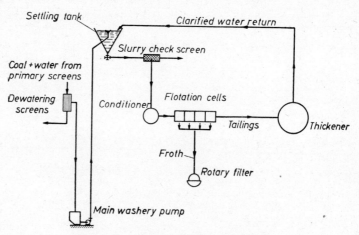

Figure 16. Typical coal flotation plant circuit

flow diagram of a combined froth flotation and flocculation plant. Settled slurry from the bottom of the settling cone is passed over a screen for rejection of any oversized material from the flotation plant. The effluent from the screen is then passed into the conditioner where suitable reagents are added from one of the feeders, the other feeders being available for the introduction of reagent into individual cells. The conditioned raw slurry is piped into the flotation cells where the clean coal is separated from the bulk of the high ash material with which it is contaminated.

The slurry passes from the first cell into the second cell and so on, until it has reached the last cell in the unit, where it emerges as an effluent containing the bulk of the high ash material. Variations in the arrangements of the flow of slurry through the plant can be arranged, depending upon the nature of the raw slurry and the quality of cleaned product required. Recleaning is often necessary where very high ash material is met with in the raw slurry.

The concentration of the material emerging from the cells varies from cell to cell, the highest concentration being usually found in the first cell and the lowest in the last cell of the unit. This is illustrated in *Table 6* which shows the concentrations of typical cell froths and it will be noted from this table that the ash content of the clean coal concentrate extracted from the cells increases from the first to the last cell of the series.

The froth is dewatered by a vacuum filter, the cake being discharged by means of an air blow or string discharge—the latter method is particularly suitable in the case of cakes which are difficult to filter. The clean filter cake is mixed back with the washed products from the main coal preparation plant, usually by blending the two products intimately in a paddle mixer. Depending upon the proportion of minus 200 mesh material in the filter cake, the moisture content of the froth can vary from 20 per cent to about 40 per cent.

Table 6. *Concentration of Cell Froths.* (*From* J. L. LEWIS, *by courtesy of* The Institution of Mining Engineers)

Mesh	No. 1 Cell		No. 2 Cell		No. 3 Cell		No. 4 Cell		No. 5 Cell	
	% Wt.	% Ash	% Wt.	% Ash	% Wt.	% Ash	% Wt.	% Ash	% Wt.	% Ash
Plus 20 B.S.	13.0	1.6	25.4	2.6	18.7	3.4	12.2	3.0	7.5	3.3
20–40	15.7	2.4	14.6	4.0	14.6	6.8	11.1	7.0	7.8	8.2
40–60	14.7	3.7	14.0	6.6	12.4	11.5	11.1	11.6	9.6	15.3
60–100	10.8	4.4	7.4	8.0	8.2	11.9	6.3	13.0	6.5	19.6
100–200	11.0	4.0	9.8	6.7	10.8	8.5	12.2	9.3	12.3	14.0
Minus 200	34.8	8.7	28.8	9.0	35.3	9.2	47.1	10.7	56.3	12.7
% Average ash	—	5.07	—	6.0	—	8.2	—	9.4	—	12.5
% Solids concentration in froth	39.2		31.6		29.8		26.3		22.1	

Because of their high clay content the tailings cannot usually be filtered and the normal method of disposal is by pumping to the colliery spoil heap at a concentration of 5–15 per cent solids. If they are flocculated and thickened before disposal, they are usually pumped to settling ponds or to the spoil heap at a concentration of 30 per cent solids. *Table 7* shows, by way of illustration, the performance of a froth flotation plant working at a British colliery.

Table 7. *Performance of a Typical Froth Flotation Plant*

Grading	Raw feed		Filter cake		Tailings	
	% Wt.	% Ash	% Wt.	% Ash	% Wt.	% Ash
Plus 16 B.S.	0·00	—	—	—	—	—
16–30	10·38	5·29	16·72	2·90	9·82	6·87
30–60	18·87	8·64	26·59	6·11	7·23	21·30
60–120	16·27	18·25	17·11	13·89	4·65	64·83
120–240	9·67	22·78	9·22	13·28	5·43	73·83
Minus 240	44·81	33·33	30·36	13·57	72·87	78·35
Totals	100·00	22·29	100·00	9·8	100·00	68·42

Moisture of filter cake 26·8 per cent.

If the coal contains a very high clay content, particularly in material below 200 mesh, it is often necessary to reclean the froth to obtain a reasonable yield with low ash. An example of the results obtained by recleaning is given below.

Single Separation
    Clean coal   74·0 per cent yield with 15·2 per cent ash
    Tailings   26·0 per cent yield with 81·6 per cent ash

Double Separation after Recleaning
    Clean coal   69·7 per cent at 9·8 per cent ash
    First tailings   25·1 per cent at 83·3 per cent ash
    Second tailings 5·2 per cent at 72 per cent ash

Slurries which require recleaning, usually have a very high ash content in the minus 200 mesh fraction of the raw slurry, in the region of 40–50 per cent. To achieve the best flotation results, it is better to feed the plant with this type of slurry at a concentration of about 8 per cent solids.

Mention has previously been made of the difficulty of cleaning low rank coals by froth flotation. When this is attempted, using conventional oil mixtures, reagent consumption is often excessive and the process uneconomical. No appreciable improvement is gained by altering the pH and temperature of the pulp or by increasing the conditioning time. However, it has been shown that a notable reduction in reagent consumption and cost can be achieved by conditioning in two stages. In the first stage, an auxiliary oil such as a medium fuel oil is used to impart to the coal the required degree of hydrophobicity and this is followed by second stage conditioning using a modest quantity of expensive frother such as cresylic acid. This method cuts the cost of frothing these low rank coals considerably, although the cost is still higher than that realized with flotation of normal rank coals[11].

Among the special applications of froth flotation applied to coal is the production of superclean coal. Most of this is used, in the form of coke, in the chemical industry for the manufacture of aluminium and calcium carbide, although quantities of it are used in the making of electrode carbon. The coke required for this purpose should have an ash content of less than 1·5

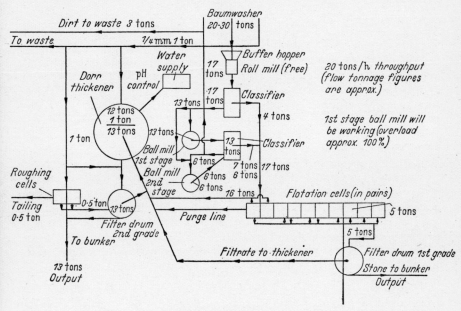

Figure 17. *Flow-sheet of flotation plant producing superclean coal.* (From E. HINDMARCH, and P. L. WATERS, *by courtesy of* The Institution of Mining Engineers)

per cent, a volatile matter of less than 0·2 per cent, and a moisture content of not more than 2·0 per cent. In the preparation of aluminium, much of the ash remaining in the coke enters into the metal and thus the quantity and quality of the ash is particularly critical. The raw coal should not, therefore, have an ash content exceeding 1·0 per cent.

Coals from which such a low ash product can be prepared are not readily available and only a few seams in Great Britain are suitable for this purpose. It is necessary to crush these down to very fine limits to liberate the ash-free material and to separate this by froth flotation. All the normal factors which affect froth flotation, such as pH value, quantity of dissolved solids in the water, pulp density and temperature, have to be carefully controlled.

A typical design of a plant producing superclean coal is illustrated by the flow diagram *Figure 17*. This installation has a throughput of superclean coal of 20–30 tons/h and the stages are briefly as follows:

(a) A small Baum washer regulates the quality of the raw coal feed to ensure consistent operating conditions of the fines plant.
(b) The necessary size grade is obtained by the use of two-stage ball-mill and

classifying screens which further ensures the minimum production of fines.

(c) The superclean coal is obtained by flotation. The coal not floated passes as tailings to a thickener and vacuum filter, ultimately to be used for the production of foundry coke in coke ovens. Only sufficient reagent to float the superclean coal is required since the tailings require no further cleaning for the production of foundry coke.

SELECTIVE FLOTATION

Selective flotation takes advantage of the different floatabilities of minerals and is concerned with floating one mineral ahead of or in preference to another and, if desired, subsequently floating a second mineral. In these cases the differences in floatability are much finer than for ordinary collective flotation and the choice of suitable reagents is therefore more precise. Further, the whole operation is more delicate and the results not so definite. Selective flotation can be applied to an untreated ore pulp or to a concentrate from a previous collective flotation operation. If the result is achieved by the careful regulation of the reagents only, without any prior treatment of the ore particles, the operation is said to be normal selective flotation. Sometimes, however, advantage has to be taken of other properties of the mineral constituents, *e.g.* their tendency for oxidation. In such a case it may be possible to promote sinking of the more floatable mineral and floating of the less floatable one, although such irregular flotation, which often makes use of costly procedures, is only normally applied to collective concentrates.

There are so many applications of selective flotation that it is only possible to illustrate a few of the principles involved in this section. Each case usually presents its own particular problems and its treatment is therefore unique. For a full description of some typical applications, consultation of some of the references at the end of this section should be made.

In the separation of galena from blende, the galena particles are sometimes finer than the blende particles, probably because the greater part of the coarse galena particles may have been removed in a water concentration process. For this reason, and because galena possesses a greater natural floatability than blende, the conditions are such that it is comparatively easy to set the flotation plant to float the galena particles and sink the blende particles. The conditions for flotation in this case include the use of a minimum of frothing agent, using a neutral circuit and a careful control of the air supply, if this is possible. These conditions produce a weak yet voluminous froth out of which the blende particles, initially entrained, readily fall. In addition, the froth area at the overflow is constricted and thus enables the galena to crowd out any competing blende. With the galena removed, more intense flotation conditions, *i.e.* more oil and greater violence of agitation, raise the blende.

Galena has also been separated from blende by manipulating the froth overflow from pneumatic flotation machines. With the Callow machine the froth is so frail and its depth so great that from a mixed galena–blende froth building up at the bottom, only the galena will arrive at the top. A skimmer is then arranged to take off the top two or three inches and therefore most of the galena. The galena that is not removed will fall back and eventually be recovered in a separate cell.

If the two minerals have approximately the same degree of fineness and selective flotation is not assisted by a difference in size, some other means is necessary to promote the separate sinking of the less floatable mineral. This can be accomplished by the action of sulphur dioxide gas on the blende. By the suitable control of this soluble gas, the blende can be sunk and the galena still floated. The mechanism of the process is by no means certain, but it is supposed that the sulphur dioxide being both a reducing agent and soluble, diffuses into and displaces the air film which contaminates the sulphide particles. This temporarily destroys, in the case of the blende particles, their affinity to attach themselves to air bubbles.

The surface of the more floatable mineral may also be modified to float the less floatable mineral by fractional roasting. This process makes use of the fact that some sulphides are more readily oxidized by roasting than others. A particular case is that of galena which is more readily oxidized than blende, and is applicable where the lead–zinc concentrate is so fine that separation of the two sulphides by water is ineffective. Roasting converts the galena to sulphate, whilst the blende remains practically unaltered. During flotation of the two minerals, the blende is floated and the oxidized galena sinks, although a small fraction of the blende is oxidized and goes into solution. All the methods involving fractional roasting require careful control of the roasting and cooling temperatures.

A further method of selective flotation makes use of the fact that various chemical solutions can react with one or other of the sulphides to be separated, thereby rendering them incapable of flotation. For instance, if ferric chloride were added in solution to a feed of galena and blende under conditions which left the blende unaffected, the galena would be covered with a thin film of lead chloride and would not be floated with the blende. A further process utilizes an aerated solution of ammonia which promotes the oxidation of metallic sulphides. The oxides of zinc, copper and lead are soluble in most ammonia salts and therefore the floatability of their sulphides remains unimpaired. Pyrite and pyrrhotite, however, are rendered unfloatable by a film of insoluble iron oxide.

Selective Flotation Applied to Coal

It is possible to utilize flotation as a means of separating the petrographic constituents of coal. If kerosene is used as a reagent, it is possible to produce a froth containing 76 per cent of vitrain and clarain, and 24 per cent of durain. With phenol, a separation of bright coal from dull coal is possible. Fusain, which is the least valuable carbonaceous constituent of coal, can be selectively depressed by starch when cresol and petroleum oils are used as flotation agents.

Starch, glue, tannin and albumen have been used to depress dull coal, while it has been stated that starch or dextrin if pretreated with hydrochloric acid would depress all the coal constituents except fusain.

In the Ekof method[12], a fusain concentrate can be made by depressing all the other petrographic constituents with polyhexoses treated with hydrochloric acid or suitable chlorides. Then by making the pulp slightly alkaline and using petroleum oil as a collecting agent, the remaining oil can be floated away from the refuse.

Another method of selectively floating fusain which has been developed by BIERBRAUER and POPPERLE[13] depends upon the differential surface oxidation of the petrographic constituents. The humates present in the clarain, vitrain and durain are oxidized to humic acid by the addition to the coal pulp of a small amount of oxidizing agent such as nitric acid, hydrogen peroxide or potassium permanganate. The humic acid forms a surface layer which is hydrophilic and therefore the clarain, vitrain and durain constituents are depressed. Fusain, because it has a lower content of humates is not affected by the oxidizing agents to any large extent and can be readily floated with the acid of a frothing agent. After removal of the fusain, the remainder of the coal can be floated away from the refuse by the addition of a collector powerful enough to overcome the hydrophilic property caused by the oxidation.

The success of this method depends upon the careful choice of reagents, since the humate content varies with the rank and type of coal.

The Flotation and Depression of Pyrite in the Presence of Coal

During experiments performed by YANCEY and TAYLOR[14], it was found that during flotation, the pyrite content of the froth increased as the pyrite content of the coal increased. A higher proportion of pyrite was also floated than in the absence of coal particles and it was concluded from this that the floatability of pyrite alone is not the only consideration, but at high concentrations mechanical entrainment takes place in the large number of coal particles entering the froth. To reduce this entrainment, it is necessary to reduce the pulp density of the feed to the flotation cells and also to reclean the coal froth.

If pyrite is not depressed it tends to concentrate in the froth; for this reason a pyrite depressant should be used in coal flotation and reagents that act as collectors for pyrite should be avoided. In practical cases the use of lime is sufficient to depress pyrite. Its advantages are that it is cheap, maintains the circuit in an alkaline condition, and assists in the settling of solids in thickeners and settling tanks. It has been shown, however, that the depressing effect of pyrites by the use of lime increases as the pH of the pulp increases, but at high pH values the coal is also depressed.

If it is desired to recover pyrites from coal as a practical commercial proposition, this can be accomplished by employing the usual collecting agents for sulphides as in normal ore dressing flotation procedure.

REFERENCES

[1] TAGGART, A. F., TAYLOR, T. C., and INCE, C. R., *Trans. Amer. min. (metall.) Engrs*, 87 (1930) 285

2 DEL GUIDICE, G. R. M., *Engng Min. J.*, 137 (1936) 291
3 WARK, I. W., and COX, A. B., *Engng Min. J.*, 137 (1936) 641
4 RALSTON, O., *U.S. Bur. Min., Inform. Circ.*, 3397 (1938)
5 HORSLEY, R. M., and SMITH, H. G., *Fuel, Lond.*, 30 (1951) 54
6 GAUDIN, A., and VINCENT, K., *Amer. Inst. Min. Engrs T.P.*, 1242 Nov. (1940)
7 'Improvements in or Relating to Froth Flotation Apparatus', *Pat. No.* 695107, Aug. 1953
8 *U.S. Pat.* Nos. 1,346,817/1920; 1,346,818/1920 and 1,401,598/1921
9 'An Improvement in Pneumatic Flotation', *Engng Min. J.*, 122, 874 (1926)
10 *U.S. Pat.* No. 2,647,629, Aug. 4, 1953
11 ALLUM, WHELAN, 'Froth Flotation of Low Rank Coals', *J. Inst. Fuel*, vol. XXVII, No. 158
12 SCHAEFER, W., and MERTENS, W., *U.S. Pat.* No. 1,944,529, Jan. 23, 1924
13 BIERBRAUER, E., and POPPERLE, J., *Brit. Pat.* No. 450,044, July 9, 1936
14 YANCEY, H., and TAYLOR, J., *U.S. Bur. Min.*, 3263 (1935)

BIBLIOGRAPHY

EVANS, L. F., and EWERS, W. E., 'Bubble-mineral attachment', *Instn Min. Metall. Symp.* on 'Recent developments in mineral dressing'

BANERJI, B. K., 'The theory of contact angles', *Min. Mag., Lond.*, Jan. 1953, 23–5

PHILIPOFF, W., and COOKE, S. R. B., 'Contact angles and surface coverage', *Trans. Amer. Inst. min. (metall.) Engrs Tech. Publ.* No. 3266B (1952) 283–6

MORRIS, T. H., 'Measurement of equilibrium forces between an air bubble and an attached solid in water', *Trans. Amer. Inst. min. (metall.) Engrs*, vol. 187

MOMOSAKI JUNJIRO, 'Attachment of minerals to bubbles in flotation', *J. Min. Inst., Japan*, 50, Oct. 1954, 467

WROBEL, S. A., 'Power and stability of flotation frothers', *Mine & Quarry Engng*, Aug. 1953, 275–80; Sept. 1953, 314–19; Oct. 1953, 363–7

WROBEL, S. A., 'Selectivity of frothers', *Mine & Quarry Engng*, 20, June 1954, 267–70

BOOTH, R. B., and DODSON, J. E., 'Frothing agents for the flotation of ores and coal', *U.S. Pat.* No. 2,695,101 to American Cyanamid Company

LEJA, J., and SCHULMAN, J. H., 'Molecular interactions between frothers and collectors at solid-liquid-air interfaces', *Min. Engng, N.Y.*, vol. 6, No. 2,221–8

GAUDIN, A. M., and COLE, R. E., 'Double-bond reactivity of oleic acid during flotation', *Amer. Inst. min. (metall.)*, vol. 196

BAARSON, R. C., and PARKS, J. R., 'The application of fatty chemicals to flotation', *J. Amer. Oil Chem. Soc.*, vol. XXXI, No. 6, June 1954, 261–6

BROWN, D. J., and SMITH, H. G., 'Continuous testing of frothers', *Colliery Engng*, vol. XXXI, June 1954, 245–50

CARPENTER, J. E., 'Condensation products of fatty acids', *U.S. Pat.* No. 2,668,165 to American Cyanamid Company

'Flotation of sulphide minerals', *Brit. Pat.* No. 702,957 American Cyanamid Co.

FISSHER, A. H., 'Xanthogen compounds', *U.S. Pat.* No. 2,608,573 to Mineral Corp. *Chem. Abstr.*, American Chemical Society, Washington, vol. 47, 5961h

MORRIS, T. M., 'Rate of flotation as a function of particle size', *Min. Engng*, vol. 4, Aug. 1952

'British unsaturated acids for flotation', Ore Dressing Notes. *Min. Mag., Lond.*, Oct. 1950, 223

Barium

HALL, C. L., and WHELAN, P. F., 'Laboratory tests on the concentration of witherite from the Northern Pennines by froth flotation', *Trans. Instn Min. Metall. Lond.*, vol. 62, pt. 6

Cassiterite

EDWARDS, G. R., and EWERS, W. E., 'Adsorption of sodium cetyl sulphate on cassiterite' (flotation of cassiterite), *Aust. J. sci. Res.*, A.4 (1951) 627–43; *Chem. Abstr.*, American Chemical Society, Washington, vol. 46, 7973

THOMPSON, A. G., 'Experiences with the flotation of cassiterite', *Min. J.*, vol. 234, June 16, 1950

KLASSEN, V. I., 'Cassiterite flotation', *U.S.S.R. Pat.* No. 69,978, Dec. 31, 1947; *Chem. Abstr.*, American Chemical Society, Washington, vol. 44, 105c

NEDOGOVOROV, D. I., 'Method for separation of scheelite and cassiterite by flotation', *Chem. Abstr.*, American Chemical Society, Washington, 43 (1949) 2139

Chromite Ores

WEINIG, A. J., 'Selective flotation of chromite ores', *U.S. Pat.* No. 2,469,422, May 10, 1949; *Chem. Abstr.*, American Chemical Society, Washington, 43 (1949) 499e

Coal

GISLER, H. J., 'Froth flotation recovers marketable coal from washery rejects and middlings', *Deco Bull.* No. F.10–B.37, Sept. 1949

NELSON, H., 'The selective separation of super low ash coal by flotation', *Deco Trefoil*, June 1950

JONES, W. I., JONES, D. C. R., and GREGORY, D. H., 'Concentration or cleaning of minerals by froth flotation', *Brit. Pat.* No. 655,905, Aug. 8, 1951; *Chem. Abstr.*, American Chemical Society, Washington, vol. 45, 8823c

SCHAFER, W., 'Preparation of fine coal up to 10 mm size in flotation apparatus', *Germ. Pat.* No. 800,387, Nov. 2, 1950; *Chem. Abstr.*, American Chemical Society, Washington, vol. 45, 358

HORSLEY, R. M., and SMITH, H. G., 'Principles of coal flotation', *Fuel, Lond.*, 30 (1951) 54–63; *Chem. Abstr.*, American Chemical Society, Washington, vol. 45, 3577

HINDMARCH, E., and WATERS, P. L. 'Froth flotation of coal', *Trans. Inst. Min. Engrs, Lond.*, 111 (1951–2) 221–34

HORSLEY, R. M., 'Oily collectors in coal flotation', *Colliery Guard.*, 184, April 24, 1952, 509–12

CRAWFORD, A., 'Preparation of ultra-clean coal in Germany', *Trans. Inst. Min. Engrs, Lond.*, 111 (1951–2) 204–18

HORSLEY, R. M., and SMITH, H. G., 'Xanthates in coal flotation', *Fuel, Lond.*, 31, July 1952, 302–11

BAILEY, R., WHELAN, P. F., 'Influence of pulp temperature of four British fine coals', *J. Inst. Fuel*, 25 (1953) 304–7; *Chem. Abstr.*, American Chemical Society, Washington, vol. 47, 5094h

GREGORY, D. H., SIMPSON, D., and WHELAN, P. F., 'Low-cost froth flotation', *Coal Age*, 58, Nov. 1953, 94–5

RILEY, H. L., and GANDRUD, B. W., 'Vacuum flotation for coal fines', *U.S. Bur. Min. Bull.* R.I. 5071

Lead-Zinc

FLEMING, M. G., 'Effects of soluble sulphide in the flotation of secondary lead minerals', *Inst. Min. Metall. Symp.* on Mineral Dressing, Sept. 23–5, 1952

HITOSI JAGIHAR, 'Mono and multilayer adsorption of aqueous xanthate on galena surfaces', *J. phys. Chem.*, 56, May 1952, 616–21

BLASKETT, K. S., 'Lead-zinc ores—continual challenge to flotation', *Chem. Abstr.*, American Chemical Society, Washington, vol. 45, 4616d

HENDRICKSON, T. A., STICKNEY, W. A., and WELLS, R. R., 'Selective flotation concentration of lead-zinc from the Musick and Helena Mines, Bohemia District, Oregon', *Chem. Abstr.*, American Chemical Society, Washington, vol. 45, 6549e

IWAYA, T., and KASHIWAYA, K., 'Studies on the differential flotation of lead-zinc ore', *J. Min. Inst. Japan*, Oct. 1943, July 1944

Uranium

GINOCCHIO, A., 'Flotation of phosphated uranium minerals in fatty acids without desliming', *Chem. Abstr.*, American Chemical Society, Washington, vol. 48, 8709b

Pyrrhotite

CHANG, C. S., COOKE, S. R. B., and IWASAKI, I., 'Pyrrhotite flotation characteristics with zanthates', *Min. Engng, N.Y.*, 6, No. 2, 209–17

WICHMANN, A. P., and BHAPPU, R. B., 'Effect of fine particle sizes on sulphide flotation', *Colo. Sch. Min. Quart.*, 50, No. 2, April 1955

Iron Ores

KECK and JASBERG, 'A study of the flotative properties of magnetite', *Amer. Inst. min. (metall.) Engrs, Tech. Publ.* 801 (1937)

Non-metallics

TAGGART, A. F., 'Flotation—application to non-metallics', *J. Min. Engng*, 137 (1936) 90–1

Manganese

DE VANEY and CLEMMER, 'Floating of carbonate and oxide manganese ores', *J. Min. Engng*, 128 (1929) 506–8

Gypsum

KECK and JASBERG, 'A study of the flotative properties of gypsum', *Amer. Inst. min. (metall.) Engrs, Tech. Publ.* 762 (1927)

9

SEDIMENTATION

R. FORBES STEWART

THE CLARIFICATION AND THICKENING PHASES

To the engineer, sedimentation generally means gravity settling or subsidence of solids suspended in liquids. Corollary to sedimentation are the terms thickening and clarification.

In thickening, the emphasis is on the solids as the end product of the separation–the compacting of the suspended solids into a dense slurry or sludge in order to facilitate subsequent processing or disposal of the solids.

The term clarification, on the other hand, implies a primary interest in the character or quality of the liquid as the product of a gravity separation in a highly dilute suspension, as in the treatment of a water supply carrying a few hundred parts per million of suspended matter.

HISTORICAL

Sedimentation, the act or process of depositing sediment, has been used by man for generations to clarify liquids and concentrate solids. Conceivably, the first application of the process might have occurred when early man discovered that muddy water, when allowed to settle, became clear enough to drink. However, the use of sedimentation as a unit operation was first recorded by AGRICOLA[26] in his classic book, *De Re Metallica* published in 1556. His woodcut diagrams definitely show crude progenitors of modern sedimentation equipment.

By the end of the nineteenth century the process of separating solids and liquids by sedimentation had been developed as an art to the extent that non-mechanical settling tanks and cones were being used in metallurgical work. Operation of these units consisted merely of a fill-and-draw procedure utilizing the swing syphon, as illustrated in *Figure 1*. Development of various discharge valves for cones later provided near-continuous operation of these non-mechanical settlers. However, due to intermittent building up and sloughing off of settled solids on the cone sides, steady discharge of a desired density was unobtainable. The proportional increase of height with diameter made large cones prohibitive in cost and size, while the operation of a battery of small cones in parallel, with the attendant problem of synchronization, proved troublesome.

Thus, it was not until the invention, by J. V. N. Dorr, of the *classifier* in 1904 and the *thickener* in 1906 that the process of sedimentation was successfully operated on a continuous basis (see *Figure 2*). The use of slowly moving angled plough blades on the thickener arms introduced a positive mechanical means for discharging settled solids and permitted the use of flat-bottomed tanks, thus overcoming both the plugging and constructional limitations of the cone.

From that beginning, progress in the art of thickening and clarification has been measured primarily, (1) by increased knowledge and understanding growing from the constant search for ways of promoting and accelerating the work of gravity; (2) by a continuing improvement of the basic designs and construction of mechanisms; and (3) by development of special designs and types of machines to keep pace with the expanding scope of applications.

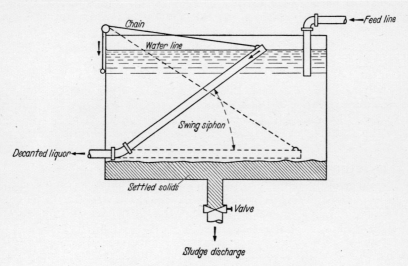

Figure 1. Batch settling tank with swing syphon

The widespread applications of thickening and clarification are generally well known. Neither gives an absolute separation of solids from liquids or of liquids from solids. In many instances, clarification alone affords adequate removal of the solids for all practical purposes. In others, where absolute clarity of the liquid is required, the last traces of suspended matter are removed by filtration through appropriate media in gravity or pressure type filters. This step is sometimes called 'polishing'.

SCOPE OF SECTION

The process of sedimentation may be defined as the movement of solid particles through a fluid due to an imposed force, which may be gravitational, centrifugal, or some other force. In this presentation only those processes will be considered in which (a) the fluid is a liquid and (b) the force is that of gravity. The process is applicable only to systems which separate into two products: substantially clear liquid and thickened solids. Industrial applications will be taken from the chemical, metallurgical, and water and sewage treatment fields, which provide examples of practically all wet processing of finely divided solids suspended in liquids.

Before any practical use can be made of the sedimentation process in solid–liquid separations, two very important preliminary steps must be considered: (1) destabilization of colloidal suspensions, and (2) aggregation of the separate particles into clusters or flocs. Both steps are discussed under the heading 'Flocculation' before taking up the study of the sedimentation process.

As shown in *Figure 2*, the subject of sedimentation covers a very broad field. In general, the mixtures of solids and liquids, hereafter referred to as 'pulps', can be classified as either dispersed or flocculent, depending on whether the particles are discrete or aggregated. The in-between pulps, in most cases, can be treated so that they fall into one or the other category. Practical applications of sedimentation are divided into two distinct groups: the first includes the operation of classification, which is best performed on dispersed pulps; the second, clarification and thickening, best performed on flocculent pulps. The general types of sedimentation equipment commercially available to perform these operations are listed under appropriate sub-headings in *Figure 2*. This section is concerned only with the clarification and thickening phases of sedimentation and the equipment commercially available for performing these operations.

Use of Terms

The usage of the terms 'clarification' and 'thickening' may at first cause some confusion. Whether a given sedimentation process is called one or the other generally depends on whether the liquid or the solid is the desired product. Thus, in an operation involving a dilute pulp (*e.g.* 1–5 per cent solids by weight) where a clear overflow is desired, the term clarification is used and the sedimentation unit is called a 'clarifier'. In an operation involving a more concentrated pulp (*e.g.* 15–30 per cent solids) where the main object of the treatment is to remove as much liquid as possible from the settled solids, the term thickening is applied and the unit is called a 'thickener'. It should be understood, however, that the solids in a dilute pulp, after settling, go through all the thickening phases, and, conversely, the liquid in a concentrated pulp, after separating from the solids, may be said to be clarified. Listed below are a few examples of typical thickening and clarification operations:

Clarification (liquid desired product)
 1. The removal of mud from hot limed and carbonated sugar solution before evaporation and crystallization.
 2. The clarification of water for municipal and industrial use.
 3. The clarification of brines used for the electrolytic production of caustic soda.

Thickening (solids desired product)
 1. Dewatering of cement slurry before being fed to the kiln.
 2. Thickening of flotation concentrates ahead of filtration.
 3. Dewatering of fine, classified lithopone.

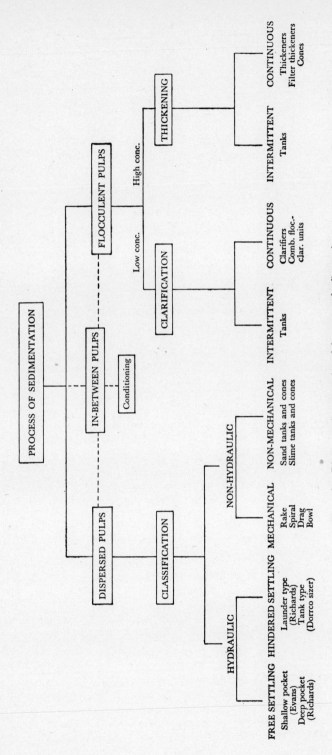

Figure 2. Diagrammatic presentation of the process of sedimentation

Limitations of Clarification and Thickening Processes

In chemical and metallurgical practices, suspensions of solids coarser than $74\,\mu$ (200 mesh) are called sands, while finer fractions are called slimes. Most pulps encountered in these fields consist mainly of slimes with some sands and relatively little colloidal matter ($0 \cdot 1 – 0 \cdot 001\,\mu$ mean diameter).

In clarification and thickening operations, due to certain pulp characteristics, some very definite limitations must be considered. If thickened coarse solids are to be removed from the sedimentation equipment they must contain sufficient fine particles or lubricating slimes to permit pumping. Coarse particles alone pack to such an extent that discharge cones and underflow lines become plugged. Thus, the final practicable density of a thickener underflow is limited to about 60 per cent solids by weight (based on a $2 \cdot 7$ specific gravity solid in water). On the other hand, some pulps, such as magnesium hydroxide, contain solids which are of such a light and fluffy nature that the final pulp density reached, even after long settling, is only a few per cent solids by weight. With such pulps, if further concentration is required, filtration equipment must be used.

Clarification has its limitations, too. The clarifier cannot make a 100 per cent removal of solids in a reasonable length of time, although some operations produce overflows containing as few as three or four parts per million (p.p.m.). As discussed later, mechanical flocculation and flocculating agents are often employed to aid in obtaining a clearer effluent. Where a perfectly clear overflow is required, further treatment, such as filtration or centrifuging, must be used. Thus, every thickening or clarification operation must be considered individually and extreme care be taken in applying data obtained on one particular sample under a given set of conditions to another sample taken under seemingly similar conditions. While some progress has been made in attempting to establish thickening and clarification as exact engineering sciences, at present there is still considerable art involved in determining many factors. Subsequent discussion will point out some of the pitfalls in the path of the inexperienced design engineer.

Since the objective of all clarification–thickening problems is to obtain the desired separation with the smallest possible settling area, any method of increasing the effective particle size of the suspended solids will produce a corresponding reduction in the required settling area.

FLOCCULATION

Importance

Flocculation is a term used to designate the combined operations of destabilization and aggregation. As previously mentioned, these important preliminary operations must be performed before the separating action of the sedimentation process can be effectively utilized. Some pulps are naturally in a separable condition and, for them, flocculation is of little importance. Others require special treatment to destabilize the suspension and induce aggregation so that the separation can be obtained. Of these pulps, some need only gentle agitation to promote flocculation, while others require treatment with special chemicals before agitation. The subsequent discussion is concerned mainly with the latter group.

The importance of the flocculation step lies in the fact that it promotes the coalescence of separate particles and small flocs into clusters, with sufficient diameter to settle at an economic rate. Later it is shown that the rate of subsidence of the slowest settling particles in a solid–liquid suspension is an important controlling factor in determining the size of a sedimentation unit. Thus, if the settling rate of the finest particles can be increased, the required area can be correspondingly decreased. Control of the settling rates of these very fine solids is, therefore, of prime importance.

Treatment of Colloidal Material

The coarse particles in a suspension are controlled mainly by gravity, but as the particles become finer the effect of gravity is reduced and is eventually overbalanced by the forces of surface energy and Brownian movement[10]. Finally, a colloidal state is reached in which the dispersed particles (in the range of $0\cdot1$–$0\cdot001\,\mu$ in mean diameter) remain in permanent suspension in the liquid. Thus, any substance can be brought into the colloidal state if reduced to a sufficiently fine state of subdivision that the Brownian movement will keep the particle suspended, provided the proper conditions are present to prevent aggregation and to maintain stability.

Colloidal suspensions may be divided into two groups, based on their sensitivity to the action of electrolytes. Since most separation processes deal with suspensions of solids in water, the term 'hydrophobic' is used to designate non-hydrated colloids which are extremely sensitive to electrolytes, and 'hydrophilic' for hydrated colloids not affected by electrolytes[10].

According to theory, the stability of the hydrophobic colloid is brought about by the preferential adsorption of charge-carrying ions on the surface of the dispersed particle. If the particles are charged with like sign they will repel each other and remain dispersed throughout the liquid medium. In order to destabilize the suspension it is necessary to neutralize the charge on the particle by adding a charge of opposite sign, usually in the form of an electrolyte. If the charged particles become completely neutralized (the isoelectric point), the probability of adhesion, if they collide, is greatest[10]. Since the effectiveness of the added electrolyte increases greatly with the increase in its valence, polyvalent electrolytes are used wherever possible. The flocculation of dilute suspensions is best carried out by the formation *in situ* of a gel-like substance (*e.g.* aluminium hydroxide formed by the addition of aluminium sulphate to the pulp) which envelops the particles and holds them in clusters. This method is commonly used in the clarification of water, sewage, and trade wastes, and is known as chemical dosing.

In the treatment of hydrophilic colloids stability is a function of both electric charge and hydration. The most effective flocculating electrolytes for such colloids are those which dehydrate the suspended material and also help to adjust the charge to the isoelectric point. After dehydration, a hydrophilic substance will react to treatment as a hydrophobic colloid.

After the charge on the particles has been adjusted to a point which would permit cohesion, the rate of aggregation becomes a function of the probability of collision. Without an outside energy source this probability depends on the concentration of the colloid and the temperature, the latter being a factor because of its effect on the Brownian motion of the particles. In general, for the dilute suspensions encountered in commercial applications, the probability of collision due to Brownian movement and sedimentation is not sufficient for an economical treatment rate. Thus, mechanical means of increasing the probability of collision must be used. For more detailed information on this subject see Reference 10. Various mechanical devices to accomplish this are discussed later under the heading 'Combined Flocculation–Sedimentation Units'.

Having first considered pre-conditioning of solid–liquid suspensions to effect destabilization and aggregation, we are now prepared to discuss at length the separation process itself.

THEORY OF SEDIMENTATION

Basic Considerations

The general principles of settlement of solids are discussed under 'Laws of Settling' in this volume, pp. 162–171.

The equation for uniform slow motion of a rigid sphere in a viscous fluid of infinite extent is presented below in the form due to Stokes[23] and used since 1850.

$$u_m = \frac{a^2 g (\varrho_0 - \varrho)}{18\eta} \quad \ldots \ (1)$$

where u_m = maximum velocity of the sphere
g = acceleration due to gravity
ϱ_0 = density of the solid
ϱ = density of the fluid
a = diameter of the sphere
η = viscosity of the fluid

The Newtonian form of equation, as applied to flow which is fully turbulent, is written:

$$u^2_m = \frac{4(\varrho_0 - \varrho)ga}{3\varrho C_D} \quad \ldots \ (2)$$

C_D is the drag coefficient, being a function of the Reynolds number, R_e.

Combining equations (1) and (2) results in

$$C_D = \frac{24 \cdot \eta}{a u_m \varrho} = \frac{24}{R_e} \quad \ldots \ (3)$$

For values of R_e up to about 2, laminar flow conditions prevail and the free-settling velocity of a given particle may be determined by trial and error from a logarithmic plot of C_D against R_e. To obtain a direct solution a separate plot of $C_D R_e$ against R_e, similar to that shown in *Figure 3*, may be drawn. The expression $C_D R_e^2$ is independent of u_m and can be calculated directly. From the corresponding value of R_e taken from the separate plot,

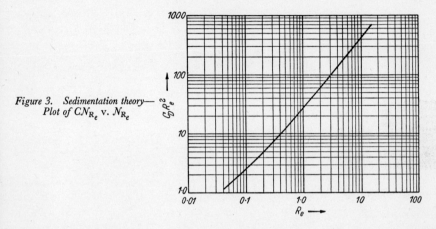

Figure 3. Sedimentation theory—Plot of $C.N_{R_e}$ v. N_{R_e}

u_m may be determined. In the clarification and thickening phases of sedimentation, laminar flow conditions are encountered in most cases since the fluid is generally water and the particles are usually in the range of 0·1–100 μ.

Table 1. Character of Subsidence of Various Types of Pulps*

Operation	Character of subsidence	Description	Example
Clarification	Independent	Particles or flocs settle independently. No definite line of subsidence. Settling rate dependent upon size and density of particle or floc.	Turbid water, sewage and many trade wastes
Intermediate, Class 2	Phase	Upper zone of independent subsidence. Lower zone of collective subsidence. Line of demarcation not sharp.	Chemical and metallurgical pulps
Concentrated, Class 3	Collective	Definite line of subsidence. Settling rate decreases with increasing concentration of solids. Settling rate retarded by particle or floc interference.	Chemical and metallurgical pulps
Thickening compact, Class 4	Compression	Flocs and particles in intimate contact. Subsidence due to pressure of particles or flocs on those below them.	All pulps eventually pass into this zone through sedimentation

* Based on a classification originally suggested by DEANE[9].

The results of several studies of the settling of spheres in various solutions indicate deviations from Stokes' law[23] ($C_D = 24/R_e$) begin around a Reynolds number of 2 and an intermediate region is encountered up to about 400, where the turbulent region begins. Studies of wall and end effects, non-spherical particles, effective viscosity, and density differences have been made and are reported in the literature, some of which work is listed in the references at the end of this section. In general, the relationships developed are useful in studying the character of fluid resistances and, consequently, the probable effects of changes in fluid conditions.

Pulp Characteristics—In addition to the variables expressed in Stokes' law, settling behaviour is also influenced by particle shape and pulp dilution. As shown in *Table 1*, the character of the subsidence from high to low dilutions will vary as the solid phase becomes more concentrated.

Since this section stresses practical rather than theoretical considerations, further pursuit of the subject is left to the student.

PRACTICAL APPROACH TO SEDIMENTATION

Although the work of COE and CLEVENGER[5] in establishing methods for determining the capacities of slime settling tanks was published as early as 1916, it is still recognized as the most important contribution to the understanding of the basic relationships of sedimentation. By carefully observing the process of batch sedimentation in glass cylinders, these investigators were able to study the fundamental principles involved and describe the various settling phases which took place. Correlating the data obtained from their small-scale batch settling tests with those from commercial-sized equipment enabled them to develop methods for determining the clarification capacity of continuously operated thickeners.

Much of the following descriptive material has been taken from their original paper on the subject.

Mechanics of Clarification

One of the most important concepts in the study of clarification problems is that of dilution ratio or the weight ratio of liquids to solids. With any given suspension of solids in liquids, the character of the settling or subsidence will vary as the dilution ratio is changed. For instance, if a very dilute metallurgical slime is thoroughly mixed in a glass cylinder and allowed to settle, the following observations can be made[21]:

Classification takes place, in which the coarsest particles settle at a comparatively rapid rate while the finest particles, settling at the slowest rate, remain on top, with graduation in size between the two limits. All particles are free to settle at a constant velocity expressed mathematically by Stokes' law, equation (1).

Clarification takes place gradually without a clear line of demarcation between the settling solids and the supernatant liquid. This is called 'independent particle subsidence' and is typical of sewage and trade wastes.

Colloidal material, if present and not destabilized, often remains suspended in the clarified zone even after long periods of settling.

If the pulp density is gradually increased by mixing in additional solids, the following will be noted:

1. A concentration is soon reached in which, after a short period, the fastest settling particles begin to form into a zone and descend collectively, but at a slower rate. Where the particles in the upper portion are settling independently and, in the lower portion, collectively, the settling phenomenon is called 'phase subsidence'.

2. As the pulp density increases, this collective zone forms at progressively earlier periods until a point is reached where the initial subsidence of solids is *en masse* with a sharp line of demarcation between the solids and the supernatant liquid. This is known as 'collective subsidence' or 'line settling' and is typical of nearly all concentrated chemical and metallurgical pulps. The rate of settlement of the line is called the settling rate of the pulp and is usually expressed in feet per hour. For most pulps this rate will be constant for a time and will then slow up considerably. Such pulps are designated Type I and produce settling curves similar to that shown in *Figure 4*.

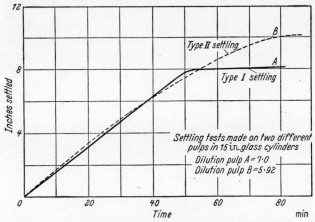

Figure 4. Settling rates of Types I and II pulp. (After COE and CLEVENGER[5])

3. At the point of concentration where the marked falling off in rate occurs, called the 'point of compression' or 'critical point' of the pulp, the settling phenomenon changes from collective to compact subsidence. All line-settling pulps eventually pass into this zone, the so-called compression zone, where further subsidence can take place only by compression of the settled material.

Exactly how a given pulp will behave when mixed and allowed to stand depends not only on the dilution ratio, but also on the factors of size distribution, and degree of flocculation. If the pulp contains particles smaller than 5 μ in diameter it is essential that it be in a flocculent condition in order to have these particles gather in clusters and settle in flocs. If coarse particles are present and the effective viscosity of the pulp is not sufficiently light to entrain them, the particles will immediately settle to the bottom of the cylinder.

The behaviour of pulps during settlement may be described in terms of four zones, as illustrated in *Figure 5*. In the first cylinder the pulp, thoroughly mixed, is at the feed concentration (zone B). As soon as the settlement begins (second cylinder), the other three zones begin to form: zone A, the substantially clear, supernatant liquid; zone C, a transitional phase between collective and compact settling; and zone D, the compression zone, the lower layer of which is composed of the coarsest particles, and the upper layer of settled flocs.

As noted in the next two cylinders, the pulp is able to maintain a constant rate of settlement, the reason being that the vertical column of settling solids descends as a whole and the lower end apparently folds up as it settles on the rising compression zones. No gradual increase in concentration along the length of the column occurs, only the rather sudden one at the BC interface. The extent of zone C depends on the suddenness of this change. Where quite abrupt, zone C is practically non-existent.

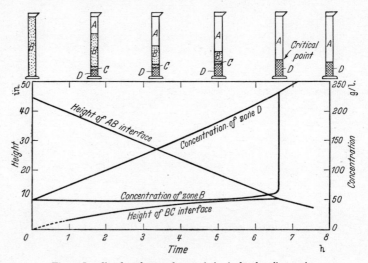

Figure 5. *Zonal settlement characteristics in batch sedimentation*

When the AB interface finally meets the rising BC interface (at the critical point), zones B and C disappear and further settling takes place only through the compaction of the settled solids, which accounts for the marked decrease in rate at this point.

Not all pulps behave as outlined above. With some, the concentration of the descending column gradually increases all along its length, causing a gradual decrease in the settling rate. This is known as Type II settling, an example of which is illustrated in *Figure 4*. In most cases the decrease is small and the critical point is easily located. However, with some pulps the curvature of the plotted data is so great that the critical point is difficult to find. The importance of locating the critical point will be made clear later.

ROBERTS[20] has proposed that a Type I pulp be considered as a subsiding teeter column of discrete, uniformly sized flocs and suggests the following equation for the settling of fine flocs in water at 20° C, in the low Reynolds number range:

$$v = 826\,(\varrho-1)\,D^2\,\frac{F\varrho}{\left(1+\dfrac{1}{F\varrho}\right)^2} \quad \ldots\ (4)$$

where v = settling velocity of floc in feet per hour
ϱ = density of floc
D = diameter of floc
F = weight of free water divided by weight of floc

The teeter behaviour was first noted at a percentage of solids by volume of about 45, and occurred at higher and lower dilutions. When the dilution was increased to a point where the per cent solids by volume fell 10, the pulp no longer settled with a definite line. By using settling rates from laboratory tests, Roberts solved the above equation for D and found by calculation that the effective diameter of the individual flocs changes with F. Thus, the equation does not hold over the whole system and therefore area determinations at various dilutions are still necessary.

NICHOLS[18] and FREE[13] have shown that, on pulps containing 2–40 per cent solids, a rise in temperature increases the collective settling rate in exact proportion to the decrease in viscosity of the liquid. This may indicate that Stokes' law is being obeyed, but another, and possibly preferable, viewpoint, is that the water rising through many fine channels between settling solids is subject to the law of POISEUILLE[7], wherein the rate of flow is inversely proportional to the viscosity of the liquid. STEWART and ROBERTS[23] suggest the latter concept may prove to be more fruitful in the theoretical study of collective settling than the standard approach using Stokes' law.

The general concepts of batch sedimentation obtained by studying settling in glass cylinders may now be applied to the problem of determining the clarification capacity of a commercial-scale, continuously operated unit.

Settling Capacity

The settling capacity of a sedimentation unit is directly proportional to the area of the tank and is determined by the free-settling* rate of the suspended solids, a rate which is independent of the depth of the liquid. The pulp in settling passes through zones of various dilutions between the dilution of the feed and that of the final discharge. If the feed dilution, for example, were 10 and the final discharge 1·5, then intermediate zones of various depths would exist between these limits with different settling rates

* Used in this sense, 'free' settling means all non-compression settling.

for each zone. Consequently, the zone having the lowest settling rate in proportion to the liquid separated would control the size of the unit, since all solids must eventually pass through this zone.

The relationship between the settling rates of various zones and the tank area required may be expressed by the following formula[5]:

$$A = \frac{2000\,(F-D)}{62\cdot 4 \times 24 \times R\,(\text{sp. gr.})_z} = \frac{1\cdot 333\,(F-D)}{R\,(\text{sp. gr.})_z} \quad \ldots\ (5)$$

where A = area in square feet per ton of dry solids per 24 h
 F = weight ratio of liquid to solids in pulp tested
 D = weight ratio of liquid to solids in discharge
 R = settling rate in feet per hour of pulp with dilution F
 $(\text{sp. gr.})_z$ = specific gravity of liquid

By applying this formula to pulps ranging in concentration from that of the feed to that at compression, the maximum value of A in the free-settling zone is determined, which will then be the proper number of square feet per ton of dry solids per 24 h to be provided in the sedimentation unit.

Mechanics of Thickening

General design of sedimentation tanks does not, for the most part, follow strict geometric proportions. The depth–diameter relationship is significant only to the extent that the projected tank volume will provide the requisite minimum detention or holding time, taking into account efficiency and mechanism design factors. The design and size proportions of feed wells, on the other hand, can be quite critical in affecting the performance of sedimentation units, clarifiers in particular.

As pointed out earlier in this section, the thickening properties of different pulps vary considerably. Some pulps reach a low final dilution within a few hours, while others retain as much as 95 per cent moisture even after days of settling. These variations are largely due to differences in the specific gravity of the solids and in the physical and structural characteristics of the flocs.

Again consider the batch sedimentation of a line-settling pulp in glass cylinders. After the pulp has reached the critical point it is in compression and further subsidence can take place only by the gradual consolidation of the settled particles. The liquid being displaced can be seen to channel up through overlying flocs until an equilibrium condition is finally reached wherein the thickening zone is at its minimum or final dilution.

In a continuously operating thickener, compact subsidence has been described by Roberts[23] as a process during which the settled flocs deform and rearrange themselves so as to reduce the percentage of voids. Contrary to their behaviour in the free-settling zones, the flocs settle in the compression zone at a rate which is dependent on the depth of the thickener. The concentration of the underflow depends on the detention time in the compression zone and, therefore, for a given area, on the depth. Sufficient capacity must be provided in the design of the unit so that ample storage space is available to permit retention until the desired density is reached.

An increase in the final concentration of the underflow may be obtained by the action of rakes moving slowly through the lower section of the compression zone. By breaking up the floc structure and providing channels

Table 2, Effect of Raking on Final Dilution
(Reported as percentage of solids in sludge)

Time h	Sludge from raw water		Sludge from dosed raw water	
	Without rakes	With rakes	Without rakes	With rakes
0·0	2·80	2·94	3·26	3·26
5·0	6·4	13·3	10·3	15·4
9·5	11·9	18·5	12·3	19·6
20·5	15·0	21·7	14·1	23·8
30·8	16·3	23·5	15·4	25·3
46·3	18·2	25·2	17·2	27·4
59·5	20·0	25·8	18·5	27·4
77·5	21·1	26·3	19·6	27·6

for the upward displacement of the void water, higher density underflows are obtained. The data of BULL and DARBY[2] shown in Table 2 illustrate the rake action in assisting thickening.

Thickening Capacity

In 1920, DEANE[9] published a paper entitled 'Settling Problems', in which he presented the following formula for the calculation of the volume of the thickening zone:

$$v = \frac{1 \cdot 33\, T\, (G_S - G_L)}{G_S\, (G_P - G_L)} \quad \ldots (6)$$

where v = volume of compression pulp in cubic feet per ton of solids per 24 h
T = detention time in hours
G_S = average specific gravity of the solids
G_L = average specific gravity of the liquid
G_P = average specific gravity of the pulp in compression

In order to obtain the necessary data to use equation (6), batch sedimentation tests must be performed on the compression pulp, observing the height of the compression zone as a function of time and then calculating the average values needed above. From v and the unit area found by using equation (5), the depth of the compression zone can be calculated.

Using the same batch settling data, Roberts[20] has developed a graphical method for determining the compression depth and for locating the true critical dilution. If settling data from these tests are plotted against time, the shape of the curve suggests a logarithmic decrement type of behaviour, the mathematical expression for which is:

$$-\frac{dD}{dt} = k(D - D_\infty) \qquad \ldots\ (7)$$

where D = dilution at any time t
D_∞ = dilution at infinite time
k = rate constant

Equation (7) indicates that a semilogarithmic plot of $D - D_\infty$ versus t should result in a straight line relationship, provided that the correct value of the unknown, D_∞, is selected. This type of plot is illustrated in *Figure 6*, where, by trial and error, a value of 0·455 for D_∞ was found to produce a straight line relationship.

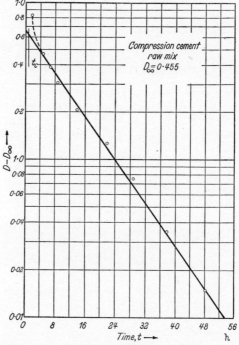

Figure 6. Semilogarithmic compression plot of raw cement mix settlement. (After Roberts[20])

If the settling test is started with the pulp above the critical point and continued into the compression zone, a curve similar to that shown in *Figure 7* for cement rock raw mix will result. The discontinuity is also shown at the upper end of the curve in *Figure 6*. To determine the true critical dilution, compression curve b is extended to zero time with the aid of the straight line portion of *Figure 6*. Time t_c is then located where the ordinate of curve a is just halfway between that of curve c and the pulp

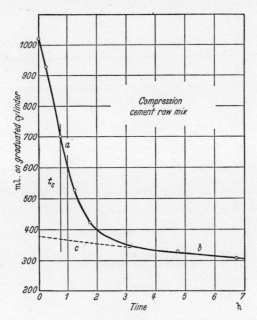

Figure 7. *Rectangular co-ordinate plot of raw cement mix.*
(*After* ROBERTS[20])

height at zero time. Since part of the pulp actually went into compression before this time and part went after, t_c represents an average value at which, in effect, all of the pulp entered compression. The critical dilution, D_C, can be read from the semilog plot at t_c.

Practical Considerations

In endeavouring to present a general conception of the process of sedimentation, it is not practical nor, indeed, possible to point out the many exceptions to the general rules and theories discussed. Before passing on to the sedimentation equipment now being used to effect solid–liquid separations, it might be well to consider briefly some of the problems faced by the design engineer due to the unpredictable behaviour of certain pulps.

Capacity is essentially a function of the settling characteristics of the solids which, in turn, involve a number of factors: particle size and distribution; shape, specific gravity and flocculence of the solids; specific gravity and viscosity of the liquid; liquid–solids ratio and temperature. Other factors which may affect capacity in varying degrees are: design of feeding arrangements, speed of the mechanism, convection currents and, in some locations, wind disturbance and evaporation in uncovered tanks of large diameter.

Settling characteristics of solids of the same composition but of a different origin may vary greatly. For example, magnesium hydroxide precipitated from brine requires from 60 to 100 ft.^{2} of thickening area per ton of solids (dry basis) per day, while magnesium hydroxide precipitated from sea water requires from three to five times as much area.

Table 3. *Thickener Unit Area Requirements for Various Pulps**

Type of pulp	Composition of pulp	Feed %	Underflow % solids	Unit area ft.^{2}/ton/24 h
Copper flotation concentrates	H_2O + 48 mesh porphyry ore	14–50	40–75	3–20
Beet sugar mud	11–15% sugar juice containing ppt. $CaCO_3$	8–20	17–22	4–20
Gold cyanide process slimes	1% NaCN soln. + 200 mesh quartz	16–33	40–55	5–13
Lead flotation concentrates	Alkaline H_2O + 65 mesh PbS ore	20–25	60–80	7–18
Recausticizing lime mud	Weak NaOH soln. + ppt. $CaCO_3$	8–10	32–40	14–18
Cement rock slurry	H_2O + 200 mesh limestone + shale or clay	16–20	60–70	15–25
Lime–soda process carbonate mud	15–20% NaOH soln. + ppt. $CaCO_3$	9–11	35–45	15–25
Water floated whiting	H_2O + 300 mesh $CaCO_3$	3–5	30–50	47–75
Water floated clay	H_2O + 300–325 mesh clay	1–4	15–45	50–225
Magnesium hydroxide from brine wells or bitterns	$Mg(OH)_2$ in 15–20% Ca or Na chloride or sulphate soln.	8–10	25–50	60–100
Magnesium hydroxide from sea water	$Mg(OH)_2$ in NaCl or $CaCl_2$ soln.	0·3–0·6	11–22	200–500
Titanium dioxide residue	H_2O + 200 mesh calcined TiO_2	5–10	40–45	60–110
Bauxite residue after H_2SO_4 digestion	30°Be $Al_2(SO_4)_3$ soln. + fine silica	5–9	20–30	75–150

* The above figures are general averages for illustrative purposes only, since each material must be checked by tests before the determination of the size of the machine required.

Generally metallurgical pulps settle faster and require lower unit areas than pulps of chemical origin. Particle size and specific gravity usually account for the difference. *Table 3*, while being far from a complete list, shows the range of unit areas for some typical thickener applications.

The variation in settling rate and unit area requirements of seemingly identical pulps can best be illustrated by citing specific examples from actual industrial operations:

Settling tests were made on samples of slaked lime from four different plants. The unit areas indicated by the tests were 50·2, 64·8, 114, and 142 ft.^{2}/ton of solids per 24 h. Even though similar production techniques had been used, due to chemical and physical dissimilarities, the required areas of the thickeners varied by a factor of over 2·8 to 1.

A similar divergence was noted in the following series of pigment plants: three different blanc fixe plants yielded areas of 1·68, 14·5, and 60 (a factor of almost 36); four clay samples indicated areas of 11·6, 36·2, and 330 (a factor of 28); three lead pigment tests showed 1·39, 13·2 and 25·1 (a factor of 18); and four lithopone plants gave areas of 5·82, 6·45, 15·9, and 30·9 (a factor of 5).

Likewise, four different samples of precipitated calcium carbonate from the lime–soda process of caustic soda manufacture indicated required areas of 3·26, 6·7, 17·4, and 32·2 (a factor of 10).

These selected examples serve to point out the dangers involved in attempting to predict the settling behaviour of some pulps. Most eccentricities can be controlled to a large extent by the application of physico-chemical principles. Mechanical or chemical flocculation, correction of the pH, control of the temperature, elimination of convection currents, or recirculation to promote crystal growth are a few of the techniques which may be used to alter the settling characteristics of troublesome pulps. Through the use of special picket-fence type rakes, which release liquid occluded in the settled solids, a more compact discharge may often be obtained. These and other special techniques have helped to make possible the wide application of sedimentation equipment in the chemical, metallurgical, and related fields.

For additional information on more recent studies of sedimentation principles reference should be made to the work of such investigators as COMINGS[6], EGLOF and MCCABE[12], GERY[15], KAMMERMEYER[16] and STEINOUR[22].

TYPES OF EQUIPMENT

Intermittent Settling Tanks

The simplest and oldest device for thickening is the batch settling tank which is still in limited use today. The thickened material is removed through a discharge valve in the bottom and the clarified solution is withdrawn either by a swing syphon, as shown in *Figure 1*, or through draw-off connections located at suitable intervals along the side.

Cones [24]

These non-mechanical settlers have a 40°–60° conical tank equipped with a manually or automatically controlled discharge valve at the bottom, and a centrally located feed well and peripheral overflow weir at the top. They can also be used as classifiers by increasing the feed rate to give a desired mesh of separation in the overflow.

One of the earlier cones is the 'Caldecott' which embodied the first successful device for the continuous discharge of thickened sand. This device consists of a disc diaphragm near the apex of the cone and so supported as to leave an annular space between the edge of the diaphragm and the cone wall. The diaphragm serves to slow down the flow of sand towards and through the spigot thus preventing bridging and plugging. The usual dimensions of the Caldecott cone are 6 ft. diameter by 9 ft. depth and 8 ft. diameter by 10 ft. depth.

On the Allen cone, the discharge valve is controlled by a link mechanism actuated by a float in the centre of the cone. By changing the weights on the actuator, or by the position of the weights, the density of the discharged sludge can be changed. Allen cones are available in diameters of 3·5 ft., 4·5 ft., 6 ft. and 8 ft. with corresponding depths of 5 ft. 2 in., 6 ft. 2 in., 7 ft. 8 in., and 9 ft. 11 in.

Table 4. Sizes of Callow Cones

Outside diam. (diam. tank top)	Inside diam. (diam. overflow weir)	Depth (tank top to bottom spigot)	Overflow gal./min
2 ft. 9 in.	2 ft. 0 in.	3 ft. 4 in.	—
3 ft. 3 in.	2 ft. 6 in.	3 ft. 9 in.	—
3 ft. 9 in.	3 ft. 0 in.	4 ft. 2 in.	—
4 ft. 9 in.	3 ft. 0 in. and 4 ft. 0 in.	5 ft. $\frac{1}{4}$ in.	6–8
5 ft. 9 in.	4 ft. 7 in. and 5 ft. 0 in.	6 ft. 0 in.	10–12
6 ft. 9 in.	5 ft. 5 in. and 6 ft. 0 in.	6 ft. $10\frac{1}{4}$ in.	14–18
8 ft. 9 in.	7 ft. 5 in. and 8 ft. 0 in.	8 ft. 7 in.	25–30

Another type of cone is the 'Callow'. Here the feed enters through a submerged well at the top of the tank, the overflow is collected in a peripheral trough, and the density of the discharge is controlled by the use of a syphon of the adjustable goose-neck variety. The greater the height of the syphon above the apex opening, the greater will be the density of the discharge sands. Callow cones are available in the sizes indicated in Table 4.

Continuous Mechanical Thickeners and Clarifiers

Single Compartment Units—These units are available commercially in round, square, and rectangular tanks. The most commonly used are shallow, cylindrical settling tanks with a centrally located drive mechanism which provides positive, mechanical removal of the settled solids, similar to the unit shown in Plate I.

For round tanks up to about 60 ft. in diameter, the drive mechanism is usually supported on a steel superstructure or on I-beams which span the top of the tank. For tanks over 60 ft. steel or concrete centre piers are generally used. Feed is admitted through a central feed well, overflow is collected in a peripheral trough, and the settled solids are slowly raked to a centrally located, pump regulated discharge outlet by means of revolving radial arms equipped with plough blades. (Plate I.)

The Dorr thickener mechanism consists of a central vertical shaft carried by bearings and a supporting bracket. A worm gear, keyed to the vertical shaft, is driven by a worm on a tangential shaft mounted on the supporting bracket. This worm shaft is provided with a pulley, sprocket, or gear for driving by belt, chain, or directly connected motor. At the lower end of the vertical shaft is attached a spider from which are extended four radial arms, usually two long and two short, equipped with plough blades. These arms are inclined from the horizontal, which for flat-bottomed tanks, results in the formation of an inverse cone of settled solids under the blades. The plough blades are mounted in such a manner that the entire area of the bottom is swept by them with each revolution and the underflow solids are moved inwardly to a central discharge outlet.

An automatic (or manual) lifting device permits the vertical raising of the mechanism a foot or two to relieve the load when starting up after a shut-down or operating interruption. The thrust of the worm shaft on a spring-loaded bearing actuates a pointer which indicates to the operator the presence and degree of an overload. If the overload increases to the point where a mechanical breakage may occur, an electrical contact is made through a mercury switch which shuts down the drive motor and sounds a warning.

A unit especially designed to handle periodic overloads is the Torq thickener, which features a torque-actuated, automatic lifting construction that causes the rake arms to pivot upward and rearward when an overload is encountered, and return to normal position when the overload is passed (see *Figure 8*). In the traction thickener the driving power for the rakes is applied at the end of a radical truss by means of a motor-driven carriage which runs along the top of the tank sides. A stationary, bridge-type truss, extending from the tank periphery to the centre pier, supports the feed trough and electrical conduits and also serves as a walkway for operators.

The Hardinge thickener uses an auto-raise device which consists of two concentric torque tubes, the outer and shorter one being entirely above the liquid level. A yoke at the top of the inner torque tube has extended rollers which normally rest at the bottom of two diagonally opposite sloping slots in the outer torque tube. When an overload occurs, the resistance to the blades causes the rollers to move along and up the sloping slots, which shortens the length of the combined torque tubes by telescoping one inside the other. When the overload has decreased, the scrapers automatically lower to their normal position. At maximum position an alarm and cut-off switch is actuated. On Hardinge units a spiral rake is sometimes used in place of the plough blades on radial arms.

When square sedimentation tanks are used, usually in the clarification of water, trade wastes and sewage, one of the radial raking arms is equipped with a special corner blade which automatically reaches into the four corners of the tank and moves the settled sludge to a point where it can be picked up by the regular raking arms. Two Squarex clarifiers are shown in *Figure 9*.

For rectangular tanks a straight line raking motion is utilized. The drive mechanism is either mounted in a stationary position at one end of the tank or is carried on a car which travels back and forth along the entire length of the tank (see *Figure 10* showing the Hardinge rectangular clarifier). The stationary drives may operate an endless chain with attached scrapers or a cable-driven car with an attached single, transverse blade. These tanks have been built to over 300 ft. long and 100 ft. wide for some of the larger installations.

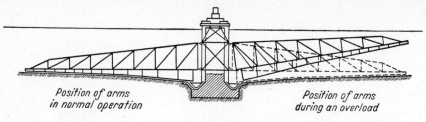

Figure 8. Diagram of Dorr Torq thickener mechanism. (*By courtesy of* Dorr-Oliver Co. Ltd.)

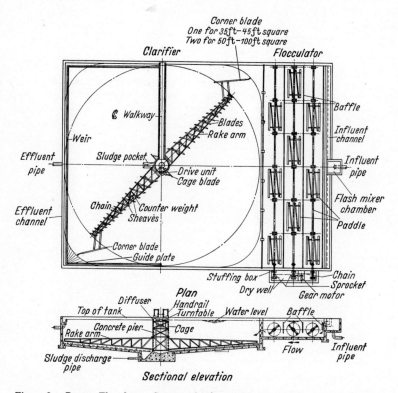

Figure 9. Dorrco Flocculator—Squarex clarifier combination. (*By courtesy of* Dorr-Oliver Co. Ltd.)

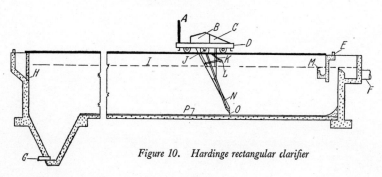

Figure 10. Hardinge rectangular clarifier

Multiple Compartment Units—Where floor space is limited and ample headroom is available, multiple or 'tray' type sedimentation units are employed. These consist of a number of shallow, superimposed settling compartments, rarely over 90 ft. and usually about 25–45 ft. in diameter. Each compartment is equipped with a raking mechanism which is attached to a common central rotating shaft. The feed, overflow and underflow arrangements vary according to the material being treated and the operation being performed. In the open type, all the feed enters the top compartment and is distributed equally among the other compartments through an annular opening around the central shaft. Overflow is taken from each compartment, but the underflow is raked to the central opening of each tray where it is joined by part of the feed. The sludge accumulates as it works its way to the bottom compartment, where it thickens further and is then discharged.

In the balanced tray unit (see *Plate II*) each compartment has individual feed and overflow connections. The solids settled in each compartment are discharged by gravity into the compartment directly below, through a central downcast boot which extends below the sludge bed in the bottom of the next lower tray, thus effectively sealing the tray and preventing intermingling of sludge and solution. The depth of the sludge bed in each compartment is regulated by raising or lowering the height of the overflow column, which is balanced by the sludge bed, as in a U-tube.

In the washing type tray thickener, the compartments operate in series, the feed entering the top compartment and the solids progressing downward while the wash water enters the bottom compartment and flows countercurrent to the solids. Integral with the sludge seal is a downcast boot which projects into the next lower compartment and serves as a mixing well. Wash solution enters the well to repulp and dilute the settled sludge prior to its being thickened again. This type of unit is especially adapted to relatively small countercurrent decantation (c.c.d.) operations handling hot solutions, since it requires relatively little floor space and can be easily insulated. It is regularly supplied in diameters up to 60 ft. and depths up to 40 ft. (5 compartments). A 'combination' type is also available in which one or more settling trays in parallel are followed by a series of two or more washing trays.

A special type of unit is the filter thickener, which consists of a cylindrical or rectangular tank, feed and discharge connections, and one or more submerged filter elements for vacuum or gravity withdrawal of overflow. The filtration medium, in some cases, consists of a number of fabric filter elements immersed in the pulp and, in other cases, of a layer of sand or other granular substance laid on a false bottom. The submerged cake which forms on the filter medium is periodically removed by the application of either low-pressure air or water on the reverse side of the medium. In some cases, a mechanical scraping device is used. In all cases, the solids are discharged from the bottom of the tank as thickened sludge and not as a filter cake.

Combined Flocculation–Sedimentation Units—In the treatment of suspensions requiring destabilization and agglomeration before clarification and thickening, it has been the practice to pre-condition the pulp in a single unit called a 'Flocculator', which consists of a rectangular tank containing rotating paddles, to provide the necessary shear for flocculation. More recent developments employ combination units, which, in a single tank, provide for flocculation in separate cells preceding the clarification and thickening steps.

An example of the latter type of unit is the Clariflocculator, which consists of a circular tank with an inner flocculating zone suspended centrally within a conventional clarifier. The flocculating mechanism consists of a series of V-shaped vertical blades mounted on two horizontal rotating arms. Inter-meshing with these blades are four sets of downwardly extending stationary V-blades that are supported from beams above the liquid surface. After sufficient detention has been provided, the flocculated solids pass through a central opening in the floor of the inner zone to the outer zone, where they settle to the tank bottom and are raked to the central discharge outlet by means of conventional sludge raking blades. The clarified effluent is discharged over a weir at the periphery of the outer zone. The unit is constructed in standard sizes from 12 to 150 ft. in diameter with the inner zones ranging from 6 to 70 ft. in diameter.

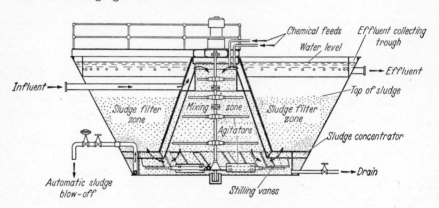

Figure 11. *Permutit precipitator.* (*By courtesy of* Permutit Co. Ltd.)

Another important type of combination unit is the high-rate, sludge blanket type, which performs the steps of blanket filtration, clarification, and sludge collection and thickening in three distinct zones. The mechanism, which may be installed in either round or square tanks, consists of a centre shaft which supports a steel drum, sludge scrapers, and four hollow distributing arms. Each arm is equipped with a series of baffled orifices which control the feed to the blanket. Two of the arms have plough blades which rake the solids to an annular thickening well for removal. Overflow is peripheral on units under 45 ft. diameter and by means of both a peripheral and an inner annular launder in larger units.

The trade name of the unit just described is the Hydro-Treator, a cutaway view of which is shown in *Plate III*. Its use is mainly in the softening of municipal and industrial water. All chemical reactions take place in the blanket, resulting in the formation of dense, fast settling particles. A flocculating agent, such as aluminium sulphate, is generally used to assist coagulation. A machine working on a similar principle is the Permutit precipitator shown in *Figure 11*.

In the Dorr Multifeed clarifier, widely used in the clarification of sugar cane juice, a top conditioning compartment is provided to permit flocculation of the suspension before it is fed to the several clarification compartments and a skimmer is used to remove floating scum. The performance of this machine has been greatly improved recently for the treatment of cane juice by the development of the RapiDorr Clarifier, which allows faster processing of hot, thin juice with consequent reduction of losses. It can handle equivalent tonnages and achieve comparable results with 30 per cent less volume than the older unit. The proved pattern of juice and mud flow in the RapiDorr is the same as in the previous Multifeed. Juice is introduced at the centre and flows outward at a decreasing velocity and increasing quiescence. Settled muds are moved from all points on the trays inward towards the centre for better concentration. A cut-away view is shown in *Plate IV*. The RapiDorr design can be applied to existing units thereby doubling working capacity with only a small volume increase at moderate cost.

SUPPLEMENTARY DESIGN DATA
Materials of Construction

Sedimentation units are normally constructed of mild steel, concrete, or wood. In special cases where corrosive solutions are handled, stainless, lead-lined, and rubber-covered steels, silicon irons, Pioneer metal, and bronze may be used. For acid-resisting duty wooden tanks may be constructed and then lead lined.

Standard trays are usually fabricated from $\frac{3}{16}$ in. steel plate and arranged for either steel-welded construction or field riveting. All trays are suitably reinforced against upward pressure and steel tanks, generally, are provided with manholes constructed in the sides to permit access to each compartment.

Where thickeners are handling chemical pulps at high temperatures the tanks are always covered and are usually of special design. In the causticizing section of a paper pulp mill, for example, the trays are constructed of $\frac{1}{4}$ in. plate, and the overflow boxes of $\frac{3}{8}$ in. plate. Both overflow and feed boxes are provided with hinged covers. In some plants these boxes are also lined with concrete. The discharge cone is made of heavy steel plate and is usually field welded to the bottom of the tank. The tanks are of heavy welded construction with bottom and side plates $\frac{3}{8}$ to $\frac{1}{2}$ in. thick, and the cover $\frac{1}{4}$ in. thick. The drivehead mechanisms are usually constructed of Meehanite or cast iron. A special stainless steel sleeve is used at the air–liquid interface on the shaft in the top compartment of tray units handling hot caustic liquor.

Selection of Type

After the areal and depth requirements for a particular separation have been ascertained through the settling tests previously described, the chemical engineer is next concerned with the selection of the most efficient yet economical type of sedimentation unit to use. While certainly no hard and fast set of rules can be established, a few pertinent considerations may be listed for general guidance:

1. *Floor Space Available.* If this is limited, the tray type should be strongly considered as it gives the greatest capacity per unit of floor space occupied.
2. *Conservation of Heat.* If this factor is important, the tray type is most suitable, being easily covered and insulated, and giving the least temperature drop between feed and overflow.
3. *Amount of Underflow Solids.* For this, the traction type has the greatest raking capacity per unit area, since the power is applied at the end of a long raking arm and deeper and larger plough blades can be used than on other types, a point worthy of consideration if the underflow is large.
4. *Periodic Overloads.* If these are frequent and severe the torque type with automatic self-raising and lowering arms or the auto-raise type units should be considered.
5. *Corrosive Material.* A single compartment, central-drive type unit should be used if possible, since the portion of the unit below the liquid level can be constructed of special materials, as mentioned in the previous section.

Special designs of the standard sedimentation units have been made for such specific problems as the treatment of trade wastes, sewage, turbid water, cane and beet sugar juice, and others, the names and manufacturers of which may be found in the trade magazines.

Ancillary Equipment

Satisfactory operation of the major sedimentation units could not be obtained without the use of certain accessories. A brief discussion of the more important of these will now be given.

Pumps—The most important accessory for continuous operation of sedimentation equipment is the pump used to withdraw the settled solids, and of these the most common is the diaphragm pump, which is available for either suction or pressure operation. Suction types are usually capable of lifting sludge against a total head equivalent to about 15 ft. of water, but in handling thickener underflows, this figure should be adjusted by dividing it by the specific gravity of the pulps. Pressure type pumps can be used against a pressure head equivalent to 45 ft. of water. Most manufacturers supply a stroke adjustment device which permits the stroke, and hence the capacity of the pump, to be changed while the pump is in operation. This feature permits close regulation of the moisture content of the discharge, which is a function of the rate of withdrawal. The pumps are supplied in sizes ranging between 1 in. and 4 in. and in simplex to quintuplex types arranged for either belt or direct-motor drive. The rated horsepower of the motor for the pressure types ranges from 1 h.p. for the 1 in. simplex to 10 h.p. on the 4 in. quintuplex; slightly lower ratings are quoted for the suction types.

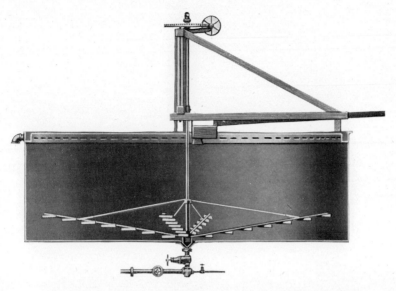

Plate I. Early Dorr unit thickener. (*By courtesy of* Dorr-Oliver Co. Ltd.)

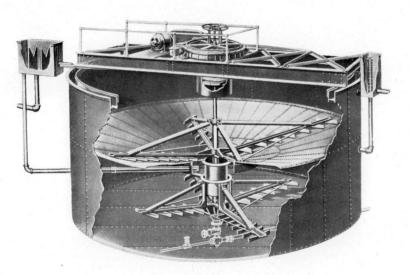

Plate II. Two-compartment balanced type tray thickener

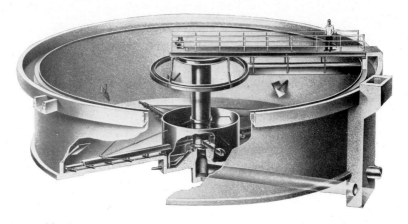

Plate III. Dorrco Hydro-Treator. (By courtesy of Dorr-Oliver Co. Ltd.*)*

Plate IV. RapiDorr multi-compartment cane juice clarifier. (By courtesy of Dorr-Oliver Co. Ltd.*)*

The usual speed is 50 strokes/min. For a sludge containing 50 per cent solids, the displacements in cubic feet of sludge per minute for the same range of sizes given above are from about 0·5 to 22.

Density Controls—By attaching a recording device to the spring taking the thrust of the worm shaft in the driving mechanism, a continuous record of the load on the raking mechanism can be available to the operator. Any building up of resistance can be detected and corrected before an overload occurs. If one does occur which trips the alarm and stops the machine, a record of the conditions leading up to the time of overload is available for study. By connecting the torque-indicating device so that it actuates the discharge valve on the unit, automatic density control of the underflow can be obtained. In a few installations gravity discharge has been used, thereby eliminating the regulatory pumps. In some clarification operations the amount of settled solids is not sufficient to permit continuous withdrawal of solids. In such cases the torque recorder may be used to indicate to the operator when sufficient solids have settled to require pumping. Electric timers may also be used for periodic operation of pumps.

All of the devices operating from the shaft of the raking mechanism are inherently crude, but to date no better method has been successfully demonstrated.

Underflow Line Features—Certain pulps exhibit a tendency to pack after settling for short periods. When an interruption in sludge withdrawal occurs, solids may settle out in the underflow lines of the unit. A stoppage of the raking mechanism due to an overload or breakdown may also allow the pulp time to pack in the discharge cone. To circumvent this difficulty a high-pressure water or steam line is usually attached to the underflow line directly below the discharge cone. By closing a valve in the underflow line, water or steam under pressure may be forced upward through the cone to dislodge the settled material. If the line itself is plugged, a valve in the line to the cone can be closed so that the high-pressure water or steam can force open the line leading away from the unit.

A recirculating line which returns all or a portion of the thickened solids to the feed well of a unit will also aid in preventing plug-ups during a shutdown of the thickener mechanism. If all of the underflow is being recirculated, the unused portion of the line leading from the unit must be flushed out to prevent plugging in the line due to settling.

Other Special Devices—The overload alarm for sedimentation units has already been described. Other devices which aid in smooth operation include sprays to break up foam, skimmers for removing surface scum and foam, and screens in the feed boxes to protect the units from oversized particles. Sampling devices are installed on some tray units so that the sludge level in each compartment can be determined and samples obtained for checking pulp densities.

Equipment Costs

Costs of complete installations are subject to so many variables, even when prices are stable, that it is practically impossible to arrive at a dependable factor of unit costs. Mechanism costs in themselves can vary widely for any given diameter, depending upon the type of construction. However, the delivered price of mechanisms of standard design and construction at the present time will probably represent on the average from 20 to 30 per cent of the total cost of the installation.

Shown in *Table 5* are rough unit cost estimates based on single compartment thickeners of standard construction in concrete tanks in a range of diameters from 24 to 100 ft. These estimates cover complete, ready-to-operate units and include excavation and back-fill, tank construction, all mechanical equipment, supports, weirs, erection of mechanism and wiring.

Table 5. *Cost Estimates for Concrete-basin, Single-compartment Thickeners*

Thickener diam. ft.	Cost per ft.^{2} of settling area shillings
24–35	120–180
35–40	100–120
40–50	80–100
50–60	72–86
60–70	66–80
70–80	60–72
80–90	60–66
90–100	52–66

The unit cost of thickeners in diameters less than 24 ft. increases rapidly with decreasing diameters and may well reach £16–£25 per square foot of area at 10 ft. Above 100 ft. diameter the cost per unit of area decreases very little. If a clarifier rather than a thickener mechanism is employed, installed cost is reduced 3–8 per cent. Similarly, heavy duty mechanisms will increase costs by approximately the same amount. Special mechanisms and tanks will increase the estimated unit costs substantially.

The initial cost of a steel tank with substructure may run appreciably less than for concrete in diameters up to about 80 ft., but the cost of maintaining and painting steel tanks may offset any initial advantage over the years.

Operating Costs

Power consumption of thickeners is a small item indeed, as indicated in *Table 6*. Power requirements of clarifiers run even less than for thickeners of the same diameter.

Table 6. *Motor Horsepower and Rake Speed of Typical Unit-compartment Thickeners*

Tank diam. ft.	Drive motor h.p.	Rake speed min/rev
20	1	4
60	2	10
100	3	13
150	5	19
200	7½	23

On the whole, repair, maintenance and operating labour costs are comparatively modest, due to the very slow speeds of the mechanism and the absence of submerged machine parts and bearings. Overall operating costs for metallurgical thickeners handling large daily tonnages of solids rarely exceed 4d. per ton; generally ranging from 1d. to 4d. depending upon the size of the operation.

CONTINUOUS COUNTERCURRENT DECANTATION

Definition

Continuous countercurrent decantation (c.c.d.), an important washing operation involving a series of continuous thickeners, is a system of washing finely-divided solids to free them from liquids containing dissolved substances. Chemical precipitates, ground metallurgical ores, and residues from leaching operations are commonly treated in c.c.d. plants. The purpose of the operation is to attain a high separation of soluble materials from the solids with a minimum number of decantations and with the use of a minimum amount of wash liquid. The solids being washed move in a direction countercurrent to the wash solution and are progressively impoverished of soluble material by coming in contact with progressively weaker wash solutions, ending finally with fresh water. Similarly, the wash water, flowing in the opposite direction to that of the solids, becomes progressively stronger by coming into contact with larger and larger amounts of soluble material. The efficiency of the system is dependent on (*a*) removing the settled solids from each thickener with the least amount of solution, and (*b*) on the thorough repulping of the sludge and wash solution between stages.

Industrial Applications

Classical examples of the use of this method in chemical engineering processing are found in (1) the c.c.d. treatment of sodium carbonate solution

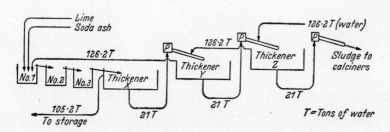

Figure 12. Flow sheet of washing by continuous countercurrent decantation

which has been causticized with lime (see *Figure 12*), and (2) the c.c.d. treatment of a slurry containing aluminium sulphate solution and a silica residue. Typical data resulting from good plant operation are presented in the *Table 7*.

Table 7. *Typical C.C.D. Operating Data*

	Aluminium sulphate	Caustic soda
Solids	Bauxite ore	$CaCO_3$ & $Ca(OH)_2$
Liquid	50°Be H_2SO_4	Na_2CO_3
Extraction in agitators, %	98·5	91·5 (% causticity)
Strength of finished solution, °Be	35 (hot)	15·8 (hot)
Temp. of finished solution, °C	90	74
Number of agitators	4	3
Number of thickeners	3	3
Washing efficiency, %	99	99·3
Overall recovery, %	97·5 (of available Al_2O_3)	—

Some of the obvious advantages of the c.c.d. system over a number of intermittent agitation and settling tanks are:

1. Less operating labour.
2. Less steam for heating.
3. Higher washing efficiency.
4. Higher concentration of finished liquor.
5. Fewer tanks and less floor area occupied.

Economic Considerations

The annual savings made when a 100 ton/day alum plant changed from batch digestion of rock and washing of residue to the c.c.d. system are shown in *Table 8*.

Table 8. *Annual Savings of C.C.D. v. Batch Process in Alum Plant*

	Batch plant	C.C.D. plant	Annual savings £
Labour	3 men per shift	1 man per shift	8,100
Steam in reaction	—	—	3,125
Overall recovery, % available Al_2O_3	90	97·5	3,133
Temperature of finished liquor, °C, saving of evaporator steam	30	90	2,400
Concentration of finished liquor, °Be, and saving of evaporator steam	30	35	3,250
		Total savings	20,008

Basis:
  Alum, 17 per cent Al_2O_3, 100 ton/day
  Labour, 10s. per hour (total cost)
  Bauxite, 50 per cent available Al_2O_3, at 66s. per ton
  Steam at 5s. per 1,000 lb.

The washing type thickener, previously described, provides a small, compact c.c.d. plant in a single unit. Considerable savings in floor area, heat loss by radiation, pumps, *etc.*, over the multiple unit plant may be realized.

C.C.D. Calculations

The calculations involved in determining the washing recovery of a typical c.c.d. operation are illustrated in the sample problem given below.

Sample Problem

Soda ash solution, when causticized with lime, yields a solution of caustic soda in which is suspended precipitated calcium carbonate. This pulp is treated in the c.c.d. system shown in *Figure 12*. Using the data given below and in the flow sheet, calculate the washing efficiency. Assume:

1. Ten tons of NaOH produced per 24 h.
2. 1·4 tons lime mud per ton of NaOH produced.
3. Lime mud discharged from all thickeners at a water–solids ratio of 1·5:1.
4. The overflow should contain approximately 190 lb. NaOH per ton of water.

Solution

Let X, Y and Z equal the pounds of dissolved NaOH per ton of water in the respective thickeners. Equate the pounds of NaOH into and out of each unit as follows:

$$\text{Thickener } X \quad 105 \cdot 2\,X + 21\,X = 126 \cdot 2\,Y + 20{,}000$$
$$\text{Thickener } Y \quad 126 \cdot 2\,Y + 21\,Y = 126 \cdot 2\,Z + 21\,X$$
$$\text{Thickener } Z \quad 126 \cdot 2\,Z + 21\,Z = 126 \cdot 2\,(0) + 21\,Y$$

Solving by substitution:

$$X = 189 \cdot 2 \text{ lb. NaOH per ton of water}$$
$$Y = 30 \cdot 8 \quad ,, \quad ,, \quad ,, \quad ,, \quad ,, \quad ,,$$
$$Z = 4 \cdot 4 \quad ,, \quad ,, \quad ,, \quad ,, \quad ,, \quad ,,$$

The 105·2 tons of water leaving Thickener X contain 189·2 lb. of NaOH per ton, or 9·953 tons of NaOH. Thus, the washing efficiency is

$$\frac{9 \cdot 953}{10{,}000} \times 100 = 99 \cdot 53 \text{ per cent}$$

SAMPLE CALCULATION OF SEDIMENTATION EQUIPMENT SIZE

A complete sample problem is presented below in which the solution has been worked out step by step. The problem chosen is not typical of either straight clarification of a dilute pulp or the thickening of a concentrated one, but is an in-between type which requires that the factor of detention and area be studied to determine which is controlling. The first portion of the example may serve as an illustration of the procedure used in handling a typical dilute pulp, and the last portion, a typical concentrated pulp. The student should plot the various data and note the shape of curves obtained.

Example Problem

A coal company has 3,500 gal./min of washer water containing 3·81 per cent solids (fine coal) by weight which requires clarification to 12 gr./gal.

before it can be reused in the plant. The fine coal to be removed is 100 per cent − 20 mesh. From representative samples of the feed determine the diameter and depth of the unit needed to produce the desired effluent. The specific gravity of the solids is 1·4.

Solution

Preliminary Settling Test

A sample of the feed pulp was thoroughly mixed and allowed to settle in a graduated glass cylinder. At first the solids at the top of the cylinder settled independently, but after a few minutes a line appeared which was quite indistinct but gradually became more definite. The pulp entered the line-settling phase at this point and settled collectively.

The above behaviour indicated that both detention and area must be considered in determining the size of the unit. Tests on both the independent particle subsidence phase (clarification) and the collective subsidence phase (line settling) must be performed.

Clarification in a Long Tube

In clarification problems involving dilute pulps where the particles or flocs settle independently, the clarifier area is determined by the settling rate of the slowest settling particles or flocs. In order to determine the overflow rate which will produce an effluent containing about 12 gr./gal., a sample of the feed (3·81 per cent solids) was allowed to settle in a 6 ft. tube. After various detention periods, samples were removed from the tube at the 5 ft. level and the percentage of solids determined with the results shown in *Table 9*.

Table 9

Detention time h	Overflow rate ft./h	Effluent	
		% solids	gr./gal.
0·00	—	3·81	2030
0·50	10·0	0·049	29
0·75	6·7	0·021	12
1·00	5·0	0·017	9·9
1·25	4·0	0·016	9·3
1·50	3·3	0·016	9·3

Thus, according to the laboratory tests, an overflow rate of 6·7 ft./h with a detention of 0·75 h will produce the desired effluent. This rate must be adjusted in order to apply it to a commercial-sized unit. Assuming a safety factor of 20 per cent for illustrative purposes, the plant rate would be $(6·7)/(1·2) = 5·6$ ft./h.

Determination of Compression Point

A sample of line-settling pulp was prepared, thoroughly mixed and allowed to settle in a graduated glass cylinder. Readings of the descending pulp line were taken every few minutes and recorded as shown in *Table 10*.

Table 10

Elapsed time min.	Reading of pulp line ml.	Elapsed time min.	Reading of pulp line ml.
0	995	8	285
2	810	10	210
4	608	12	180
6	435	14	150

These data were then plotted and the break in the curve (indicating a Type I pulp) was obtained by extrapolation, as illustrated in *Figure 7*. The critical point was found to be at the 245 ml. mark, indicating the critical dilution, $D_C = 4\cdot 8$, calculated as follows (the pulp contained 44·4 g solids):

$$\text{Dilution} = \frac{\text{parts by weight of liquid}}{\text{parts by weight of solids}} = \frac{245 \text{ ml. pulp}}{44\cdot 4 \text{ g solids}} - \frac{1}{1\cdot 4 \text{ g solids/ml.}}$$

$$= 4\cdot 8 \frac{\text{ml. H}_2\text{O}}{\text{g solid}} \left(1\cdot 0 \frac{\text{g H}_2\text{O}}{\text{ml.}}\right) = 4\cdot 8 \frac{\text{g H}_2\text{O}}{\text{g solid}}$$

Final Dilution Test

In order to determine the dilution of the discharge from the clarifier, a 700 ml. sample containing 114·5 g of solids was allowed to settle to a point near the critical dilution. Readings of the pulp level were then taken until the subsidence practically ceased. Occasional stirring of the settled pulp

Table 11

Pulp reading ml.	Dilution	Avg. dilution	Time interval h	Time int. × avg. dilution
700	5·40			
		4·29	0·16	0·69
445	3·18			
		2·59	0·25	0·65
310	2·00			
		1·89	0·25	0·47
285	1·78			
		1·68	0·50	0·84
262	1·58			
		1·52	0·50	0·76
247	1·45			
		1·41	1·00	1·41
238	1·37			
		1·33	1·00	1·33
229	1·29			
		1·25	1·00	1·25
220	1·21			
		1·19	1·00	1·19
215	1·17			
		1·15	1·00	1·15
210	1·12			
			6·66	9·74

(Based on 114·5 g of solids)

to simulate the raking action of the clarifier was employed. The results of the test and the method of calculating the average dilution in the compression zone are given in *Table 11*.

From these results the final dilution, D, was found to be 1·12 after 6·66 h. The average dilution in the compression zone was $9\cdot74/6\cdot66 = 1\cdot46$. From this, the average specific gravity of the pulp in the compression zone can be calculated:

$$\text{sp. gr.} = \frac{\text{vol.}}{\text{wt.}} = \frac{1\cdot46 + 1}{1\cdot46 + \dfrac{1}{1\cdot4}} = 1\cdot13$$

Determination of Unit Areas

Settling tests were then made on pulps having dilutions ranging from that of initial line settling to the critical dilution. The rate of settling for each zone was observed and its required unit area calculated. The procedure used was as follows:

Three samples of pulp were made up containing 110, 68, and 45 g of solids, respectively, in 1,000 ml. of total pulp. Starting with the 110 g sample, the pulp was thoroughly mixed, allowed to settle, and readings of the pulp line were recorded at intervals of 2 or 3 min until a definite constant rate was established. Then 100 ml. of clear solution was decanted from the cylinder and the same procedure followed with the resulting pulp of a slightly lower dilution. By repeating the decanting procedure down to the

Table 12

Pulp dilution	Rate ft./h	Unit area ft.²/ton *solids*/24 h
21·5	6·40	4·24
19·2	5·80	4·17
17·0	5·20	4·08
14·9	4·50	4·08
14·0	4·10	4·20
12·5	3·33	4·60
11·1	3·06	4·35
9·6	3·06	3·71
8·4	2·94	3·31
7·5	1·94	4·39
6·6	1·49	4·82
5·6	1·07	5·61
4·7	0·71	6·76

Based on a final dilution of 1·12 to 1.

700 ml. mark for each of the three pulps a rate of settling was established for various pulps beginning at line-settling dilution (about 21·5) and ending near the compression point (4·8). The results of these tests were then used in the formula for area, $A = 1\cdot33\,(F - D)/R$, using as the final dilution $D = 1\cdot12$, as obtained in the test previously made on the compression pulp. These data are listed in *Table 12*.

From the above, the minimum allowable unit area was noted to be 6·76 ft.²/ton solids/24 h. Based on this figure, the required compression depth is calculated as follows:

Average dilution of pulp in compression zone	1·46:1
Average specific gravity of pulp in compression zone	1·13
Minimum allowable unit area, ft.²/ton solids/24 h	6·76
Hours detention for reaching final dilution	6·66
Pounds of pulp requiring storage per square foot	

$$\frac{2000}{24} \bigg| \frac{6 \cdot 66}{6 \cdot 76} \bigg| \frac{1 \cdot 46 + 1}{} = 202$$

$$\text{Compression depth in feet} = \frac{202}{62 \cdot 4 \times 1 \cdot 13} = 2 \cdot 86$$

Thus, from the laboratory tests, a minimum allowable unit area of 6·76 ft.²/ton solids/24 h was obtained. Using a safety factor of 20 per cent to take care of changes in the character of the pulp, variations in temperature, and the conversion from laboratory quiescent conditions to plant dynamic conditions, the unit area becomes $(6 \cdot 76)(1 \cdot 2) = 8 \cdot 10$ ft.²/ton solids/24 h.

Determination of Clarifier Area

1. *From overflow data*

Using the adjusted overflow rate of 5·6 ft./h found by the long tube clarification test, the required clarifier area is:

$$\text{Total area} = \frac{3{,}500 \text{ gal./min}}{} \bigg| \frac{60 \text{ min/h}}{7 \cdot 48 \text{ gal./ft.}^3} \bigg| \frac{}{5 \cdot 6 \text{ ft./h}} = 5{,}000 \text{ ft.}^2$$

2. *From zone tests*

Using the adjusted unit area of 8·10 ft.²/ton solids per 24 h from unit area zone tests above, the clarifier area is calculated as follows:

Tons dry solids/day =

$$\frac{3{,}500 \text{ gal./min}}{} \bigg| \frac{1{,}440 \text{ min/day}}{2{,}000 \text{ lb./ton}} \bigg| \frac{8 \cdot 34 \text{ lb. pulp}}{\text{gal.}} \bigg| \frac{0 \cdot 0381}{} = 800$$

Clarifier area = $(800)(8 \cdot 10) = 6{,}480$ ft.²

Since this figure is larger than that determined by the overflow rate test, the collective subsidence phase is the controlling factor and the figure of 6,480 ft.² must be used in the design of the unit. The diameter of the clarifier based upon the maximum unit area calculates to be:

$$d = \frac{6{,}480}{0 \cdot 786} = 91 \text{ ft. or 95 ft. in a standard size.}$$

Determination of Clarifier Depth

The total depth of the clarifier may be calculated as follows:

Overflow rate from 95 ft. diameter clarifier:

$$3{,}500 \text{ gal./min} \times \frac{60 \text{ min/h}}{7 \cdot 48 \text{ gal./ft.}^3} \times \frac{1}{0 \cdot 785(95)^2 \text{ ft.}^2} = 3 \cdot 95 \text{ ft./h.}$$

Laboratory tests indicate a detention of 0·75 h at 6·7 ft./h rate will produce the desired effluent. Since the plant rate in the 95 ft. unit is considerably lower, the effluent should contain less than the 12 gr./gal. specified. Allowing for a 1·0 h detention in the plant, the clarification depth to be provided is:

$$(3 \cdot 95)(1 \cdot 0) = 3 \cdot 95 \text{ ft. or } 4 \cdot 0 \text{ ft.}$$

Thus the total depth in feet may be calculated as follows:

Clarification depth	4·0
Compression depth	2·86
Ineffective depth due to sloping of rakes	4·64
Feed well	2·00

Total clarifier depth = 13·50 or 14 ft.

Thus the clarifier needed by the coal company is a unit 95 ft. in diameter by 14 ft. deep.

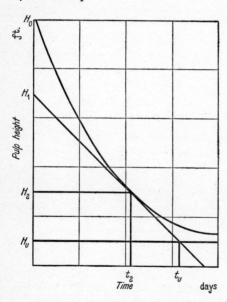

Figure 13. *Chart for calculating unit areas of thickeners*

This method of calculating the size of a clarifier or thickener required for a given job, as explained above, has been used without significant alteration for almost 35 years until the development of a new and simplified means was started during the past two years. The principles of thickening are being actively studied and a recent paper[25] shows that it is possible to dispense with the 'zone' tests described on p. 258. When a more precise method of locating the compression point is developed, this method (KYNCH[17]) will be of great value in simplifying the testing procedure. As now being applied it is only necessary to run one batch test starting with a relatively dilute pulp and continuing through compression to final concentration.

The pulp depth at various times is plotted graphically against the time required to attain that depth. (*Figure 13.*) A tangent to this plot is drawn at the point of compression and the intersection of this tangent and the line representing final pulp dilution determines the time (t_u) required to attain ultimate density.

Maximum unit area may then be calculated by substituting in the formula, $U.A. = t_u/C_0 H_0$, the graphically determined (t_u), pulp depth (H_0) and concentration (C_0) at the start of the test. Unit area is expressed in square feet per ton of solids per day, time in days, pulp depth in feet, and concentration in tons per cubic feet.

The new method eliminates as many as fifteen batch tests and has proved to be as accurate as the original method in dilute zones, and more reliable in zones of greater concentration. Comparison of actual thickener unit areas and those predicted as outlined have checked very closely in four sugar-beet plants.

NEW DEVELOPMENTS

For the most part, equipment design has been relatively stable since the early days of continuous sedimentation with improvements limited chiefly to driving mechanisms.

Redesign of Recausticizing Clarifiers

The major exception has been the recent redesign of the green liquor clarifier and dregs washer in pulp and paper chemical reclaiming system. The four compartment tray machines formerly used at both points in the flow sheet have been replaced by unit compartment clarifiers of new design equipped with special feed wells.

Tray units, while producing overflows of acceptable clarity at low rates, did not fully utilize existing tank volume. Feed was short-circuiting, streaming from the rather shallow feed well to the overflow weir. Thus green liquor was detained for only a fraction of the desired time.

A survey of a number of recausticizing systems indicated that short-circuiting could be minimized by designing clarifiers with a depth to diameter ratio greater than conventional, and with a larger diameter and deeper feed well. Unclarified new feed can then be introduced well down in the tank at a point where solids concentration is more nearly equal to feed concentration—a condition which assures maximum flocculation of solids in suspension and better clarification.

This same survey also revealed a relationship between detention time and overflow rate of the liquor. Formerly it had been thought that overflow rate was the primary factor affecting overflow clarity, but these tests clearly pointed out the fact that by providing sufficient detention time, overflow rates could be radically increased without impairing clarity. To produce the best overflow at a minimum investment cost, design of the clarifier hinges on the establishment of an optimum relationship between overflow rate and detention time.

For green liquor this optimum relationship permits use of a single compartment clarifier, somewhat deeper than conventional, and equipped with a large, deep feed well to boost the overflow rates three or fourfold without sacrifice in overflow clarity. This unit compartment design has now been incorporated into the standard recausticizing system for both the dregs washer and the green liquor clarifier.

Redesign of Metallurgical and Chemical Thickeners

Equipment redesign, however, has not been limited to the pulp and paper industry. In the metallurgical field, thickeners of special unit compartment design have been applied to the clarification and thickening of blast-furnace flue dust and, to date, results have been excellent.

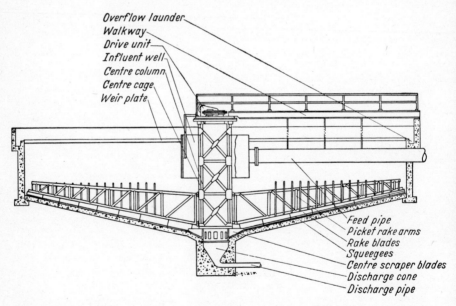

Figure 14. Dorr Densludge thickener with picket-fence type rakes. (By courtesy of Dorr-Oliver Co. Ltd.)

With extremely deep units, overflow rates have been increased four to sixfold and overflow clarities simultaneously have been substantially improved. It is expected that operating results from shallower units now being erected will also show marked, but not as spectacular, improvement.

The chemical industry also has been affected by this change and unit compartment clarifiers of this general design are soon going into operation. Should these applications prove as successful as those mentioned above, clarifiers of the new unit compartment design may well be applied to many chemical problems formerly served by tray units.

Thicker Underflows

For a number of years equipment manufacturers have experimented with several different methods of producing a thicker underflow. Vertical pickets (see *Figure 14*) have probably been the most successful of these devices, but they have increased underflow density of some solids only a few per cent in most cases. The most striking application of the so-called 'picket-fence' type thickener has been in the Dorrco Densludge System for the production of a thicker sewage sludge. By continuously withdrawing dilute sludge from the raw sewage clarifiers for thickening under controlled conditions in the Densludge thickener, the volume of sludge in the underflow has been decreased by as much as 50 per cent compared to conventional clarifier underflows, thereby greatly increasing sewage digester capacity. More recently, however, other types of mechanisms have been utilized in experimental work. In the near future more dense pulps—which may be unpumpable and removable only by screw conveyor—may become a reality.

REFERENCES

[1] ANABLE, A., and KNOWLES, C. L., *Chem. metall. Engng*, 45 (1938) 260
[2] BULL, A. W., and DARBY, G. M., *Trans. Amer. Inst. chem. Engrs*, 18 (1926) 365
[3] BROWN, G. G., et al, *Unit Operations*, John Wiley & Sons, New York, 1950
[4] CAMP, T. R., *Proc. Amer. Soc. civ. Engrs*, 71 (1945) 445
[5] COE, H. S., and CLEVENGER, G. H., *Trans. Amer. Inst. min. (metall.) Engrs*, 55 (1916) 356
[6] COMINGS, E. W., *Industr. Engng Chem. (Industr.)*, 32 (1940) 663
[7] DALLAVALLE, J. M., *Micromeritics*, Pitman Publishing Corp., New York, 1943
[8] DARBY, W. A., *Waterwks & Sewerage*, 86 (1939) 209
[9] DEANE, W. A., *Trans. Amer. electrochem. Soc.*, 37 (1920) 659
[10] DORR, J. V. N., and LASSETER, F. P., 'Solid–Liquid Separations', Vol. VI, p. 782–799, in *Colloid Chemistry*, Ed. J. Alexander, Reinhold Publishing Corp., New York, 1946
[11] DORR, J. V. N., and ROBERTS, E. J., *Trans. Amer. Inst. chem. Engrs*, 33 (1937) 106
[12] EGLOF, C. B., and MCCABE, W. L., *Trans. Amer. Inst. chem. Engrs*, 33 (1937) 620
[13] FREE, E. E., *Engng Min. J. (-Press)*, 101 (1916) 243, 429, 509, 681
[14] FISCHER, E. K., and GANS, D. M., Vol. VI, p. 286–327, in *Colloid Chemistry*, Ed. J. Alexander, Reinhold Publishing Corp., New York, 1946
[15] GERY, W. B., *Chem. Engr*, 62 (1955) 228
[16] KAMMERMEYER, K., *Industr. Engng Chem. (Industr.)*, 33 (1941) 1484
[17] KINCH, G. J., *Trans. Faraday Soc.*, 48 (1952) 161
[18] NICHOLS, H. G., *Trans. Instn Min. Metall., Lond.*, 17 (1908) 293
[19] PERRY, J. H., *Chemical Engineers' Handbook*, McGraw-Hill Book Co., New York, 1950
[20] ROBERTS, E. J., *Trans. Min. Engrs*, 1 (1949) 61
[21] ROBERTS, E. J., *Trans. Amer. Inst. min. (metall.) Engrs*, 112 (1934) 178

[22] STEINOUR, H. H., *Industr. Engng Chem. (Industr.)*, 36 (1944) 618, 840, 901
[23] STEWART, R. F., and ROBERTS, E. J., *Trans. Instn chem. Engrs, Lond.*, 11 (1933) 124, 126
[24] TAGGART, A. F., *Handbook of Mineral Dressing*, John Wiley & Sons, New York, 1945
[25] TALMADGE, W. P., and FITCH, E. B., *Industr. Engng Chem.*, 47 (1955) 38
[26] AGRICOLA, G., *De Re Metallica*, transl. by H. C. Hoover, and L. H. Hoover, Dover Publications, Inc., New York, 1950

BIBLIOGRAPHY

DONALD, M. B., 'Sedimentation and Flocculation', *Trans. Instn chem. Engrs, Lond.*, 18 (1940) 24–28

DAVIES, C. N., 'The Sedimentation of Small Suspended Particles', *Instn chem. Engrs* and S.C.I. Symposium on Particle Size Analysis, 1947

TOMALIN, E. F. J., 'The Theory and Practice of the Treatment of Coal Washery Effluents', *Trans. Instn chem. Engrs, Lond.*, 16 (1938) 231

Torq, Squarex, Clariflocculator, Flocculator, Multifeed, Monorake, and Hydro-Treator are trademarks of Dorr-Oliver Incorporated.

10

WET CLASSIFICATION

R. FORBES STEWART

CLASSIFICATION[1-3, 6, 8, 12, 16] may be defined as the separation of solids into two or more fractions from their suspension in a fluid medium, usually water. The separation is seldom, if ever, complete, mainly because of variations in the shape and density of the particles, but the extent of its completeness is determined by sizing the products on a series of standard screens down to 325 mesh per inch (0·043 mm opening) and below this by various subsieve sizing methods based on Stokes' law for rates of settling of solid particles in a fluid. (See 'Size Analysis' and 'Laws of Settling'.) In general, wet classification methods are: non-mechanical classification; mechanical classification; and hydraulic classification.

The largest use of classification is in the dressing of metallic ores and non-metallic minerals. Other important applications of classification methods and machines are:

Draining, in which particles are separated from suspension together as one fraction without regard to size or specific gravity. This may be defined as the dewatering action occurring when wet particles of a granular nature are placed on or dragged over a flat or inclined surface.

Washing, in which the main purpose is the removal of already dissolved substances from particles collected from suspension in one fraction without regard to size or specific gravity by one or more applications of water or other liquid, each application being followed by a dewatering step to displace the bulk of the dissolved substances.

Leaching, which may be defined as a process of removing, by the application of a solvent, that constituent of the substance being treated which is readily soluble in the solvent applied.

Closed circuit grinding, the classification in which the main object is the separation of partly ground material into finished and unfinished fractions in order to obtain increased capacity and savings in grinding costs.

This section is limited to a discussion of classification in an aqueous medium, where the size separations referred to are accomplished by employing the difference in the rates of settlement in water between particles of different sizes. These rates follow Stokes' law of viscous resistance in the case of very small particles and Newton's law of eddying, or turbulent resistance, for particles above about 1 mm diameter (sp. gr. 2·65) in water. The intermediate rates are best found by a dimensionless equation devised by CASTLEMAN[8], from which it is possible to predict the behaviour of a particle of any size. By using a modification of this method it is also possible to predict the settling rate of any size range in a mass of particles where the effect of 'crowding' or interference of other particles becomes pronounced.

Where conditions are such that the particles are relatively far apart their movement is said to be free settling; where crowding occurs and the particles are interfering with one another the state is described as one of hindered settling. Under ideal conditions the relative diameter of two particles for constant conditions of viscosity, flocculation, *etc.*, determines their relative

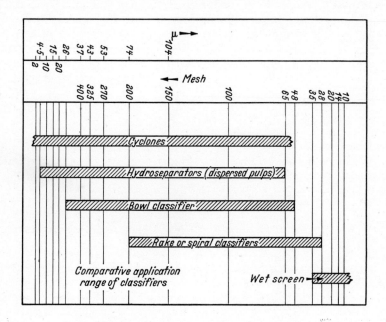

Figure 1. *Comparative classification range of classifiers*

rate of fall, but in practice particle shape and density are also involved. The settling ratio (both free and hindered) is defined as the ratio of the diameter of the light particle to that of the heavy particle which is equal-settling with it; where heterogeneous mixtures are involved, this becomes a factor of importance in making size separations on a commercial basis.

When a body falls in a liquid medium it encounters a resistance that is a function of its velocity, which increases until the resistance force equals the pull of gravity. Thereafter the body falls at this constant terminal velocity.

The state of flocculation of the very small [minus 200 mesh (0·074 mm)] particles in a pulp has a profound effect on their classification characteristics, if the flow rates and detentions involved permit the formation of flocs or clusters, since these settle at very much higher rates than the individual particles of which they are composed. For the most efficient classification of such fines a well-dispersed pulp is usually desirable, and is often essential, whether dispersion be obtained by chemical or mechanical means. For a detailed discussion see 'Laws of Settling', pp. 162–171 and reference 12.

Aqueous classification[7] may be accomplished in a number of wet type mechanical devices such as the reciprocating rake and spiral classifiers for separations in the coarser size ranges, and by the use of machines such as the bowl classifier, centrifuge, hydroseparator, or liquid cyclone for finer sizes. Non-mechanical hydraulic classifiers are also still used to a certain extent.

Classification is affected by the following factors: specific gravity of particles, size of particles, shape of particles, possible selective attachment of air bubbles, magnetism of particles, density of fluid medium, and viscosity of medium. Thus it seems that unless the material to be sized is uniform in shape and specific gravity, there is little chance of obtaining a true size separation by classification methods alone. Screening, however, will give a size separation which is independent of the above factors, although particle shape, magnetism, and viscosity of the fluid may have some effect, especially when the finer sizes of screens are used. (See Section 6, 'Screening Machines', p. 147.)

Classifiers have been designed and built in a wide variety of types which may be divided into two main groups: hydraulic and non-hydraulic, depending on whether or not an upward current of 'hydraulic' water is used as a sorting column. The latter may be further subdivided into mechanical and non-mechanical classifiers.

In addition to the above there are also centrifugal classifications, pneumatic classification and wet screening separations. By far the largest number of classifiers now employed in metallurgical and industrial sorting operations are those of the mechanically operated type and *Figure 1* shows the comparative particle size separation range of the different forms of this type of classifier.

NON-HYDRAULIC CLASSIFIERS

Non-mechanical Classifiers

The simplest type of classifiers consist of tanks, usually conical in form, placed in a stream of pulp, the feed flowing in on one side and overflowing on the other. The overflow contains the fine material, which does not settle, while the coarser material sinks to the bottom of the tank and is discharged by a spigot. Several of these cones may be placed in a series of increasing size from which may be obtained a series of graded products determined by the velocity of the surface stream and upward eddy currents in each tank and the law of equal falling particles. Rittinger's Spitzkasten[8] is such a series. Devices of this type were at one time widely used for desliming operations in ore dressing.

Among the other non-mechanical classifiers is the Caldecott cone which gave results considered to be satisfactory before the advent of more efficient mechanical classifiers. It consists of a sheet-iron conical tank with a central feed pipe and a peripheral overflow launder with the sides of the cone sloping 60° or 70° from the horizontal. A similar classifier to the Caldecott is the Allen cone, which is distinguished by an automatic discharge. This has been used to a considerable extent in phosphate rock washing plants. Other cone classifiers are the Callow, Boylan, Wuensch and Nordberg-Wood. For a discussion of non-mechanical classifiers see reference 8, pp. 35–43, which includes illustrations of all types. Cone classifiers range in size from 4 to 8 ft. in diameter and up to 8 ft. in depth and have capacities ranging from 100 to 1,000 tons/24 h depending upon the character of the feed. At the present time their use is limited to special applications, generally of limited size.

These devices may be used for both classifying and thickening by varying the rate of feed. For thickening, the feed is retarded to a rate whereby essentially all solids settle out and are discharged in the underflow—usually intermittently.

Mechanical Classifiers

In mechanical classifiers of the rake and spiral types[1, 6, 8, 12] such as the Dorr, Akins, Wemco, Hardinge, drag or Esperanza, classification is a function of surface currents and rising velocities as in the non-mechanical machines, but in addition mechanical agitation is employed. The agitation due to rakes, surge, plunge of the feed *etc.* is adjusted to produce a condition of turbulence sufficient to prevent the settlement of the undersize particles, thereby making a cleaner sand product than can be obtained from non-mechanical machines.

In any of these classifiers used for desliming of pulps the size of the particle overflowed is a function of its settling velocity. With a fixed size of unit and quantity of feed the settling velocity can only be changed by variations in the density of feed or overflow to produce a separating medium of high density and high viscosity, thereby greatly reducing the rate of settlement of the individual particles.

Functions[16]—Primarily, a classifier must function as a container in which the solids segregate. This could be a tub or tank of almost any shape. Although batch separations can be made in such a container, for modern commercial operation the classifier should fulfil one or more, and in many cases all, of the following functions: (1) It must collect the undersize and convey it to a convenient point by means of a launder. (2) It must separate the oversize from the undersize and dewater it as much as possible. This is usually accomplished by rakes or screws in inclined troughs whereby the oversize is removed from the pool of overflow pulp and drained. (3) In closed-circuit grinding, which is the principal application of classifiers, the oversize must be conveyed from the discharge to the feed end of the mill. This is performed by the rakes or screws. (4) Changing operating conditions of tonnage, temperature and character of solids necessitate some flexibility of control in obtaining the separation desired. Within limits, this can be done by moisture control, but it is also advantageous to control both the actual and effective area of the classifier pool. In certain classifiers this can be done by changing the overflow weir, the slope and the raking speed.

With correct dilution and rake speed, a pulp fed to the classifier segregates rapidly into (*a*) the coarse, quick-settling material, larger than the mesh of separation, and (b) the fine, slow-settling material, finer than the mesh of separation. The coarse portion sinks to the bottom of the pool and is transported up the inclined deck by the action of the rakes. The fine portion, held in suspension by the specific gravity of the pulp and the degree of agitation of the rakes, flows to the overflow lip and leaves the classifier.

With the proper size and type of classifier, the mesh of separation can be controlled by means of various adjustments including pulp dilution, tank slope, and rake speed. Broadly speaking, the greater the rake speed, the higher is the concentration of solids in the pulp; and the steeper the slope, the coarser is the mesh of separation and *vice versa*. All factors, including speed, dilution, slope, separation mesh, and nature of materials, remaining the same, the overflow and raking capacity are proportional to the classifier width. The finer the material to be raked, the slower must be the raking speed, and the less the tank slope, resulting in reduced capacity.

Reciprocating rake classifiers[13] are exemplified by the Dorr classifier (see *Plate I*), which is essentially an inclined rectangular settling tank in the form of a trough open at the upper end for the discharge of oversize grains and furnished with a weir at the lower end where fine particles and slimes are overflowed. In various installations the inclination of the bottom of the tank is $1\frac{1}{2}$ in. or more per foot, depending on the size at which a separation is to be made. A reciprocating raking mechanism carries settled solids up the inclined deck to the sand discharge lip. Feed enters through either end of a transverse launder near the overflow. The launder is equipped with splitter vanes for obtaining an even distribution across the width of the tank.

The modern type H mechanism is supported at only three points. At the discharge end it is carried by the support tube which spans the tank and rests on two pedestal bearings. A support shaft and bearing attached to the hydraulic lifting device by a hanger rod provides the third point of support near the overflow end. All bearings are above the pulp level. The machine is designated as simplex for one rake, duplex for two rakes, and quadruplex for a machine with four rakes. The type H mechanism may be driven by a constant or variable-speed drive, connected by conventional V-belts to a pinion shaft, and integral with this shaft are two helical pinions of opposite hand which drive the gears and crank shafts. To each crank shaft is mounted a connecting rod which drives the rakes forward and backward.

The vertical lifting of the rakes at the end of each stroke and subsequent lowering at the beginning of the next is activated by a cam, cam roller arm, and pivot shaft contained in an oiltight housing at the head end of the unit. The pivot shaft is bolted to the upper rocker arm so that the oscillation of the cam roller causes the entire torque tube, rocker arm and slide assembly to rock from side to side on its centre-line axis. This motion alternately raises and lowers each set of rakes in synchronization with their forward and backward movement. The motion of the reciprocating rakes aids the release of undesired fines from the sand bed through agitation and slight attrition. Natural drainage, which may be assisted by water sprays applied above the pulp level, further aids in the removal of fines and entrained liquid. Removal of a considerable portion of accompanying free moisture from the sands is accomplished by compression on the upper end of the inclined drainage deck above the pulp level.

Rake-type classifiers are divided roughly into three groups: light, medium, and heavy duty, depending upon the amount and characteristics of the materials which they are to handle and the mesh of separation.

Light-duty machines range in size from 1 ft. 6 in. wide × 18 ft. long to 5 ft. × 30 ft. and their raking capacities vary from 100 tons/24 h for a 2·70 sp. gr. ore at 200 mesh to 2,800 tons for a 4·50 sp. gr. ore at 28 mesh. The mechanisms are built in simplex and duplex units. Medium-duty classifiers range in size from 6 ft. × 18 ft. to 16 ft. × 30 ft. with an extreme variation in raking capacity from 750 to 10,500 tons/24 h depending upon the above-mentioned conditions. For heavy-duty machines, the corresponding figures are 5 ft. × 24 ft. to 16 ft. × 32 ft. with rake capacities ranging from 2,200 tons to 25,000 tons/24 h. The medium and heavy-duty classifiers are made in duplex and quadruplex units. Bowls for bowl classifiers range in diameter from 6 ft. to 25 ft., depending upon the size of the reciprocating raking mechanism.

Power consumption by classifiers is notably low compared with other process machinery. It will range from 1 h.p. motors for the smallest light-duty machine, to 15–20 h.p. for medium duty, and 30 h.p. for the largest heavy-duty machines. The bowl mechanisms will require 1 h.p.–7½ h.p. motors over the extreme range of sizes.

Another reciprocating rake classifier, the Geco, was made by General Engineering Co. In this machine two sets of blades act independently, one moving up the slope as the other is travelling back over it. Thus, the raking movement is practically continuous, the sands are not allowed to slip back, and the pulp is agitated continuously. The design allows a steeper rise which facilitates the return of sands to the mill in closed-circuit grinding and reduces floor space. This machine is no longer on the market.

The Dorr bowl classifier consists essentially of a straight Dorr classifier upon the lower end of which there is superimposed a shallow circular bowl with a revolving raking mechanism. Feed enters through a central loading well, overflow takes place across a peripheral weir, and coarse solids are raked to an opening in the centre of the bowl and pass thence into the reciprocating rake compartment below. Bowl-type classifiers are also made by the Hardinge Company. They are used when a cleaner rake product is desired, where the overflow product is to be of extremely fine size, and where the overflow capacity is large in comparison to the raking capacity.

Spiral rake classifiers[13] belong to another group of mechanical classifiers which have applications similar to those of reciprocating rake machines, the largest fields being closed-circuit grinding and minerals washing. A typical model is the Akins (see *Plate II*) which consists of the usual sloping trough where the pulp is kept gently agitated by one or more spiral ribbons mounted

Plate I.   Dorr Type HX reciprocating rake classifier.   (By courtesy of Dorr-Oliver Co. Ltd.)

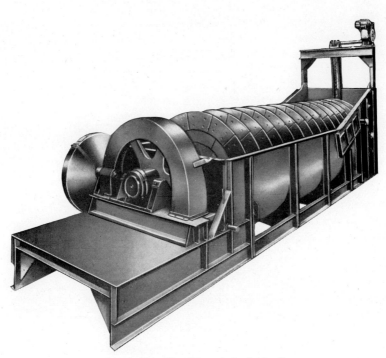

Plate II.   Akins spiral rake classifier

Plate III. Dorr sand washer. (By courtesy of Dorr-Oliver Co. Ltd.)

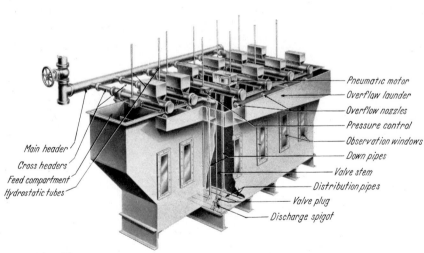

Plate IV. Dorrco jet sizer (9½ pocket unit). (By courtesy of Dorr-Oliver Co. Ltd.)

on a shaft. When a mixture of water or solution and solids is fed into the settling pool at the lower end of the trough, the coarse particles settle out and are carried up the incline by the revolving spiral. Drainage is affected at the upper end of the inclined trough or deck, and at the lower end the discharge of fine particles carried in suspension with slime and water is accomplished by an overflow weir. The mesh of separation, capacity, power consumption, and other functions of the Akins are controlled by the same factors as those for the Dorr classifier. The slope is usually from 3 to 4 in./ft., and several sizes in both the 'high weir' and 'submerged spiral types' are furnished. Wash sprays can be used above the pulp level to aid in the removal of entrained fines from the sand load. Spiral raking classifiers similar to the Akins are manufactured by Denver Equipment Co. and Western Machinery Co.

The Hardinge countercurrent classifier differs from the regular type of spiral classifiers in that the continuous spiral rake is rigidly attached to the inside of a rotating drum mounted on tyres and rollers at a slight slope from the horizontal. The material to be classified is fed at the lower end of the drum, and, as the machine revolves, the coarse particles that settle out are carried up the slope and are discharged at an elevation above the pulp bath level. The fines are held in suspension and washed back towards the overflow end by the wash water introduced at the sand or oversize discharge end. Control of size separation is obtained through the speed of rotation, pulp dilution, and depth of pulp pool. Its principal application is in the closed-circuit grinding of ores.

Hydroseparators[7] in action are essentially overloaded thickeners in that all of the feed particles are not settled out and the portion of the feed solids carried off in a cloudy supernatant liquid will be composed of the finest particles in the original feed. The practice of deliberately overloading a thickener so that it classifies the feed particles roughly according to their size is known as 'hydroseparation'. In practice, the mechanism for hydroseparation does not differ materially from that suitable for thickening (see Section 9, 'Sedimentation'). The main points of difference between sedimentation machines functioning as hydroseparators (classifiers) and thickeners are discussed briefly below.

Hydroseparators have a higher ratio of overflow to underflow capacity. Thickeners are usually rated on the basis of unit areas—square feet of settling area per ton of solids per 24 h. On the other hand, hydroseparator capacities are expressed as overflow rates, that is, vertical upward displacement. For example, to produce a fine overflow product (98·5 per cent minus 325 mesh) on a water-floated clay, a hydroseparator has to provide space to overflow a 26:1 pulp at the rate of 3 ft./h. On a thickener basis this is equivalent to a unit area of 11·6 ft.^{2}/ton/24 h. The underflow must be removed at a higher dilution than in other types of classifiers, which means that not only more water must be subsequently removed from the product but that the hydroseparator underflow will carry a larger amount of undersize as void solids. This may necessitate an additional machine, such as a rake classifier, if a clean product is required.

When hydroseparation is made at about 60–200 mesh, as in sand or phosphate rock washing plants and in many other industrial applications, the amount of settled material to be raked can be much greater in relation to the size of the machine than with thickeners operated as such. Moreover, the action of hydroseparation itself removes from the sands most of the slime normally acting as lubricant so the rake product or underflow is more difficult to handle. To meet these conditions, heavier mechanisms rotating at a higher than normal speed must be used in place of the standard thickener mechanisms. The field for the application of hydroseparators may be divided roughly into three classes: (1) for extremely fine separations where large areas and very quiescent settling conditions are essential; (2) for roughly sloughing off slimes ahead of subsequent treatment where the expense of bowl classifiers is not warranted; and (3) for rough removal of fines and dewatering a very dilute, large volume feed. Slush from anthracite coal breakers is typical, where this step is primarily for dewatering, but the elimination of fine slimes with high ash content is also accomplished.

Principal makes of hydroseparators are the Dorr hydroseparator, the Hardinge hydroclassifier and the Denver hydroclassifier.

Washing Classifiers[12]

Although log washers are not ordinarily ranked as classifiers they are included in this section because they are used largely for the separation of valuable minerals from fine wastes. These washers, which might be termed rough washing classifiers, usually consist of one or more 'logs' or shafts carrying metal blades at the surface and placed in an inclined trough at a slope of about $1-1\frac{1}{2}$ in./ft. A machine of this type is the Allis-Chalmers log washer.

The Dorr washer is a standard classifier adapted for breaking up a tenuous bonding of dissimilar grains such as a sand–clay mixture. It is substantially a single-stage Dorr classifier with a perforated washing trommel partially submerged in the classifier bath across its overflow end. This arrangement makes it possible to obtain three products from a mixed feed: (1) a scrubbed coarse oversize as trommel discharge, which may be used as coarse concrete aggregate; (2) a granular rake product (fine aggregate); and (3) a fine overflow consisting of silt and slime.

Another mineral washing machine is the Dorrco sand washer (see *Plate III*), which is designed primarily for the desliming of common sand for concrete aggregate use. It is popular in this industry because of its high raking capacity per unit of space. However, a certain lack of flexibility, as compared with a Dorr rake classifier, limits its field of use mainly to desliming and dewatering large tonnages of sandy material in the range of 28–80 mesh where the feed is predominantly coarser than 100 mesh and where the ratio of water to solids is usually less than 2:1 by weight as in dry pit operations, but it also has some applications as a rough classifier in the coarser range, either in open or closed circuit for general industrial work.

Light scrubbing and washing of coarse sand is also performed in screw washers similar to spiral classifiers and operating in the same manner, with the moist sands discharged at the upper end of the screw trough.

Drag classifiers were once quite widely used, but have now been largely superseded by the modern mechanical classifiers. They consist simply of a continuous belt or link chain which carries a series of flights and runs on pulleys or sprockets so located that part of the travel of the belt is above and part below the level of pulp in the tank. Feed is delivered at the lower end of the tank, slimes overflow a weir, and sands, after settling, are dragged up the sloping bottom by flights and discharged at the upper end. The Esperanza is a typical example of these machines.

In a number of industries where washing, leaching, or the dissolution of chemicals is an important operation, the Dorr Washing or Multideck Classifier[1] has performed very efficiently. This machine is, in effect, a

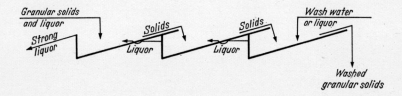

Figure 2. Flow sheet showing operation of multi-deck classifier

group of interlocked, series connected classifiers in which the raking mechanisms and raking compartments may range in number from two to six, or more, depending upon the characteristics of the problem. The long mechanism is equipped with a series of individual rakes, one for each compartment, and is driven from a single head motion. Wash water, solution, or lixivant added to the last or top compartment flows progressively through all compartments via external connecting launders to the first compartment whence it overflows enriched in solubles. Feed solids added in the first compartment are raked upwards through the others in the opposite direction to the last compartment whence they are discharged with a minimum of solubles. *Figure 2* is a flow sheet of the countercurrent decantation operations which are accomplished in this machine. The Akins spiral rake classifier can be arranged in a similar manner.

HYDRAULIC CLASSIFIERS OR SIZERS

These classifying devices are characterized by the use of water in addition to that of the feed pulp, so introduced that its course of flow opposes that of the settling particles. It is termed hydraulic water. Its quantity and resultant velocity constitute the principal means of controlling operations and results.

Hydraulic classifiers are of two types, the grouping being based on the degree of crowding of the particles in the separating zone. If this zone is relatively uncrowded so that collision between particles is relatively infrequent, the machine is free-settling. If the zone is crowded, so that no particle can pass through without material hindrance, the device is hindered-settling.

Free-settling Hydraulic Classifiers[8]

The fundamental principle of free-settling hydraulic classifiers is to utilize the varying settling rates of grains of different sizes or densities to obtain a series of graded products. The particles, during their forward movement by the carrying current, are subjected to a series of rising hydraulic currents or eddies, which usually decrease in velocity through a series of pockets. The grains, which are sufficiently heavy to settle against the upward current in the first pocket, do so, and are withdrawn through an orifice at the bottom, while the lighter grains are carried on to the pockets with successively lower velocities where in turn they are collected and withdrawn.

An example of free-settling hydraulic classifiers is the Richards deep-pocket or launder-vortex machine, which consists of a shallow trough with four pockets of successively increasing cross section. Each pocket is provided with a 9 in. long sorting column composed of a section of 3 in. pipe with a vortex fitting at the bottom. The vortex imparts a whirling motion to the entering hydraulic water which carries light grains upward into the next pocket and allows the heavy ones to drop into the spigot. This classifier will handle about 60 tons/24 h of pyrite ore containing about 75 per cent of quartz gangue ground to $\frac{3}{16}$ in. size and in a 10:1 pulp of water and solids. Other machines of this type are the Evans, Calumet, and Richards annular vortex.

Hindered-settling Hydraulic Classifiers[7, 8]

One of the most widely used classifiers of this type is the Dorrco sizer which is a modern form of its prototype, the Fahrenwald sizer, the principal improvement being the adaptation of the 'pressuretrol' and motor combination actuated by a hydrostatic tube within the pocket, which makes its operation completely automatic. *Plate IV* shows a Jet sizer which, in appearance, varies considerably from the Dorrco sizer. The sizer consists of a trapezoidal tank divided into three or more communicating pockets with their individual controls and valves.

The pockets are separated from each other by submerged baffle plates and are provided at the bottom with constriction plates which have holes with diameters and spacing designed to produce a teeter condition (alternate rapid rising and falling) in each pocket correlated with the approximate size and gravity of the particles to be retained in that pocket. Water under hydraulic pressure from a header is introduced through an adjustable pinch valve into the hydraulic compartment under each pocket, and flowing upward through a constriction plate, furnishes a uniform rising current in the sorting zone of each compartment. A discharge valve, automatically controlled by a special mechanism, is provided for each pocket, the automatic control mechanism functioning in accordance with conditions of pulp density in the sorting compartment. Associated with this mechanism is an indicator which enables the operator to detect easily and at all times the pocket density conditions. Feed is introduced at the narrow end of the sizer into the feed compartment which is equipped with a constriction plate, and in which preliminary sorting occurs. When the machine is adjusted the position of the discharge mechanism in each pocket reaches a balance determined by the hydrostatic pressure established by the control instrument. When no change occurs in the screen analysis or composition of the feed the product of each individual spigot is held constantly within the desired range of sizes and is discharged continuously. Whenever a change occurs in the size distribution or rate of the feed, the machine adjusts automatically to such new conditions, and the discharge is again continuous until a further change in feed takes place. No manual adjustment is necessary except in the case of abnormal changes in the feed or where the product requirements are altered. For such requirements the operator can adjust the machine readily to meet the new conditions.

A radical redesign of the Dorrco sizer called the 'Jet sizer' has recently been introduced to the metallurgical, chemical and non-metallic fields. The unit makes more effective use of hydraulic water to produce clean deslimed fractions sized within very narrow limits. It is particularly applicable where hydraulic classification is indicated for large tonnage scale sizing or where grading of minus 8 mesh pulps is required. The machine has wide flexibility of cell arrangement whereby $1\frac{1}{2}$ to $21\frac{1}{2}$ pockets can be set up in series or in banks in virtually any combination. *Plate IV* shows a $9\frac{1}{2}$ pocket machine.

Hydraulic classification water is introduced by means of an overhead piping arrangement rather than through a constriction plate. Water from a constant head tank travels through a large header and into a horizontal pipe directly above each bank of pockets. Two vertical pipes into each compartment convey the water to four removable horizontal distribution pipes along the bottom of the pocket and nozzles at selected spacings in the horizontal members admit water to the sizing pockets. Product is discharged continuously through spigots in the centre of the floor of each pocket, control is by a pneumatically operated system in which the action of a hydrostatic head is magnified to operate a valve plug.

Some of the major fields of application of such machines are: lead and zinc ore sizing or preliminary concentration; ilmenite, rutile, fine coal and phosphate rock sizing or concentration; specification sand and abrasive sizing.

There are now a number of other hindered-settling classifiers in commercial use which are described briefly below.

The Deister 'Concenco' classifier (supersorter) consists of a series of sorting compartments of uniform size. By regulating the amount of hydraulic water to each compartment a series of products is obtained, grading from coarse at the feed end to fine at the overflow end.

The Deister cone baffle classifier comprises a cast iron truncated cone which is bolted to the bottom of a pointed box, conical settling tank, or launder. The interior of the cone contains a series of conical diaphragms slotted to permit the passage of pulp. The hydraulic water passes into the classifier through radial holes in the side and on rising encounters the descending ore grains. The lighter particles to be separated are washed over the top, while the heavier ones settle and are discharged through an orifice in the bottom of the cone.

There is also the Bunker Hill, which is similar to the Fahrenwald, the Pellett, the Delano and the Richards-Janney, which have limited applications. They are described in detail on pp. 48–55 of reference 8.

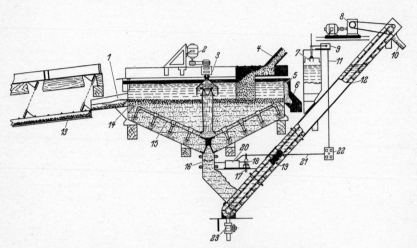

Figure 3. Wilmot Hydrotator hindered-settling classifier

1. Clean coal discharge
2. Pump motor
3. Pump
4. Raw feed
5. Pool
6. Overflow
7. Float
8. Conveyor drive
9. Limit switch
10. Refuse
11. Cable to limit switch
12. Refuse conveyor
13. Dewatering shaker screen
14. Baffle
15. Agitator
16. Automatic Butterfly valve
17. Water supply
18. 0-0 valve
19. Refuse dewatering conveyor
20. Electric valve operator
21. Cable to electric operator
22. Wilmot automatic control panel
23. Slush valve

The Hydrotator classifier[17] made by the Wilmot Engineering Co. is a partially hindered-settling machine which is used almost exclusively for coal cleaning (see *Figure 3*). It differs from the previously mentioned devices in that it has a hindered-settling zone at the bottom which supplements a free-settling zone at the top of the classifier. These zones are not brought about by a constriction of the bottom of the classifier, but by an increase in velocity of flow in the bottom zone without increase in the top zone. The cardinal feature of the Hydrotator is the continuous circulation through the machine of a relatively large quantity of fine mineral particles by the use of an outside pump discharging through an arrangement similar to a revolving lawn sprinkler. Discharge of the sediment occurs whenever the pulp density exceeds a predetermined figure. Control is by means of a float which actuates a valve.

The Dorr Hydroscillator[7, 13] is a comparatively new machine which combines the features of both mechanical and hydraulic classification. Sharper separations and cleaner rake products are obtained than with conventional rake and spiral classifiers. Compared with multiple-pocket hydraulic classifiers, this unit uses extremely small quantities of water and delivers an overflow of high percentage solids and a dewatered oversize product.

In spite of their wide use mechanical classifiers have certain limitations. For instance, the rake product entrains fines from the incoming feed, and the various advantages of hindered-settling classification, including improved hindered-settling ratios mentioned above, cannot be utilized. Although not critical in many applications, such as in most closed-circuit grinding and desliming operations, there are instances where sharper classification and a deslimed sand is desirable.

The multiple-pocket hydraulic sizer or classifier, which is described earlier in this section, operates in such a manner as to meet these requirements. However, this machine has a number of disadvantages, including (as compared with the mechanical rake classifier) the use of relatively large volumes of water, considerable loss of head room, and limited capacity per square foot of floor space.

Therefore, to combine the advantages of both the 'mechanical' and 'hydraulic' type classifiers the Hydroscillator has been developed. The fields of operation have not been widely explored as yet, but it has shown satisfactory results where sharp two-product separations are required as in iron ore beneficiation, silica sand processing, and closed-circuit grinding of ores. However, its considerably increased cost over conventional rake and spiral classifiers has militated against its extensive adoption.

Centrifugal Classification (Hydraulic)

Centrifugals—The theory and operation of the various types of centrifugal machines are discussed elsewhere (see 'Centrifugals', Vol 6), and only the functioning of centrifugals in the field of hydraulic classification will be considered here[3, 5, 10, 11]. They have the advantage of requiring a much smaller floor space than gravity machines, since the effective settling area under centrifugal force, which may reach 9,000 times gravity, is greatly increased. However, most types of centrifuges have high first cost and relatively high power and maintenance requirements.

A representative machine of this type is the Bird continuous centrifugal, which consists of a truncated cone fixed to and rotating at high speed on a horizontal shaft. Rotating at a slightly slower speed is an internal spiral to remove continuously solids deposited on the inner surface of the cone or 'bowl'. Feed enters the cone by means of the hollow centre shaft; overflow leaves the cone or bowl through ports at the large end, and oversize solids, moved by the spiral, exit through ports at the small end of the cone.

Another centrifugal separator which finds many applications in the field of classification is the Merco continuous centrifuge which rotates at such a high r.p.m. that it generates centrifugal separating forces as high as 9,000 times gravity. It has a unique return flow principle which recirculates a considerable portion of the underflow combined with wash liquor. Simple external adjustment of the wash liquor counterflow creates any desired particle separation equilibrium. Wash liquor counterflow along with feed rate and concentration are adjusted to exceed the settling velocity of the smaller particles, forcing them back to the overflow. A faster counterflow classifies at a larger size and a slower rate gives a smaller size. Major applications are in bentonite refining, pigment classification in the low micron range, and separation of starch from gluten based on specific gravity differentials.

Continuous centrifugals functioning as classifiers are used in the processing of very fine-grained pigments such as titanium dioxide and lithopone, which require sizing at 100 per cent–5 μ. Other materials such as clay, calcium carbonate, and fillers are similarly classified. Also, in closed circuit grinding in the cement industry, the Bird can handle a thick slurry ground to 85 per cent–200 mesh, which is suitable for direct blending and calcination.

Liquid Cyclones [5, 9, 10, 15, 19, 20]—The hydrocyclone is a compact cylindro-conical classification unit utilizing centrifugal force rather than gravity. It was developed at the Dutch State Mines, Limburg, in 1939 and was first used in fine coal washing plants. Somewhat later, manufacturing rights were acquired by the Dorr-Oliver organization and made under the name of DorrClone from 1948 onwards.

It soon became evident that liquid cyclone processing had unique advantages in many instances compared with other methods of classification and since that time an intensive research programme has been carried on by several equipment organizations in the development of machines and their applications to various classification operations. Other models of hydraulic cyclones having similar characteristics to the DorrClone are the Centriclone, Centri-cleaner, Heyl and Patterson Cyclone, Whirlcone of Georgia Iron Works, the Prinz Streamcleaner, and the Krebs Cyclone of Equipment Engineers, Inc. *Plate V* shows a cutaway view of the DorrClone showing construction and the relation of principal parts.

In recent years the hydrocyclone has been gaining rapidly in popularity in the mining, minerals processing and chemical processing fields, and it should be considered wherever it is desired to make a size separation in the range of from 2 to 200 μ or for dewatering when the recovery of fines is not important. Where it is applicable, the cyclone offers the advantages of high capacity (10–1,500 gal./min) for the space required and moderately low initial cost.

The cyclone consists mainly of a short cylindrical section mounted on an inverted conical section, as shown in *Plate V*. In operation, feed furnished by a centrifugal pump enters tangentially into the upper section of the liquid cyclone at sufficient pressure, ranging between 5 and 120 lb./in.^{2}, to support the vortex action of the pulp in the unit. This action is somewhat similar to that occurring in dry cyclones serving as dust collectors. The centrifugal forces in the vortex throw the coarser particles, contained in the feed, to the walls of the cone where they collect and pass downward and out of the unit through the apex valve and tailpipe. The diameter of this apex

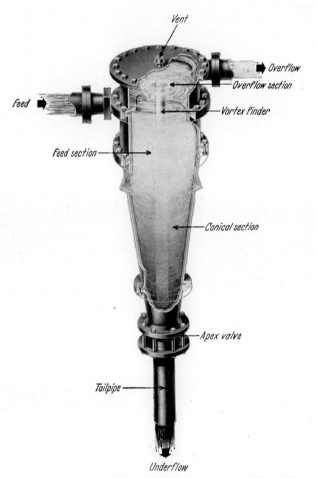

Plate V. DorrClone liquid cyclone. (*By courtesy of* Dorr-Oliver Co. Ltd.)

valve opening controls the consistency of the underflow—as it decreases the consistency increases, and vice versa. Fine particles move to the inner spiral of the vortex and, together with most of the water, are displaced upward into the vortex finder. This overflow goes directly to a 90° elbow attached to the vortex finder.

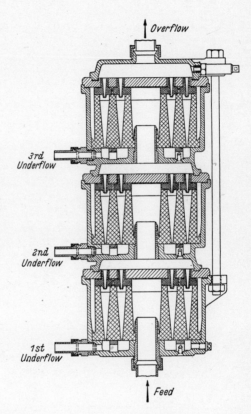

Figure 4. Dorr Type TM3 multiple cyclone. (*By courtesy of* Dorr-Oliver Co. Ltd.)

One of the cyclone's best applications is as a desliming classifier for the removal of fines ranging between $147\,\mu$ (100 mesh) and $10\text{--}20\,\mu$ from a mixed feed. It classifies particles in almost any pulp which can be pumped through the unit—even in those pulps of such high plasticity and solids content that their classification is impossible in gravity sedimentation machines.

The main action in the cyclone is one of classification[19] rather than thickening. For effective thickening it is desirable to cause the fine particles to form flocs or agglomerates which have much higher settling rates than the individual particles. In the cyclone flocculation does not take place for two reasons: (1) the large shear forces caused by differences in tangential velocities quickly break up any existing flocs and prevent the formation of new ones, and (2) the tangential force on the particle is so much greater than the inter-particle force that flocs cannot form during their brief stay in the cyclone. Also, the same characteristic which makes the cyclone a poor thickener increases its effectiveness as a classifier. For good classification clean separation of particles is required and this can be obtained in the cyclone without the addition of dispersion agents, which is frequently required in gravity classifiers.

Various forms in the design of cyclones have made their appearance. One type called the Centriclone is equipped with a motor-driven impeller in the cylindrical section of the casing which obviates the necessity of an auxiliary pump. Another, the Krebs Cyclone, is a two stage arrangement where primary separation takes place in a cylindrical cyclone, from which the underflow is discharged tangentially into a smaller, conical secondary cyclone. The overflow from the secondary can be recycled if desired. A third variation is the Whirlcone, in which the whirling motion of the feed is actuated by a spiral chamber near the inlet.

Dorr-Oliver Inc. have introduced several modifications of the DorrClone, among which is the type TM multiple-cyclone which may be arranged in one, two or three stage assemblies. *Figure 4* shows the three stage arrangement for the 24 cyclone rubber block unit. A 32 cyclone unit is moulded in bakelite.

The net throughput capacity of a TM3 three stage unit depends upon the type and number of cyclones used and the volume of the recycle streams. For feed pressures in the range of 120 lb./in.^{2}, a unit equipped with twenty-four 15 mm cyclones per stage has a total capacity in the range of 22–26 gal./min.

Applications: These small size units were originally developed for the concentration of dilute starch suspension, but they have also been found suitable for other operations, such as making separations in the range of 2–20 μ, in the fine sizing of clays for china, whiting and other fillers for paper, pigment production, grinding powders, and pharmaceuticals.

Large size single liquid cyclones[20] ranging in diameter from 3 to 24 in. are gaining favour as classifiers in closed-circuit grinding, especially where grinding to a fine size is required as in raw cement mills, cyanidation mills, and flotation plant regrind circuits. In iron ore grinding circuits cyclones have the added advantage of minimizing difficulties due to magnetic flocculation and full-scale installations are being made.

One of their most important applications, as mentioned earlier, has been in desliming operations, three of which are listed below:

1. Desliming ground ores of iron, zinc, magnesite, phosphate rock, *etc.*, prior to flotation, tabling and other concentration processes in which the slime fraction is detrimental.
2. In the mining industry for desliming mill tailing to make them suitable for mine backfill. This application has reached large proportions.
3. Removing undesirable minus 200 mesh fines from glass sand.

Though liquid cyclones have a very low initial cost compared with other forms of classifiers, they have much higher operation and maintenance charges due to the heavy power consumption of the feed pumps and the rapid wear resulting from the abrasive effect of most solid particles at the high rotational speeds developed within the cyclone. For these reasons the adoption of cyclones must be considered very carefully as no two cases are ever exactly alike and often seemingly identical applications will prove radically different economically. Also, with the use of larger cyclones, the cost of pump and piping often approaches 70 per cent of the total cost of the installation.

The problem of abrasion prevention or reduction is being studied very intensively by several manufacturers and considerable progress has been made to date. Trials have been made of hardened steel, heat treated Meehanite, and Ni-Hard alloy. Cone liners composed of two types of ceramically bonded, abrasion-resistant materials, Linatex and a great many combinations of both natural and synthetic rubber have been thoroughly tested. Some of these materials give satisfactory results under certain service conditions and fail completely in others.

It is clear that within the past few years the wet cyclone has earned its place as a useful classification unit. Its enormous capacity for size, its simplicity and its relatively modest initital cost have wide appeal, and its practical advantages have been proved beyond doubt in many applications. But for all its strong points it should not be considered as a cure-all for classification problems, but rather as another valuable tool for the practising engineer. The user must be aware of its limitations.

Tables 1 (*a*) and (*b*) show the operating characteristics and fields of application for various types of aqueous classifiers.

CLASSIFIER DESIGN AND PERFORMANCE
Fundamentals of Design

Classifier design is based first upon the rate of volumetric displacement of incoming feed (solids plus fluid) and the rate of settlement of the finest particle to be retained in the coarse fraction, and second upon the rate at which the products must be discharged from the machine.

Within limits, the lower the feed dilution, the smaller the machine for a given tonnage of solids and required settling rate, but as feed dilution is lowered the viscosity and/or density of the solid–fluid mixture tends to increase, thus reducing the settling rate. A more serious disadvantage results from the fact that the voids between the particles of the coarse fraction carry a higher percentage of feed solids, thereby lowering the separating efficiency. These relationships are shown in *Figure 5*.

The settling rate of a solid particle in a fluid depends upon its diameter, density, shape factor, and the density and viscosity of this fluid (see Sedimentation section). The effect of increased numbers of particles (crowding) is roughly that of increasing the density and viscosity of the

Table 1(a). *Operating Characteristics and Fields of Application for Aqueous Classifiers*
(*From* J. V. N. Dorr *and* F. L. Bosqui[7])

Type of classifier	Normal size range		
	Width	Diameter	Maximum length
Straight or unit type	14 in.–20 ft.	—	40 ft.
Bowl classifier	18 in.–20 ft.	4–28 ft.	38 ft.
Bowl desiltor	4 ft.–16 ft.	20–50 ft.	38 ft.
Hydroseparator	—	4–250 ft.	—
Hydraulic classifiers:			
Sizer	Varies with no. of pockets	—	Varies with no. of pockets
Super sorter	6 ft.	—	About 40 ft.
Hydroscillator	4 ft.–12 ft.	4–14 ft.	30 ft.
Centrifugal classifiers:			
Cyclone	—	3–30 in.	9 ft.
Solid bowl centrifuge	—	18–54 in.	70 in.
Cone classifier	—	2–12 ft.	—
Sand washer	—	7, 9, 12 ft.	—

Table 1(b)

Type of classifier	Normal feed density range %	Underflow or rake product % solid	Drive motor range h.p.
Straight or unit type	Not critical	80–83	½–25
Bowl classifier	Between 10–75 solids	75–83	Bowl: 1–7½ Recip. rakes: 1–25
Bowl desiltor	Not critical	75–83	Bowl: 1–10 Recip. rakes: 5–25
Hydroseparator	5–20 (not critical)	30–50	Fractional to 15
Hydraulic classifiers:			
Sizer	40–60	40–60	1–2 for air pressure
Super sorter	30–60	40–60	1 to operate pincer valves
Hydroscillator	40–80	75–83	Bowl: 3–10 Rakes: 5–20
Centrifugal classifiers:			
Cyclone	1–30	55–70	Power or pressure head
Solid bowl centrifuge	1–30	40–70	10–150
Cone classifier	Not critical	35–60	None
Sand washer	About 30–35 solids	80–83 solids	5–10

Table 1 (a)—continued

Classification		Normal feed tonnage range	Maximum oversize	Normal overflow % solids
Average separation ranges	Relative classifier efficiency			
20–150 mesh	Medium	1–350 tons/h	1–1½ in.	5–65
65–325 mesh	Medium	1–300 tons/h	½ in.	5–25
100–325 mesh	Medium	5–250 tons/h	½ in.	1–15
Usually 100–325 mesh*	Low	A few lb/h to 500–700 tons/h	½ in.	1–20
8–100 mesh	High	2–100 tons/h	½–3/16 in.	5–20
½–150 mesh	High	40–150 tons/h	½ in.	5–20
20–200 mesh	High	5–250 tons/h	½–1 in.	15–30
35 mesh to 3 μ	Medium	10–1,500 gal./min	14–20 m	5–30
100 mesh to 5 μ	Medium	Up to 500 gal./min	½ in.	5–30
28–325 mesh	Low	Up to 100 tons/h	½ in.	5–30
28–65 mesh	Medium	25–125 tons/h	1 in.	5–20

* Special cases 35 mesh to 5 μ.

Table 1(b)—continued

Hydraulic water required	Typical applications
Spray wash optional	Closed circuit grinding. Washing and dewatering. Process feed control
Water added with feed or in classifier for correct overflow density	Closed circuit grinding—mostly in secondary circuit but sometimes in primary. Washing and dewatering
Spray wash optional	Recovery fine sand, limestone, coal, fine phosphate rock from large flow volumes
All water enters with feed	For fine separation where large feed volumes are involved
Up to 4 tons/h feed	Gravity concentration of metallic ores. Preparation of table feed. Sizing of homogeneous materials
Up to 4 tons/h feed	Gravity concentration, preparation of table feed, coal. Sizing of homogeneous materials
Up to 1·5 tons/h/ton feed	Used where exceptionally clean rake sands are needed. Closed-circuit grinding
None	For fine separations in variety of fields
None	For fine fractionating to 5 μ
None	For desliming and primary dewatering
None	For desliming and dewatering large tonnages of bulk material

fluids. At 20° C, for instance, the settling rate of a quartz sphere (sp. gr. 2·65) of 0·147 mm diameter is theoretically 3·0 ft./min in water, whereas in a pulp of 25 per cent solids the rate falls to 0·4 ft./min, and under hindered-settling conditions at 40 per cent solids (volume basis) it drops to about 0·30 ft./min, or roughly one-tenth the rate at infinite dilution.

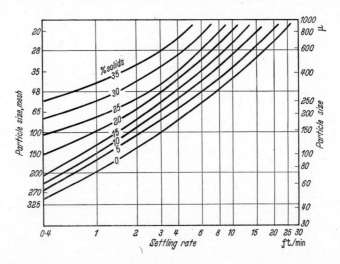

Figure 5. *Typical settling rate curves for normal classifier operation on solids of 2·65 sp. gr. in water at 20° C*

For mechanical classifiers, in the case of metallurgical pulps, the overflow dilution should not be less than 3:1 for coarse (up to 28 mesh separations) and not less than 5:1 for 200 mesh separations. For limestone and clays twice this dilution may be necessary. Having determined the settling rate at the size of separation, usually from actual test, a unit area (in square feet per ton of overflow solids per 24 hours)[1] is calculated. This means that it is necessary to provide a certain total settling area in one or more classifiers to handle the tonnage indicated. In practice, various area factors which may range from 1·4 to 5·0, depending upon the machine, are applied in calculating 'actual' as compared to the 'theoretical' areas indicated. Raking capacities vary with the particular design of machine, the rake type being available with single and multiple (up to quadruplex) raking units, so that most rake to feed ratios encountered in practice can be handled without difficulty.

Hydroseparator and stationary cone sizes are calculated in a similar manner, except that in estimating the area required the volume of pulp discharged through the bottom spigot must be subtracted from the volume of incoming feed to determine the net upward displacement. The size of classifiers should be specified within close limits since it is apparent that oversize units will fail as seriously as undersize units to give the desired performance. In the case of hindered-settling hydraulic classifiers, the number of pockets required will depend upon the particle size range in the feed and the maximum particle size overflowed. Not more than a 1–2 mesh drop (largest sized particle not more than 1½–2 times the size of the smallest particle present) should be allowed per pocket, otherwise the water rate used in that pocket will be too low for the coarser size and too high for the finer size present.

Table 2. *Typical Screen Analyses for Hydroscillator versus Standard Classifier*
(From J. V. N. DORR and F. L. BOSQUI, *by courtesy of* The Institute of Mining and Metallurgy)
Screen analyses reported as per cent cumulative plus mesh

Sizing			Dorr Hydroscillator		Dorr classifier	
Mesh	Microns	Feed	Overflow	Rake	Overflow	Rake
Example 1*						
35	417	40.9	—	51.7	—	46.8
48	295	57.9	0.1	73.7	—	66.3
65	208	70.3	1.5	89.0	1.7	80.5
—	199†	71.2	2.0	90.5	2.0	81.7
100	147	77.6	11.8	96.7	8.3	88.0
—	141	78.2	14.2	97.2	9.7	88.3
150	104	83.1	29.4	98.4	23.5	92.3
—	100	83.6	32.8	98.5	25.5	92.6
200	74	87.7	50.4	98.8	39.4	94.8
Dry tons		200.5	46.5	154.0	26.5	174
Per cent solids		—	22.4	79.2	22.4	79.2
Per cent eliminations:						
1st critical				34		14
2nd critical				80		35
Below 2nd critical				93		61
Example 2						
35	417	58.2	0.1	77.0	—	68.0
48	295†	70.6	2.0	92.5	2.0	82.0
65	208	77.0	14.4	98.1	8.9	88.3
100	147	83.0	34.6	99.5	24.8	92.7
150	104	87.4	49.7	99.7	40.6	95.2
200	74	90.0	61.3	99.8	51.4	96.4
Dry tons		206.5	49.4	157.1	29.4	177.1
Per cent solids		—	19.6	78.2	19.6	78.2
Per cent eliminations:						
1st critical				43		16
2nd critical				83		38
Below 2nd critical				98		63

* 0.70 ton hydraulic water per ton Hydroscillator oversize.
† Mesh of separation in microns.

From numerous empirical data the settling rate or upward displacement of hydraulic water for particles of various sizes and density at various void ratios (that is, ratio of voids to solids on a volumetric basis) have been worked out by manufacturers of this equipment. Constriction plate densities of about 0.6 voids and cleaning with a 3:1 hydraulic water to solids ratio appear to give satisfactory motility. It is then possible to calculate the square feet of settling area required for each ton of material within a given mesh size and to determine the area of each pocket required. While these areas do not usually conform exactly to manufacturers' designs, it is nearly always possible to find a particular model or combination of models that will meet specifications.

The sharpness of separation is to some extent dependent upon the ratio of hydraulic water to solids, but, since the areas required are proportional to the hydraulic water used, it is usually not economical to exceed the 3:1 ratio mentioned. Feed dilutions should be in the range of 30–40 per cent solids.

Comparative Performance of Classifiers

To compare the performances of a standard rake classifier, a Hydroscillator classifier, and a hindered-settling classifier, tests of these three machines operating on the same materials in pairs were made; the results are given in the accompanying *Tables 2, 3* and *4*. In these tables the term 'critical size'

Table 3. *Typical Screen Analysis for Hydroscillator versus Four EX–8 Sizers**

(*From* J. V. N. DORR *and* F. L. BOSQUI, *by courtesy of* The Institute of Mining and Metallurgy)
Screen analyses reported as per cent cumulative plus mesh

Sizing		Dorr Hydroscillator†			Dorrco sizers		
Mesh	μ	Spigot	Feed	Overflow	Coarse	Feed	Fine
28	—	33.6	12.9	2.04	36.8	12.9	3.74
	475	46.5	18.5	4.0	37.5	—	—
35	580	57.6	25.1	6.8	62.7	25.1	12.3
	336	74.5	37.0	16.0	—	—	—
	410	—	—	—	63.0	—	12.7
48	—	83.0	49.9	26.7	85.5	49.9	38.4
	238	93.2	66.5	50.0	—	—	—
	290	—	—	—	86.2	—	38.8
65	—	96.5	77.6	66.0	96.5	77.6	71.3
100	—	99.6	94.4	91.0	99.4	94.4	91.5
Per cent solids		50	42	23	—	42	—
Short tons/h		5.9	15.5	9.6	36.5	140	103.5
Ton coarse/ton feed		0.38	—	—	0.26	—	—
Ton water/ton coarse		2.55	—	—	16.4	—	—
Per cent eliminations:							
1st critical				41		49	
2nd critical				75		76	
Below 2nd critical				92		93	

* Superelevation 6.4 at 10 in. depth; water rate 60 gal./min over 7.8 ft.² pool area: oscillator 225 stroke/min; stroke 3¼ in.
† 42 in. diameter Hydroscillator classifier with spigot discharge.

refers to $\sqrt{2}$ screen size ranges below the mesh of separation. In classification practice the mesh of separation is defined as that mesh on which approximately 1·5–2·0 per cent of the sample is retained under standard screening conditions. The 1st critical designates the proportion remaining on the next finer mesh sieve; the 2nd critical is that remaining on the next finer, *etc.* The percentage elimination of these various criticals is then a measure of the effectiveness of the classifying operation and of the cleanliness of the separated product.

The data shown in *Table 2* illustrate the relative performance of a standard Dorr classifier and the Dorr Hydroscillator classifier on a homogeneous ore, the mesh of separation being 65–100 in Test 1, and 48 in Test 2. The improved elimination of 1st and 2nd critical sizes from the rake sands of the Hydroscillator classifier should be noted. In *Table 3* the data represent comparative tests between the Dorr Hydroscillator and a hindered-settling 8-pocket Dorrco sizer operating on phosphate rock. The results show the great similarity in the performance of these two classifiers. In *Table 4*, which shows comparative tests between a standard Dorr classifier

Table 4. Tests Illustrating Difference in Size Distribution between Products from an Hydraulic Sizer and a Standard Rake Classifier*

Screen analyses reported as per cent cumulative plus mesh

Sizing		Feed	Dorr classifier rake products	Dorrco sizer spigots 1–5	Dorr classifier overflow	Dorrco sizer spigot 6 + overflow
Mesh	μ					
28	475	0·28	0·5	0·5	—	—
35	336	9·65	13·5	19·4	—	—
48	238	29·3	33·0	47·8	—	—
65	208	46·1	63·0	73·0	2·4	2·4
100	147	66·5	86·0	95·1	22·0	19·4
150	104	80·7	94·6	99·8	47·0	47·7
200	74	86·8	97·1	100·0	62·0	72·0
Weight per cent		100·0	70·3	66·2	29·7	33·8

* Separation in each case at 2·4 per cent +65 mesh.

and a Dorrco sizer, the greater elimination of fines below 65 mesh from the coarse fraction of the hydraulically sized product is indicated both by the higher cumulative percentage figures in the 100 mesh and finer fraction, and also by the greater percentage weight of material in the overflow product. The figures are the results of tests on grinding sand.

AUXILIARY EQUIPMENT AND MATERIALS OF CONSTRUCTION

Most of the machines described in this article require for their normal operation certain items of auxiliary equipment which are discussed below.

Feeders

Many of the machines are fed directly by delivery chutes or pipes but some of them require mechanical feeders of which many types are available including wet, dry, proportioning, weighing, disc, chain, vibrating, screw, and belt-operated. Detailed discussions of these various types may be found in the section 'Weighing and Gauging Solids' and in references 13 and 18.

Pumps

Mechanisms such as hydroseparators, which operate continuously in tanks, usually discharge through a valve-controlled spigot, but in some instances it is desirable to have their underflows controlled by a pump which may be for either suction or pressure operation. Diaphragm pumps are most commonly used for this purpose (see Sedimentation section); underflow lines and density controls are also covered in the same article. Pressure pumps, usually of the centrifugal type, are required for feeding liquid cyclones such as the DorrClone and Centri-cleaner. The delivered pressure of these pumps ranges normally from 15 to 40 lb./in.^{2} with an extreme range of 3–120 lb./in.^{2} The extreme capacity range is from 8 to 1,600 gal/min on cyclones varying in size from 3 to 24 in. diameter.

Miscellaneous

There are a number of attachments for mechanical classifiers to increase their efficiency and to control their operation. Among these are a hydrometer density indicator for the control of the dilution of classifier overflow, a floating chip-removing device, and a suction box to reduce the moisture of the rake product. It is occasionally necessary to use an auxiliary sand scoop or wheel to complete the circuit between a classifier and the ball mill. For bowl classifiers there is a 'critical size' control to improve classification efficiency by eliminating surging in the pool. Coarse screens are usually placed ahead of classifiers to scalp-off heavy oversize. To prevent damage to drive mechanisms of hydroseparators, torque recorders and automatic overload alarms are available.

Materials of Construction

Where severe abrasion is present manganese steel and other hard alloy steels are employed and rubber linings installed at certain critical locations to protect the machine. In hydraulic classification iron and steel are used for construction, since water is the usual medium, but, if separations are made in corrosive solutions, rubber, lead-covered, stainless steel or wooden submerged parts are employed. Abrasion is quite severe in hydraulic cyclones and its effects are minimized by using rubber and other materials such as Carbofrax at critical locations. The same practice applies to other types of sizing equipment.

ACKNOWLEDGMENT

The author acknowledges with gratitude the invaluable help of Mr. D. C. Gillespie and other members of the Dorr-Oliver organization in preparing this contribution.

REFERENCES

[1] ANABLE, A., 'Classification', *Chemical Engineers' Handbook*, McGraw-Hill, N.Y., 1950, pp. 922–37
[2] ANABLE, A., and KNOWLES, C. H., 'Technique of Settling Separations', *Chem. metall. Engng*, 45 No. 5 (1938) 260–63
[3] BROWN, G. G. et al., *Unit Operations*, Wiley, N.Y., 1950, Chap. 8, pp. 84–98; Chap. 19, pp. 258–68
[4] CRAMPTON, F., 'Cost Estimating in Mineral Engineering', *Amer. Inst. chem. Engrs*, Biloxi, Miss., Mar. 10, 1953

[5] DAHLSTROM, D. A., 'Fundamentals and Applications of Liquid Cyclones', *Amer. Inst. chem. Engrs*, Biloxi, Miss., Mar. 10, 1953
[6] DORR, J. V. N., and BOSQUI, F. L., 'Classification', *Handbook of Nonferrous Metallurgy*, McGraw-Hill, N.Y., 1945, pp. 101–25
[7] DORR, J. V. N., and BOSQUI, F. L., 'Recent Developments in Classification and Fluidization', *Miner. Dress. Symp. Inst. Min. & Metall.*, Lond., Sept. 23, 1952
[8] DORR Co. Engrs, 'Classification with Water', *Taggart, Handbook of Mineral Dressing*, Wiley, N.Y., 1945, Sect. 8, pp. 1–61
[9] FISCHER, A. J., and FORGER, R. D., 'Current Status of Cyclones as Classifiers', *Min. World*, 16 (1954) 44–8
[10] FITCH, E. B., and JOHNSON, E. C., 'Operating Behavior of Liquid–Solid Cyclones', *Min. Engng*, 5 (1953) 304–8
[11] FLOWERS, A. E., 'Centrifuges', *Chemical Engineers' Handbook*, McGraw-Hill, N.Y., 1950, pp. 992–1012
[12] GAUDIN, A. M., *Principles of Mineral Dressing*, McGraw-Hill, N.Y., 1939, Chap. 8, pp. 165–201; Chap. 9, pp. 202–30
[13] HITZROT, H. W., 'Guide for Proper Application of Classifiers', *Min. Engng*, 6 (1954) 534–9
[14] HYDE, R. W., 'Feeders and Feeding Mechanism', *Chemical Engineers' Handbook*, McGraw-Hill, N.Y., 1950, pp. 1370–6
[15] LEWIS, F. M., and JOHNSON, E. C., 'Liquid–Solids Cyclone as a Classifier *in C.C.G.*', *Min. Engng*, 6 (1954) 620–1
[16] ROBERTS, E. J., 'Sizing in Water Classification', *Amer. Inst. chem. Engrs*, Biloxi, Miss., Mar. 10, 1953
[17] SWARTZMAN, E., 'Progress in Coal Cleaning', *Canad. Min. metall. Bull.*, 45 (1952) 518–23
[18] TAGGART, A. F., *Handbook of Mineral Dressing*, Wiley, N.Y., 1945, Sect. 18, pp. 97–107
[19] TANGEL, O. F., and BRISON, R. J., 'Wet Cyclones', *Chem. Engng*, 62, No. 6 (1955) 234–8
[20] WEEMS, F. T., 'Metallurgical Applications of the DorrClone', *Min. Engng*, 3 (1951) 681–90

BIBLIOGRAPHY

ARMSTRONG, D. G., 'The Design Construction and Operation of a Laboratory Classifier', *Byll. No. 592 Instn Min. Metall.*, Lond., Trans. 65 (1956) 229–39
JOWETT, A., 'A Simple Mathematical Treatment of the Operation of the Hydrocyclone', *J. Leeds Univ. Min. Soc.*, 29 (1954) 39–46
KELSALL, D. F., 'A Study of the Motion of Solid Particles in a Hydraulic Cyclone', Symposium on Mineral Dressing, *Instn Min. Metall. Pap. No. 17*, London, 1952
FERN, K. A., 'The Cyclone as a Separating Tool in Mineral Dressing', *Trans. Instn chem. Engng, Lond.*, 30, No. 2 (1952) 87–108
SPEAKMAN, J. E. D., 'Improved Methods of Cleaning Coal', *Industr. Chem., Lond.*, April (1950) 171–4
BELHAM, R. L., 'The Eldorado Gravity Plant. Port Radum, N.W.T.', *Canad. Min. metall. Bull. No. 528*, Trans. 59 (1956) 154–61
HIRST, A. A., 'Gig Washers in Theory and Practice', Symposium on Coal Preparation, *Leeds Univ.*, 152, 61–82
PRYOR, E. J., 'An Introduction to Mineral Dressing', Mining Publications, Ltd., London, 1955
ANDREWS, L., 'Classified Grinding Research', *Trans. Instn Metall.*, Lond., Vol. 48, 141–207

11

DENSE MEDIUM COAL WASHING

R. SYMINGTON

PRESENT day mining conditions tend to produce coal having more and more complex washability characteristics and therefore lead to an increasing demand for cleaning processes which will separate coal and refuse with a high degree of efficiency and thus produce the maximum yield of saleable products. Such a separation is accurately accomplished by the simple floating and sinking of the raw coal in a dense medium, the specific gravity of which may be varied and controlled within very accurate limits to suit the nature of the coal to be treated. Organic liquids and inorganic solutions have been employed as dense media but the dense medium processes at present in general commercial use employ a suspension of finely divided solids in water as the dense medium.

The accurate separation accomplished is not materially affected by variation in the rate of feed or in the size range of the raw coal. Any range of sizes from $\frac{1}{8}$ in. or less up to 24 in. or more can be treated and, in fact, it is sometimes possible to treat entire outputs of run-of-mine coal without any hand cleaning, thus gaining substantial economy in labour costs. The accuracy of separation and range of specific gravities enable the full recovery of coal products, a feature which has not been fully realized in other processes.

PRINCIPLES

Dense medium coal washing, in common with jig washing, depends for its success on the fact that clean coal with a low proportion of ash-forming material has a lower specific gravity than low grade coal with a higher ash content. *Table 1*, which is an analysis of a British coal according to specific gravity fractions, illustrates this relationship.

Table 1. Sink and Float Analysis

Specific gravity of coal fraction	Yield per cent	Ash content per cent
Less than 1·25	5·0	0·8
Between 1·25 and 1·30	60·0	2·5
,, 1·30 ,, 1·35	10·0	8·0
,, 1·35 ,, 1·40	4·0	16·0
,, 1·40 ,, 1·50	3·0	25·0
,, 1·50 ,, 1·60	3·0	35·0
Greater than 1·60	15·0	70·0

It will be evident that if a sample of raw coal is introduced into a solution of a predetermined density, a clean coal containing a required ash content will be obtained as a float product while the residue of the coal, the discard, will sink. In practice, solutions of carbon tetrachloride, toluene and bromoform, which are miscible one in the other, are commonly used in the laboratory to analyse coals in the manner illustrated in *Table 1*.

HEAVY LIQUIDS

In commercial practice, organic and inorganic solutions have been used in the past to achieve such a separation, and in 1858 Sir Henry Bessemer patented the first process using solutions of inorganic chlorides, one such plant being built at that time in Germany. In 1928 the LESSING[1] process, using calcium chloride, was introduced and three plants were built in Yorkshire and South Wales which achieved high efficiencies of separation. Other processes of the same type were the BERTRAND[2] (1934), MARTELL and MOORE[3] (1933), BELKNAP[4] (1935) and Crescent coal washing plants.

These processes did not, however, achieve great commercial success on account of the corrosive nature of the separating medium, the difficulty and expense of recovering the solutions from the products of separation and, principally, the development of the dense medium processes. For the same reasons, the DU PONT[5] process introduced in 1936 and using solutions of chlorinated hydrocarbons was used at only one colliery in the U.S.A.

DENSE MEDIA

All the processes now in general commercial use employ a suspension of finely divided solids in water, termed the 'dense medium', and it has been found that such suspensions exhibit many of the properties of true solutions and in particular can support particles of coal with a lower density than the suspension density provided that the mean size of the medium solids are smaller than the coal.

Since the medium solids are larger than the colloidal size range, though generally below 200 B.S., some degree of agitation is required to maintain a suspension of uniform density. The rate of settlement and consequently the degree of agitation are, of course, dependent on the size grading and the specific gravity of the solids, together with their volume concentration in the suspension.

The following equation determines the density of a suspension:

$$D = \frac{100}{(100-C) + \frac{C}{d}}$$

D = density of the suspension (gm/c.c.)
d = density of the solids (gm/c.c.)
C = weight concentration of the solids in suspension (gm/c.c.)

From this equation, it will be seen that the density of a suspension can be varied at will by adjusting the proportion of medium solids present in the suspension. There is, however, a natural limit to the proportion of solids imposed by the volume concentration since it is important that the suspension should have a relatively low viscosity to permit the free travel of coal and dirt in it. This limiting volume concentration is similar for all suspensions with particles of similar size range, but can be modified to a limited extent by changes in the size grading of the medium solids.

The equation also shows that the density of the suspension is a function of the specific gravity of the medium solids and it naturally follows from this that media composed of higher specific gravity solids will have a higher density at the limiting volume concentration than suspensions composed of lower specific gravity solids.

The term 'stability' is used in dense medium practice to indicate the rate of settlement of a suspension, and a stable medium is one in which settlement occurs slowly and uniformly throughout the suspension. Unstable media settle rapidly and there is evidence of differential settlement in the body of the suspension. In practice, there is a further class of media, which we shall term semi-stable, comprising higher density solids whose settling characteristics are modified by the addition of low density solids at a controlled level to obtain a uniform rate of settlement over a wide range of operating densities.

Table 2. *Settling Rate of a Semi-stable Medium with Added Slimes (Suspension Density—1·50)*

Proportion of slimes Wt. per cent	Settling rate min/in.
0	1·78
10	3·98
15	6·31
20	9·55
25	14·13
30	21·38
35	31·62
40	47·86

The most stable media in commercial use are those in which the solids are very finely divided and are of low specific gravity (2·0–3·0 sp. gr.) such as loess or shale. Such a suspension permits separation of the coal under relatively quiescent conditions with a high degree of accuracy of separation, but is limited by the low specific gravity of the solids to a suspension density of 1·60. Above this limit, the volume concentration becomes so high that the suspension exhibits thixotropy. For the same reasons, in the case of such suspensions it is extremely important to avoid contamination of the medium with fine coal or shale particles of low specific gravity which are introduced with the raw coal, and considerable care must be taken to ensure the removal of this material in the medium recovery process.

High density solids such as magnetite (5·0 sp. gr.) and barytes (4·2 sp. gr.) give suspensions which are generally unstable but which permit higher suspension densities up to 2·0 or higher.

Fortunately, however, the stability of such media can be considerably modified by control of the size grading and by the amount of coal and shale slimes which are present in the suspension, and it is found that by fine grinding and in the presence of 30–40 per cent of slimes, semi-stable suspensions can be achieved at densities as low as 1·30–1·35.

The influence of added clay to an unstable medium can be seen from *Table 2* which shows that the settling rate is rather more than halved with each 10 per cent increase in the slimes concentration, and that with the addition of 30 per cent of slimes the settling rate is so far reduced that such a medium might well be considered to be stable. At higher concentrations of slimes, however, the viscosity rapidly increases and it is not generally possible to operate with more than 50–60 per cent of slimes.

BASIC DESIGN OF A DENSE MEDIUM PLANT

The design of any dense medium coal washing process may be considered to consist of four interrelated parts, namely:

The separating bath.
The apparatus for feeding the raw coal into the separating bath and removing the products of separation.
The circulation of the separating medium through the bath.
The recovery, cleaning and densification of the medium carried out of the bath on the products of separation.

The schematic flow diagram illustrated in *Figure 1* may be regarded as representative of any dense medium coal washing process.

The Separating Bath

The separating bath is primarily a vessel in which the raw coal feed is introduced into the medium so that a separation into floats and sinks can take place. The size of the bath is selected so that the coal is permitted to sink or float freely without mechanical interference, and so that the near gravity coal or middlings remain in the bath sufficiently long to achieve their correct placing with the sinks or floats.

It is usual to classify the separating vessels as shallow or deep baths. The features of the shallow baths are that they contain a relatively small volume of medium and that when a stable or semi-stable medium is used circulation rates can be comparatively low. They cannot, however, be used to carry out a three product separation and they have a limited capacity for overload. Deep baths have a higher capacity for overload and permit a three product separation to be effected, but unless a stable medium is used the rates of medium circulation require to be relatively high.

The capacity of a coal washing bath may be related to the surface area on which the coal floats. While it is usual to design the length of the bath so that the coal travels 7–10 ft. in passing from the feed point to the coal discharge, it is the width of bath which mainly controls the capacity. The width is arranged so that the coal is able to move freely on the surface of the medium and so to place itself correctly with the sink or float material. The coal generally travels at 20–40 ft./min across the surface of the bath,

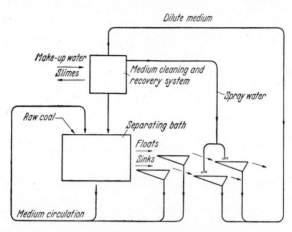

Figure 1. *Typical D.M. flow diagram.* (By courtesy of University of Sheffield)

and under these conditions it is usual to allow a foot of width for each 10–15 tons/h of floating coal in the 1–$\frac{1}{2}$ in. size range and the same width for each 15–25 tons/h in the 3–2 in. range.

Raw Coal Feed to the Separating Bath and the Removal of the Products

Since the efficiency of a dense medium coal washing bath falls off when it is overloaded, and this is particularly true of shallow baths, it is usual to endeavour to maintain a constant rate of feed. The coal will normally be fed to the bath by conveyor belt, shaker or other mechanical means, and, in some systems, the coal is pre-wetted in a stream of medium before entering the bath.

Floating material may be mechanically assisted across the surface of the bath by means of scrapers, paddles and star wheels, or the flow of medium to a weir discharge may be employed. The sinking material may be removed by air lifts, bucket elevator, scraper, rubber belt, or by means of a drum. In the Chance process an automatic gate is used.

Medium Circulation

As already stated, it is necessary in all systems to maintain some circulation of the medium in the separating bath in order to obtain stability, the actual rate of circulation being governed by the properties of the medium used and the design of the separating bath.

A second, and in many cases more important, consideration in selecting the correct medium circulation rate, is the necessity for removing from the medium fine coal and dirt in the 30 mesh to $\frac{1}{4}$ in. size range. This fine material may be due to inefficient pre-screening of the raw coal feed, or it may be formed by breakage in the bath, but in any event if it is allowed to accumulate in the medium, it will eventually raise the viscosity to a level at which a separation can no longer be made effectively. It is usual, therefore, to pass the circulating medium through a 30 mesh screen before returning it to the bath.

Medium Cleaning and Recovery System

The products of separation leaving the coal washing bath carry on their surface a coating of the medium suspension, which for reasons of economy must be recovered and it is the general practice to pass these products over vibrating desanding screens to remove the majority of this medium, returning it directly to the medium circulation. It is not possible to remove all the adhering medium in this way and the products are therefore subsequently washed with water on vibrating spraying screens or shakers to remove the remaining medium. It is usual to employ water from the medium recovery system for the initial washing and to finish with a spray of fresh make-up water to give the products a good appearance.

As might be expected, the quantity of medium adhering to the products after desanding bears a relationship to the specific gravity of the medium in which they have been separated. In fact, it is found that $1\frac{1}{2}$ in. coal from a 1·50 sp. gr. separation carries 50 lb. of medium per ton of coal on its surface after desanding, together with a proportion of slimes which were of course present in the bath and adhere to the products.

It is the function of the medium recovery and cleaning system to densify the medium contained in the spray water so that it can be returned to the circulation system, and to free it from the associated coal and clay slimes which are also washed off the products of separation. These slimes, as distinct from the fine coal and shale produced by breakage, are formed in part by the disintegration of clay present in the raw coal feed and, being very finely divided, they cannot be removed by screening. In common with the fine coal and shale they will, if allowed to accumulate in the medium circulation, eventually increase its viscosity to a point at which separation becomes ineffective.

In most cases it is found that the amount of slimes leaving the bath on the products of separation, and which is subsequently eliminated from the medium in the cleaning system, is sufficient to maintain the viscosity of the medium in the separating bath at a satisfactory level. In certain cases, however, where there is a large proportion of disintegratable clay present in the raw coal feed, the equilibrium concentration of slimes in the bath may become too high, and in such cases a part of the medium circulation is continuously bled to the medium cleaning system. The design of the medium recovery and cleaning system is arranged to suit the particular medium employed and the degree of cleaning which is necessary.

In systems which employ media whose specific gravity or size range is close to that of clay slimes, such as loess or froth flotation tailings, it is usual to employ Dorr thickeners to recover and thicken the medium. More recently, however, the Dutch State Mines have developed the cyclone thickener for the recovery and thickening of loess and froth flotation tailings media. Where the medium has a specific gravity or size range widely different from the slimes, such as pyrites, magnetite and other magnetic media, it is customary to employ settling cones to recover the medium.

In all systems, some degree of medium cleaning is effected in the thickener or cone used for the recovery of the medium, and in the Chance, Ridley-Scholes, and Tromp systems all the cleaning is effected at this stage by so selecting the size of the settling cone that the coarse and heavy medium particles settle to the base of the cone, while the finer slimes overflow the rim. A 12–15 ft. diameter 60° cone is normally used for a 100 ton/h washer. Such a separation is not entirely successful as a proportion of the slimes in the larger size range settles with the medium, while the finer medium particles tend to overflow the rim. In the Tromp and Ridley-Scholes process the spray water is screened at 30, 60 or even 150 mesh to reduce the proportion of large slime particles. In the Barvoys and Dutch State Mines systems it is usual to employ froth flotation to clean the underflow from the Dorr thickener and to remove associated coal and clay slimes by this means.

The American Cyanamid Co. system of medium cleaning takes advantage of the magnetic properties of the medium and of the fact that when a magnetic medium is magnetized the particles flocculate and achieve a greatly increased rate of settlement. The spray water, after passing through a vibrating screen to remove oversize slimes, is magnetized and its rate of settlement in the thickener is thereby increased. Underflow from the thickener is then fed to one or more Crockett type magnetic belt separators in series and a separation of the slimes is effected. The cleaned medium from the separators is thickened in an Akins classifier and is demagnetized before its return to the medium circulation.

In the Simon-Carves system of magnetic medium cleaning and recovery processes, the three stages of recovery, cleaning and densification of the magnetic medium are carried out in one specially designed cone. It has been found that magnetic medium magnetized under streamline flow conditions flocculates to a much greater extent and has a higher rate of settlement than medium magnetized under turbulent conditions. This increased rate of settlement makes it possible to obtain a substantially complete recovery of the magnetic medium fed to the cone and at the same time, by virtue of the greater upward flow rates which are possible in the cone, to clean the medium to the degree necessary to permit of its return to the circulation after demagnetization.

The published literature shows that processes which use magnetic media without employing magnetic cleaning and recovery systems report a medium consumption of 4–5 lb. or higher per ton of coal washed. When, however, magnetic recovery is practised, as in the Cyanamid and Simon-Carves processes, the medium consumption is $\frac{1}{2}$–2 lb. per ton, and a large part of this difference can be attributed to the improved efficiency of recovery made possible by magnetic flocculation.

PROCESSES COMMERCIALLY USED IN GREAT BRITAIN

These comprise the CHANCE[6], BARVOYS[7], TROMP[8], RIDLEY-SCHOLES[9], DUTCH STATE MINES[10], NELSON DAVIS[11] and SIMCAR[12, 13] systems and have been introduced into Britain in the order given. These processes, between them, cover the full range of suspensions in use and illustrate the different types of medium recovery systems and separating vessels used. In many cases, variations in design are practised but, since these seldom affect the general principles used in each process, only one type of medium system or separator will be discussed.

Chance System

This process employs a suspension of sand in water, the sand having a specific gravity of about 2·6 and a size range of 30–100 B.S., and is the oldest of the commercially successful systems, being first used for cleaning anthracite in the U.S.A. in 1921[14] and later introduced to the United Kingdom in the early 1930's where it is now widely used, particularly in South Wales.

A pictorial layout of the Chance system is shown in *Figure 2*. The vessel in which the actual separation is carried out is conical in shape with water inlets at various levels to ensure an even distribution through its depth. Each of the water inlets has its own valve for controlling the specific gravity of the fluid. A slow-moving agitator is provided to give the fluid a rotary movement around the cone to enable the floated coal to be carried from the point of feed to the overflow weir and on to the desanding screen. The base of the cone is provided with two air operated slide gates in between which is mounted a refuse chamber for receiving the sinking refuse. These gates are automatically controlled and operate as described.

The chamber is first supplied with water from the filling receiver, *via* the air operated filling valve, closing a pressure switch when full. The closing of this switch automatically opens the top slide gate which allows the refuse that has accumulated in the bottom of the cone to fall into the refuse chamber. After a given time, usually 15–20 sec, the top gate closes and automatically opens the bottom gate which allows the sand, water and refuse to fall on to the desanding screen. The bottom gate closes after the predetermined time and the cycle is recommenced by the opening of the air operated filling valve.

Both the clean coal overflow from the top of the cone and the refuse extracted from the bottom of the cone are desanded and dewatered on screens and the sand and water are laundered to a main sand and water sump where, due to the low rising velocity of the water, the sand is retained in the bottom of the sump and the water overflowed into an outer compartment. From this sump sand and water are pumped back to the top of the cone and the water required for gravity control is returned through the control valves to the cone water inlets.

Another function of the sand sump is to hold a reserve of sand. If the cone has been operating at a high gravity and is suddenly changed to a low one the excess of sand automatically leaves the cone and is retained in

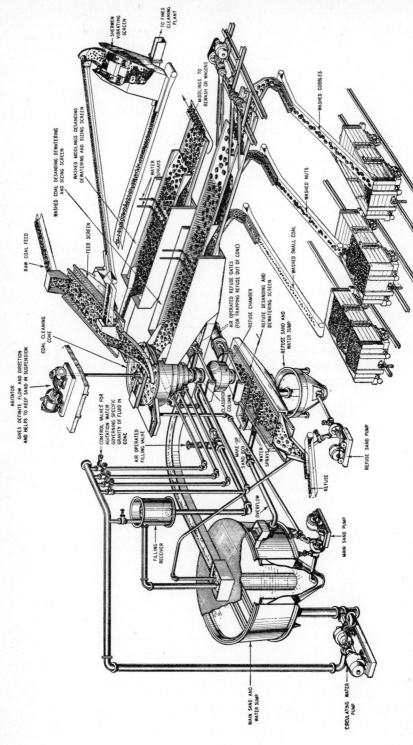

Figure 2. Pictorial layout of Chance system. (By courtesy of The General Electric Company Limited, Fraser & Chalmers Engineering Works)

the sump. This reserve of sand is, likewise, available when more sand is required in the cone. The sand losses in operating the Chance process are in the neighbourhood of 1½–2 lb./ton of raw coal treated, and to compensate for this it is usual to add make-up sand once a shift. This make-up sand is also stored in the sand sump thus avoiding the inconvenience of adding sand continuously.

Separation of Middlings

A Chance cone may be designed for either two product or three product separations. The normal two product design for producing only clean coal and refuse, as already described, is illustrated in *Figure 2*. When it is desired to extract separately a middlings product as well as coal and refuse, the type of separator is used, in which a middlings column is connected to

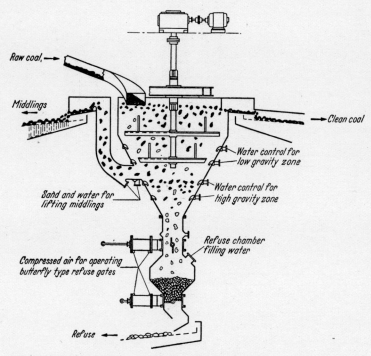

Figure 3. Chance separator with middles column. (From B. M. BIRD and D. R. MITCHELL, 'Coal Preparation', by courtesy of the American Institute of Mining and Metallurgical Engineers)

the side of the cone. (See *Figure 3*.) This column is in direct communication with the sand suspension in the cone, and is operated by either increasing the gravity at the bottom of the cone or lowering the gravity at the top by adjusting the water control valves and thus allowing the middlings product to be floated off on the high gravity zone up the column to the desanding screen as shown in *Figure 3*. It is of interest to note here that this type of cone can be operated as a single gravity separator or a two gravity separator merely by opening or closing the valves controlling the sand and water for lifting the middlings up the column.

Barvoys System

The Barvoys system, developed at Sophia-Jacoba Colliery in Holland by De Vooys about 1931, was introduced into England about 1936 and is now well established. The system uses suspensions of higher stability which approach very close to those of a true liquid, and achieves separation under non-turbulent conditions. The Barvoys medium consists of finely divided

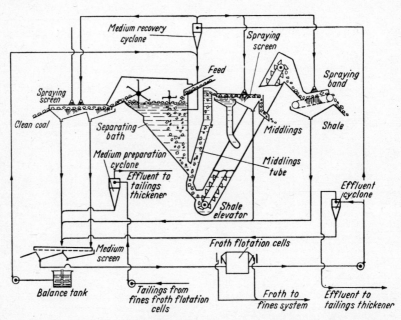

Figure 4. Barvoys bath with middle tube. (*By courtesy of* Head, Wrightson Colliery Engineering Limited)

barytes and clay, the size of the barytes being of the order of 25 μ while the clay, which is generally obtained from disintegrating shale in the coal treated, is still finer in size. A suspension of this type is virtually stable when the solid concentration is high and settles reasonably quickly when diluted with water. The specific gravity of barytes is 4·2 and of shale about 2·2–2·6 and their proportions may be adjusted so that the suspension has the required stability at specific gravities ranging from 1·30–1·60. The concentration of solids in the suspension is usually between 25 and 30 per cent by volume. Above this value the medium tends to become viscous, and at lower concentrations becomes unstable.

When the products of separation are rinsed with water to remove adhering medium, the resulting dilute suspension of medium can be recovered and densified by settlement in thickeners, but these are of relatively large diameter. As in other processes, arrangements must be made to remove coal slimes originating from the coal being treated, and for this purpose, a portion of the recovered medium is treated in froth flotation cells to separate the coal slimes and surplus clay from the barytes, which is then returned to the washing circuit. Medium consumption is stated to be 2–3 lb./ton of coal cleaned.

The Barvoys bath is of the deep type which can be used for any suspension of suitable character and the coal is transported mechanically across the surface of the separator either by a scraper or by rotary paddles while the shale is removed by bucket elevator (*Figure 4*). The middlings which sink in the separator together with the discard are separated from the discard by an upward current of medium arranged in the lower part of the vessel and a hydraulic classification of the middlings is thus effected. This separation by hydraulic current is not accurate since the upward force exerted on the middlings depends on their size and shape, but where the coal treated is in a relatively narrow size range, this may be unimportant in practice.

Dutch State Mines System

The Dutch State Mines system operates like the Barvoys system with a relatively stable medium and employs hydraulic cyclones to recover the medium. In this case, the medium itself was originally natural loess which is a very finely divided siliceous material similar to clay but containing very little colloidal material and not subject to decomposition in water. The specific gravity of loess is about 2·6 and the particles are about 300 mesh in size. Due to these physical characteristics and, in particular, to the absence of colloidal material, the loess medium may contain a high proportion of solids, up to 40 or 50 per cent by volume, without unduly affecting the viscosity of the medium, and so, despite the low specific gravity of the solids, suitable suspensions of up to 1·6 sp. gr. can be employed.

The dilute medium obtained by spraying the products of separation with water was found to be susceptible to thickening in the hydraulic cyclone and, further, that these cyclones provided an effective method of rejecting the finest coal and clay slimes with which the medium became contaminated during use. More recently, the cyclones have been used to prepare a medium from natural clays found in the shale discard and such clays are freed from slimes before being introduced into the washing process. In plants which include froth flotation equipment, the rejected tailings provide a source of dense medium and the necessity for using relatively expensive materials like barytes is avoided. However, with such low density solids, extreme care must be taken to ensure that the suspension does not become fouled with slimes and, in particular, froth flotation plant has to be included to remove coal slimes. Medium losses are of the order of 4–6 lb./ton but this is unimportant where the medium is obtained from the discard and costs no more than its preparation charges.

Figure 5 is a flow diagram of the Dutch State Mines process which employs a shallow type of separator about 4–6 ft. deep through which travels a continuous scraper, the top flights transporting the floating coal across the surface of the separator while the lower strand scrapes the refuse along the bottom of the separator in the opposite direction. It is claimed that the agitation provided by the scraper is itself sufficient to agitate the medium and no circulation of medium through the separator to maintain stability is necessary. It is, however, generally considered that some circulation

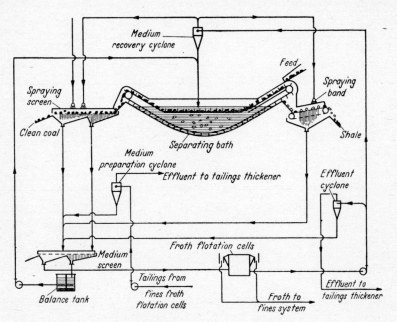

Figure 5. Flow diagram of Dutch State Mines process. (*By courtesy of* Head, Wrightson Colliery Engineering Limited)

must nevertheless be employed in order to ensure that the medium is continuously removed from the bath for cleaning on fine screens in order to remove associated coal.

Tromp Process

This process, which was developed by K. Tromp in 1938 at Domaniale mine in Holland, was the first to employ magnetite commercially as medium and the process is now used widely to treat coals sized down to ¼ in. at specific gravities from 1·40–1·90. The magnetite, and in later practice, mill scale, spathic ore or sinter, is ground to 100 B.S. mesh in wet ball mills before introduction to the plant to obtain a medium which is unstable and tends to settle in the separator.

The Tromp system, *Figure 6*, employs a deep bath in which the magnetite medium tends to settle in layers of increasing density. Layer density is controlled by maintaining a number of horizontal streams of medium across the bath, each of which has its own separate inlet to, and outlet from, the bath. On leaving the bath, each stream passes through a grid to retain coal and is then elevated by air lift to a mixing tank in which its density is corrected before return to the bath. In this way, the settlement of medium from one layer to the next is corrected and the density of the streams maintained constant.

The density of the respective layers is a function of the stability of the medium and of the rate of medium circulation in each stream. In practice, the sizing of the medium, once set, is maintained at a constant value and only the rate of circulation is varied. This is most simply effected since air lifts are used and an infinite variation in flow rate is possible.

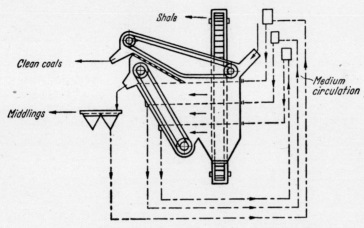

Figure 6. *Tromp bath and medium circulation system.* (*From* F. F. RIDLEY *and* H. Y. ROBINSON, *by courtesy of* The University of Leeds)

In such a bath, it will be appreciated that it is possible to separate more than two distinct products and, in fact, the material sinking at the surface is separated into a middles and a dirt product. The floating coal in the separator is conveyed across the surface of the medium by a scraper conveyor which also lifts the coal from the bath over an inclined draining screen fitted with wedge wire sieves on which the medium is drained from the coal. The middles are removed by a similar scraper which is fully immersed in the medium and collects middlings which have migrated to the grid protecting the medium outlet points and to the bath overflow which is the outlet point for the other circulating stream. The dirt is removed from the base of the separator by a bucket elevator. It will be appreciated that the wear on this conveying equipment is relatively high, but good maintenance makes it possible to keep the equipment in satisfactory condition.

The products leaving the separator are drained of medium on screens and are subsequently sprayed with recovered water in the usual way. The dilute medium is screened at 60 mesh to remove coarse coal breakage and is then pumped to a conical settling tank in which the medium is recovered and thickened by settlement for return to the bath. Automatic control of the density of the medium return to the bath is obtained by an automatic weighing device and good density control is possible. The area of the cone is selected so that most of the medium is recovered in the cone, while a proportion of the lower density coal and clay slimes overflow the rim and can be bled from the system to maintain the medium viscosity at a satisfactory level. The simplicity of this type of recovery system is attractive and is due to the relatively high settling rate of the coarser magnetite used in the Tromp process.

Ridley-Scholes Process

This process has the distinction of being the only widely used commercial process to be developed in Great Britain, and it was developed by F. F. Ridley and first applied at Derwenthaugh in 1947.

A magnetite suspension is used in this system but it is applied in a more stable form than in the Tromp system and has a medium of uniform density to make the conventional two product separation between surface floats and settled sinks. The stability of the working medium is obtained by the degree of fineness of grinding the magnetite, by suitably proportioning the coal and clay content of the suspension, and by upwardly inclined currents in the separating bath.

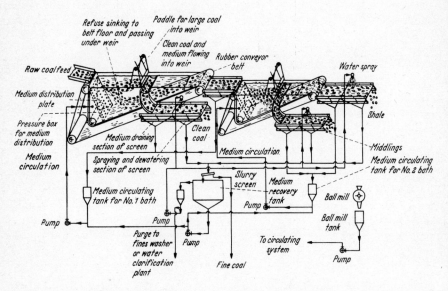

Figure 7. Ridley Scholes bath. (From B. M. BIRD *and* D. R. MITCHELL, 'Coal Preparation', *by courtesy of the* American Institute of Mining and Metallurgical Engineers)

The Ridley-Scholes bath, *Figure 7*, which is adapted to the use of any semi-stable suspension, is essentially a shallow-type separator. It comprises a wedge-shaped tank which tapers longitudinally from a depth of 4–5 ft. to zero. The medium is fed through inlets distributed over the cross section at the deep end and flows as a stream along the bath to surface overflow weirs in the shallow zone. Due to the diminishing depth of the tank, the flow of medium converges to the surface with upwardly inclined currents. In a shallow bath of this form, and of relatively small volume, the whole of the medium can be kept moving at a rate sufficient to maintain it stable and of uniform density during its period of passage through the bath.

The raw coal is fed into the deep end and travels along the bath in the same direction and at about the same speed as the stream of medium. Thus the separation occurs progressively with the minimum of relative motion between the coal and shale and the medium, and disturbance and displacement currents are thus minimized.

The other distinctive feature of the system is that the tank has virtually a moving floor which is provided by a troughed rubber conveyor belt which entirely covers the inclined bottom of the tank and travels along it from the deep end to the surface of the medium at the shallow end. The floating coal is moved along the surface of the medium by reciprocating paddles, which make a forward stroke in the medium and retract above it, and as it travels the shale and middlings, which are denser than the medium, sink and settle on to the belt at various points according to their rates of fall. Thus the belt is open to the medium and to the reception of sinking material over the whole area of the bath and, similarly, any material which is not dense enough to rest on it can rise at any point in its travel. Since the conveyance of the sinks is due to their frictional grip on the belt, only material denser than the medium can remain upon it.

In the more recent installations the belt enters the tank through multi-stage flexible seals in the end wall of the tank. These seals progressively reduce the hydrostatic pressure and pass only a small leakage which joins the circulating medium in the pumping tank. The belt passes round head and tail drums exterior to the tank, and the return side is below the tank.

At an intermediate position along the top of the tank is a transverse trough which is set below the surface of the medium and has inclined weir plates on its forward and rearward sides. The coal and part of the flowing medium is delivered into the trough over the forward weir, with the aid of a paddle, and part of the medium passes under the trough and discharges into it over the rearward weir. This maintains some flow over the greater part of the belt and assists deposition of the sinks of lesser density.

Coal passing into the trough is floated along it in the discharged medium and is delivered to a drainage screen at the side of the bath which drains out the medium. The latter then flows to the surge and storage tank from which it is pumped back to the bath. The coal then passes over the spraying screen where adhering medium is washed off it. The shale is similarly drained and sprayed.

The diluted medium is screened on a fine mesh wire cloth screen to remove particles of coal and shale and is then thickened in a settling cone, as in the Tromp process, which also classifies out a sufficient proportion of coal and clay slimes in the overflow. The cone overflow water is used for spraying the products before the clean water sprays are applied and an effluent is bled from this overflow to purge the system of slimes.

The thickened suspension is discharged from the cone through an oscillating ball valve which delivers it into a continuous weighing mechanism which registers its density and also operates an electrical circuit with timing relays. This circuit controls the operation of the ball valve and interrupts it when the medium is not of the required density and, after a time interval, restarts the cycle. The density of the medium returned to the system is thus automatically controlled.

Nelson Davis Process

This system, manufactured by the Nelson L. Davis Co. of Chicago, in common with all the more recently developed processes, employs magnetite as the medium together with magnetic means for recovering the medium sprayed off the products. The first plant of this type in the United Kingdom was installed during 1955 at Yorkshire Main Colliery.

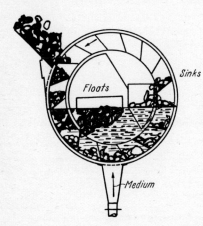

Figure 8. Nelson Davis separator. (From F. F. Ridley and H. Y. Robinson, by courtesy of The University of Leeds)

The separator, *Figure 8*, is cylindrical in shape and has an inner revolving drum, fitted with flights, which serves to introduce the coal to the separator and to remove the sink material which is elevated to a discharge trough, while the floating coal is carried over a weir by the circulating medium. The diameter of the drum is about 10 ft. and its length, which is designed to suit the type and difficulty of coal being treated, is about 8 ft.

The medium cleaning and recovery system, *Figure 9*, is of the type developed by the American Zinc Lead and Smelting Co. and comprises a thickener to recover the medium following magnetic flocculation to promote its settlement, a magnetic separator to clean the recovered medium, and a densifier and demagnetizer to control its density and to deflocculate it prior to its return to the separator. This type of recovery system is relatively efficient and medium losses can be reduced below 1 lb./ton of coal feed and is particularly valuable where the coal contains a high clay content or when a high separating density is required and extreme cleanliness of the medium is essential. It is, however, necessary to use a relatively coarse magnetic medium to ensure good magnetic pick up and this may lead to unstable conditions in the separator unless high flow rates are employed.

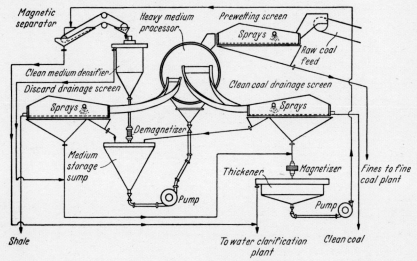

Figure 9. *Nelson Davis recovery system.* (From F. F. RIDLEY and H. Y. ROBINSON, by courtesy of The University of Leeds)

Simcar Process

This process is a combination of the Drewboy separator developed by Preparation Industrielle des Combustibles in France and a recovery system developed by Simon-Carves Ltd., England, and was first introduced into the United Kingdom at Rossington Colliery. The recovery system itself was, however, first employed in 1951 at Williamthorpe Colliery and superseded a standard recovery system of the Tromp type there.

The Drewboy separator, *Figure 10*, is a bath of the shallow type making a two product separation and is characterized in the fact that there are no wearing parts in contact with the medium and in that the volume of the bath is low. Coal introduced at the surface of the bath with the medium is carried across the surface of the bath and discharged over the overflow weir by two slowly rotating star wheels with chain flights. The sinks fall to the foot of the separator and are picked up in a perforated dirt wheel divided into compartments in which they are lifted out of the medium and discharged through the dirt wheel.

In certain applications, the medium may also be fed into the separator at its base and this is particularly useful when there is a high proportion of near gravity coal which teeters in the medium and is thus directed to the floats. Similarly, an underflow can be taken from the separator to direct these teetering particles to the dirt. The medium cleaning and recovery circuit is primarily of the magnetic type, though it has the distinction that it can also be used with non-magnetic medium when it is used in the same way as in Tromp and Ridley-Scholes processes.

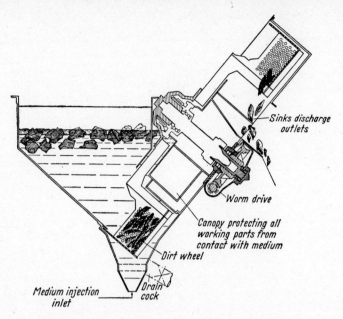

Figure 10.   *Drewboy separator.*   (*By courtesy of* Simon-Carves Ltd.)

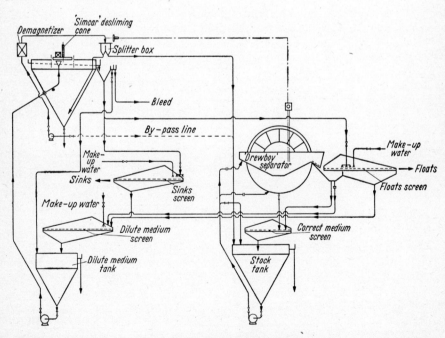

Figure 11.   *Flow diagram of Simcar dense medium plant incorporating Drewboy two-product separator.*
(*By courtesy of* Simon-Carves Ltd.)

Referring to *Figure 11*, the products of separation are sprayed on vibrating screens equipped with $\frac{1}{2}$ mm stainless steel wedge wire decks and the 'dilute medium' obtained is screened at 30–60 B.S. to remove fine coal breakage. The dilute medium is then pumped to a Simcar Desliming Cone in which the magnetic medium is recovered, cleaned and densified for return to the separating bath, after demagnetization. The desliming cone, *Figure 12*, comprises a conventional 60° cone with launder and the feed pipe terminating in a tun dish set inside a circular skirt. Immediately over the tun dish and just above the level of the liquid in the cone is a deep field electromagnet,

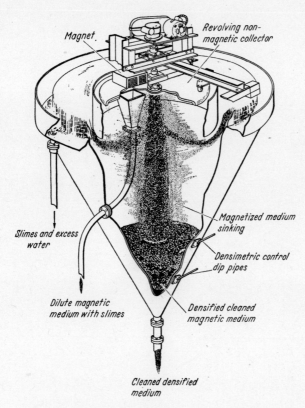

Figure 12. Simcar desliming cone. (*By courtesy of* Simon-Carves Ltd.)

and between the tun dish and the electromagnet is a non-magnetic spinner plate, with vanes on its underside, rotating at 12 r.p.m. The dilute medium entering the desliming cone is thus discharged into a strong magnetic field which magnetizes and flocculates the particles of magnetic medium, the magnetized particles adhering firmly to the underside of the spinner plate until its rotation carries them out of the magnetic field. The magnetic flocs formed are of large size and settle very rapidly to the base of the cone while the non-magnetic slimes and fine coal particles remain in suspension and overflow the cone at the launder. This overflow is used as spray water and a bleed may be taken from it to purge the system of slimes.

The remarkably high settling rate of the magnetic flocs make it possible to wash with very finely divided magnetic medium and this in turn results in a higher degree of stability in the separating vessel than in any other system using magnetic medium. As a further consequence, the flow rate of medium through the separator can be exceptionally low and this allows coal separation under virtually ideal conditions.

The Simcar process employs pneumatic type control instrumentation to control the density of the medium leaving the desliming cone and hence the specific gravity of separation in the separator.

Processes widely used commercially outside Great Britain

Descriptions of the HUMBOLDT DEUTZ [15], Link Belt [16] and S.K.B. dense medium processes are included here since, while they are not in commercial use in the United Kingdom, such processes are widely used commercially in Europe and the United States.

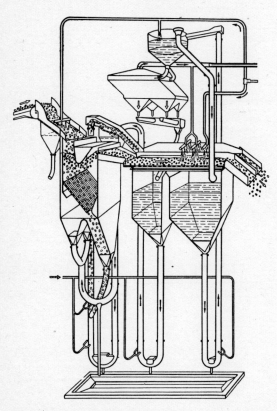

Figure 13. Humboldt Deutz deep bath. (By courtesy of Klöckner-Humboldt Deutz, Cologne)

Humboldt Deutz Process

The Humboldt process (1949) is marketed by Klöckner-Humboldt Deutz A.G. of Cologne and employs a deep bath designed for use with magnetite or other high density media such as pyrites. The raw coal is fed to the separator proper at a point below the liquid level into an upward stream of the medium which carries the floating material to an overflow weir. The sinking material passing down the bath body is deflected by a perforated baffle plate from the medium inlet point and settles to the foot of the bath where an air lift is employed to elevate the discard with medium to a drainage screen. On this screen, the medium is separated from the discard and is returned in part to the separator and partly to increase the flow of medium up the discard pipe. The medium overflowing the separator with the floats is similarly separated from the floats on a drainage screen and is returned to the separator. The distinguishing feature of the separator is the total absence of moving parts in it and the use of air lifts for circulating the medium to the bath.

Figure 13 is a flow diagram of the process in which the coal and discard are drained on a divided screen or on two parallel screens thus maintaining unidirectional flow of the products. The products are sprayed with recovered spray water in the usual way and the dilute medium is elevated by air lift to a head tank to control the flow rate and thence to the Humboldt thickener in which the medium is recovered and thickened for return to the separator.

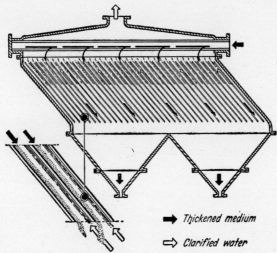

Figure 14. Humboldt thickener. (*By courtesy of* Klöckner-Humboldt Deutz, Cologne)

The Humboldt thickener (*Figure 14*) is of a novel design and is fitted with baffle plates to accelerate the settlement of the medium and to ensure a uniform flow distribution throughout the area of the unit. It is claimed that the thickener has a larger equivalent area for settlement than any other vessel of comparable size.

Link Belt Process

The link belt process (1946) was developed by the Link Belt Co. in co-operation with the American Cyanamid Co. and others. The process is distinguished by the design of the separator (*Figure 15*) which was the first of the drum separators to be brought into commercial use.

Raw coal is fed axially to the separator with a medium flush and the floating coal is carried across the surface to the discharge weir by the flow of medium; additional medium as required being fed to the base of the separator for this purpose. The discard sinking in the separator falls into a rotating perforated dirt wheel with lifting flights which elevate the dirt and discharge it into the dirt flue.

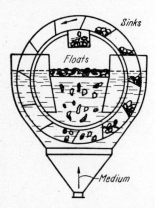

Figure 15. Link belt drum separator. (*From* F. F. RIDLEY *and* H. Y. ROBINSON, *by courtesy of* The University of Leeds)

The link belt separator has also been adapted to make three products by dividing the dirt wheel into two sections by means of an annular ring. Dirt settles rapidly in the bath and falls into the first section, while middlings settle less rapidly into the secondary section.

The recovery system, *Figure 16*, is of the conventional electromagnetic type as developed by the American Zinc Lead and Smelting Corporation and permits of low medium consumption. This type of separator, in common with the Nelson Davis unit, is claimed to handle large coal up to 10–12 in., but inevitably the large medium flow rates required to discharge coal over the floats weirs cause some degree of misplacement of smaller sized coal.

S.K.B. Process

The S.K.B. Process (1953) now in operation in Germany is marketed by Schüchtermann & Kremer-Baum of Dortmund and is adapted for use with unstable medium.

Figure 17 is a part section of the separating vessel which is again of the drum type with axial feed of raw coal. The floating clean coal is discharged at the surface of the separator over a weir to the drainage screen on which the coal is separated from the medium. The sinking material settling in the drum is carried forward by a spiral ring attached to its inner surface and is discharged to a bucket wheel which is compartmented to lift the dirt out of the separator. The medium cleaning and recovery system is of the non-magnetic type.

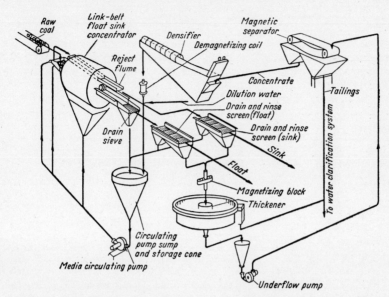

Figure 16. Flow diagram of recovery system incorporating link belt separator. (From F. F. RIDLEY and H. Y. ROBINSON, by courtesy of The University of Leeds)

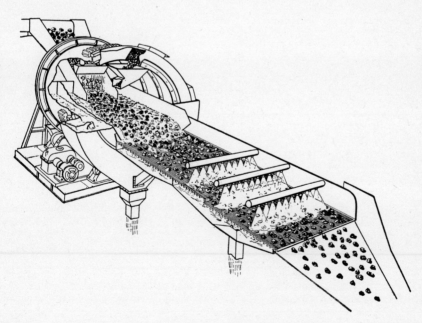

Figure 17. Part section of S.K.B. drum separator. (By courtesy of Schüchtermann & Kremer-Baum, Dortmund)

REFERENCES

[1] BIRD, B. M., and MITCHELL, D. R., 'Coal Preparation', *Amer. Inst. min. (metall.) Engrs* (1950) 471–2

MOTT, R. A., 'The Cleaning of Coal Using Dense Media', *Inst. Fuel*, Feb. 1936

[2] BIRD, B. M., and MITCHELL, D. R., 'Coal Preparation', *Amer. Inst. min. (metall.) Engrs* (1950) 472–3

BERTRAND, M. F., 'Pure Coal and Its Applications', *J. Inst. Fuel*, 8 (1935) 328–336

[3] BIRD, B. M., and MITCHELL, D. R., 'Coal Preparation', *Amer. Inst. min. (metall.) Engrs* (1950) 473

[4] BIRD, B. M., and MITCHELL, D. R., 'Coal Preparation', *Amer. Inst. min. (metall.) Engrs* (1950) 474–5

[5] BIRD, B. M., and MITCHELL, D. R., 'Coal Preparation', *Amer. Inst. min. (metall.) Engrs* (1950) 475–9

[6] BIRD, B. M., and MITCHELL, D. R., 'Coal Preparation', *Amer. Inst. min. (metall.) Engrs* (1950) 480–94

[7] BIRD, B. M., and MITCHELL, D. R., 'Coal Preparation', *Amer. Inst. min. (metall.) Engrs* (1950) 471–2

MOTT, R. A., 'The Cleaning of Coal Using Dense Media', *Inst. Fuel*, Feb. 1936

[8] BIRD, B. M., and MITCHELL, D. R., 'Coal Preparation', *Amer. Inst. min. (metall.) Engrs* (1950) 498–9

SHAFER, O., 'Coal Preparation by the Tromp Dense Liquor Process', *Colliery Engng*, Jan. 1939

RIDLEY, F. F., and ROBINSON, H. Y., 'Dense Medium Processes for Coal Cleaning', Symposium on Coal Preparation, Nov. 1952. *Univ. Leeds Min. Dep.*, 32–4

[9] BIRD, B. M., and MITCHELL, D. R., 'Coal Preparation', *Amer. Inst. min. (metall.) Engrs* (1950) 502–4

GROUNDS, A., 'The Ridley-Scholes Coal Washing System', *Trans. Inst. Min. Eng.* May 1947, 106, 8

RIDLEY, F. F., and ROBINSON, H. Y., 'Dense Medium Processes for Coal Cleaning', Symposium on Coal Preparation, Nov. 1952. *Univ. Leeds Min. Dep.*, 35–6

[10] BIRD, B. M., and MITCHELL, D. R., 'Coal Preparation', *Amer. Inst. min. (metall.) Engrs* (1950) 499–502

DRIESSEN, M. G., 'Cleaning of Coal by Heavy Liquids with Special Reference to the Staatsmijnen Loess Process', *Inst. Fuel*, May 4, 1939

RIDLEY, F. F., and ROBINSON, H. Y., 'Dense Medium Processes for Coal Cleaning', Symposium on Coal Preparation, Nov. 1952. *Univ. Leeds Min. Dep.*, 28–32

[11] BIRD, B. M., and MITCHELL, D. R., 'Coal Preparation', *Amer. Inst. min. (metall.) Engrs* (1950) 510–11

RIDLEY, F. F., and ROBINSON, H. Y., 'Dense Medium Processes for Coal Cleaning', Symposium on Coal Preparation, Nov. 1952. *Univ. Leeds Min. Dep.*, 42–3

[12] SYMINGTON, R., and HAMILTON, J., 'The Development of the Simcar Dense Medium Washer', *Gas World*, Coking Section, June 1951

[13] SYMINGTON, R., 'Dense Medium Coal Washing Processes with Reference to the Simon-Carves Dense Medium Coal Washing Process', *Sheffld Univ. Min. Mag.*, 1952

[14] BIRD, B. M., and MITCHELL, D. R., 'Coal Preparation', *Amer. Inst. min. (metall.) Engrs* (1950) 480–94

[15] RIDLEY, F. F., and ROBINSON, H. Y., 'Dense Medium Processes for Coal Cleaning', Symposium on Coal Preparation, Nov. 1952. *Univ. Leeds Min. Dep.*, 46

[16] RIDLEY, F. F., and ROBINSON, H. Y., 'Dense Medium Processes for Coal Cleaning', Symposium on Coal Preparation, Nov. 1952. *Univ. Leeds Min. Dep.*, 40

12

AIR FLOW SELECTION

R. A. SCOTT

In certain processes particles are separated into discrete fractions by the action of forces arising from the flow of air past the particles. In all such separators or classifiers the force of the air current on the particle is opposed or modified by gravitational and inertial forces associated with the mass of the particle. Individual particles move in a manner which depends on the balance between those forces due to impact of the stream and those due to the mass of the particle. Such processes separate particles in a manner which reflects differences in terminal velocity rather than in size.

The general class of air flow selection processes may be considered to embrace at one extreme, laboratory air elutriators used for the measurement of terminal velocity and, at the other extreme, industrial dust and particle collectors intended for the complete removal of particles from an entraining stream of air. Whilst some reference is made to such devices where they are useful in illustration of principles, the main attention of this section will be directed towards methods for selecting from a stream of particles those which lie in a particular range of terminal velocity. Further information on laboratory elutriators is to be found in Section 2, 'Size Analysis', and particle separators are dealt with in the subsection 'Dust Collectors', p. 426.

It has been shown in the subsection headed 'Laws of Settling' that the drag force on a particle can be described quantitatively in terms of the dimensions of the particle and the viscosity and density of the fluid. A general formula (6) for terminal velocity is given on page 167 in terms of the drag coefficient.

Where particles of sufficiently small size fall in air, the flow about the particle is streamline and the drag force depends on the viscosity of the air and the size of the particle as expressed by a typical length dimension l. The terminal velocity is then given, if the buoyancy of the air is neglected, by

$$u = mg/k\eta l \qquad \ldots . (1)$$

where the symbols have the same significance as in formula (2) of page 163. The terminal velocity in air of a small particle of given shape (and therefore constant k) is directly proportional to the ratio of its mass to its width: m/l.

Large particles create turbulence in the air as they fall; the drag force, in this case, depends on the density ϱ of the air and the area of section A of the particle normal to the direction of fall. The terminal velocity is then given by

$$u = (mg/\tfrac{1}{2}\varrho A C_D)^{\tfrac{1}{2}} \qquad \ldots . (2)$$

as in formula (6) of page 167, but with C_D taking a roughly constant value for all particles. The terminal velocity in air of a large particle is proportional to the square root of the ratio of its mass to its area of section: $(m/A)^{\frac{1}{2}}$. All particles of common density and like shape, whether large or small, have distinct terminal velocities according to their size; air flow separation may, therefore, be used for the grading of such particles according to size. Curve A of *Figure 1* shows the variation of terminal velocity for a wide range in size of spherical particles of unit density, falling

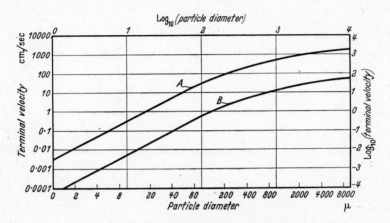

Figure 1. Terminal velocity of spherical particles

in air. In practice, the use of air flow separators for sizing purposes is ordinarily restricted to processes for which sieving is inefficient, as for example where the majority of particles are smaller than 100 μ in mesh size.

Air flow separations may also be used to make characteristic separations according to shape. Among particles of definite screen size and common density, those of compact form have higher terminal velocity than particles of lamellar form; granular particles are, for example, less easily lifted by an upward current of air than are fibrous or flake-like ones. Air flow selection may also be used, at least in principle, for separation according to density. Among particles of closely similar size and shape, the denser ones have the higher terminal velocity. However, moderate differences in density are obscured by small differences of shape or size and liquid flow selection methods, with their extra buoyancy and therefore greater discrimination, usually give a better separation.

Winnowing or wind-sifting of ground corn to remove husk must rank as one of the earliest applications of air flow selection. Subsequently developed forms of air flow separators fall into one of two principal classes, characterized respectively by actions which depend on using the force of air resistance on the particle, either

(a) to modify the shape of trajectory of the moving particle, or
(b) to draw the particle in a direction directly opposed to the gravitational or a centrifugal force.

The 'catch-box' separator or settling chamber shown in *Figure 2* illustrates the first principle; the air column elutriator in *Figure 3* illustrates the second principle. The action of these separators is described in more detail below.

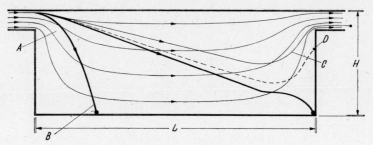

Figure 2.   *Air and particle movement in simple settling chamber*

In a typical settling chamber, the incoming stream of air falls in velocity as it enters the chamber A. The entrained particles are therefore initially projected horizontally into the chamber with a speed substantially greater than that of the air. Particles of very large terminal velocity of fall under

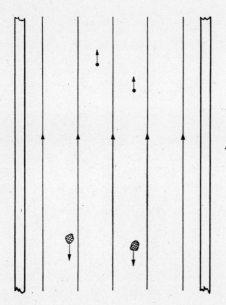

Figure 3.   *Air and particle movement in simple air column elutriator*

gravity, sink to the bottom of the chamber in an arc of roughly parabolic shape, B, while those of very small terminal velocity, C, are quickly retarded in their forward flight, relative to the air stream, and are thereafter sufficiently fully entrained by the stream to emerge from the chamber with the outgoing air. Particles of intermediate terminal velocity move across the chamber at much the same speed as the air, but lose height under the action of the downward pull of gravity. If the air stream takes t seconds to cross the chamber and the particle is small enough to obey Stokes' law it may be said with sufficient accuracy that the particle falls a distance ut in flight where u is the terminal velocity of the particle. A particle which is near the top of the inlet duct as it enters the chamber and which falls fast enough just to reach the bottom of the stream during the time t is trapped in the chamber. Thus, if H is the depth of the chamber, all those particles for which

$$u < H/t \qquad \ldots\ (3)$$

are separated from the air stream. The course of a particle which critically satisfies this condition is shown by the dotted curve D of *Figure 2*.

In contradistinction to the settling chamber, the flow of air in the air column elutriator is vertically upwards (see *Figure 3*). The air drag on each particle exerts a force which opposes the weight of the particle. When the upward velocity of the air is equal to the terminal velocity of a particle which enters the air stream, the air drag force is just equal to the weight of the particle, and the particle remains suspended in the stream. All particles with terminal velocities greater than the upward velocity of the air fall through the air column and are separated from the air stream; others are lifted by the air stream.

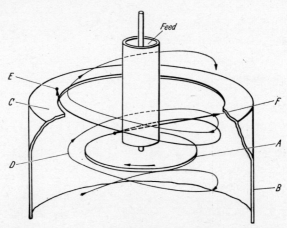

Figure 4. Air and particle movement in centrifugal separator with helical air flow

The settling chamber and the air column separator offer the simplest examples of the trajectory separator and the elutriator respectively. The various varieties of centrifugal separator fall into two well-defined classes which may be regarded as analogues of the settling chamber and of the elutriating column. In the first of these classes the particles are introduced into a stream of air which is moving with a rapidly rotating helical motion; in the second class the particles are introduced into a stream of air moving with a rapidly rotating spiral motion. In each case the place occupied by gravity in the simpler sort of air separator is taken by the centrifugal acceleration of the particle as it is whirled in a curved path by the rotating body of air.

Figure 4 illustrates the essential features of the separating zone of what may be called the centrifugal settling chamber. The particles are fed to the upper surface of a spinning disc, A, and are thrown tangentially from its edge. A cylindrical casing, B, surrounds the disc and terminates in a shallow ledge or lip, C, which extends a short way inwards from the upper rim. The disc lies a short distance below the rim. A system of guide vanes or impellers causes the air to follow an upward helical path, D, past the edge of the disc and ultimately out through the top of the casing. The spinning current of air tends to entrain the particles so that the finest, for example at E, tend to follow the course of the air and emerge through the top of the casing. The entraining action is opposed by the inertia of the particles with the result that relatively massive particles, or more particularly those of high terminal velocity, are not much deflected from their initial tangential flight. These particles strike the wall of the chamber at some point, such as F, below the lip and are separated from the air.

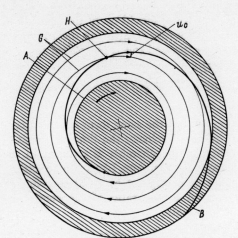

Figure 5. *Outward drift of particles in helical flow separator*

A somewhat simplified view of the action of this class of separator is illustrated in *Figure 5*, which shows a plan view of the separation region immediately above the spinning disc. At this point the traces of the streamlines of the air flow are circles, G. Small particles such as H are very nearly entrained by the air flow. Apart from a small, steady, outward drift across the stream, these particles move with the upward and rotating velocity components of the air. The resistance force which acts on the particles by virtue of the component of drift velocity is just sufficient to offset the centrifugal force due to the flight of the particle in its curved path. All particles which enter the helical stream of air pass through the aperture provided that their speed of outward drift relative to the stream is so slow that they have not reached the wall B by the time they have risen to the plane of the aperture.

The outward drift velocity u_c may be determined by equating the centrifugal force on the particle with the drag force associated with the drift of the particle across the air stream. For a small particle of mass m and width l the drag force F, calculated from the streamline formula (2) of page 163, is equal to $k\eta l u_c$. If the particle moves in a path of radius of curvature r with tangential velocity V, the centrifugal force (neglecting buoyancy effects) on the particle is mV^2/r. Thus, by equating forces:

$$u_c = mV^2/k\eta l r.$$

By comparison with formula (1) on page 337, it may be seen that the value of the drift velocity u_c is N times larger than the terminal velocity u of free fall of the particle, where $N = V^2/rg$, and is therefore equal to the centrifugal acceleration expressed in terms of the acceleration due to gravity. In practical separators, particles of such size that they are just separated move with a velocity, V, very nearly equal to the air velocity v in a path of radius only slightly larger than that of the air flow. The centrifugal acceleration of the particle, and therefore N, is approximately equal to that of the air flow just inside the lip of the aperture.

The condition that a particle should just be separated from the air stream is substantially that its drift velocity u_c should carry the particle to the wall of the chamber during the period of time t', in which the air stream rises to the level of the lip. Otherwise expressed this is equivalent to

$$u_c < H/t'$$

where H is in this case the annular radius of the gap between the rim of the disc and the wall of the chamber.

It should be noted that this expression has the same form as (3), except that to separate a particle of given terminal velocity the time of transit of the air through the separating section of the chamber is reduced in the ratio of u_c/u. The centrifugal separator acts, therefore, like a simple settling chamber in which the mean velocity of passage of air, and therefore of fines, through the chamber is increased in the ratio of the centrifugal acceleration to the acceleration due to gravity.

The calculations given above are based on drag forces given by Stokes' law; the nature of the separation is similar in cases where the drag is associated with turbulent flow, but the calculations are more complicated. In practice, the Reynolds number of the drift flow relative to the particle usually takes values substantially greater than unity for all particles larger than a few microns in diameter.

Figure 6 illustrates the essential features of the separating zone of what may be called the centrifugal elutriator. The particles are fed to a region, A, in which the air is made to move inwards towards a central discharge, B, whilst it rotates about the axis of the chamber. The air tends to entrain fine particles, C, so that they move spirally inwards with the air. The inertia of the particles opposes the entraining action and large particles, as at D, which enter the chamber tangentially, tend to remain on the same course until they meet the outer wall of the chamber.

A particle, E, of intermediate terminal velocity will remain circling at a fixed radius, r, with substantially the same velocity, v, as the air if the centrifugal force mv^2/r is exactly balanced by the drag F caused by the radially inward velocity component u_c of the spiral motion of the air.

If the drag force is assumed to follow Stokes' law

$$F = mv^2/r = k\eta l u_c \qquad \ldots \ (6)$$

where k, l and u_c refer to the critical particle and η is the viscosity of the air. Comparison of this formula with (1) shows that the particle is in equilibrium when the inward velocity component u_c is as many times greater than the terminal velocity of free fall under gravity, as the centrifugal acceleration of the particle is greater than that due to gravity. Particles of smaller terminal velocity than that of the critical particle move inwards from the radius and may be discharged with the air through the outlet port B; particles of higher terminal velocity spiral towards the outer regions of the chamber. Once again, the separator behaves like the analogous elutriating column but with the velocity of transport of air through the chamber increased in the ratio of the centrifugal acceleration to the acceleration due to gravity.

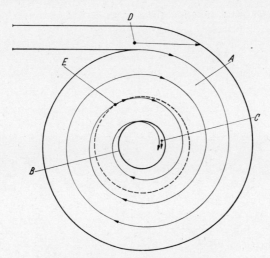

Figure 6. Air and particle movement in centrifugal separator with spiral air flow

Air flow selection is widely used in two distinct fields. The simpler sorts of machine such as the settling chamber or the elutriator are used for separation of particles according to characteristics of shape or of density in applications where only a small proportion of particles have terminal velocities below a few hundreds of feet per minute (say, less than a Stokes' diameter of 500 μ for particles of about unit density). The centrifugal forms of machine are used for particles of substantially lower terminal velocity. In the majority of applications such machines are used to separate streams of particles of common shape and density but of diverse particle size. In such cases, the centrifugal separator acts primarily as a size grader. Its special value lies in its capacity for separating moderately large quantities of material in the size range below 100 μ, for which the alternative methods of fine screening are unduly cumbersome.

The advantage of centrifugal classifiers over their simpler counterparts in the separation of fine particles is conveniently illustrated by consideration of the factors which determine throughput in classifiers generally. The mass rate of transport of, for example, the finer particles through a separating chamber of given size in any of the principal forms of separator depends on the product of the volume concentration of the particles and the mean rate of transport of particles in the chamber. For maximal capacity each factor must be as large as possible. An upper limit to the volume concentration is set by the need to ensure that the motion of individual particles is not too greatly affected by the presence of near neighbours. The capacity of a separator of given size is therefore directly proportional to the mean rate of transport of particles through the separating region.

In a simple settling chamber particles of some given size are just collected in the chamber, when the forward velocity of the air flow in the chamber is roughly L/H times the terminal velocity of the critical particle (see *Figure 2*). This ratio is usually only a few times greater than unity. Where particles of low terminal velocity are to be separated, they move through the chamber at low velocity. The capacity of a settling chamber of given size is therefore low for particles of low terminal velocity.

In a typical centrifugal separator the centrifugal force on a particle may be 500 to 5,000 times the force due to gravity. For a particle of critical size the air stream through the separator must exert a correspondingly large drag force. To produce such a force, at least on particles of a few microns in diameter, the air stream must move through the chamber with a velocity 500 to 5,000 times that in a non-centrifugal classifier of analogous type. The capacity of a centrifugal separator may, therefore, be maintained at a relatively high level even for particles of as low a terminal velocity as 1 ft./min (say, a Stokes' diameter of 10 μ for particles of unit density).

While the principles of action of the centrifugal elutriator are conveniently described by analogy with the gravity elutriator, the detailed behaviour of practical forms of the separator presents several special features. In all separators of this class the air possesses both rotatory and inward moving components of velocity, therefore the path of an air particle is an inwardly directed spiral.

For any chosen position in the flow pattern the capacity of the flow to centrifuge a particle of given terminal velocity from the throughgoing stream of air depends on maintaining a high ratio of centrifugal acceleration to radial velocity. In consequence, even in the simplest classes of air flow pattern the capacity of the air flow to separate particles of given terminal velocity varies with radial position across the flow pattern.

Three important main classes of flow pattern are distinguishable in practical separators. These are respectively: free vortex flow, modified free vortex flow, and forced vortex flow. The characteristics of these flow patterns are given in outline below.

FREE VORTEX FLOW

Where air is set in rotation towards the periphery of a separating chamber and is drawn radially through a central discharge tube the air pattern tends to form a free vortex in which the angular momentum is conserved at all radii.

Where the angular momentum is conserved the tangential component of velocity varies inversely with distance from the central axis of the chamber. As the air moves in towards the centre of the pattern the rotating velocity steadily increases. The centrifugal acceleration, which depends on the ratio of the square of the velocity to the local radius, therefore increases sharply towards the centre of the pattern.

Figure 7 illustrates the variation with radial position across the free vortex of the rotating component of velocity, and the centrifugal acceleration, v^2/r, of the air. The radial component of velocity u_c has the same form of variation as v, except where the flow enters the discharge tube.

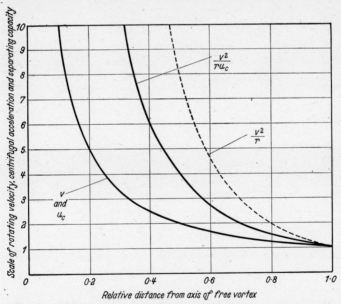

Figure 7. *Velocity, acceleration and separating capacity at various radii in a free vortex*

The graph also shows the ratio of centrifugal acceleration to radial velocity v^2/ru_c. For particles small enough to obey Stokes' law this latter quantity is a measure of the severity of the separating action: the greater this quantity, the lower the terminal velocity of a particle just separated at the corresponding position. Since the centrifugal acceleration rises much more steeply than the radial velocity as the radius decreases, the separating action of this type of flow is therefore markedly greater towards the central outlet than at regions remote from the outlet.

Modified Free Vortex Pattern

The free vortex pattern, referred to above, exists when the angular momentum of the air is conserved throughout its inward flow. However, the angular momentum of the air remains constant at all radii only if the confining walls of the chamber exert a negligible drag on the flowing air and if the shearing of the air in the body of the chamber leads to no appreciable transfer of energy from layer to layer. In practice, the wall drag and the shearing stresses (often of turbulent, rather than viscous origin) accompanying the progressive change in angular velocity across the chamber, disturb the velocity pattern. Where the effect of the dissipatory forces is considerable a modified free vortex flow is to be found.

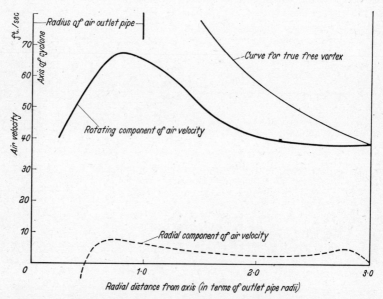

Figure 8. *Typical velocity pattern just below rim of outlet pipe of cyclone.* (*After* A. J. TER LINDEN [1])

A simple example of this type of flow is provided by the reverse flow cyclone separator (see *Figure 21* and also the subsection on particle and dust separation from gases). Towards the outer regions of the flow pattern the velocity increases steadily with decreasing radius. The velocity continues to rise until the radius of the path falls to about half that of the outlet pipe; thereafter, the velocity falls steeply to zero at the axis of the cyclone. *Figure 8* shows the relationship between the rotating component of air velocity and the radius of path in a diametrical section immediately below the mouth of the outlet pipe of a typical cyclone separator. The data are drawn from TER LINDEN's[1] classical account of cyclone behaviour.

The outer portion of the pattern is commonly known as the free vortex, although the rotating component of velocity increases at a significantly lower rate than in a true free vortex. For example, if v is expressed as proportional to the nth power of the radius of the path, n has a value of -1 for a free vortex. In the outer regions of a typical cyclone a value of $n = -0.5$ is generally considered to give the best fit (see discussion on ter Linden's paper, *loc. cit.*).

Towards the axis of the cyclone the air flow departs radically from the free vortex pattern. A free vortex of small diameter involves a very high rate of shear of the air and therefore a high rate of dissipation of energy to maintain the flow in the presence of the viscosity (and also of the 'eddy viscosity' associated with turbulence[2]) of the air. The throughgoing air is clearly limited in its capacity to supply such energy. Towards the axis, where more and more energy is withdrawn from the stream to overcome the viscous effects, the air tends to assume a constant angular velocity and therefore a vanishingly small rate of shear. This inner region of the flow is commonly known as the forced vortex.

In spiral flow air separators, the throughgoing air is generally withdrawn from the separating chamber by way of an axial outlet pipe. The air tends to pass laterally from the chamber through the outer annular region of the pipe since in that manner the violence of the shear in the central core, and therefore the total drain of energy from the system, is minimized. In the modified free vortex flow of a typical cyclone separator, the centrifugal acceleration, and also the ratio of this acceleration to the radial component of velocity, are maximal in the region of the rim of the outlet pipe (see *Figure 8*). The separating action for particles immersed in the flow is therefore greatest in this region (see, however, ter Linden, *loc. cit.*).

Forced Vortex Pattern

Where the inward spiral flow is maintained, for example, between closely spaced, rotating, diametrical plates, and particularly where the inward component of velocity is much lower than the rotating component and therefore carries little store of energy, the effect of the drag of the flanking walls tends to cause the air to move with the common angular velocity of the rotor. In the extreme case of a completely core-like motion the rotating component of velocity diminishes towards the centre linearly with radius of the path, while the radial velocity increases towards the centre inversely as the radius of the path. The centrifugal acceleration is greatest towards the outside of the pattern. The separating capacity, as determined by the ratio of the centrifugal acceleration to the radial component of velocity, is correspondingly greatest towards the outside of the pattern. *Figure 9* shows a cross sectional view of a small-scale particle classifier (Bahco Centrifugal Dust Classifier[3, 4]) in which the condition of constant angular velocity is at least approached. The separating chamber A is defined by closely spaced, rotating discs attached to a common rotor system. The rotor contains a fan blade system with vanes B which draws air in towards the lower central cavity through the lower annular aperture C. The air moves outwards through the gaps of the spaced set of discs which impart a smooth symmetrical rotating motion to the air. The air spirals inwards through the separating chamber to the root of the fan impeller and emerges at the periphery of the impeller.

Dust particles are fed through the central hole D in the upper face of the rotor and are drawn along the duct E so as to enter the separating chamber at F. Coarse particles are thrown out tangentially to collect ultimately in the bowl G. Fine particles of sufficiently low terminal velocity are drawn in by the entraining action of the air current and pass through the fan impeller. Much of the fine material is deposited by centrifugal action on to the inside of the bowl H.

Since the air flow between the discs tends to move with not much greater angular speed than the discs, the separating capacity is no longer greatest towards the inner radii of the separating space. The discs of the separating chamber have diameters of 12 cm and are driven at 3,000 r.p.m. The centrifugal acceleration of the air at the rim of the discs is, therefore, about 700 g, and at such high acceleration it is possible to make an effective separation at a critical terminal velocity corresponding to particles of a few microns in size.

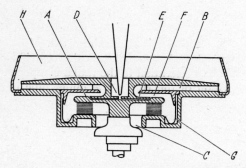

Figure 9. *Spiral flow laboratory classifier.* (*By courtesy of* Bahco, Sweden)

The size of a particle at the separation limit is controlled by alteration of the width and therefore throttling action of the annular gap through which the air enters the rotor. For example, spherical particles of unit density and $3 \cdot 7\,\mu$ in diameter have a terminal velocity of free fall in air of about $0 \cdot 05$ cm/s. If the air quantity through the classifier is controlled so that the radial velocity of the air at the rim of the chamber is 700 times this value, then, on the assumption of the continued operation of Stokes' law, particles of $3 \cdot 7\,\mu$ are just separated. Larger air quantity gives a larger separation limit or 'cut size'. A separator of this sort is capable of separating a few grammes per minute of particles of a few microns in cut size.

AIR FLOW CLASSIFIERS

For purposes of description, particle classifiers for industrial use are divided below into two groups. The first group consists of non-centrifugal classifiers such as settling chambers and air column elutriators, in which the air resistance forces are comparable with the weight of the particle. The second group consists of centrifugal classifiers in which the forces are very considerably increased by centrifugal means associated with rotation of the mass of air in the body of the classifier.

Non-centrifugal Classifiers

Separators Using Differences in Shape of Trajectory—The simple settling chamber already illustrated in *Figure 2* classifies entrained material of the feed into two classes. Those particles which have high terminal velocities fall through the air stream and are collected in the main chamber. Particles which have sufficiently low terminal velocities leave the chamber with the outgoing air.

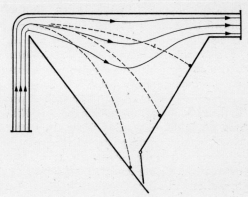

Figure 10. Simple catch box separator with hopper discharge. (*By courtesy of* Henry Simon Ltd.)

Figure 10 shows a typical catch-box such as is commonly used in the cereal industries for the separation of grain from light particles of husk or bran. Particles of high terminal velocity fall in smooth trajectories to meet the sides of the chamber and are removed from the bottom of the hopper.

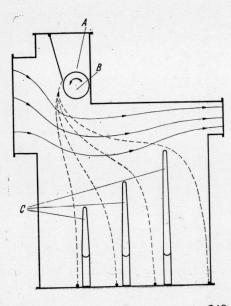

Figure 11. Separator depending on characteristic differences of trajectory shape. (*By courtesy of* Henry Simon Ltd.)

Such separators are suitable for the separation of particles of density about unity and diameter of, say, 1 mm and larger. Such particles have sufficiently large terminal velocities (1,000 ft./min and more) to allow them to fall rapidly out of the main air stream of the box. An incoming feed containing a high concentration of particles therefore thins out soon after entering the chamber. A substantial throughput of separated material can be collected without the volume concentration of particles in the chamber becoming so great that one particle interferes with the course of its neighbours.

Figure 11 illustrates a separator similar in form to the catch-box, but capable of yielding more than one separated product. Particles are fed from the hopper A by means of a rotating feed roll B, into a stream of air drawn transversely through the main chamber. Particles of low terminal velocity are strongly deflected by the air stream; those of high terminal velocity are weakly deflected. Adjustable partitions, C, canalize the deflected particles into streams which leave the machine by separate outlet ports. Separators of catch-box form are simple in construction but not very exact in operation. The lack of precision in separation is partly due to the difficulty of introducing the particles with sufficiently controlled velocity and direction—especially when the feed rate is high and the particles are close together—and partly to the disturbing effects which arise from cross currents and eddies of the air in the chamber. In such separators there exists also a strong tendency for some fine or flaky particles to be entrained by adhesion to the coarse particles.

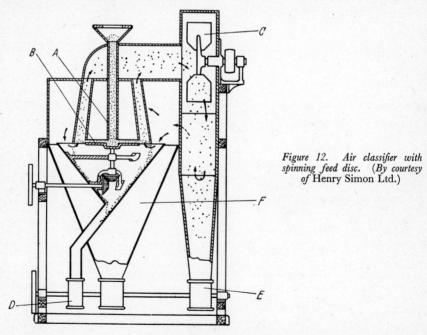

Figure 12. Air classifier with spinning feed disc. (By courtesy of Henry Simon Ltd.)

Figure 12 shows a different form of trajectory separator in which the particles are projected from the edge of a spinning disc into an upwardly moving air stream. Feed enters through the pipe A and falls on to a spinning disc B, which throws the particles tangentially from its edge. A fan, A, draws air upwards past the edge of the disc. Particles of high terminal velocity fall through the air stream to an outlet spout D. Particles of lower terminal velocity are drawn into the fan. The heavier of the lifted particles fall to the bottom of the expansion chamber E, while the lighter ones are separated from the air in the cyclone F. The cleaned air stream returns to the separating region to complete its circuit.

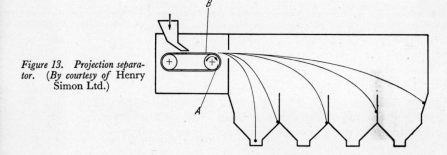

Figure 13. *Projection separator.* (*By courtesy of* Henry Simon Ltd.)

Alternative forms of separator make use of characteristic differences in the shape of the trajectories of particles when projected into relatively still air, (see *Figure 13*). The particles are thrown forward into a chamber with a velocity acquired by ejection from the end roller A of a fast moving feed belt B. The particles of low velocity are quickly retarded by the still air and, under the action of gravity, fall short of those with high terminal velocity. The throughgoing material may be separated into discrete classified fractions by the appropriate placing of collecting hoppers below the falling streams of particles. For low feed rates, particles of 0·1 mm and upwards can be separated into sharply defined fractions. When, however,

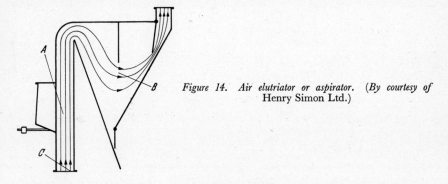

Figure 14. *Air elutriator or aspirator.* (*By courtesy of* Henry Simon Ltd.)

the feed rate is increased beyond a certain level the mass of particles projected from the belt tends to entrain a considerable mass of air which moves forward to disturb the paths of the individual particles and causes a deterioration in the accuracy of the classification.

Air Elutriators—The air elutriator or 'aspirator' has a range of application similar to that of the trajectory separator. Its general form is illustrated in *Figure 14*. Air is drawn upwards through a vertical channel A into which the stream of particles is fed; those particles with terminal velocities less than the velocity of the upward flow are lifted and deposited in an upper expansion box B while those with greater terminal velocity fall to a discharge port C.

Such air elutriators are commonly employed in certain branches of the food industry as alternatives to trajectory shape separators for the separation of husk and of small foreign seeds from grain. Simple air elutriators generally act more efficiently as classifiers than do separators relying on trajectory shape. Their chief advantage over the latter lies, however, in the possibilities offered for an improved efficiency resulting from repeated treatment in successive aspirator columns.

The efficiency of trajectory shape classifiers is seriously limited in practice by lack of uniformity of air flow and by complications arising from the concentration of particles in the stream. The various forms of air elutriator suffer from the same disturbing factors. Thus, eddies arise at bluffs in the

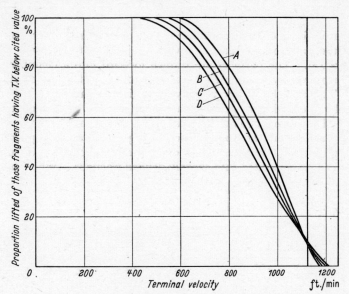

Figure 15. *Effect of increasing feed rate on efficiency of air column elutriator.*
(*By courtesy of* Henry Simon Ltd.)

channel and carry heavy particles upwards, while light ones which find their way to the region of low velocity near the wall, may fall. Likewise, the disturbing effect of near neighbours on the lifting force on each particle, may, according to circumstances, cause heavy particles to rise or light ones to fall.

The ill effects of non-uniformity of the air flow may be reduced by keeping at least one cross dimension of the channel small and the length of the channel moderately large. The ill effects of proximity of particles may be reduced by the use of low feed rates and by the even spreading of feed over the cross section of the channel. The combined effect of increasing feed rate on the separating efficiency of an elutriating column designed for the separation of the smaller fragments or 'screenings' from whole wheat grains is shown in *Figure 15*. The graphs show the position of screenings with terminal velocities less than the figure indicated on the horizontal scale, which are lifted in a typical case by a fixed air current as the feed rate of the wheat stream is increased from a very low value (curve *A*) to successively higher feed rates of 100, 200 and 300 lb./in./min respectively (curves *B*, *C* and *D*). Where the rate of feed is high the efficiency of separation falls severely, particularly when the material is so concentrated in the channel that each particle is spaced, on average, only a few diameters from its neighbours. Even where the rate of feed is low, however, particles which have terminal velocities lower than, but close to, the mean air velocity have a substantial chance of falling and particles with higher terminal velocities have a substantial chance of rising.

The air elutriation method of separation may be considerably refined by arranging two separate aspirator channels in a retreatment cycle. Such an elutriator for separating husk from wheat is shown in *Figure 16*. The air velocity in the first channel (1) is set about 10 per cent larger than the terminal velocity of the lighter particles which are to be removed; a high proportion

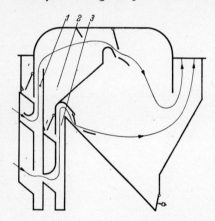

Figure 16. Two stage aspirator for wheat cleaning. (By courtesy of Henry Simon Ltd.)

of these lighter particles is therefore lifted together with a small proportion of heavy ones. The material carried upwards by the air current is separated from the air in the expansion chamber (2) and is fed to the second aspirator channel (3). Here the air velocity is set at such a (lower) value that every heavy particle in the channel falls out of the air stream; these heavy particles are then returned to the feeding point of the first channel. In this dual system such wheat particles as are lifted in the first channel, because they arrive by chance in a local region of high air velocity, are returned to the feed for retreatment by way of the second channel. The chance that a heavy particle be lifted in the first leg twice in succession is relatively low;

a substantially improved efficiency is therefore provided without an unduly large 'circulating load'.

Figure 17 illustrates a form of multiple treatment elutriator designed to separate only the fraction of highest terminal velocity from a substantial mass of particles of lower terminal velocity. The machine is designed especially for the separation of the relatively smooth and compact wheat germ particles from a carefully graded fraction of an intermediate ground

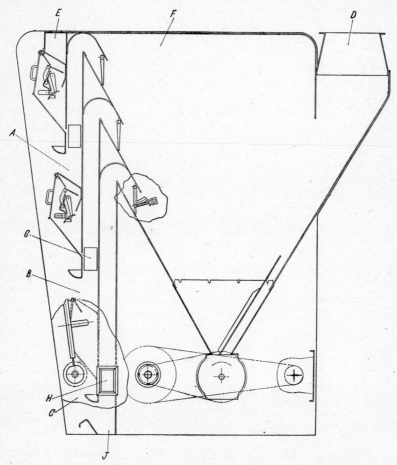

Figure 17. Triple aspirator for wheat germ separation. (*By courtesy of* Henry Simon Ltd.)

product of the grain. The unwanted particles are of much the same size as the germ particles; their terminal velocities are, however, lower, since the unwanted particles are generally more angular in shape and offer a higher ratio of area to weight. In the machine illustrated, the first two stages of the separation are concerned with the removal of those particles which are substantially 'lighter' than the germ so as to concentrate the germ into a sufficiently small fraction for the final separation of the third stage.

The three separate streams of air enter at ports A, B and C under the action of suction applied at D. Separate valves are provided to control the air flow in each column. Feed enters at E; light material is carried over to the catch-box F and is deposited there. The heavier fraction falls and enters the second column at G and once again the heavier fraction falls to form the feed to the final column H. The final product of separated wheat germ leaves the machine at J. The feed rates of material to the first two stages are necessarily substantially greater than to the third stage; the air velocity is, therefore, set at lower values in these columns so that little of the germ is lifted to the catch-box. The quality of the product depends critically on the setting of the air valve of the third stage.

In each of the elutriating column separators so far described, considerable care is taken to distribute the feed uniformly over the cross section of the air channel. The spreading is accomplished by means of feed gates of special design. A typical machine contains channels with cross sections 25 in. long by $3\frac{1}{2}$ in. wide. The mass of feed accumulates in a compartment alongside the channel and issues from the lower part of the compartment into the channel through a shallow opening which extends across the full breadth of the channel. A head of stock, sufficient to ensure that the stock spreads fully across the machine and yet is never great enough to cause a 'choke', is maintained by a self-regulating mechanism which causes the discharge gap to increase when an excess of feed enters the compartment. For free flowing material such as whole wheat grains, the base member of the compartment is pivoted and counterbalanced so that the direct weight of accumulating stock opens the feed gate. For particles which do not flow sufficiently freely, a hinged gate forming the upper extremity of the feed gap is connected by levers to an inclined plate which projects into the compartment at a level somewhat above the base. The inclined plate is depressed to an extent proportional to the weight of stock above its surface and its deflection causes the hinged gate to open appropriately. The material is fed through the gap by, for example, the forward thrust of a mechanically oscillated shoe which forms the lower extremity of the feed gap. Elutriators of the types described operate satisfactorily only if the feed spreading mechanism is efficient and the air channel is carefully constructed to give an even upward flow of air.

Centrifugal Air Classifiers

The Fan Separator—A centrifugal fan presents a simple means for the rough classification of coarse and fine particles suspended in an air stream. Air enters the eye of the fan and, under the action of the fan impeller, spins in the casing. The main stream of air emerges from the fan outlet, carrying the lighter particles in suspension. Heavy particles are thrown towards the wall immediately on entering the fan casing and these may be collected in a separate portion of the emerging air stream. *Figure 18* illustrates a fan separator of this general form. A fan casing, A, contains a long rotor which is formed of two separate impellers. The air stream, with its burden of particles, enters the eye of the fan at B and is set in rotation by the first impeller C. Particles of higher terminal velocity concentrate just beyond the periphery of this rotor and leave the casing at D, together with some of the throughgoing air. The main body of the air is drawn through the second impeller E and leaves the fan casing with the remainder of the particles at F. The path of particles into the fan separator and the air flow pattern in the fan casing are not sufficiently under control to allow this form of separator to make sharply defined classifications. Nevertheless, the fan separator is sometimes used for rough separation of coarser material from a suspension of particles in an air stream[5].

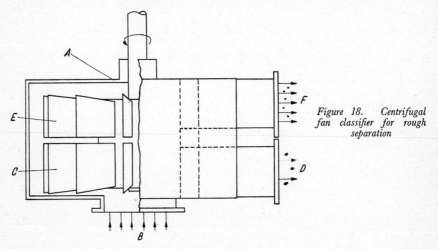

Figure 18. Centrifugal fan classifier for rough separation

Centrifugal Classifiers with Helical Flow of Air—Figure 19 illustrates a common form of air separator suitable for the separation of particles of terminal velocity greater than a few centimetres per second (say, greater than 40 μ in diameter for spherical particles of unit density). Particles enter the feed hopper at the top of the machine and fall on to the horizontal spinning plate A and are projected tangentially from the rim of the disc. The main casing of the classifier is divided into a central chamber surrounding the spinning plate and an outer annular passage. A fan rotor, B, draws air through the opening in the top of the central chamber and returns it to the bottom of the central chamber by way of the annular passage and the gap between the vanes C. The air current which moves generally upwards through the central chamber is made to rotate by the action of the blades of a lower rotor D which lies just below the rim of the central chamber.

The particles projected from the spinning plate enter a stream of air moving upwards with a helical motion, and those of high terminal velocity are thrown to the wall and spin down the inside of the central hopper to be delivered at the outlet port E; others of lower terminal velocity are more fully entrained by the air stream and, if fine enough, follow the course of the throughgoing air and are thrown to the wall of the outer casing of the classifier to be delivered ultimately at the outlet port F. The severity of the classifying action is greatest when the air stream in the separating region spins at high velocity and moves upwards at low velocity. The spin of the air is determined by the number (and size) of the blades fitted to the lower fan rotor; the greater the number of blades the more effective is the impeller and the finer the product—that is, until the stream of air about this lower rotor is moving substantially at the same angular speed as the rotor.

The throttle valves G control the upward rate of flow of the air. Restriction of the upward flow of air leads, at least for low rates of feed, to a smaller 'cut-size'. Where, however, moderate and high feed rates are used a high

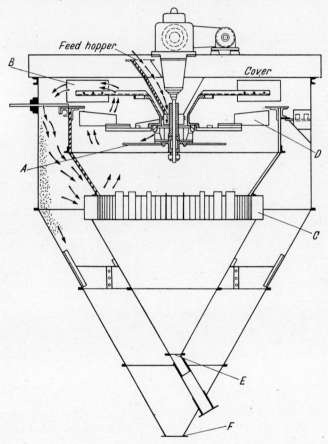

Figure 19. Centrifugal air classifier (Sturtevant). (From RIEGEL, by courtesy of Reinhold Publishing Co.)

rate of upward flow is needed to give a sharp separation. In such circumstances the volume concentration of particles in the separating region is kept at minimal value and the paths of individual particles are least affected by the presence of close neighbours.

Figure 20 illustrates an alternative form of the type of classifier described above. This differs mainly in the detail of the spinning feed plate and lower impeller. A is the feed plate, B the fan producing the main flow.

Helical flow air classifiers range in size from laboratory models with casings of about one foot in diameter, capable of classifying a few pounds of particles an hour (*e.g.*, for particles of unit density and a cut-size of 20 μ), to large-scale models with casings 20 ft. in diameter, capable of classifying a few tons an hour of particles of similar size.

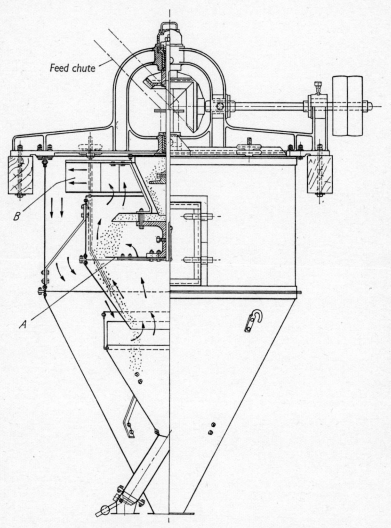

Figure 20. Centrifugal air classifier. (By courtesy of **Edgar Allen & Co.**)

Centrifugal Classifiers with Spiral Air Flow—Many classifiers contain a separating region through which the general movement of the air is along an inwardly directed spiral. Probably the best known example is to be found in the cyclone dust collector (see *Figure 21*). The critical region of this type of separator is, as already stated, near the entry to the central outlet tube *A*. At this point the throughgoing air rotates at high angular velocity. Very fine particles which pass into the region near the outlet tube are carried out of the chamber with the air. Coarse particles spiral outwardly from the central region since their inertia is large and the entraining action of the inward component of air flow is not great enough to draw them inwards against centrifugal force.

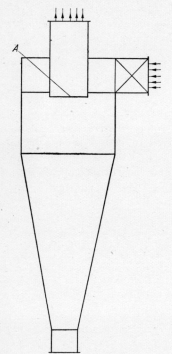

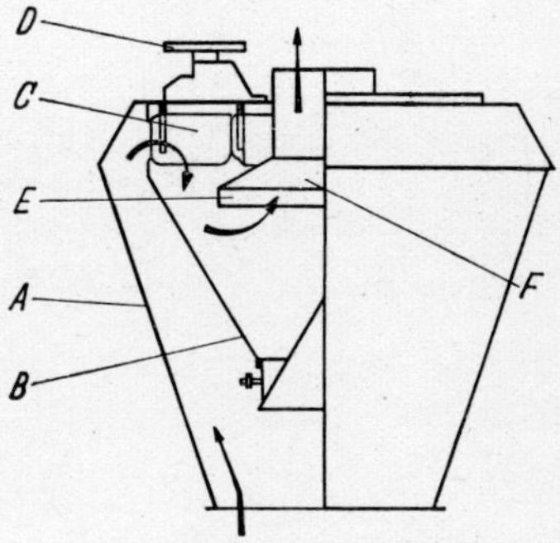

Figure 22. Double cone classifier. (By courtesy of British Rema Manufacturing Co. Ltd.)

Figure 21. Reverse-flow cyclone. (From A. J. TER LINDEN, *by courtesy of* The Institution of Mechanical Engineers)

Conventional cyclones have a positive classifying action and commonly operate to give a separating limit corresponding to particles with a terminal velocity of a few centimetres per second (10–20 μ for a spherical particle of unit density). The sharpness of separation is not good, however, since particles which are initially close to the wall as they enter the chamber experience little inward entraining action and pass with coarse ones to the lower discharge port. *Figure 22* shows a typical example of a modified form of cyclone classifier which is commonly used to separate coarse particles for regrinding in closed circuit grinding installations.

The particles from the grinder are carried upwards by the air stream into the bottom of an outer conical chamber, *A*. An inner concentric conical chamber, *B*, serves as the main cyclone chamber. The air stream with its entrained particles is directed into the top of the inner chamber through the passages between the vanes *C*. The inclination of the vanes may be altered from outside the classifier by adjustment of the hand wheel. The severity of the rotation imparted to the air in the inner chamber may be controlled by adjustment of the inclination of the vanes with respect to radii. An external control wheel, *D*, moves all vanes simultaneously. Coarse particles which pass into the inner chamber are thrown to the wall and return, for example, to the grinding mill through a port at the bottom of the chamber. Fine particles pass with the air between the baffle *E* and the wall of the inner chamber and are drawn in a generally spiral path towards the bottom of the outlet pipe. Air and fine particles issue from the top of the central outlet pipe *F*.

Several varieties of impact grinder (see section on Grinding) contain spiral flow air classifying elements, built integrally with the grinding chamber.

Figures 23 and *24* show the form of grinding chamber and discharge port respectively of the Herbert Atritor mill and of the Micronizer, to which reference has already been made in Section 3. In each of these markedly different types of grinder, the comminution of particles takes place towards the periphery of a flat cylindrical grinding chamber through which an air current is drawn in an inwardly spiralling path. In the grinder illustrated

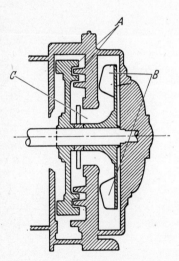

Figure 23. Impact mill incorporating air classifier section (Atritor Pulverizer). (By courtesy of Alfred Herbert Ltd.)

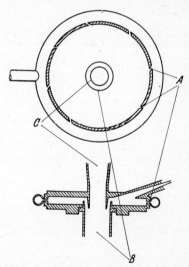

Figure 24. Jet mill incorporating air classifier section (Micronizer). (From M. F. Dufour *and* J. B. Chatelain, *by courtesy of* Mining Engineering)

in *Figure 23*, the air is set in rotation by the fast moving impact members, *A*, of the grinder rotor and is drawn radially inwards by the suction induced by the fan rotor, *B*, mounted on the common shaft. When individual particles of sufficiently small size are produced in the grinding zone, the inward component of the air flow draws them towards the central outlet port, *C*. Coarser particles remain towards the periphery of the chamber and continue to be subjected to grinding impact until reduced in turn to such a size that they may be drawn to the outlet port.

The rotating component of air flow in the grinding chamber of the Micronizer (*Figure 24*) is produced by the tangential setting of the air nozzles *A*. The air supplied to the nozzles is allowed to escape only from the region of the axis of the chamber. Fine particles produced by impact in the grinding zone move inwards with the air. Near the axis of the chamber the angular velocity of the air rapidly increases (as in the free vortex region of the cyclone) and the fine particles are centrifuged from the air stream and pass with some of the throughgoing air to the 'fines' outlet *B*. The main part of the air stream, largely cleaned of particles, is discharged through the air outlet pipe *C* in the upper face of the grinding chamber. The high concentration and violent turbulence present in practical grinding machines necessarily leads to lack of precision of the integral air classifier. Nevertheless, such classifiers usefully remove finely ground particles from the grinding chamber as soon as they are formed, and leave the coarser ones for further grinding action.

The principles involved in spiral flow and helical flow air classifiers are commonly used in separation plant intended for the complete removal of particles or of fine dust from air streams. Direct flow and reverse flow cyclone particle collectors are typical of such separators and are described in fuller detail in a later section. Such separators have a limited separating efficiency for particles smaller than, say, 10 μ. Centrifugal separators of much more severe action are constructed in cyclone form but with mechanically driven cage-like rotors lying in the path of the throughgoing air. This construction leads to the formation of an especially intense vortex within the body of the rotor and to a correspondingly high efficiency for the separation of small particles from the air. In smaller models of certain separators[6] of this class, centrifugal acceleration of about 20,000 g may be maintained and a separation limit of as small as 2–3 μ (spheres of unit density) can be achieved.

REFERENCES AND BIBLIOGRAPHY

[1] TER LINDEN, A. J., *Proc. Instn mech. Engrs, Lond.* 160 (1949) 234
[2] RICHARDSON, E. G., *Dynamics of Real Fluids*, p. 45, Edward Arnold, London, 1950
[3] GUSTAVSSON, K. A., *Tekn. Tidskr., Stockh.*, 78 (1948) 667
[4] WOLF, K., and RUMPF, H., *V.D.I., Verf-Techn.*, 2 (1941) 29
[5] LOCKWOOD, J. F., *Flour Milling*, Northern Publishing Co., Liverpool, 1948
[6] Brit. Pat. 740997

BROWN, R. L., *Some Aspects of Fluid Flow*, Edward Arnold, London, 1951
TAGGART, A. F., *Handbook of Mineral Dressing*, Chap. 10, John Wiley and Sons, New York, 1945

13

MIXING OF SOLIDS

R. A. SCOTT

The blending and mixing of solid and quasi-solid materials occupies an important place in numerous industries. The wide variation in form of available machines is partly due to special requirements of each industry and partly to the difficulty of specifying and testing the mixing performance. Progress has, however, been made in recent years in formulation of theories of mixing, and these theories afford some help in providing measures of degree of mixing and in differentiating between the various types of machine.

In discussion of mixing processes a clear distinction must be made at the outset between a perfect mixture, on the one hand, and a regularly ordered arrangement of particles on the other. *Figure 1* illustrates this distinction. In *Figure 1* (a) two varieties of ingredient present themselves in exact alternate order throughout the assembly, as with atoms in the face of a cubic crystal; in *Figure 1* (b) the particles form a random assemblage with no discernible order among the particles. The random assemblage is the typical product of continued reshuffling of the position of particles, which is the essence of mixing action. The ordered array may conceivably be produced by hand but cannot be obtained or even approached by mixing action.

In the sense appropriate to industrial mixing operations, a perfect mixture is one in which the initial arbitrary grouping of the elementary particles has been replaced by a fully disordered arrangement of particles throughout the mass. Where the irregular tumbling, stirring and rearranging motions of a mixing machine act without discrimination on all elements, the process tends, if continued long enough, to produce in the end a perfect mixture in the sense defined.

In practice, mixing processes are seldom complete either because the operation is not continued for a sufficient time or because some secondary mechanism intervenes to cause, for example, segregation. Various authors[1, 2] have applied themselves to deriving measures which define the degree of imperfection of a mix. Such estimates are based on the departure of the mix from fully disordered arrangement.

At an intermediate stage of a mixing process the arrangement of ingredients may differ from full disorder in many ways and the manner of the deviation depends on the charactertisic actions of the mixer. Ingredients may be well mixed on a large scale and yet fully segregated on a small one. Large samples drawn from the mass may then contain closely similar proportions of the ingredients, while small samples may often be composed of an undisturbed aggregate of only one ingredient. It is clear therefore that a single number cannot fully describe a state of partial mixture.

Irregular transposition of the initially segregated components leads ultimately to any of an extremely large variety of random arrangements, of which an overwhelmingly large proportion have no clearly discernible grouping. Quantitative estimates of the state of a mixture must, moreover, be based on measurements of samples or sub-batches drawn at random from the mass. The techniques which have been formulated for defining the extent of mixing are therefore drawn from statistical method, which characteristically deals with such variable data.

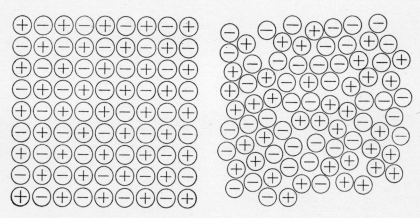

Figure 1. Distinction between (a) regular array and (b) typical perfect mixture

The intermediate stages of mixing may be characterized by the pattern or texture of particles. Thus, in a ribbon mixer, blocks of medium size in comparison with the mass are dispersed by the ribbon or scroll through the body of the machine. The large coarse aggregates tend, therefore, to be broken up and give place to a 'mosaic' of finer scale; under certain circumstances this fining-up of the texture may be defined by a characteristic length such as, for example, the average distance between the blocks of the new mosaic. This approach to the definition of extent of mixing has been developed by DANCKWERTS[1] in terms of a parameter called the scale of mixing and is referred to in a following section.

Alternatively, the intermediate stages of mixing may be characterized by a measure of the variability in composition to be found among samples of given size. This type of index is generally, though not always, a function of the size of the sample. Thus, in a ribbon mixer, a 'sample' embracing the whole mass must have by definition the 'average composition'. The deviation from average composition of a sample of medium size, that is, one containing very many blocks of the 'mosaic', is also likely to be small compared with the deviation for a sample as small as the average block of the mosaic. A suitably defined measure of the intensity of deviation from average composition therefore reflects the relationship between size of sample and texture of the mix, and provides a measure of the extent of mixing complementary to 'scale of mixing'.

EXTENT OF MIXING ASSESSED IN TERMS OF CONSISTENCY IN COMPOSITION OF SUB-BATCHES

For many industrial purposes the most important function of a mixing process is to reshuffle the ingredients of the whole batch sufficiently to allow arbitrary sub-batches of some assigned size to be regarded as having substantially the same composition as the whole. Danckwerts[1] and also Lacey[2] have derived indexes of mixing which may be used to indicate how far a mixing process has progressed from this particular point of view.

Figure 2. Grouping of like particles by chance in small samples

If the product of a mixing operation is sampled by drawing from it sub-batches of some fixed size, the concentration of a selected component will, in general, vary from sample to sample. The extent of the variation may be reckoned in terms of the standard deviation of the concentration of the component as estimated by classical statistical technique[3].

For samples of particulate material in which only a small number of particles of the selected component are present, some variation of concentration between samples occurs even for a mixture which is perfect in the sense of complete random arrangement. This innate variability of small samples is illustrated in *Figure 2* which, again, displays a random pattern of two varieties of ingredient. Reshuffling of the particles cannot lead to any further disarrangement and the particles are therefore 'perfectly mixed'. Nevertheless, in certain regions, for example those ringed, substantial patches may readily be discerned which contain particles of a single sort only. The concentration of any one component in arbitrarily chosen patches of, say, 10 particles would be likely in a small number of trials to cover the extreme range from unity to zero. The larger the number of particles included in the patch, the less likely is the concentration to assume extreme values; for sufficiently large patches of randomly arranged particles the concentration of either component will differ inappreciably from the 'proper' average value of 0·5.

Materials which are imperfectly mixed exhibit larger variations between samples of a given size than does the perfect mixture. For any given size of sample or sub-batch drawn from the mix, the extra component of variability provides an index of the effectiveness of the mixing operation.

Suppose the concentration of the selected component in a sample is expressed in terms of the proportion by volume it occupies in the sample. Individual samples drawn from the mixture will have values of concentration a_k ranging about the mean value for the whole mix. The variation between samples can conveniently be expressed in terms of the standard deviation σ estimated from the individual values a_k and the mean value $\bar{a}$ of n samples according to the formula:

$$\sigma^2 = \sum_1^n (a_k - \bar{a})^2/(n-1)$$

For a perfectly mixed sample containing very many particles of the selected component, σ may be small. For smaller perfectly mixed samples σ may have a value σ_r which is dependent in general on the smallness of the sample. Partially mixed samples of the same size have a larger standard deviation, say σ_m. The value of σ, or rather of the *variance* σ^2, may be regarded generally as compounded from two independent factors: firstly, that due to the smallness of the sample, and secondly, that due to the imperfection of mixing. Thus

$$\sigma^2 = \sigma_r^2 + \sigma_m^2$$

and therefore

$$\sigma_m^2 = \sigma^2 - \sigma_r^2$$

If no mixing at all has taken place σ_m^2 has the large value $\sigma_0^2 - \sigma_r^2$ characteristic of the unmixed state. Where, however, the mixing process has so far moved towards completion that variations in composition of samples of the size under review are not affected by prolonging the mixing operation, then σ_m^2 tends to zero. Lacey has defined a useful index of mixing M, in the following form:

$$M = 1 - (\sigma^2 - \sigma_r^2)/(\sigma_0^2 - \sigma_r^2)$$

For a partially mixed charge, the value of M changes with size of sample or sub-batch in a manner characteristic of the type of mixing process and of the extent of the mixing. The value of M is generally zero for the initital unmixed state and unity for all sub-batch sizes in cases where the mix is perfect in the sense that further prolonging of the mixing operation produces no increase in consistency of sample composition.

EXTENT OF MIXING ASSESSED IN TERMS OF TEXTURE OF MIX

In certain mixing operations the progress of mixing is most usefully assessed by some measure of texture which relates to the amount of patchiness or mottle which remains in the mix. Examples of industrial processes of this sort are to be found where mechanical constructional materials are formed by compounding or cementing together discrete ingredients, or where paints or powders are manufactured.

Danckwerts[1] has suggested a useful index which he calls the scale of segregation of the mix. In form it is analogous with the quantity defined as the scale of turbulence which describes spatial characteristics of turbulence in fluid motion. The technique involved is one of constructing a measure of the relationship which on average is observed to exist between the local deviations in concentration at separated regions.

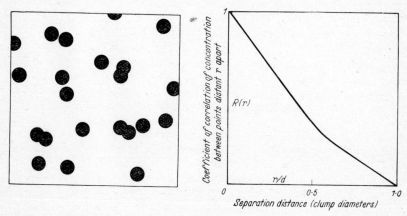

Figure 3. Scale of mixing: (a) random arrangement of circular clumps; (b) corresponding correlogram of concentration at separated points. (From P. V. DANCKWERTS, by courtesy of Applied Scientific Research)

Consider two regions adjacent to one another in a mosaic of partially mixed materials. Where the regions are part of the same agglomerate the local concentrations of one chosen component are identical: a large deviation from average concentration in one region is then associated with a correspondingly large deviation in the other region. For widely separated regions, however, deviations from average concentration in one region are likely to be almost unrelated with those in the other region. This state of affairs may be described by saying that the deviations from average concentration are strongly correlated for closely spaced regions, but uncorrelated for widely separated regions. The distance over which the correlation remains substantial, expressed in an appropriate manner, is defined as the scale of segregation. Physically, the scale of segregation for a mosaic of agglomerated particles is generally related to the size of the agglomerates, provided that the individual regions for which the deviations are measured are smaller than the agglomerates themselves. Suppose the concentration by volume of a component of the mix is a_1, at an arbitrary small local region, and a_2 at a second local region distant r from the first; the quantity:

$$R(r) = \frac{\overline{(a_1 - \bar{a})(a_2 - \bar{a})}}{\overline{(a - \bar{a})^2}}$$

is known as the correlation coefficient[4] between concentrations at points r apart. The separate bracketed terms are the deviations of the concentrations at the local points from the mean value of concentration $\bar{a}$ for the whole mix. The numerator is the value of the average of the product of these terms for a large number of pairs of points mutually spaced by the distance r. The denominator is the variance of the locally determined concentrations.

For the mosaic of agglomerated particles already referred to, $R(r)$ tends to the value of unity for values of r smaller than the size of the average component blocks of the mosaic and to zero for values of r appreciably greater than the size of the blocks. For a partially mixed mass, in general, $R(r)$ falls off with increasing r in a manner characteristic of the state or texture of the mix. A graphical illustration of the application of this technique, as given by Danckwerts, is reproduced in *Figure 3*. *Figure 3 (a)* displays a two-dimensional array of circular clumps (black) randomly spaced in an excess of a second component (white background). The curve of *Figure 3 (b)* shows how $R(r)$ varies with increasing separation of the points. In this typical example, the value of $R(r)$ falls to a very low figure when the separation of the points exceeds the diameter of a clump.

Danckwerts has suggested the use of the following function as a linear measure of the effective size of agglomerates remaining in an imperfect mix:

$$S = \int_0^\infty R(r) \, dr$$

The linear scale of segregation S, so defined, is the area below the curve of the correlogram. In the example of *Figure 3 S* has a value 0·42 times the diameter of the individual clump and is therefore comparable with the radius of the clump.

MECHANICAL MIXING ACTION

The various mechanical motions which characterize mixing processes combine to disturb the arrangement of the whole mass of particles. The nature of the disturbances has been conveniently classified by Lacey (loc. cit.) into three principal categories:

(*i*) convective mixing, in which groups of particles are bodily transposed;
(*ii*) shear mixing, in which adjacent layers of particles slide with respect to each other as the result of the creation of slip planes within the mass;
(*iii*) diffusive mixing, in which particles adjust their mutual positions by individual (usually small scale) movements.

Mixers generally combine two or more of these actions. Thus all elements which involve movement of the mass tend to increase the internal mobility of the individual particles and lead to some degree of diffusive mixing. Nevertheless, the conveying elements of ribbon or trough mixers have a predominantly convective action, while rearrangement of the main body of the charge in a tumbling mixer involves a large element of shear mixing.

Convective and shear action are jointly responsible for most coarse-scale rearrangement in mixers; diffusive action is mainly responsible for the fine-scale rearrangement. Coarse and fine-scale mixing do not proceed at the same rate: diffusing mixers are slow to produce uniformity in composition of very large sub-batches; convective mixers are slow to produce uniformity in small samples. The combination of mixing elements proper to a given mixing process depends on the degree of segregation of the components in the initial charge, the free-flowing characteristics of the particles of the charge and on the degree of homogeneity required of the final mixed product.

Continuous Mixing

In many otherwise continuous manufacturing processes the mixing operations are performed in batch mixers. This is in part because the composition of the total mass in the mixer may then be closely predetermined by charging the mixer with accurately weighed components, and in part, because of the ease with which the degree of mixing may be controlled from batch to batch by adjustment of the mixing time. In many industries, however, streams of material are continuously mixed without interruption of the mean flow of the stream.

There are two distinct forms of continuous mixing operation. In the first, the quantity of material in the machine is generally small and the mixing operation is confined to the intermingling of individual particles. Such mixers are often similar in general form to grinding machines in which the respective impact and shear elements are set to provide a maximum of intermingling. Where continuous mixers of this class are used the coarse-scale blending of ingredients is carried out as a separate process.

The second form of continuous mixing operation involves simultaneous coarse- and fine-scale blending of the components and replaces batch mixing where the process which follows requires an unbroken stream of material. Continuous mixers of this second type contain the same sorts of mixing element as are found in batch mixers; indeed, batch mixers of certain sorts are frequently used for continuous mixing. Thus, continuous mixers contain agitating and conveying mechanisms which rearrange the individual particles, displace discrete volumes of the material and move the charge generally forward from inlet to outlet of the machine. Where batch mixers fitted with mixing ribbons or with agitating members are adapted for continuous mixing, some sort of restricting device usually exists near the outlet port to provide a controlled, steady outflow. This may take the form of a worm feeder of limited capacity or of a physical barrier equivalent in effect to a weir. For the latter case outflow commences when the level of material has risen to a definite height; increase in level above this height tends to increase the rate of outflow so that an equilibrium level is reached for which the outgoing flow equals the incoming flow.

In continuous blending the composition of the outflowing stream reflects both the degree of small-scale intermixing of the components in the mixer and also the extent to which the mixer unifies the composition of the gross contents of the mixer. As shown by DANCKWERTS[5], the averaging effect of the mixer on the fluctuating composition of the incoming flow can be described in terms of the characteristic times for which elements of the stream reside in the machine. He defines a function (F function) to describe the distribution of 'residual times' and shows how to evaluate the function from experiment and how to use it to predict the extent to which fluctuations of inflowing composition are reduced in the outflow. For a detailed treatment the reader is referred to the original paper; the main conclusions which derive from this argument are that effectiveness of the blending requires a moderate, as distinct from a too broad or too narrow, distribution of residual times. Thus, if the mean volumetric rate of inflow (or outflow) is v and the mixer volume is V, residual times should be variously distributed about a value of V/v. Only a small proportion of elements should possess any substantially larger residual time. These requirements may be interpreted as meaning that paths of the various elements should be diverse in length and in characteristic speed of transit, but that dead patches with indefinitely long hold-up times must be avoided.

MIXING MACHINES

Almost all industrial mixing machines involve some element of convective, diffusive and shearing action. Machines may nevertheless be grouped roughly into the following categories according to which types of action predominate:

ribbon, worm, spiral and fountain mixers; mainly convective and diffusive in action:
sifter mixers and high speed beater mixers; mainly diffusive in action:
drum mixers and tumblers; mainly shearing in the body of the material together with diffusion at the free surfaces:
ball mills, mullers and colloid mills; mainly shearing in action, often augmented by convective and diffusive action induced by moving scrapers or rakes:
dough mixers, banbury mixers and mixing rolls; mainly shearing and convective in action.

The proper choice of type of mixer depends especially on the flow characteristics of the charge and on whether coarse-scale intermixing or fine-scale homogeneity is the aim of the mixing operation. Dry materials which are moderately free flowing may be mixed in many varieties of mixer. The main mass is mixed on a coarse scale by the direct action of agitators and rakes or by the irregular falling and sliding movements induced by rotation of the mixer body. At the same time, the general movement of material makes the mass sufficiently mobile for individual particles to migrate from their original positions in respect of their immediate neighbours so that the mass is mixed on a fine scale by diffusion.

Drum mixers and tumblers are especially suitable for dry, free-flowing materials; they are generally mild in action and lead to very little breaking up even of friable materials. Ribbon mixers and trough mixers provide more positive displacement within the body of the charge and are suitable in addition for the coarse mixing of fibrous and lumpy materials.

Sifter mixers and mixers in which the material is flung by fast moving beaters against a wall, provide good mixing action on the intermediate and fine scale, especially for those particles which tend otherwise to remain in agglomerates. Ball mills and mullers and other mixers of grinder-like construction are primarily designed for providing homogeneity of composition and texture on a small scale and frequently are required to provide a grinding as well as a mixing action. Mixers which produce good fine-scale mixing frequently confine the mixing action to a small local region in which the action is particularly intense and, in consequence, tend to produce fine-scale homogeneity without causing much averaging-out of variations in the mean concentration. Mullers and ball mills are therefore usually fitted with additional scrapers and rakes to assist in the intermixing of the main mass.

Dough mixers, Banbury mixers and mixing rolls are designed for the mixing of pastes and plastic masses rather than for dry particulate solids, and therefore form a distinct class of their own.

Drum Mixers and Tumblers

Mixers which operate by a tumbling action of free-flowing material in a drum or chamber are commonly used for the mixing of granular material. A tumbling drum in its simplest form is illustrated in *Figure 4*. The cylindrical container A, partly filled with granular material, is supported so that it may be rotated about a horizontal axis by an external drive. The charge tends to be carried round with the container until the free surface in the upper part, B, of the container tilts steeply forward. The

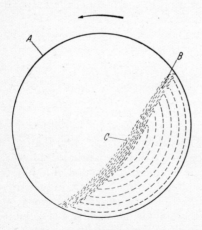

Figure 4. Simple tumbling drum

upper layers of the pile of free-flowing ingredients then become unstable and slide towards the foot of the pile along planes of shear which develop in the material. These planes make an angle with the horizontal which is rather steeper than the angle of repose of the granular material. As the sliding mass C moves downward in the container, the motion becomes increasingly irregular. The movement of material may be regarded as composed of two phases. In the initial phase, substantial volumes of

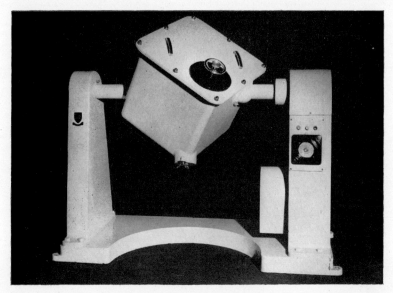

Plate I. Rotocube tumbling mixer. (*By courtesy of* Foster, Yates and Thom, Ltd.)

Plate II. Drum type mixer. (*By courtesy of* Sturtevant Engineering Co. Ltd.)

Plate III.   Mix-Muller.   (By courtesy of August's Limited)

Plate IV.   Dough mixer.   (By courtesy of Morton Machine Co. Ltd.)

material are detached from the mass by the shearing action; in the second, the individual particles are intermixed by 'diffusion' in the tumbling layer as it moves towards the foot of the pile.

Tumbling mixers may be constructed as simple horizontal cylinders, but are commonly of rather more elaborate geometrical form. *Plate I* illustrates a mixer in the form of a cubical box, having one face removable for loading and set skew on a horizontal spindle so that the material tumbles in turn from one face to another. Alternative forms of chamber are illustrated in *Figure 5*; all these forms have been developed with the aim of ensuring good

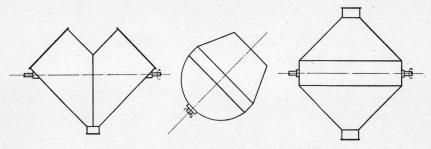

Figure 5. *Other typical forms of tumbler mixers*

mixing, not merely in a single vertical plane, but also from side to side of the container. Simple tumbling mixers are fast in operation for free-flowing solids and are sufficiently gentle to cause little breaking up of fragile grains. The absence of moving parts leads to relative ease in obtaining a complete discharge of ingredients, and to good accessibility for cleaning.

Drum mixers are frequently fitted with internal agitators or scrapers, or with lifting blades which form part of the drum lining; these turn over and generally disturb the body of the material in the drum. The agitators and lifters are usually set to displace material laterally in the drum and therefore to increase the rate of mixing in the direction of the length of it. Conventional cement mixers provide a well-known example of this class of mixer.

Plate II illustrates a drum mixer suitable for batches of up to about 1 ton of free-flowing granular material. The charge is fed through a gate in the centre of one end face of the rotating drum. Scoops attached to the inside of the drum casing lift the material and ultimately shed it so that the stream of particles falls back to the bottom of the drum by a path lying close to the feed gate. Material which spreads laterally along the axis of the drum is therefore continually drawn together to a common point before it is redispersed in its next fall. When mixing is complete, a central chute at the entry port can be extended into the interior of the drum to catch the stream falling from the scoops and direct them out of the mixer.

Spiral Mixers

Spiral or fountain mixers consist of a vertical cylindrical casing carrying a central continuous Archimedean screw which tends to lift the material through the height of the casing so that it descends in the regions nearer the wall. In certain forms of the fountain mixer the screw is tapered in section so that its swept area steadily increases with increasing height in the chamber. In consequence material is drawn into the upward conveying stream at all heights and this action assists the coarse-scale intermixing. Such a fountain mixer is shown in *Figure 6*. Feed enters at the top A, and is ultimately discharged when the batch is mixed through the side port, B, towards the lower end of the hopper section. Agitating arms C, attached to the bottom of the screw shaft, assist both in mixing and in discharging.

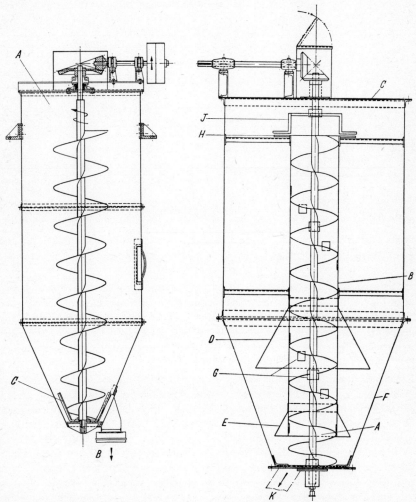

Figure 6. Fountain mixer. (*By courtesy of* W. S. Barron & Son Ltd.)

Figure 7. Spray mixer. (*By courtesy of* Henry Simon Ltd.)

Figure 7 shows a modified form of spiral mixer in which the central screw is contained in an inner cylindrical casing. Feed enters at the top of the casing and falls freely through the main height of the chamber; two cowls attached to the inner casing assist in dispersing the material. Material is drawn into the screw at the foot of the hoppered section of the chamber and as the column of material moves upwards some portion is shed laterally through side ports situated at intervals through the height of the casing. These falling sprays of material mix intimately with other material which has been elevated to the top of the chamber and swept from the ledge by the action of vanes which rotate with the main shaft. When mixing of the batch is complete the charge leaves the machine by way of the lower port.

Trough Mixers

In their simplest form, trough mixers consist of a U-shaped trough carrying a longitudinal shaft fitted with agitating and propelling blades which lift or displace parts of the mass so that they fall back irregularly to diffuse over the remaining material. *Figure 8* illustrates a trough mixer with a pair of agitating shafts A and B which run in a casing which takes the form of a double trough C. The shafts are geared together and driven

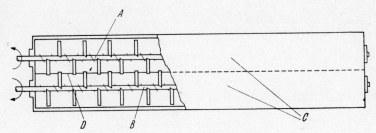

Figure 8. Double worm trough mixer

so that the motions of the separate agitators combine to lift the material upwards in the region of the central ridge D. The mixer is continuous in operation: the agitators are set at an angle to give a general forward movement to the stream, and the effective mixing time being controlled by adjustment of the angles of the agitator blades. In a typical machine the shaft speed is usually in the region of 150–300 rev./min.

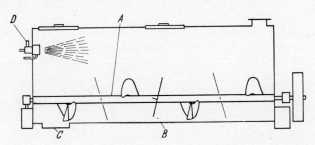

Figure 9. Cascade trough mixer. (*By courtesy of* Henry Simon Ltd.)

Figure 9 shows a modified form of trough mixer intended for continuous operation. The shaft A, and agitators B, occupy the lower part of a deep casing. Size of agitators and speed of the shaft are chosen so that the material is thrown upwards to form a diffuse cloud in the upper part of the casing. Feed enters towards one end of the casing and leaves through the port C, at the other end. The pronounced dispersing action of the fast-moving blades assists the intermixing and also provides suitable conditions for the addition of a liquid ingredient through a spray nozzle D, suitably located in the casing.

Ribbon and Paddle Mixers

Ribbon mixers and paddle mixers consist of a stationary chamber containing moving agitators designed to give a complicated pattern of continuous movement to the elements of the charge. Such mixers may be used for free-flowing solids, but are particularly suitable for the batch

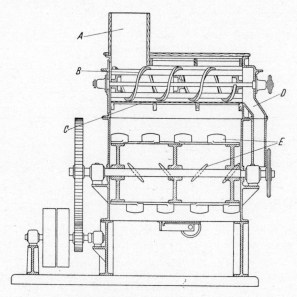

Figure 10. *Mixer with sleeve and agitating chamber*

mixing of fibrous or coherent granular material. The agitator system usually consists of rings of paddles or of ribbon-like helical conveying scrolls which are revolved by a single central shaft. The various sections of the agitators which combine to make the assembly are inclined at a variety of angles and the parts of the mass therefore move with complicated and erratic movements. In addition to the coarse-scale intermixing by convecting action and shear induced by the agitators, some fine-scale mixing is provided by diffusion as the parts of the mass lifted clear of the general free surface by the agitators fall back on to the remainder of the mass.

A paddle mixer is shown in the section below the sifting screen of *Figure 10*. The typically diverse inclinations of the paddle elements, E, are evident. In the ribbon mixer the agitator is formed of two or more separate rings of helical conveying ribbons with adjacent ribbons set to convey in opposing directions. In agitator mixers, generally, the outer ring of agitating elements usually runs close to the wall of the casing and at the end of the mixing operations sweeps the mixing chamber clear of material through an exit port.

The forced movements imposed by the agitators lead to rapid coarse-scale mixing. Mixers with close-spaced driven agitators are therefore substantially faster in operation than spiral or tumbler mixers, especially where the material is naturally inclined to be coherent. Agitator mixers are, however, not as easy to clean nor are they as economical in power.

Sifter Mixers

Intimate mixing of fine powders is sometimes carried out with the aid of a screen mounted immediately above a chamber containing a revolving agitator assembly. *Figure 10* illustrates a sifter mixer of this type. Material is fed into the machine through the port A. The revolving brush B assists in working the particles through the sieve cover C. A spout D carries away lumps which do not pass through the sieve apertures. The tumbling action induced by the revolving brush mixes the material by shear and diffusion. The sieving action disperses loose agglomerates, which exist in the feed, and diffuses the individual particles regularly over such material that has already penetrated the sieve. The revolving agitators E complete the mixing operation.

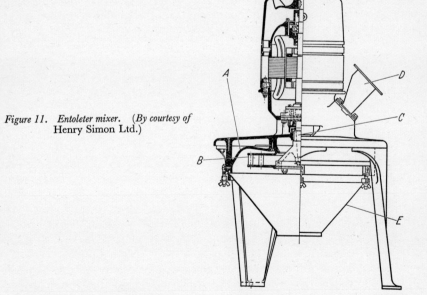

Figure 11. Entoleter mixer. (*By courtesy of* Henry Simon Ltd.)

High Speed Beater Mixers

Particles can be intermixed on a continuous scale by feeding the stream of proportioned particles into some form of high speed rotary impeller which spreads the stream into a spinning curtain of umbrella form. The particles are separated and intermixed by impact and by the highly turbulent course of the air motion induced by the impeller. *Figure 11* shows a mixer of this sort. A casing A contains a pegged-disc rotor B driven through a vertical shaft C. Proportioned material enters through ports D in the upper casing and is drawn into the chamber with the help of the air stream induced by the rotating rotor. The material is thrown outwards by the disc to meet the wall of the casing and is intermixed by diffusion both in the turbulent air stream and as it swirls down the wall of the hoppered delivery section E. The rotor in a typical case has a diameter of 17 in. and is rotated at a speed of 1,500–3,000 rev./min according to the severity of action required.

Ball Mills, Mullers and Colloid Mills

The intensive shearing action which takes place in certain types of grinding machines makes them particularly suitable for the fine-scale mixing of particles. Ball mills, mullers and colloid mills all exhibit this intense shearing action.

The action of ball mills has already been described in the section on grinding machines. The severity of the mixing action and particularly the severity of such grinding action as accompanies it, can be controlled by a suitable choice of size, weight and number of balls and also by choice of scale of the mill. The use of ball mills for mixing is generally restricted to hard and moderately hard materials: an undue proportion of fines produced by grinding retards mixing by rendering the material less free flowing. Mixing takes place mainly through the action of shearing and tumbling, but is sometimes assisted by the convective action of agitators or rakes which move relatively to the charge.

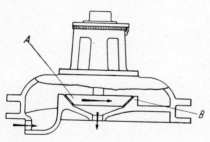

Figure 12. Conical-rotor colloid mill. (From L. CARPENTER, 'Mechanical Mixing Machinery', *by courtesy of* Ernest Benn Limited)

The muller or pan mixer (illustrated in *Plate III*) makes use of the mixing action of heavy rollers which are made to ride over a bed of material. Above the base of the main chamber lies a pair of heavy rollers which run on separate horizontal spindles. A central vertical shaft assembly is driven so that the rollers ride over material which lies in the chamber. Rolling movement of the wide rollers is accompanied by some skidding action where the rollers engage with the mass of material. These movements induce a pronounced local shearing which mixes the material intensely on a fine scale, and a further coarse-scale mixing of the material is produced partly by the displacing action of the forward moving rollers and partly by the ploughing and redirecting action of scrapers carried round by the roller assembly.

Apart from ball mills and mullers many other forms of mill may be used to combine the grinding or dispersing of aggregates with intense mixing action. The class of machines known as colloid mills is frequently used for fine-scale mixing of moist solids or pastes. Such mills usually consist of a pair of closely spaced surfaces which move relatively to one another at high speed. Feed passes in a continuous stream into the narrow gap and is violently sheared as it is carried through the gap. *Figure 12* shows a form of colloid mill in which a moving conical rotor A, mounted on a vertical spindle is set so that it just clears a fixed conical seating B. The solid material, in this case suspended in a liquid medium, is drawn into the gap from above and after dispersion is discharged in a steady stream below.

Dough Mixers, Banbury Mixers, Mixing Rolls

Where materials to be mixed are of a plastic nature, as in the case of pastes, paints and rubber-like masses, mixers must involve the physical distortion of the ingredients. The action of such mixers is primarily of

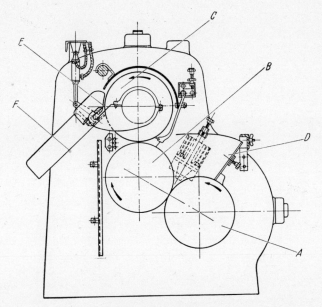

Figure 13. Three-roll mixing mill (Simon-Lehmann). (*By courtesy of* Henry Simon Ltd.)

disruption by shear usually followed by the reformation of the mass under the action of either intertwining beaters which knead the plastic mass, or of mixing rolls which wipe new layers of mixed material over earlier layers which have been carried back to the nip by recircling the surface of the roll.

Plate IV shows a typical dough mixer containing a double agitation system mounted generally in a double U-shaped casing to give an irregular kneading action when driven through gearing from outside the casing. Such mixers are also applicable to the mixing of dry or moist granular material.

Two or three-roll mills of suitable design are used for the mixing of solid particles into plastic and viscous fluid media. *Figure 13* shows a three-roll

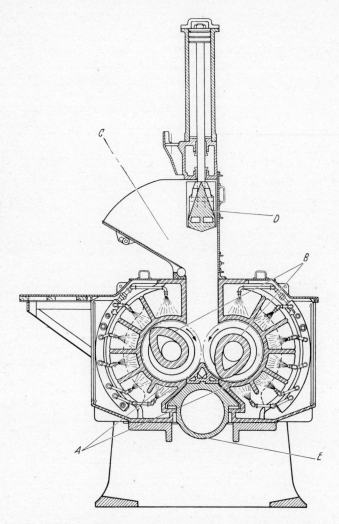

Figure 14. Banbury mixer. (By courtesy of David Bridge & Co. Ltd.)

mill designed for the final stage of dispersion of solid pigments in paint and printing ink. The lower, middle and upper rolls, A, B and C, are geared together to run at speeds in the ratio $1:3:6$ and are supported and adjusted so that a controlled pressure exists between the engaging faces of the rolls during operation. The generally proportioned and coarsely mixed ingredients are introduced to the hopper D which is mounted above the lower roll. The material is picked up as a film on the upper surface of the lower roll and the film is violently sheared as it passes through the nip. The emerging film is carried by the rear surface of the faster moving middle roll to be further sheared in the nip below the upper roll. The medium with its dispersed pigment is finally carried upwards on the front surface of the upper roll, which runs at about 350 rev./min, and is withdrawn by the knife E, and delivered at the apron F. The main action of the machine is to disperse the solid aggregates of pigment so that the individual particles are evenly distributed in the medium.

A Banbury mixer, commonly used for compounding chemicals with raw rubber or plastics, is shown in *Figure 14*. The chamber itself is formed of two interconnecting cylindrical spaces A, which each house a rotor B of spiral shape. The chamber is loaded from above, C, and a floating weight, D, is brought into position to confine the charge completely to the mixing zone. The rotors are driven at slightly different speeds in opposite directions and continually distort and transpose the elements of the plastic mass. The mixing process consumes much energy and provision is made to cool both the main walls of the casing and the body of the individual rotors. The mixed mass is finally discharged through the lower door E.

REFERENCES

[1] DANCKWERTS, P. V., *Appl. sci. Res., Hague*, A 3 (1952) 279
[2] LACEY, P. M. C., *J. appl. Chem.*, 4 (1954) 268
[3] FISHER, R. A., *Statistical Methods for Research Workers*, Oliver & Boyd, Tweeddale Court, Edinburgh, 1948, Chap. 3
[4] FISHER, R. A., *Statistical Methods for Research Workers*, Oliver & Boyd, Tweeddale Court, Edinburgh, 1948, Chap. 6
[5] DANCKWERTS, P. V., *Chem. Engng Sci.*, 2 (1953) 1

BIBLIOGRAPHY

ADAMS, J. F. E., and BAKER, A. G., 'An Assessment of Dry Blending Equipment,' *Trans. Instn. chem. Engrs., Lond.*, 33 (1956)
BEAUDRY, *Chem. Engng*, 55 (1948) 112
BUSLIK, D., *Bull. Amer. Soc. Test. Mat.*, 66 (1950) 165
COULSON, J. M., and MAITRA, N. K., *Industr. Chem. Mfr.*, 26 (1950) 55
RILEY, D. F., *Chem. & Process Engng*, 34 (1953) 67
CARPENTER, L., *Mechanical Mixing Machinery*, Benn, London, 1925

14

STORAGE AND HANDLING OF SOLIDS

J. A. W. HUGGILL

THE storage and handling of materials must always be regarded as an inevitable liability in any productive process, adding appreciably to the cost of manufacture of a product without increasing its intrinsic value. It is only natural, therefore, that commercial enterprises should require that the expenditure relating to storage and handling must be as small as possible. In the last decade or so, greatly increased production coupled with rising labour costs has forced manufacturers to mechanize their production processes to a much greater extent than previously and, with this, to re-assess their storage and handling facilities to bring them into line. It is now widely recognized that a considerable amount of capital sunk in an efficient storage and mechanical handling system will quickly pay dividends and at the same time will allow a throughput which would have been virtually impossible with only manual methods in use. It is significant also that mechanical handling can substantially reduce the time taken to load and unload ships, trucks, *etc.*, saving demurrage and releasing transport for its proper job.

It is important that storage and handling in the general sense should be considered together, since storage will always require means of transporting material into and out of storage, while handling implies movement from one point to another, and even though the points in question may not deserve the title of 'storage', they represent a static state of the material and must be taken into consideration. To determine the most efficient system the whole flow-line should be analysed, and the reader who is making plans for a comprehensive storage and handling system is strongly urged to consult a reputable firm specializing in this work, rather than attempt a piecemeal solution of his own, followed by the purchase of individual items of equipment which may not be ideal for the job. The account published here must be regarded as no more than a brief survey of principles and current practice in this field and not as an authoritative textbook. Most of the methods and equipment described have resulted from years of practical development and it is asking for trouble to ignore advice given from first-hand practical experience. A fuller account of the general principles and economic aspects of materials handling has been published by IMMER[1] and a more comprehensive technical treatise on conveyors and related equipment has been published by HUDSON[2]. Several specialized works and technical papers will be referred to in the following subsections.

From the point of view of storage and handling, materials can be classified in two groups: (*i*) granular or powdered material in bulk, (*ii*) material in the form of, or made up into, units, *i.e.* bags, sacks, cartons, packages or individual objects. Unit handling was inevitable when only manual labour was available and the unit (*e.g.* a sack of grain) was of a weight appropriate to the strength of a man. Modern times call for the predominance of bulk handling for practically all materials, except the final product of manufacture, which may be bagged for ease of sale and distribution. A few particularly objectionable materials are only reluctantly considered for bulk handling methods for reasons such as toxicity, corrosion, extreme dust nuisance, or contamination from all ordinary constructional materials. Such substances would in all probability be handled in lesser quantities than those of the commoner materials such as grain, coal and similar minerals, where methods other than bulk handling are now considered impracticable.

Even though the internal storage and handling arrangements in a process plant may be highly efficient, co-operation between supplier and user can often be of mutual benefit where transport costs may be reduced by suitable equipment for loading and unloading bulk material from road and rail wagons. Many materials such as flour, sugar, cement and hydrated lime are, by long tradition, handled in bags, but at the present time there is a distinct move over to bulk handling in transport. These materials may be subject to contamination or are too dusty for movement in the usual tipping lorry, and special closed wagons holding 10 tons or more are now available with self-contained equipment (*e.g.* pneumatic) for unloading into the customer's storage bin. The disadvantages of bagged material is that labour is required for filling and emptying the bags, also, bags are often not returnable and have to be paid for.

STORAGE AND HANDLING OF UNIT MATERIALS

With the increasing availability of powered lifting tackle and cranes there is a trend to increase the size of units to be stored and handled. Special containers for bulk materials, and trays and racks for stacked packages and similar articles, can be lifted quickly on or off trucks for transport, furthermore, such a unit load may be conveniently left intact for short term storage if necessary. Of special importance is the almost world-wide use of pallets with fork lift trucks. The pallets are flat trays made of wood or metal in various sizes from three feet square upwards. In their thickness (about 5 in.) are two holes to take a pair of prongs projecting from the front of a powered truck. The prongs are lifted a short distance to raise the pallet above the ground for transport, or lifted several feet for economical stacking (of palletized material). Considerable progress has been made on standardization[3] of pallets, with the result that a unit load made up of bags, drums or similar articles strapped on a pallet need not be broken up during many stages of transportation and storage, thereby saving much of the labour required for handling the material as individual articles. The fork lift truck can handle many types of large packages, containers and articles without pallets, and some makes of truck can be adapted to handle drums and various other shaped objects. Although only economical for transportation over short journeys, the fork lift truck is so adaptable and manœuvrable that it deserves consideration in any plant where unit loads may be handled and stored. *Plate I* shows a typical fork lift truck.

The efficient use of available floor area for unit load storage means that stacks must be as high as possible, consistent with stability, and the number of aisles should be the minimum consistent with ready access to different varieties of materials stored. Fork lift trucks can lift loads of 1 ton or more to heights of about 15 ft. and they have the advantage of requiring little headroom, which makes them particularly suitable for use in multi-storey buildings. A survey of specifications of fork lift trucks (petrol, diesel and electrically driven) has been published in *Mechanical Handling*, October 1955. Mobile cranes have considerable use in open air stores, but the headroom required by them restricts their use indoors.

The storage of a large variety of miscellaneous goods which must all be readily accessible—a situation which occurs in warehousing activities and most works' general stores—is usually accomplished by using racks built up on a tubular or angle-iron framework. According to the nature of the goods, they are held in containers or made up into unit loads on pallets which are placed in or removed from their allotted position in the racks by fork lift trucks travelling in the aisles between the racks.

The stacking of units, such as bags and packages which are small enough to be manhandled, can be greatly facilitated by portable roller, belt or slat conveyors, along a path from the receiving bay where, for example, a wagon is being unloaded, into the storage area where the final conveyor is inclined to lift the articles to the top of the stack. The only manual labour required is the unloading of the article from the wagon on to the adjacent conveyor and the positioning of it on top of the stack.

Handling of small units from one process or machine to another is frequently done with conventional types of continuous conveyors. The advantage of this being that no labour is required, since the machine or process operator can load or unload the conveyor, while in some cases it can be made automatic, or the process (*e.g.* filling bottles or cartons) can be done without removing the article from the conveyor. Flat belt conveyors and slat conveyors can handle practically any kind of article, and roller conveyors can carry any object with a flat base. *Plate II* shows bags of plastic granules being conveyed to a road vehicle in a loading bay.

THE PROPERTIES OF BULK MATERIALS

Although the storage capacity of a bunker or the conveying rate of handling equipment is almost invariably specified in terms of the weight of material stored or conveyed, the design is based largely on the volume of the bulk material. The property of first importance is therefore the bulk density. Unfortunately it is a property which can vary quite widely, even between different samples of the same material, since it is dependent on size grading (where small particles can fill the pores between the larger particles, thus increasing the bulk density), on the degree of compression, particularly with fibrous materials like sawdust, and on such processes as vibration, which causes arched groups of particles to collapse to form a more closely packed arrangement. With very finely powdered materials the slow release of air trapped between the particles when the bunker is filled can appreciably increase the bulk density and show up as a lowering of the level in the bunker. A low estimate of the bulk density (loosely packed material) is appropriate in most cases of storage and handling, since this corresponds to the usual condition of the material. However, the highest likely value should also be known because this will show if there is any chance of a bunker becoming overloaded with excess weight of material.

The second important property is the angle of repose. This determines directly the height of the cone of material under the point of its entry, giving the volume wasted in the upper corners in the case of a flat-topped closed bunker, or the possible amount of overfilling in an open bunker. The angle of repose is a measure of the apparent friction between layers of material, and it is therefore of importance in calculating the pressures on the walls and base of a bunker. Related to it, and usually of about the same value, is the minimum angle of slope down which the material will slide. The sloping bottom of a bunker must be designed so that the angle of the cone of a circular bunker, or the valley between adjacent slopes of a rectangular bunker, are considerably greater than this angle of flow to ensure complete emptying.

Finely powdered materials are often found to have an indeterminate angle of repose, readily forming vertical cliffs, while moisture appears to emphasize the effect. JORDAN[4] has shown that with fine particles the cohesive force due to intermolecular attraction at any point of contact can be comparable to the weight of the particle so that particles may be inclined to cling together in any orientation. A strongly polar surface layer such as that which would be created by adsorption of moisture would increase the attractive force. Surface moisture also appreciably affects the flow of coarser granular materials.

Table 1 gives values of the bulk density, particle density (*i.e.* the density of the solid substance of which the particles are composed) and angle of repose for a few substances. They should be regarded as approximate only, since there are many factors which influence the values, as outlined above. The particle sizes where given are average values regardless of whether the material is relatively uniform, for example wheat, or covers a very wide range of particle sizes as in most crushed materials.

The design of handling equipment is influenced by the rather ill-defined flow properties of a material in bulk, and the angle of repose is only a partial measure of this. In practice it is found that many of the less free-flowing materials tend to choke wherever there is the slightest impediment, such as at a bend in a chute or at the outlet of a hopper, and the material arches across the passage and further material piling up behind packs it more firmly. Many devices consisting of vibrators, agitators and mechanical arch breakers are in use to overcome this condition. With fine coherent powders considerable success can be attained by the technique of fluidization. An upcurrent of air supplied under pressure through a porous pad over which the material flows partially lifts the material and aerates it, greatly reducing the friction between the particles. With full aeration the material flows almost like a fluid with an angle of repose approaching zero. The fluidizing pads can be arranged close to hopper outlets and in the base of chutes. A chute in which the pads extend the whole length is known as an 'airslide' and can convey efficiently at slopes of only a degree or so from horizontal with materials like cement or alumina.

In general, free-flowing materials have particles of fairly uniform size and rounded shape. The poorer flowing materials have irregular shaped particles which tend to interlock, with a wide size grading which allows the larger particles to rest in a bed of smaller ones.

Table 1. *Typical Values for Bulk Density, Particle Density and Angle of Repose*

Material	Average particle size	Bulk density lb./ft.^{3}	Particle density lb./ft.^{3}	Angle of repose degrees
Alumina (sandy)	60 μ	64	230	35
Alumina (floury)	60 μ	54	230	45
Ammonium sulphate	600 μ	45	110	40
Asbestos fibre	—	35–50	150	45
Bauxite (crushed)	—	105	160	45
Calcium fluoride	50 μ	95	200	40
Cement	20 μ	90	195	(35)
China clay	6 μ	54	165	45
Coal (lumpy)	—	55	100	40
Coke dust	15 μ	60	120	43
Lime (hydrated)	15 μ	31	105	—
Phosphate rock (lumpy)	—	90	145	45
Quartz (crushed)	15 μ	58	165	45
Rubber pellets	3 mm	30	70	40
Sand	—	90–100	165	35
Sawdust (dry)	—	15–18	40	43
Soda ash (heavy)	—	64	165	39
Soda ash (light)	30 μ	40	140	45
Wheat	3·5 mm	50	90	25

The friability of lumps is of importance in the design of equipment, since if a lump should cause jamming it is preferable for the lump to fracture and clear itself than for a machine element to break. For handling very hard materials the equipment must be so designed that either jamming is virtually impossible or that a safety device or self-clearing mechanism will come into operation.

All materials produce a certain amount of dust when disturbed by any handling process, and dust which is fine enough to be airborne for an appreciable time can give rise both to health hazards either by contact with the skin (dermatitis) or by inhalation (pneumoconiosis) and also to explosion hazards if the material is combustible. Practical action to reduce health hazards consists of enclosing the equipment as much as possible and removing the dust by an exhaust system. Particle sizes of 5 μ and below are the most important in constituting a pneumoconiosis hazard, since larger particles either settle out of the atmosphere too quickly or are adequately trapped in the nose and bronchial tubes before reaching the alveoli. Many dusts are apparently inert, but silica and a number of chemicals are known to produce serious disability. Explosion hazard is difficult to eliminate, and practical steps are based on reducing the possibility of ignition by careful maintenance of flameproof electrical equipment, adequate earthing of any equipment likely to generate static electricity, siting of plant relative to any source of ignition, followed by minimizing the effect of any explosion should one occur. Bursting panels on vessels, storage bunkers and ducts discharging into open air, with baffles along any path over which an explosion wave might be propagated serve to confine the explosion to a small area of the plant. It should be noted that often the most damage and loss of life arises from a secondary explosion due to the ignition of dust blown off ledges, roof-trusses, *etc.* by the primary explosion. This can be avoided by proper design of explosion reliefs on bunkers and ducts with attention to cleanliness of the building. The degree of explosion hazard depends on the dust concentration, which must be between certain limits (for a particular dust) for an explosion to be possible, while the intensity of the explosion increases with the fineness of the dust and the calorific value of the material. A full account of dust hazards has been published by the Society of Chemical Industry[5], and the Factory Acts[6] lay down certain specific requirements. A comprehensive account of the properties of granular and powdered materials has been given by DALLA VALLE[7].

STORAGE OF MATERIALS IN BULK

Open Air Storage

Applicable to materials which are not affected by a wide range of climatic conditions, open air storage is an economic proposition if the average handling rate in stocking out and reclaiming is small in relation to the amount in storage. Otherwise, although there is no capital cost for bunkers and such constructions, the expense of high capacity handling equipment becomes excessive. These conditions apply in most cases where a reserve of material must be kept available to take care of seasonal or other inevitable fluctuations in the regular supply and demand. Power stations and gas-works use this system on a wide scale to stock coal in summer when their requirements are small and reclaim it in winter during the peak demand period.

The site for such a stock-pile should be well drained but reasonably level. The ground should be firm, and with advantage can be paved or surfaced with a well-consolidated layer of hard core or ashes.

The elaboration of distributing and reclaiming equipment depends on the size of the stock-pile. At large power stations permanent travelling gantries with conveyors and grabs to cover the area can be justified. At smaller power stations a drag scraper is often used to distribute coal from a pile made by the intake conveyor, or to reclaim coal into a hopper below ground level which feeds a conveyor to the boiler-house bunkers. The drag scraper is pulled by cable on a radius to or from centrally placed driving gear, and at the outer end the cable reverses over a pulley which can be moved periodically so that the working radius can cover the whole storage area in due course. At the smallest stock-piles, such as at works' boiler-houses, little permanent equipment can be justified; coal is probably received by tipping lorry, and reclaimed by a mobile bucket loader or by a fork lift truck with bucket attachment and carried to the boiler bunker, or it may be pushed by a bull-dozer to a hopper below ground.

Large stock-piles of bituminous coal are liable to spontaneous combustion, an exothermic oxidation process which increases in rate with rise of temperature and in the presence of moisture, so that if the internal temperature rises above about 60° C a dangerous situation can arise requiring immediate breakdown of the pile to avoid a fire. The danger can be reduced by ensuring that the coal is packed tightly to minimize the amount of air available, and to limit the height to 15–20 ft. so that heat can be more readily conducted away.

Indoor Storage

Protection against weather given to a pile of material can range from a simple Dutch barn to a substantially constructed building. The presence of a roof structure has the advantage that a distributing conveyor can be fitted above the pile with little extra cost, as shown in *Figure 1*. Reclamation of

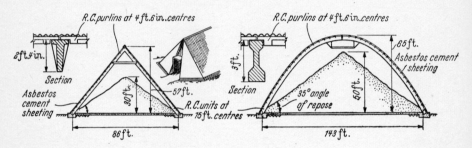

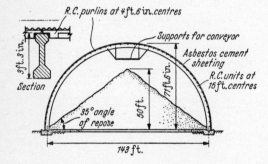

Figure 1. Storage sheds constructed of reinforced concrete and asbestos sheeting with overhead conveyors. (By courtesy of Mechanical Handling)

the material follows the same principles as for open air storage, although grabs can seldom be used owing to lack of headroom. Pneumatic conveying is a very convenient way of reclaiming many materials, particularly deserving of consideration if the expensive pumps and discharging machinery can be shared between other pneumatic conveying requirements round the plant. Easily handled flexible pipes connect the intake nozzle to the nearest tapping point on a permanent suction pipe arranged round the building.

Floor storage is occasionally used in multi-storey buildings, especially those previously used for storage of sacks of material. The quantity which can be stored is usually limited by the permissible floor loading of the building rather than the headroom available.

Bins, Bunkers and Silos

Material which must be readily available for conveying to a process at a steady rate must be stored in a container designed so that the material can be withdrawn from the bottom by a suitable conveyor without the use of manual labour. The base of the container must therefore slope down steeply enough towards the outlet aperture to ensure easy flow of the material, the rate of flow through the outlet being controlled by a valve such as a sliding plate, or if close control is necessary, by special measuring or feeding equipment. The sloping base, or hopper, alone will hold useful amounts of material and is often constructed at ground level to take a lorry-load tipped into it prior to being conveyed at a steady slower rate to more permanent storage. Further increase in stored volume is gained by vertical walls rising from the lip of the hopper base. Structures which are very tall compared with the cross section are usually termed silos, particularly in storing grain, where a number of containers, each holding 1,000 tons or more of grain may be built together as a block.

Steel is a common constructional material for shallow bins and bunkers of square or rectangular plan, the plates forming the hopper and walls being strengthened by external channel or angle sections. Reinforced concrete is more frequently used for large bunkers and silos, the principles involved in design being similar. Most of the weight of the stored material is supported by the hopper bottom in shallow bunkers, and friction with the vertical walls can only bear a proportion of the weight of the material which lies above the profile which would form if the hopper only were filled and surcharged to the angle of repose of the material. In deep silos much of the material is supported by friction against the walls and it is of importance to be able to calculate the stresses involved. Besides the downward compressive stress there is also an outward bursting pressure. The following formulae which are in common use are based on the equilibrium of a potentially sliding wedge of material. The angle of the plane of separation of the wedge of material is taken as that giving maximum outward pressure on the wall and assumes that the angles of friction of the material on itself (the angle of repose) and on the constructional material of the silo wall are known, besides the density of the stored material. Two cases are considered: the first, in which the depth is small and the plane of separation of the wedge reaches the surface of the material (assumed level) before reaching the opposite wall, the horizontal outward force P per unit length of circumference down to depth h is given by

$$P = \frac{wh^2}{2}\left[\frac{1}{\sqrt{\mu(\mu+\mu')}+\sqrt{(1+\mu^2)}}\right]^2$$

where w = density of stored material, and μ and μ' are the coefficients of friction of the material on itself and on the silo wall respectively. The angles of friction are related to the coefficients by

$$\mu = \tan \varphi$$

In the second case the plane of separation reaches the opposite wall before reaching the surface, and the corresponding force is

$$P = \frac{wb^2}{2}\left[\frac{\sqrt{\frac{2h}{b}(\mu+\mu') + (1-\mu\mu')} - \sqrt{(1+\mu^2)}}{\mu+\mu'}\right]^2$$

where, in addition, b is the breadth of the silo.

The weight of material supported by the walls at depth h is therefore the total outward force round the circumference multiplied by the coefficient of friction μ'. The difference between this supported weight at maximum depth and the total weight of material in the silo is therefore the weight supported by the hopper bottom. The design and construction of bunkers and silos in reinforced concrete has been described by GRAY[8].

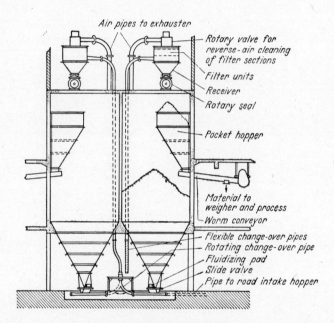

Figure 2. *Reinforced concrete bunkers for lime and soda ash with pneumatic conveying equipment*

A different kind of pressure distribution occurs in the catenary suspension bunker illustrated in *Plate III*. The curved section is nearly the natural shape taken up by a flexible membrane containing the material, so that only tensile stresses are of importance in the shell. Different levels of filling tend to distort the shape, but this is easily withstood by the steel plate construction. From rail trucks on the far side the bunker is filled with zinc ore by the pneumatic suction plant travelling along the top of the bunker. The material is withdrawn by a worm conveyor collecting from several outlet points along the bottom.

A pair of reinforced concrete bunkers for hydrated lime and soda ash are shown in *Figure 2*. The hopper bottoms in this case are of steel plate. Subsidiary storage for day-to-day requirements is held in two pocket hoppers near the top of each main bunker and a pneumatic plant provides for intake and circulation of material from the bottom of either bunker; in each case the pocket hopper is first filled until the rest of the material overflows into the main bunker. A worm conveyor and small batch weigher control the flow from either pocket hopper to the required process. Soda ash and lime are inclined to build up in cliffs and flow irregularly out of a hopper, hence fluidizing pads are placed near the bottom outlet in each case.

LEVEL AND WEIGHT MEASUREMENT

The inconvenience and possible hazard or damage caused by either overfilling a bunker or unexpectedly emptying a bunker at a critical moment during a process can be avoided by the installation of suitable level or weight detecting equipment. A simple installation would comprise high and low level alarms, but more elaborate systems may require continuous indication or recording of the level or weight.

A well established device in wide use consists of a flexible diaphragm or pivoted plate approximately flush with the wall of the bunker. The lateral pressure of material is sufficient to deflect the diaphragm against a spring and operate a switch when the level of material is slightly above the device. When the level drops, the diaphragm returns to normal. The device is highly successful for all free-flowing materials.

The following instrument is more suitable for poorly-flowing materials, and consists of a vane slowly rotated by a small electric motor. The vane is placed at the high or low level in question and the presence of material prevents its rotation. Subsequent rotation of the motor is taken up by a spring and the relative movement operates the alarm and switches off the motor. The spring remains in tension so that when the vane is released by lowering the level of material the switch returns to normal and the motor restarts. The motor does not run back under the spring tension owing to the low efficiency of the reduction gearing. A disadvantage of this device is that the vane is sometimes prone to damage from impact of lumpy material.

Various electronic methods can be used, and in a simple type now very popular the change of capacitance of an insulated electrode at the level in question, due to the presence of material, upsets the stability of a valve oscillator. The change in anode current of the valve is sufficient to operate a relay. For continuous level indication a similar principle can be used in which the capacitance between a pair of insulated conductors suspended vertically in the bunker is measured.

Bins and small bunkers can be weighed as a whole and it is possible to use a purely mechanical means of weighing, such as a steelyard, but it is far more convenient to support or suspend the bunker from a number of load cells which transmit the force to the indicating or recording instrument. A load cell may be a hydraulic capsule transmitting a pressure, or a steel stud with electrical resistance strain gauges transmitting an electric signal to a magnetic or electronic amplifier.

CONVEYING EQUIPMENT
General Remarks

It is quite impossible to be specific regarding what material should or should not be handled by a certain type of conveyor and each case should be considered individually by an experienced handling engineer. Naturally the solution should be the one which gives least capital and maintenance costs combined with convenience of operation without causing any hazard or loss of value of the material. The following paragraphs give a few of the points which influence choice.

Site Limitations—Many conveyors will convey in a straight line only with limited inclination and length. Certain conveyors require control of the feed, such equipment increasing the required headroom when taking material from a hopper or bunker. If it is necessary to use conveyors in series over an indirect route transfer points are required and it is highly advisable to have mechanical and electrical interlocks so that if one conveyor stops all the earlier conveyors in the series also stop to avoid a pile-up of material.

Average Capacity and Peak Capacity—Conveyors which are not self-feeding must be designed to handle a peak capacity appreciably greater than the rated average capacity to cater for surges in the flow of material.

Action of Material on Conveyor—Corrosion, high temperature and abrasion effects must be considered in order that maintenance should not be excessive. Hard lumpy material might cause serious breakdowns in certain types.

Action of Conveyor on Material—Loss of value of the material may occur by contamination or attrition if friable material is roughly handled. With dusty material it may be necessary to install a dust-collecting system to eliminate health or explosion hazards.

The ideal conveyor is therefore one which does not affect and is not affected by the material, takes up little space, carries the material all the way in one move, is impossible to choke or damage with lumps or overfilling, is dustless, and, of course, uses a minimum of power. It is obvious that to obtain this, one must be using an ideal material on an ideal site as well.

Conveying by Gravity

Chutes—For the controlled lowering of both bulk material and unit materials such as sacks and packages, a chute has the advantage of virtually no maintenance and no power consumption. A straight chute can also convey laterally a useful distance such as may be required for transferring material from one conveyor to another, but the slope may need to be carefully chosen to avoid excessive speed of the material and consequent damage to it, particularly in the case of packages. A spiral chute (*Plate IV*) limits the speed of units since the faster the descent the greater is the frictional force against the outer face of the chute. A long drop given to friable materials down a steep chute can cause breakage and in such cases baffles are fitted in the chute to slow down the material. The minimum angle of a chute must, of course, be appreciably greater than the angle of friction of the material.

Airslide—With fine powders it is possible to reduce the angle of friction of the material to an almost negligible value by fluidizing the material with an upcurrent of air. In this way large conveying rates are possible in a chute at an angle of only a degree or so from horizontal, and a considerable horizontal distance can be covered with a small loss of height. The base of the chute is made of a porous material (ceramic, fabric, sintered metal, *etc.*) and air under low pressure passes up through it from a duct below. The vertical air velocity required above the porous base depends on the density and particle size of the powder, and for cement and alumina is of the order of 5 ft./min. The pressure required is about $\frac{1}{2}$ lb./in.^{2} depending on the flow resistance of the porous bed and the depth of material. Bends and junctions can be made in the airslide provided the gradient is maintained. Material inlet is possible through any aperture in the top cover, and the outlet can be either over the end of the chute or through an aperture in the side wall.

Roller Conveyors—Suitable for any unit with a flat base such as a box, closely spaced rollers support the load and allow it to run downwards at an angle of about two degrees or less. The pitch of the rollers should not be greater than half the lengthwise dimension of the load so that there is no possibility of the load tipping and jamming between rollers. The acceleration of a single unit down the rollers is limited since energy is successively given to the rollers by the passing of the load, but another load following closely will accelerate more since the rollers are already revolving and it may collide with the first. Curves can be negotiated with tapered rollers, and junctions and ball tables (on which an operator can push an article in any direction) can be included. Portable conveyor units are readily obtainable, and some or all of the rollers may be power driven for conveying horizontally or up slight inclines. The dimensions of rollers and conveyor components have recently been standardized [9].

Belt Conveyors

It may be said that the belt conveyor is probably the nearest approach to a universal conveyor. It can carry a very wide variety of materials economically over long or short distances, either horizontally or at an appreciable incline either up or down. Running and maintenance costs are low provided the conveyor is properly designed to suit the material and layout. The chief maintenance expense is renewal of the belt when it is worn or damaged, and in a poorly designed conveyor the belt may have a drastically reduced expectation of life.

Essentially, a belt conveyor consists of an endless band or belt of uniform flexible material running over a wide pulley at each end, one of these pulleys being power driven. Bulk material or packages laid on the belt at some point will therefore be carried with the belt until they either drop over the end at the pulley or are scraped sideways off the belt with an inclined board at an intermediate point. The weight of the belt and material must be supported along its length either on free-running rollers placed at intervals, or for very light duties, on a smooth wood or metal deck. The returning run of the belt also needs support, but since it is not loaded, the idling rollers are more widely spaced. Various layouts are shown in *Figure 3*.

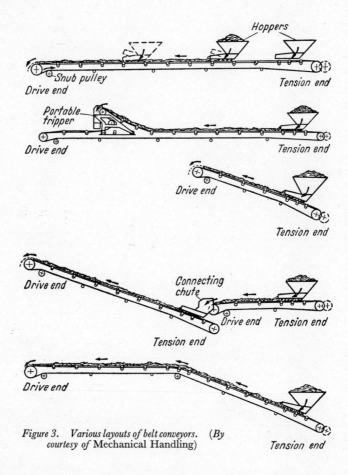

Figure 3. *Various layouts of belt conveyors.* (*By courtesy of* Mechanical Handling)

It is practical and economical to form the conveying run of the belt into a shallow trough when conveying material in bulk by dividing each supporting idler roller into three or five sections and inclining the outer sections at an angle which is commonly about 20°. This greatly increases the cross section of material which can be placed on the belt without spillage. A 3-roller arrangement is shown in *Figure 4*.

The Belt—The most widely used form of belt is built up of several plies of woven cotton fabric (duck or canvas) thoroughly impregnated and cemented together with rubber, with a layer of $\frac{1}{32}$ in. or $\frac{1}{16}$ in. rubber on the face in contact with the pulleys, and from $\frac{1}{32}$ in. to $\frac{1}{4}$ in. (or even thicker for arduous duties) on the face exposed to the abrasion of the material being carried. The fabric supplies practically all the tensile strength of the belt, but cannot withstand abrasion or exposure to humidity, nor the chemical action of many materials. The rubber impregnation and facing waterproofs the fabric and provides a resilient surface. Belts for light duty are

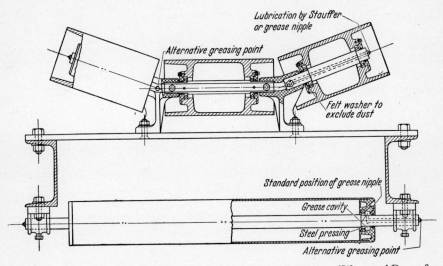

Figure 4. *Typical set of troughing idlers, showing construction.* (*By courtesy of* Westwood Dawes & Co. Ltd.)

also made with impregnation alone, either with rubber, various kinds of drying oils, bitumen compounds or balata gum. Plastics have recently entered this field to provide flameproof belts, particularly for coal mines. Belts made of natural fabric, either with the plies stitched together or solid woven to a comparable thickness, are often used for package conveyors, especially if the belt is to slide on a deck instead of being carried on rollers, since the coefficient of friction is considerably lower than that of impregnated belts. A disadvantage of natural fabric belts is that they have a much greater stretch under tension. For example, a rubber-covered or impregnated belt usually has a stretch of about 1·5 per cent under maximum working tension, while a natural solid-woven belt would have a stretch of about 6 per cent, the working tension in each case being about 10 per cent of the ultimate tensile strength.

B.S. 490:1951[10] lays down minimum standards of quality and strength, covering six weights of fabric and three grades of rubber, for rubber conveyor and elevator belts. Suggestions are given for selecting a suitable grade of belt according to the duty required and recommendations made for the appropriate minimum size of pulley. *Table 2* surveys the types of fabric covered by the British Standard, with their minimum ultimate tensile strengths and recommended maximum working tensions.

Other types of belt can be used for special purposes; notable examples are thin steel bands[11] which are often used for conveying sugar and many other sticky materials, and woven wire belts used for conveying articles through ovens and similar process equipment.

Table 2. Types of Fabric Conveyor Belts

Weight of fabric 42 in. × 36 in. ounces	Tensions lb./in. width per ply		
	U.T.S.		Maximum working tension
	warp	weft	
28	350	170	25
31*	350	200	25
32	400	185	30
$33\frac{1}{2}$*	400	250	30
36	420	200	32
42	500	240	40

* These fabrics have increased weft strength for use with bucket elevators.

The width of the belt depends partly on the largest lumps which are likely to be conveyed, and partly on the required maximum capacity, which in turn depends on the speed. If, due to the conveyance of large lumps, a wide belt is required, then the speed of the belt can economically be reduced until it is loaded to its maximum cross section of material at the required conveying rate, with an allowance for surges, and the reduced flexure and wear will considerably increase the life of the belt. On the other hand, when conveying fine material, the maximum speed is determined by the tendency to spill material when passing over idler rollers, or by dusty material blowing off at the high relative air speed, and the width is determined by the capacity required. *Table 3* summarizes present practice with regard to width, numbers of plies, maximum speeds, and average cross section of material on flat and troughed belts, though this will depend to some extent on the free running properties and angle of repose of the material.

Current trends in the development of belting include the use of synthetic fibres such as rayon, with coverings of plastics and synthetic rubbers compounded with a special purpose in mind, such as resistance to oil. Certainly Terylene and silicone elastomers will be used in due course. The other major development is the vast increase in longitudinal tensile strength given by cord construction, in which part of the fabric is replaced by continuous cords, thus making a stronger belt without reducing the lateral flexibility which ensures good troughing. In the extreme case, wire cables can be used, and HUDSON[2] reports a belt conveyor 10,000 ft. between pulleys with a working tension of 650 lb./in. width.

Supporting the Belt—At each end of the conveyor the belt must reverse round a pulley of sufficient diameter so that the continual flexing does not harm the belt. In current practice a diameter of at least 4 in. per ply for the lighter fabrics, and 6 in. per ply for the heavier fabrics is used. The width of each pulley should be 2–3 in. wider than the belt to allow a small amount of wandering without the belt overlapping the edge of the pulley and becoming unsupported. The pulley at the head, or discharge end is usually connected to an electric motor through a suitable speed-reducing unit to provide the drive for the conveyor, while the pulley at the feed-on end can rotate freely.

Table 3. *Capacities and Speeds of Belt Conveyors*

Width in.	Number of plies	Maximum size of lumps		Average cross section† of material on belt		Maximum speeds		
		Uniform in.	Mixed* in.	Flat ft.²	Troughed ft.²	Lumpy, heavy abrasive materials ft./min	Fine materials ft./min	Grain ft./min
10	3–4	1½	2	0·018	0·041	200	300	400
12	3–4	1½	2	0·030	0·066			
14	3–4	2	2½	0·043	0·096			
16	3–5	2½	3	0·059	0·131	300	400	500
18	3–5	3	4	0·078	0·173			
20	4–6	3½	5	0·098	0·22			
22	4–6	4	7	0·121	0·27	350	500	600
24	5–7	4½	8	0·146	0·33			
26	5–7	5	9	0·173	0·39			
28	5–8	5½	10	0·20	0·46	400	550	700
30	5–8	6	11	0·24	0·54			
32	5–8	6½	12	0·27	0·62			
34	6–9	7	13	0·31	0·71	450	600	800
36	6–9	7½	14	0·35	0·80			
42	6–10	8	17	0·49	1·12			
48	7–10	10	20	0·65	1·48			
54	7–10	12	23	0·83	1·90			
60	7–10	14	26	1·03	2·36			

* 'Mixed Sizes' refers to material of which not more than 20 per cent is of the maximum size stated.
† 'Average cross section' refers to material lying on the belt with its profile meeting the belt at 20° to the horizontal and in the case of a troughed belt, with the outer idler rollers inclined at 20°. With many materials, and with attention to good feed-on conditions, the cross section can be considerably increased. (Data from HETZEL and ALBRIGHT[12].)

Along the length of the conveyor the belt must be supported with as little friction as possible to economize in driving power. Free-running rollers or 'idlers' are therefore set under the belt every few feet, the spacing being such that the load of material on the tensioned belt does not cause the belt to sag between idlers sufficiently to induce any appreciable disturbance of the material. The returning run of the belt is also supported on rollers, but since this length is not loaded with material the idlers do not need to be closely spaced. For most conveyors a spacing of 3–6 ft. on the conveying run is satisfactory, and on the return run a spacing of about 10 ft. is sufficient.

The returning belt is almost invariably run flat on single cylindrical rollers, whereas the conveying run may be troughed by dividing each roller into three or five sections of which the centre one is horizontal and the remaining sections are angled to raise the edges of the belt. Angles up to

45° have been used, but the life of the belt tends to be short because of the sharp kinks formed at the junctions of the sectionalized rollers, and also a lightly loaded belt may have difficulty in conforming to the shape required by the rollers owing to its inherent lateral stiffness. Modern belt conveyors are troughed to an angle of about 20°–25° as a compromise between reduced belt life and reduced cross section of material being conveyed. Diameters of the rollers range from 4 to 7 in.

In the construction of idler rollers, fabrication from steel tube is replacing cast iron, and roller or ball bearings are replacing the old plain bearings. Most modern idlers, such as the set shown in *Figure 4*, run on a stationary shaft with seals to keep the charge of grease in the bearings, and dirt and water out. Usually there is a large reservoir of grease, and such idlers have been known to run for several years without recharging.

Belts for unit loads, *i.e.* packages, sacks, *etc.*, almost always run flat. For short conveyors it is practicable to support the belt on a polished metal or wood deck as this reduces the capital cost, but the greater friction increases the power and wear on the belt, hence running costs will be greater.

The belt will drift to one side when working unless positive steps are taken to keep it in control. Causes of the drift are chiefly:

(*a*) slight misalignment of any idlers, or pulleys;
(*b*) not loading material centrally on the belt;
(*c*) using a belt which is too stiff to trough properly;
(*d*) joints in the belt not being made quite square;
(*e*) a sticking idler roller dragging one side of the belt;
(*f*) uneven stretching of the belt at worn places.

Besides eliminating the above causes as far as possible, the head and foot pulleys should be crowned by making the central diameter greater than the edge diameter by at least $\frac{1}{8}$ in./ft. of pulley width. This provides a slight automatic steering action which tends to centralize the belt on the crown, but has the disadvantage that the centre of the belt is given greater tension than the edges. Crowning may be sufficient for the shorter belts, but for long lengths it will probably be necessary to use a number of special self-aligning idlers, of which there are many proprietary types available. These generally operate automatically by the sideways drift of the belt causing the idler frame to pivot on a vertical axis and steer the belt towards the centre.

Feed on to the Belt—The main principle to be observed is that material should, as far as possible, be given a forward velocity equal to that of the belt and laid on with as small a free drop as practicable, so that abrasion due to impact and sliding will be minimized. For many free-flowing materials all that is necessary is a chute curved at its lower end to become almost tangential to the belt and terminating just clear of it. The width of the chute should be about two-thirds the width of the belt and its side plates should be extended forward to help confine the material after it has been laid on the belt. In practice, however, it will often be found that there is too little headroom for a long chute in which adequate momentum

Plate I.   Typical fork lift truck.   (By courtesy of Coventry Climax Engines Ltd.)

Plate II.   Bags of Plastic granules being conveyed by a flat belt on to a lorry.   (By courtesy of Simon Handling Engineers Ltd.)

Plate III. Catenary suspension bunker for zinc ore. (*By courtesy of* Simon Handling Engineers Ltd.)

Plate IV. Spiral chutes for sacks. (*By courtesy of* Simon Handling Engineers Ltd.)

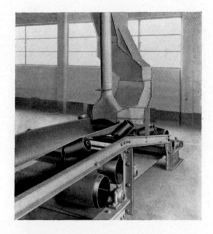

Plate V. Feed-on to a belt, showing a dust-collecting hood. (By courtesy of Simon Handling Engineers Ltd.*)*

Plate VI. Large mechanically propelled tripper. (By courtesy of Simon Handling Engineers Ltd.*)*

Plate VII. Bucket elevator for ship unloading (*in housed position*). (*By courtesy of* Simon Handling Engineers Ltd.)

can be gained, and the material may have a high angle of friction, so that after coming down a steep chute it meets the belt with a horizontal velocity which is too slow, and with a vertical velocity which is too high. To minimize sliding abrasion the belt speed should be kept low, and to minimize impact, lumpy material should meet the belt between the supporting idlers so that the belt itself can give and absorb the impact gradually. If loading over an idler is inevitable it should be of a type which provides some resilience —either with rubber covered rollers or with a resilient mounting. A feed-on station with a dust-collecting hood is shown in *Plate V*.

Discharging the Belt—The simplest form of discharge takes place over the head pulley. If the belt runs fast enough so that the centrifugal acceleration over the head pulley v^2/r is greater than gravity, then the material leaves the belt as it begins to run round the pulley and describes a parabolic trajectory (if air resistance is negligible) during which the horizontal component of velocity remains constant and the vertical component increases at a constant rate under gravitational acceleration.

For a belt of velocity v running up an incline at an angle θ to the horizontal, the horizontal component of velocity is $v \cos \theta$, and the vertical (downward) component is $-v \sin \theta$, then the horizontal and vertical distances moved at time t from the moment of leaving the belt are:

horizontally: $\qquad x = vt \cos \theta$;
vertically (downwards): $y = \tfrac{1}{2} gt^2 - vt \sin \theta$

On the other hand, a slower-moving belt will not throw off its material until it has passed round the pulley to a point where the belt subtends an angle φ below the horizontal where $v^2/r \cos \varphi = g$. The horizontal velocity component is then $v \cos \varphi$ and $x = vt \cos \varphi$; the vertical (downward) velocity component is $v \sin \varphi$ and $y = \tfrac{1}{2}gt^2 + vt \sin \varphi$. Both measured from the point at which material leaves the belt.

A proportion of dust with the conveyed material will have too high an air resistance to follow the parabolic trajectory and will merely dribble over the head pulley. The chute for the discharged material must therefore be placed sufficiently below the pulley to catch the dribble as well as the main stream. Discharge can be effected at any point along the length of the belt by a device known as a tripper or throw-off carriage (*Plate VI*). A frame carries a pulley, similar to the head and foot pulleys, and the belt is raised from its normal position and passes over the tripper pulley discharging the material into a chute directed over the side of the belt. The belt is then reversed over another pulley below and behind the point of discharge, and resumes its normal path along the conveyor. Usually the frame is made to straddle the conveyor and to move along on guide rails. Many such trippers are moved by power derived from the belt pulleys as they rotate with the belt. Flat belts can readily be discharged by a plough or inclined scraper board which diverts the material (or packages) over the side of the belt. This has the disadvantage of creating a sideways force on the belt, but for the discharge of bulk materials this can be eliminated by making the plough V-shaped and discharging equally over both sides. For discharge at various points along a flat belt conveyor a series of ploughs can be used, and these normally would be held away from the material on the belt, except for one at the required discharge point.

For discharge at a number of points in line, as for example over a row of bins, it is sometimes possible to use a shuttle belt. This is a short belt conveyor wholly mounted on a movable frame which can pass to and fro under the feed-on chute. The head pulley over which discharge takes place is positioned over the appropriate bin.

Inclined Conveyors

The maximum incline is determined by the possibility of the material being conveyed, slipping or rolling downhill when disturbed by its passage over the idlers. There is always a slight jolt in passing over the small hump caused by an idler, and a slight shifting as the belt becomes more deeply troughed. Smooth large lumps will tend to roll over each other unless cushioned by admixture with finer material, as also will small lumps to a rather less extent. The following table gives very approximate maximum angles of inclination of the belt, in degrees, for broad classifications of materials:

Table 4. Maximum Inclines (degrees) for Belt Conveyors

Material	Size		Mixed sizes
	Large	Small	
Rounded or ovoid	10	12	14
Fairly angular	12	15	18
Very angular	15	18	22
Irregular, chip-like	22	24	28

For example, washed and screened pebbles should not be conveyed above about 12°, wheat and similar grains not above 15°, and mixed coal up to 18°–20°, while wood chips can be conveyed up to about 28°. Fine powders vary widely in their flow properties according to the cohesion of the particles and the aeration of the mass. Moisture content can make a large difference.

Packages can be conveyed on inclined flat belts up to angles approaching approximately 10° less than their angle of friction on the belt provided the package is stable and does not tend to topple. Special belts have been made with deep corrugations moulded in the surface to increase the friction and they are claimed to be able to convey packages at inclines up to 45°.

Driving the Belt—The simplest way to drive is to couple a motor to either the head or tail pulley through a suitable speed-reducing device. It is preferable to drive the head pulley since the tension is then applied directly to the conveying strand of belt and the whole of the returning strand is under its least tension. The difference in tension round the driving pulley is limited by its coefficient of friction against the belt and by the wrap being only 180°. The coefficient of friction can be increased by lagging the surface of the pulley with a suitable material either riveted or bonded on, and the angle of wrap can be increased by the use of a snub pulley close to the driving pulley. If belt life is not important in relation to other considerations, as in coalface conveyors, it is possible to use a similar pulley to press the belt against the driving pulley to increase the frictional force. If even greater friction is necessary, as may be possible with long heavily laden belts, two driving pulleys can be used either both geared to a common motor or independently driven.

The tension differential required to keep the loaded belt moving can be derived from assumed values of the effective coefficient of friction of the belt supported on the idlers. The total weight of the belt (W_1) must be known, and this can be calculated from figures given in belt manufacturers' catalogues. The weight of the moving parts of the conveyor idlers and terminal pulleys can be estimated according to the number, size and type (W_2), and the weight (W_3) of material on the belt can be calculated from the length, speed and conveying rate.

The required tension differential is then (for a horizontal belt)

$$\Delta T = \mu_1(W_1 + W_2) + \mu_2 W_3$$

where μ_1 is the apparent coefficient of friction of the belt rolling over the idlers and terminal pulleys, including the effect of flexing and troughing the belt without a load of material, and μ_2 is the apparent incremental coefficient of friction as material is loaded onto the belt.

For an installation which is well maintained and has ball or roller bearings throughout, $\mu_1 = 0.015$ and $\mu_2 = 0.022$ are recommended, but the values are strongly affected by such variables as the idler spacing and design of feed-on equipment. For an inclined belt, a gravitational component $W_3 H/L$ (where H is the change in level and L is the length) must be added for a rising conveyor or deducted for a descending conveyor.

The power required at the driving pulley is, of course, $S\Delta T$, where S is the speed of the belt. With the weights in pounds and S in feet per minute, the tension is in pounds and the power in foot-pound per minute, *i.e.* the horse power is $S\Delta T/33,000$. An allowance must be made for losses in the drive between the motor and pulley shaft, and it is as well to have a good margin of power available to cover the possible inaccuracy of assessment of the coefficients of friction and the effect of the shifting of material, particularly at the loading point.

A certain minimum tension is required on the slack side of the driving pulley in order to provide sufficient frictional contact, and also some device is required to take up any increase in the length of the belt due to stretching. The simplest method is to mount the foot pulley bearings on slide rails and provide jacking screws to create the required tension and adjustment for length. Another way is to pull the frame containing the foot pulley bearings at a constant tension by means of a cable which passes over a pulley to a weight on its end. This has the advantage that stretch is automatically taken up and eliminates the chance of over-tensioning and possibly damaging the belt. The weighted tensioning and take-up principle can be applied at any place along the return run where two full size pulleys with a weighted pulley moving vertically between and below them can be accommodated.

The maximum tension in the belt is the sum of the initial tension provided by the tensioning device and tension differential required to drive the belt (except in the rare case of a steeply descending conveyor), and this must not exceed the maximum working tensions previously recommended, or the life of the belt will be curtailed.

An ascending belt conveyor can in certain cases run backwards if a power failure occurs when it is loaded, causing a pile-up of material at the foot with serious consequences. It is therefore a wise policy to provide an automatic brake to prevent this happening.

Cleaning the Belt—In many cases it may be found necessary to clean the conveying surface of the belt after the bulk of material has been discharged. This surface of the belt is in contact with the return idlers and with pulleys in throw-off carriages, double-pulley and snub-pulley drives, so that any residue of material clinging to the surface may become compacted and eventually interfere with the proper working of the conveyor and damage the

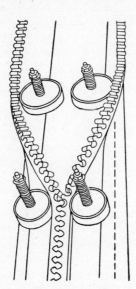

Figure 5. The method of opening and closing the zipper. (By courtesy of Mechanical Handling)

belt. Cleaning devices should be fitted close to the discharge pulley and can take the form of spring-loaded scrapers or rotating brushes driven from the pulley. More complex cleaning methods have been devised for special applications and are the subject of numerous patents. Care must be taken that the belt cleaner does not catch at the point where the ends of the belt are joined together. A vulcanized spliced joint is by far the best from this aspect, besides protecting the fabric from moisture which is a difficult problem when any type of metal clip fastener is used.

The standard textbook on belt conveyors by HETZEL and ALBRIGHT[12], the relevant chapter of Hudson's book[2], and an article by WOODLEY[13] give further details.

The Zipper Conveyor

This important development of the belt conveyor consists of a flat belt provided with a flexible rubber cover in two halves, each half attached to the outer edge of the belt and meeting at the middle where moulded teeth interlock in the same manner as the well-known 'zip' fastener. Fixed sets of rollers guide the meshing and unmeshing of the teeth as the belt passes through them (*Figure 5*). The covers are closed immediately after the material has been deposited on the belt, and the belt can then follow quite complicated paths. Besides straight runs similar to a conventional belt conveyor, the zipper belt can be guided round wide radius curves in the vertical plane and will convey even vertically when fully loaded, like an 'en masse' conveyor. The belt can also be slowly twisted in order to take a curve in some other plane, and discharge can be effected by opening the

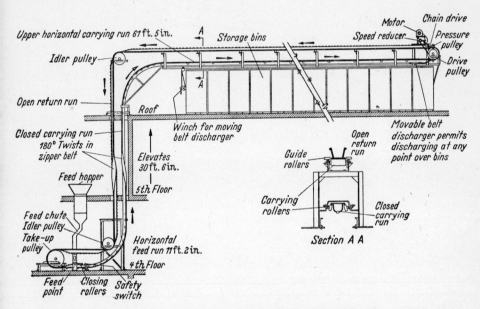

Figure 6. Layout of a zipper conveyor. (By courtesy of Mechanical Handling *and* Rownson, Drew and Clydesdale Ltd.)

covers with the belt upside-down over the receiving hopper. When the covers are opened out flat, the belt is flexible enough to pass over a conventional head pulley, as another method of discharge. These characteristics of the zipper conveyor are emphasized in *Figure 6*.

Although this conveyor has only recently become available in this country, a number of installations, of which WOODLEY[14] gives further details, are working successfully in America on such materials as carbon black.

Bucket Elevators

For the vertical lifting of bulk materials the bucket elevator is one of the most widely used pieces of equipment, which in its various forms can handle a wide range of bulk dry materials, from heavy lumpy minerals to fine powders, and also wet materials and slurries. Its obvious limitations are that the largest lumps must be able to go into the buckets without jamming, and materials must not be so sticky that the bucket will not empty even when held upside-down. A more subtle limitation is that the rather rough handling which occurs when the buckets discharge may degrade certain materials and reduce their value, also, with fine, highly aerated materials it may be difficult to fill the buckets adequately owing to spillage so that an elevator for a given capacity becomes uneconomically large.

In all forms of the elevator, suitably shaped buckets are equidistantly placed on an endless belt or chain which passes over pulleys or sprockets at the foot where the buckets are filled with material and at the head where the buckets empty as they turn over and discharge the material into a chute. Although not strictly necessary, elevators are usually enclosed in a sheet metal casing to eliminate the dust nuisance, and to confine spilled material so that it will fall into the path of the buckets at the foot and be picked up again.

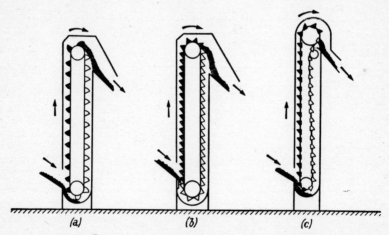

Figure 7. *Diagram showing the three basic types of elevators;* (a) *centrifugal,* (b) *continuous bucket,* (c) *positive discharge.* (*By courtesy of* Mechanical Handling)

Elevators can be classified into three groups according to the predominant principle which causes discharge, and these are shown diagrammatically in *Figure 7*.

Centrifugal Discharge—The angular velocity of the buckets passing round the head pulley is so high that centrifugal force throws the material clear of the pulley and other buckets into a suitably placed discharge chute. The high speed and short time available for discharge limits its application to comparatively free-flowing materials which will not suffer degradation on impact during discharge. It is the cheapest elevator for its capacity and is very widely used, particularly for handling grain.

Continuous Bucket—The buckets are closely spaced and are of triangular section so that as they turn over the head pulley and commence to empty, the back of the bucket ahead acts as a short chute to guide the flow of material outwards and into the discharge chute. Speeds are much slower than those used for centrifugal discharge elevators, but centrifugal action may help to clear the buckets to some extent. These elevators are used for materials such as minerals and coal, where the slower speed is necessary because of high abrasive wear from hard lumpy material or where reduction of degradation is important.

Positive Discharge—The buckets must be supported between two chains so that the chains can be snubbed under the head sprockets. Thus the buckets can actually be held upside-down over the discharge chute to give sticky or poorly-flowing materials the greatest possible opportunity of emptying.

The properties of the material to be handled control the method of filling the bucket nearly as much as they determine the mode of discharge. For free-flowing granular materials the buckets can be allowed to dig into a pile of material round the foot pulley. This could not be possible with lumpy heavy materials since the digging forces would be much greater and there would be a possibility of jamming, with consequent damage to the buckets and the chain or belt carrying them. The centrifugal discharge elevator therefore invariably has a digging or dredging feed whereas the other types usually have a means whereby the buckets are filled with material directly from a chute as they begin their upward run, and the flow of material must be carefully regulated to avoid a choke. In fact, if an attempt is made to fill the buckets of a centrifugal discharge elevator on the upward run, the high speed and spacing of the buckets causes so much scattering of the material that the buckets do not fill properly until enough spilled material collects in the 'boot' or casing round the foot pulley, to make a contribution due to dredging.

The correct speed of the centrifugal discharge elevator is important. Normally the speed is calculated so that as a bucket passes over the top of the head pulley the centrifugal force at the centre of gravity of the material in the bucket is equal to the weight of material. After this point the resultant of the centrifugal force and the gravitational force is directed forward, and the material, provided it is not sticky, begins to move forward out of the bucket and describes a parabola into the appropriately placed chute. The buckets must not be too closely spaced, otherwise the bucket ahead may interfere with free discharge.

The condition for balancing these forces is therefore:

$$\omega^2 r = g = 32 \cdot 2 \text{ ft./sec}^2$$

where ω is the angular velocity of the head pulley and r is the radius of the centre of gravity of the contents of the bucket. In terms of N, the r.p.m. of the head pulley, this becomes $N^2 r = 2940$ (with r in feet).

If the speed is too high, the resultant force on the material is directed against the sloping face of the bucket until appreciably beyond the vertical position depending on the internal angle of the bucket, and discharge can only take place by material sliding towards the lip of the bucket. With high speeds and wide angle buckets handling free-flowing materials an appreciable amount of discharge can take place in this manner before the vertical position is reached. On the other hand, with a speed which is too low, the resultant force is directed towards the back of the bucket which is attached to the belt or chain carrier, and the material will not lift clear until beyond the vertical position; the free trajectory is then probably too short to reach the discharge chute and material will spill down the return leg of the elevator.

To determine the position of the discharge chute and the minimum spacing of the buckets to prevent interference, the path of the material after leaving the bucket can be plotted to scale. The initial velocity components are:

horizontally: $\omega r \cos \theta$
vertically (downwards): $\omega r \sin \theta$

and after time t, the distances moved are:
horizontally: $\omega r t \cos \theta$;
vertically: $\omega r t \sin \theta + \tfrac{1}{2} g t^2$

The variation of r over the depth of the bucket will cause a considerable spread of the material at the discharge chute. Normally the lip of the chute is fixed just clear of the descending buckets and at an angle of 30°–35° below the shaft.

Some continental grain elevators run at a speed far higher than that required to balance gravity, and the path of the discharged material is confined and guided into the chute by the curved casing round the head pulley. The advantage is a smaller elevator for the required capacity, provided the buckets can be efficiently filled at the high speed. It will be realized from this that centrifugal force tends to prevent the filling of the buckets as they pass round the foot pulley. In fact, filling cannot be regarded as completed unless material is still being added to the bucket as it begins its straight upward run where the centrifugal force is zero.

The shape of the bucket for use with a particular material is a compromise between the requirement of a wide mouth, in relation to its depth, to facilitate filling and discharge, and an adequate volumetric capacity in relation to its projection from the plane of the belt or chain carrier. Materials with a large angle of repose need a wide internal angle between the back and front with a smooth internal contour to obtain clean centrifugal discharge. For grain and similar free-flowing substances, 40° is often used, though for finer floury materials 60° may be necessary. On the other hand, with continuous bucket elevators, the internal angle must not be too great or the chute formed by the bucket ahead of the one being discharged will not be steep enough.

Buckets are generally made from sheet metal, either fabricated with welded or riveted seams, or pressed out in one piece. Plastic buckets have also been made for light duties. Perforated buckets have been used for allowing wet material to drain in transit.

The choice of belt or chain as the bucket carrier depends primarily on the maximum tension likely to occur. When using a belt, not only is there a limit to the belt tension before damage is done, but also the frictional grip of the head pulley is limited, whereas chains have a positive drive from the head sprocket.

The majority of centrifugal discharge elevators can utilize belts. The design of the belt conforms closely to the principles used in belt conveyors and generally rubber-covered textile fabric of a suitable number of plies is used. B.S. 490:1951[10] recommends two weights of fabric for such use. It is necessary to have a rather thicker facing of rubber on both sides of the belt, as it is subjected to a good deal of abrasion due to material being trapped both between the buckets and the belt, and also between the belt and the head and foot pulleys. Due to the more arduous nature of its work, a belt in an elevator should not be expected to have such a long life as a comparable belt conveyor. The buckets are fixed to the belt by bolts through the holes punched in the belt, the heads being countersunk to present a flat surface when passing over the pulleys. There is a particularly heavy strain on the fixing when a bucket is dredging through material in the boot, and the chance of a bucket tearing loose, which can cause serious subsequent damage, must be kept as low as possible.

Chains for elevators are, on the whole, very similar to those used for chain, flight and scraper conveyors, the difference being the various forms of attachments suitable for fixing the buckets.

It is a distinct advantage for continuous bucket elevators to work slightly inclined from the vertical towards the direction of discharge. This means that the head pulley is more directly above the lip of the discharge chute and buckets with a greater internal angle can be used without the chute formed by the bucket ahead being at an insufficient slope. The backward tilt of the buckets on the upward run also means that they can carry more material without spilling. The disadvantage is that, depending on the tension and height, it may become necessary to support the belt or chain on idlers to prevent excessive sag.

Driving power is almost invariably applied at the head pulley or sprocket, while the foot pulley is made adjustable (usually with a screw take-up device as for belt conveyors). The tension in the belt or chain is greatest at the top of the upward run and is made up of the following components:
1. Weight of upward run of belt or chain with filled buckets.
2. Friction due to foot pulley.
3. Force due to dredging or change of momentum of material entering buckets.
4. Extra tension created by foot pulley screw adjustment.

The tension in the belt or chain leaving the driving pulley is only:
5. Weight of downrun of belt or chain with empty buckets.
6. Extra tension created at foot pulley [equal to (4) above].

The main power requirement is that taken to lift the load in the buckets and is

$$\frac{\text{elevating rate (lb./min)} \times \text{height (ft.)}}{33{,}000} \text{ horse power.}$$

The extra power required to overcome dredging forces is usually accounted for empirically by adding twelve *foot pulley* diameters to the actual height in the above formula. A further 5–10 per cent in power will cover friction and air resistance (in high speed elevators) giving the power required at the head pulley shaft. The motor output must also include a margin for losses in the reduction gearing. Some device to prevent reverse movement must be included in case a power failure occurs during operation.

Plate VII shows a centrifugal discharge elevator used for elevating grain from ships. The elevator pivots on the boom structure to drop vertically through the hatches, and grain flows to the buckets through holes in the boot casing. A full discussion of the design of bucket elevators has been published by HETZEL and ALBRIGHT[12].

Worm Conveyors

The principle of conveying or raising something using a helix rotating in a tube or trough has been known at least since the time of Archimedes, and although the historical development is difficult to trace, the worm conveyor has probably existed in milling practice for centuries.

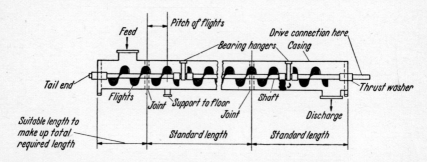

Figure 8. *General arrangement of a worm conveyor.* (By courtesy of Mechanical Handling)

The present-day worm conveyor consists of a U-shaped trough made usually of sheet steel. Granular material in the semi-circular bottom is urged along the trough by a long pitch helix rotating on a shaft co-axial with the base of the trough (*Figure 8*). With the considerable friction that exists between dry solid materials, this method of conveying is inefficient in comparison with some others, but within its limitations, its cheapness in first cost and ease of maintenance often make it the conveyor of preference. It can handle almost any material of which the lumps are not greater than about 10 per cent of the trough width. Sufficient clearance between the helix and the trough should be provided to eliminate the possibility of jamming. The more abrasive the material the slower the helix should rotate, and the larger the conveyor will be for a given capacity. Materials with a large angle of repose cannot be allowed to fill the trough as much as the more free-flowing materials, otherwise some of the material may pass over the rotating shaft and thus slip back one turn of the helix, reducing the overall capacity.

The most efficient type of helix for handling non-abrasive materials is made from cold-rolled steel strip such that the rolling forms it into a spiral having a thin outer edge and a thick inner edge to weld to the shaft. Other materials can, of course, be used if due to corrosion, contamination or similar cause steel is objectionable. The shaft is usually a tube supported at suitable intervals in bearings which must hang down through the upper part of the trough so as not to interfere with the progress of the material. A thrust bearing must be provided to take care of the considerable longitudinal force in the shaft, preferably at the discharge end so that the shaft is in tension.

The helix can also be fabricated from sheet metal segments either joined together to make a 'continuous worm' as above, or attached separately to make a 'paddle-bladed worm'. Cast iron paddle sections are often used bolted to the tubular shaft, when the conveyor is required to provide mixing during the conveying, for example, of various granular materials fed through separate apertures into the worm, or a liquid sprayed onto the material during its progress.

Feed into the conveyor takes place simply through an aperture in the top cover. The flow should be controlled so that the vertical section of the helix is only partly filled. For light, freely flowing materials, the proportion can be 45 per cent, but for heavy, poorly flowing materials, the proportion should be only 25 per cent, or even less. The conveyor can be arranged to be self-feeding from a hopper or bin by shortening the pitch of the first few turns, so that although the first part of the conveyor under the hopper works 100 per cent filled, the material spreads out to the appropriate lower proportion further along the conveyor. Discharge can be effected through any aperture in the semicircular base of the trough, and several apertures can be provided along the length of the conveyor, those not being used being closed by sliding plates although there is the disadvantage of the pocket so formed retaining material which may have to be cleaned out by hand. As a safety precaution, the furthest discharge point, should be left permanently open, so that there is no possibility of material packing hard in the end of the conveyor and causing damage.

The theoretical capacity of the worm conveyor can easily be determined in terms of the helix diameter d, the helix pitch p, the speed N revolutions per minute, and the proportion f of filling of the trough, referred to above. Then the conveying rate is $\frac{1}{4}\pi d^2 p.N.f$ ft.^{3}/min with d and p in feet. The capacity in tons per hour or similar unit can now be found using the known bulk density of the material.

Accepted practical values of conveying rates using a continuous worm are about 90 per cent of the theoretical values. For paddle-bladed worms they are rather less, depending on the shape of the blades. The pitch p is usually approximately equal to the diameter d, and maximum values of the speed N depend on the diameter and the characteristics of the material: thus, for non-abrasive free-flowing materials, with diameter 4 in., maximum recommended speed is 200 r.p.m. and with diameter 24 in., 100 r.p.m. Similarly, for heavy abrasive materials 80 r.p.m. at 4 in. diameter and 50 r.p.m. at 24 in. diameter are the maximum recommended speeds.

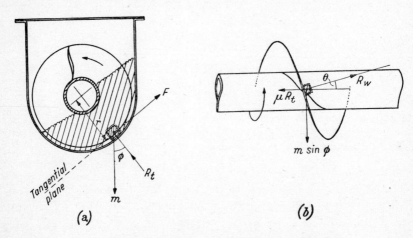

Figure 9. Forces on material in a worm conveyor

The total power requirement of a worm conveyor is mostly due to material friction. Losses in the worm shaft bearings are small, being largely dependent on the type of bearings and accuracy of alignment, and the efficiency of the reduction drive is easily taken into account.

Let us make an elementary analysis of the frictional forces on the spiral with a material whose coefficient of friction on the clean surface of the helix and trough is μ. The coefficient of friction can be determined separately by allowing the material to slide under gravity down a plate of the same substance as the conveyor. There is a certain angle α at which the material will just slide, then $\mu = \tan \alpha$.

A small mass of granular material, of weight m, is being urged horizontally along the trough by the rotation of the helix. Owing to its friction against the helix its path is not along the bottom of the trough, but at an angle φ partly up the curved side. *Figure 9* (a) shows the forces acting in the vertical plane, where R_t is the reaction force normal to the surface of the trough, and F is the tangential force component in this plane on the helix acting at radius r. For equilibrium we have the following equations:

$$F = m \sin \varphi \qquad \ldots\ldots (1)$$
$$R_t = m \cos \varphi \qquad \ldots\ldots (2)$$

In the tangential plane, *Figure 9* (b), R_t produces a frictional force μR_t parallel to the axis, and the reaction normal to the helix R_w produces the frictional force μR_w. With the weight component $m \sin \varphi$ we have the following equations for equilibrium:

$$\mu R_t + \mu R_w \sin \theta = R_w \cos \theta \qquad \ldots (3)$$

$$F = m \sin \varphi = \mu R_w \cos \theta + R_w \sin \theta \qquad \ldots (4)$$

The coefficient of friction μ is assumed to be the same for the material sliding on the helix as in the trough. Eliminating φ from equations (1) and (2) we have

$$F^2 = m^2 - R_t^2 \qquad \ldots (5)$$

and dividing equation (3) by (4) we have

$$\frac{R_t}{F} = \frac{\frac{1}{\mu} \cos \theta - \sin \theta}{\mu \cos \theta + \sin \theta}$$

whence substituting for R_t from equation (5) and squaring:

$$\left(\frac{m}{F}\right)^2 = \left(\frac{\frac{1}{\mu} \cos \theta - \sin \theta}{\mu \cos \theta + \sin \theta}\right)^2 + 1$$

Summating over the whole of the material in the conveyor, m becomes the total weight of material, and F becomes the tangential force acting at a mean radius r (slightly less than the helix radius, but we shall ignore this) supplied by the driving motor.

Table 5. *Conveying Power for Worm Conveyors*

α	μ	Values of K			
		$\frac{p}{d}=0.4$	0.6	0.8	1.0
25	0.446	2.2	1.7	1.4	1.2
30	0.577	3.2	2.3	1.9	1.7
35	0.700	4.2	3.1	2.5	2.2
40	0.839	5.3	3.8	3.0	2.5
45	1.000	6.2	4.3	3.4	2.8

Now $m = QL/NP$ where Q is the theoretical conveying rate in pounds per minute, and the horsepower $P = \pi dFN/33{,}000$.

Hence
$$P = \frac{QLK}{33{,}000}$$

where
$$K = \left[\left(\frac{\frac{1}{\mu} \cos \theta - \sin \theta}{\mu \cos \theta + \sin \theta}\right)^2 + 1\right]^{-\frac{1}{2}} \times \frac{\pi d}{p}$$

but $\dfrac{p}{\pi d} = \tan\theta$ and we can write

$$K = \pi\frac{d}{p}\left[\left(\frac{\dfrac{\pi d}{\mu p}-1}{\dfrac{\mu\pi d}{p}+1}\right)^2 + 1\right]^{-\frac{1}{2}}$$

Values of K are given in *Table 5* for various angles of sliding of the material, α, which determines μ, and for a range of values of p/d.

For an ordinary worm conveyor with $p = d$ the above values are comparable with the following empirical figures recommended by conveyor manufacturers:

Light free-flowing, non-abrasive materials, $K = 1\cdot2$.
Medium weight materials containing fines and small lumps, $K = 1\cdot4$–$1\cdot8$.
Semi-abrasive granular materials with lumps, $K = 2\cdot0$–$2\cdot5$.
Heavier, lumpy materials, $K = 3$–4.

These figures relate to actual conveying rate rather than theoretical conveying rate in the above power formula, and are higher in proportion.

The maximum length of a conveyor is determined by the torque which the shaft will safely transmit, since the power, and therefore the torque, required is proportional to its length.

Worm conveyors, discussed in more detail by HUDSON[2] and WOODLEY[15], will work up an incline, but it will be found that the capacity falls considerably, possibly by a half at an angle of 25°. Special designs can, however, be made to convey even vertically, in which case the pitch is generally made shorter, particularly the lowest few turns, so that the helix is almost full of material, and friction with the wall, now tubular, prevents the material rotating with the helix to a large extent.

Drag Chain and Flight Conveyors

A chute is limited in its application by the fact that it must slope steeply down at an angle appreciably greater than the angle of friction of the bulk material, but it is simple to imagine a method whereby the flow of material can be assisted by a chain or rope with projections or flights moving along the chute and powered by external means. Suitable wheels at the beginning and end of the chute make it possible to use an endless chain or rope continuously moving along the base of the chute and returning along any convenient path. If the flights are large enough to reach almost to each side of the chute, most granular materials will move apparently as a solid mass at nearly the same velocity as the flights, even up considerable inclines. The amount of slip between the material and the flights increases rapidly as the upward slope goes beyond about 20°, but it depends greatly on the cross section of the flight and the cohesive properties of the material.

This is the principle used in a broad group of conveyors which are suitable for handling many chemical powders and are widely used for minerals, coal and ash handling in power stations, swarf removal from machine tools, and for cereal grains and other granular foodstuffs. The chute or trough may be covered to eliminate a dust nuisance or possible contamination, or may be open over at least part of its length to allow entry of material from a number of places. The principle can also be applied to package handling, where the flights are widely spaced so that each flight can push a package along a flat deck.

A simple and commonly used design of drag-chain conveyor consists of two roller chains of pitch approximately half the width of the trough, with a bar connecting the two chains at every other link. The chains run along in the corners of a rectangular sectioned trough and the bars scrape along the bottom. At each end the chains double back over sprocket wheels engaging with the rollers, and the returning chains slide on angle supports in the upper part of the trough just clear of the surface of the material when working at maximum capacity. The sprocket wheel at the delivery end is driven by an electric motor through reduction gears to give a maximum chain speed of about 100 ft./min. Feed-in takes place through the top of the trough at any point, the material falling freely through the upper return chain. Delivery can be effected through an aperture in the floor of the trough. Several feed-in points and several delivery apertures can be provided along the length of the conveyor, those not in use being covered by slide plates.

For this design of conveyor the effective velocity of the bed of material is about 90 per cent of the chain velocity when conveying horizontally. The volume through-put of material can therefore be calculated from the cross sectional area A of the trough available for material, and the velocity V of the chain. If the bulk density ϱ of the material is known, then the mass flow rate is given by

$$0 \cdot 90 \, V A \varrho$$

The power and tension in the chain (which determines the minimum strength of the chain) depends on the effective coefficients of friction of the sliding chain and the material. For steel chain in a steel trough, the coefficient is about $0 \cdot 33$, while for the various materials to be conveyed it can range from $0 \cdot 3$ for free-flowing grains to about $0 \cdot 9$ for clinging materials like cement. The tension in the chain at the driving sprocket is thus

$$T = W_1 \mu_1 \cos \theta + W_2 (\mu_2 \cos \theta + \sin \theta)$$

where μ_1 and μ_2 are the coefficients of sliding friction of the chain and material respectively, W_1 is the total weight of the chain, and W_2 is the total weight of material in the length of the conveyor at any moment, *i.e.* $W_2 = A \varrho L$ where L is the length of conveyor between feed-in and delivery, and θ is the angle of inclination if the conveyor rises.

The power required at the driving sprocket is therefore TV. If A is in square feet, ϱ in pounds per cubic foot, W_1 and W_2 in pounds, and V in feet per minute, then T is in pounds and the power is TV in foot-pounds per minute, or $TV/33,000$ horse power. The power of the driving motor should be at least 30 per cent greater than this to allow for friction in the conveyor bearings and the inefficiency of the reduction gearing, and to provide a margin to deal with surges and starting loads.

A disadvantage of this design of chain conveyor is that the chain itself is subject to the maximum effect of abrasion due to the material. In a design to avoid this the conveying strands of chain move above the trough containing the material to be conveyed, while flights extend downward from

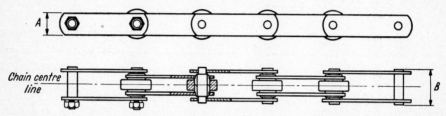

Figure 10. *Construction of precision-made conveyor chain.* (By courtesy of Renold Chains Ltd.)

the chains into the material. If a double chain is used, each chain can be supported and guided on a suitable track at each side of the trough, the flights being suspended with a good clearance, but a single chain version, as used for small conveyors, would have flights guided by the trough walls, with the chain centrally between them, as shown in *Plate VIII*.

An advantage of keeping the chains clear of the material is that it is possible to reduce the friction of the chains sliding in their guides by using larger rollers so that the weight of the chains and flights is carried on the freely revolving rollers. This reduces the coefficient of friction from 0·33 to about 0·1 with an appreciable saving in power.

Table 6. *Breaking Load and Weight of Conveyor Chains*

Size of chain		Breaking load lb.	Weight of single chain lb./ft. pitch			
A in.	B in.		2 in.	4 in.	8 in.	12 in.
0·70	0·75	3,000	1·05	0·70	—	—
1·0	1·3	7,500	2·70	1·90	—	—
1·5	1·55	15,000	—	4·05	3·00	—
2·0	2·05	30,000	—	9·15	6·25	5·25
2·4	2·88	45,000	—	—	12·5	10·1

Table 6 gives some basic data of a typical form of chain used with these conveyors and shown in *Figure 10* and Hudson [2] and Woodley [16] discuss chain conveyors in more detail than is possible here. A safety factor of six to ten should be used in calculating the safe working load from the tabulated breaking load. (Renold Chains Ltd.)

A cable can be used instead of chain, and one simple design has disc-like flights working in a U-shaped trough, the pulleys over which the cable must pass being suitably cut away to allow passage for the flights.

Chain Conveyors—Carrying Types

In this group of conveyors, the tractive effort is borne by chains in the same way as drag-chain and flight conveyors, but the flights now not only apply the propulsive force to the material, but also bear its weight and if necessary confine it to prevent spillage. In many ways, therefore, such a conveyor resembles a slow-speed belt conveyor in its application without the limitations of temperature and corrosion set by the presence of an organic material (rubber or canvas, for instance) used for conventional belt conveyors.

The extreme case of a large number of small carrying flights is the wire belt conveyor [17], which, although often driven by frictional contact with the powered head pulley as in a conventional belt conveyor, can be better constructed in the form of chain-like links engaging with a driven sprocket, and supported either on idlers or sliding on guides or on a flat deck. The open nature of the woven wire or chain link construction is of considerable importance in processes requiring material to be drained or washed during transit.

The apron conveyor (*Plate IX*) for bulk materials consists of a series of flat trays carried on one or more chains by suitable attachments to the links and which overlap to prevent leakage of material although they must be designed not to interfere with each other when the chain flexes over the sprockets. The sides may also be lipped to reduce spillage.

The slat conveyor, for packages, sacks and similar unit loads, has its carrying surface made of wooden or metal slats extending the full width of the conveyor, each attached to the chain links in the same way as the trays of the apron conveyor. Stationary side-plates or rails are desirable to eliminate the possibility of a package shifting and falling off sideways, or the carrying surface can be arranged to be flush with the floor. For conveying up steep inclines, projections can be fitted to some of the slats to engage the packages and prevent slipping back.

The usual type of roller chain used with these conveyors can only flex in one plane, although this can be either horizontal or vertical. Many conveying requirements can be satisfied by a path which is straight in plan with inclines and declines where necessary, and at the terminals the chain passes over sprockets on a horizontal axis and returns below the carrying strand. Bulk material on an apron conveyor is thus discharged over the head sprocket. Unit loads on a slat conveyor can either be tipped off at the head sprocket into a chute or ploughed off *en route* by a deflecting board, although in some applications, using slow-moving slat conveyors, a load may be removed manually as it passes a particular operator's station. A slat conveyor can be made in the form of a 'run-around' conveyor if the chain is arranged to flex in the horizontal plane, the advantage being that if the load is not removed at the proper station it will continue to travel round and not cause a dangerous pile-up in a delivery chute.

Chains are available which can flex in both vertical and horizontal planes, and with the use of these the slat conveyor becomes a very convenient way of conveying almost any kind of unit load over a wide variety of lines of flow. The working tension in the chain and the power requirements depend on the length, loading and coefficient of friction of the chain and attachments moving in their tracks, which is in turn dependent on whether the load is supported on rollers or is sliding. The length of conveyor is limited by the maximum tension permissible in the chain unless it is possible to provide the driving force at two or more spaced points working in synchronism.

The speeds of these conveyors range up to about 100 ft./min, but if any manual operations are involved, such as loading, unloading, picking or sorting of material, the speed should be kept below 50 ft./min. If the material is processed *en route* by washing, draining, heating or cooling, *etc.*, the speed may be determined entirely by the time required for completion of the process.

En Masse Conveyors

This type of conveyor[18] can be regarded as having been developed from the drag-chain conveyor, with the advantage that bulk material can be conveyed at any angle right up to the vertical. The basic principle is that the conveying strand of chain drags the material through an enclosed duct so that the material completely fills this duct, while the returning strand passes through a separate duct which may be adjacent to the conveying duct, but which may be separated and, in fact, can carry another material along a different path provided such a 'run-around' conveyor lies in one plane—either horizontal or vertical.

Such conveying ducts are usually rectangular, and the flights attached to the chain are shaped to fit the duct with a small clearance over three of its sides. The central part of the flight may be left open so that a sprocket can fit against either side of the chain. In many cases, the flights and chain links are integral castings of malleable iron, or for arduous duty with abrasive materials they are drop-forged and case-hardened alloy steel.

The materials which can be handled are similar to those for which drag-chain and flight conveyors are used, ranging from fine powders to heavy granular materials, such as ores, provided the maximum lump size is not more than about one-fifth of the width of the duct. The totally enclosed construction ensures that no dust nuisance is caused, and that contamination of such materials as foodstuffs cannot occur in transit. Corrosive materials or materials at high temperatures can be conveyed if due consideration is given to the materials of construction of the conveyor.

One great advantage is that the conveyor cannot be overloaded, except by jamming due to the entry of an excessively large lump, since as soon as the duct is full of material in the normal course of operation no further material can enter from the feed hopper. Thus when the conveyor is taking a feed from several hoppers in a row, the first hopper fills the conveyor which will not accept any more material from the following hoppers until the first hopper is either empty or the feed is stopped by closing its slide valve.

In the simple arrangement of a straight horizontal or slightly inclined conveyor, material is fed into the upper duct containing the returning strand of chain and is carried back a short distance towards the foot sprocket until it can drop into the conveying duct below, either in the foot sprocket casing or through a special transfer aperture between the ducts. If the material is reasonably free-flowing, it can be fed directly into the side of the conveying duct, but this alternative is only necessary if the return chain is being used for conveying another material in the reverse direction. As with drag-chain and flight conveyors, delivery of the material can take place through any aperture in the base of the duct.

Such a conveyor, inclined at less than the angle of repose of the material, will be satisfactory when the duct is only partly filled, but for conveying at steeper angles, or vertically, the design must ensure that a sufficient horizontal length exists before the rise for a solid seal of material to be created. This will occur in any case at maximum capacity, but at lower capacities material

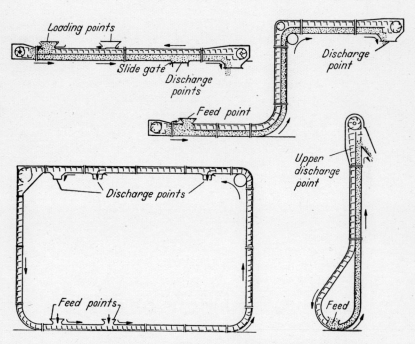

Figure 11. Typical runs of en masse conveyors with various combinations of feed and discharge methods. (By courtesy of Redler Conveyors Ltd.)

in the partly filled rising duct will fall back until sufficient accumulates at the beginning of the rise to form this seal. The net effect is that the rising duct is full of material which is moving at a lower speed than the material in the partly filled horizontal duct. It also means that the tail end of a batch of material will not be conveyed when there is too little material remaining to fill the rising duct. In certain cases where a different material is then to be conveyed, or where the material will deteriorate on standing, the duct will have to be cleared out either manually through an aperture at the lower end of the rising duct, or by incorporating a few special flights of larger cross section which are able to carry a small amount of the material, thus slowly removing the remainder of the batch of material.

A direct vertical lift from a feed point can be achieved by forming the lower ends of the duct into a loop, guiding the chain, instead of passing the chain over a sprocket, and feeding material into the top of the duct at its lowest point. This gives sufficient sealing to ensure conveying, except when conveying fine aerated materials which need to be fed in tangentially near the bottom of the down-leg. Some of the various combinations of feed and discharge methods are shown in *Figure 11*.

A neat design of *en masse* conveyor is made up of a roller chain similar to the type used for power transmission with flights attached so that they can pivot. Thus the flights stand out from the chain along the conveying strand, but along the returning strand they can be made to lie flat, considerably reducing the overall cross section of the conveyor. The maximum speed of these conveyors is about 100 ft./min, but it should be borne in mind that when using highly abrasive materials a lower speed will considerably reduce maintenance costs at the expense of a rather larger cross section of conveyor (and therefore capital cost) for the same capacity. Power requirements and chain tension are calculated on the same principles as described for drag-chain conveyors, but using a considerably larger value for the coefficient of friction to allow for the tendency for material to press against the sides of the duct as well as for its weight on the base of the duct.

A modern design which has distinct advantages in the chemical engineering field has been developed by Hapman in the U.S.A. and is now available in this country. The flights are circular and practically seal the pipe through which they pass (*Figure 12*). The links between the flights can articulate in

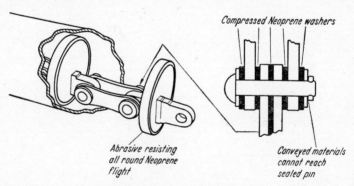

Figure 12. Circular duct and flight of the Hapman conveyor. (By courtesy of Fisher & Ludlow Ltd.)

all directions and the smooth-bored pipe is sufficient guidance for the flights round all but the sharpest bends. Material can thus be conveyed from one process vessel to another even though they may be at slightly different pressures or contain different gaseous atmospheres, and heating or cooling in transit can be provided by suitably jacketing the conveying pipe.

Vibrating Conveyors

If a tray loaded with loose material (either powdered, granular or composed of discrete objects) is moved to and fro in its own plane, the material will move with the tray if the force required to accelerate the material at the same rate as the tray is less than its frictional force with the surface of the tray. If the acceleration is too great, or the frictional force too small, the tray will slide under the material. The principle involved in vibrating conveyors is to increase the frictional force during the motion of the tray in one direction while reducing it on the return movement so that there is a net average force on the material in a definite direction along the tray. This is most conveniently done by tilting the plane of movement relative to the plane of the tray so that there is an acceleration component up and down normal to the surface of the tray. Upward acceleration therefore increases the frictional force sufficiently for the simultaneous horizontal movement to propel the material, while downward acceleration on the reverse stroke reduces the frictional force and the tray slides back under the material which still retains much of the forward momentum gained during the previous stroke. The average velocity of the material is thus nearly equal to the peak horizontal component of the velocity of the tray.

In practice the tray takes the form of a trough or tube supported by a system of parallel links which are inclined at 20° to 30° from the vertical, so that as the trough moves forward the inclined links also give it a small upward movement. The vibrating or reciprocating motion can be given in several ways, for example:

(*a*) motor driven eccentric and connecting rod,
(*b*) motor driven shaft with out-of-balance weights,
(*c*) magnetic or pneumatic vibrator.

The forces involved in accelerating the trough are large and it is a great advantage if they can be balanced out so that the reaction is not transmitted to the supporting structure and foundations. One way of doing this is to vibrate a counter-balance weight so that its momentum is always equal and opposite in direction to the momentum of the trough by means of a second eccentric and connecting rod. The forces can also be balanced out by powerful springs, and in this case the trough can only be vibrated at one particular frequency—the resonant frequency to which the driving unit must synchronize. Many vibrating units of the rotating out-of-balance weight, magnetic and pneumatic types provide a force which is in fact a reaction against a counterweight so that the elimination of other forces against the supporting structure does not arise. However, a weight on a single rotating shaft has an unwanted reaction component at right angles to that desired, which must either be absorbed by a suitable resilient mounting or balanced out by a similar weight on a contra-rotating shaft as in a 'Juby drive'.

Since the supporting links only move slightly with the vibration of the trough, their pivots can be rubber bushes with the advantage of requiring practically no maintenance. The links can take the form of leaf springs for a resonant frequency conveyor, although in this case the forces on the foundations are not balanced out.

Vibrating conveyors have numerous applications for processing operations besides straightforward conveying. With suitable construction they can handle corrosive or hot materials or pass through high temperature regions, and the vibration itself can be used for screening in transit or for such operations as shaking moulding sand out of castings. A tubular type or covered trough can be made completely dustless with flexible feed-in and delivery attachments. *Figure 13* shows a natural frequency conveyor with motor drive and balance weight to neutralize undesirable external forces.

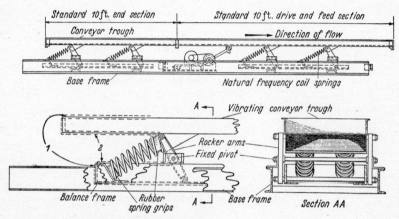

Figure 13. Natural frequency vibrating conveyor. (*By courtesy of* Redler Conveyors Ltd.)

Pneumatic Conveying

The conveyance of materials along a pipe in a moving air stream has been a practical proposition for a good many years. First developed for the unloading of grain from ships, the principle was soon used for handling coal, and in recent years has become of commercial importance in handling such diverse materials as flue dust, ashes, lime, detergent powders, plastics and many powdered and granular chemicals. Power requirements are high, but the convenience, flexibility, absence of dust and small amount of labour necessary are strongly in its favour. With a simple plant sucking grain out of ships' holds the power required may be less than 1 kWh/ton, but long pipelines conveying heavy minerals will require several times this amount. The power depends markedly on the air velocities used, and it is still an art rather than a science to choose economically low velocities for pick-up of the material and its conveying without running the risk of the material settling in the pipe and causing a choke. Some materials are also highly abrasive, and excessive air velocities lead to a short life for bends in the pipeline.

Pneumatic conveyors may be broadly classified into the two groups of high or low concentration of material in the air stream. The high concentration type is characterized by large conveying rates, high pressure gradients in the pipeline, and requires several pounds per square inch pressure or suction, provided either by positive displacement air compressors, or multi-stage turbo-compressors. With the low concentration type, the pressures required are small enough to be adequately provided by fans. The largest plants at present in use in this country can convey at the rate of about 300 tons/h, while the smallest conveying rates are those of dust collecting systems such as are used for exhausting swarf and chippings from machine tools, dust created by certain types of mechanical conveyors, and sweepings in mills and silos, not forgetting portable vacuum cleaners, both industrial and domestic.

There is no fundamental difference between sucking and blowing material. The choice is largely a matter dictated by the duty required of the plant. It is naturally convenient to suck material from a heap on the ground or in a ship's hold since only one pipe is involved and no complicated device is required at the intake nozzle. It is also more practical to suck material from various points to a central collecting point where the necessary mechanical equipment (pumps, separating and discharging devices) can be concentrated. Blowing, on the other hand, is suited to distribution of material from one point to a row of bins, for example. Combined systems are used with great effect in mobile plants (*e.g.* mounted on a road or rail truck) where portable suction and blowing pipelines can be quickly rigged up according to requirements. Closed circuit systems, in which the conveying air, or gas, is piped from the delivery point back to the intake point, are used in special circumstances where noxious fumes or dust created by the material would be difficult or uneconomic to remove, or where the conveying medium is a gas or specially conditioned air which must be conserved.

The Pressure Drop in the System—The pressure difference in the conveying pipe between the point of entry of the material and its point of discharge (either the blowing nozzle or the receiver in which the material can settle out of the airstream) is made up of a number of components. Some of these can be readily calculated, but others can at present be estimated only by empirical methods.

The first important contribution to the pressure drop is that due to the velocity of the air alone in the pipe. Its value depends basically on the Reynolds number R_e given by

$$R_e = \varrho V D / \eta$$

where ϱ is the density of the gas
V the velocity
D the diameter of the pipe
and η the viscosity of the gas.

According to the roughness of the pipe and the Reynolds number, a friction factor f has been determined experimentally giving the pressure drop from the formula

$$\Delta P_1 = f \varrho L V^2 / 2gD$$

in pounds per square foot where the length L and other quantities are in feet, with the value 32·2 ft./sec^2 for g.

Table 7 gives values of f for various sizes of commercial steel pipes at several velocities. A discussion of the data available for determining the air pressure drop has been published by WRIGHT[19], including a chart for reading off the values directly in the case of dust collecting systems where the air pressure is always close to atmospheric.

Table 7. Values of the Friction Factor, f, for Air

Equivalent air velocity at 30 in. Hg abs *and* 60° F		Pipe diameter (in.)				
ft./min	ft./sec	4	6	8	10	12
2,000	33·3	0·0215	0·0195	0·0185	0·0175	0·0168
4,000	66·7	0·0195	0·0176	0·0166	0·0159	0·0150
6,000	100	0·0185	0·0170	0·0160	0·0152	0·0145
8,000	133	0·0180	0·0165	0·0155	0·0148	0·0143
10,000	167	0·0178	0·0162	0·0151	0·0146	0·0140

The energy given to the air at the intake nozzle and after turbulence in bends must also be included. This contributes a pressure drop $\Delta P_2 = \rho V^2/2g$ at the intake and about half this value for each right-angled bend.

The energy given to the material being conveyed is a significant contribution to the pressure drop. This is similarly $\Delta P_3 = \rho_1 V_p^2/2g$ where V_p is the assumed particle velocity in the pipe and is generally about two-thirds to three-quarters of the air velocity. The density ρ_1 is now the density of the fluidized material so that $\rho_1 = 4M/\pi D^2 V_p$ and $\Delta P_3 = 2MV_p/\pi D^2 g$ where M is the rate of conveying in pounds per second. At each bend there is a further loss of energy corresponding to about half this value which must be regained in the following straight pipe.

In a rising pipe there is a contribution due to increase of potential energy; the extra pressure drop, being equivalent to the hydrostatic head of a fluid of density ρ_1 of depth h, is therefore

$$\Delta P_4 = \rho_1 h = 4Mh/\pi D^2 V_p$$

The final contribution to the pressure drop is the friction of the material on the walls of the pipe and has been the subject of a number of experiments, together with rather unsuccessful attempts to correlate the effect of variations in velocity, pipe diameter, particle size and density. The earliest experiments were made by CRAMP and PRIESTLEY[20], and by SEGLER[21], using wheat. GASTERSTADT[22], using wheat in a horizontal pipe correlated the ratio of conveying pressure drop to air only pressure drop with material-air ratio and found straight lines, of the form

$$\Delta P_{\text{total}}/\Delta P_{\text{air}} = 1 + aM/Q$$

where M/Q is the weight ratio of the conveying rates of material to air. The factor a varied from about 0·3 to 0·5 with air velocity, tending to the lower value of 0·3 for air velocities above 50 ft./sec. This simple rule is accurate enough for calculating the pressure drop due to friction in low concentration conveying systems, and in short high concentration systems where the friction component is a relatively small part of the total pressure drop.

As an example let us calculate the pressure drop in a vertical pipe of 5 in. diameter rising 26 ft. from the intake nozzle in a pile of wheat. At an intake air velocity of 100 ft./sec and conveying rate of 40 tons/h Q is 1·0 lb./sec and M is 25 lb./sec. The calculated contributions to the pressure drop are therefore:

air friction $\Delta P_1 = 0·18$ in. mercury column (in. Hg)
air entry loss $\Delta P_2 = 0·17$ in. Hg

We assume a velocity of 60 ft./sec for the rising material (rather slower than in a horizontal pipe):

material acceleration $\Delta P_3 = 2·4$ in. Hg
material static head $\Delta P_4 = 1·1$ in. Hg

Assuming Gasterstadt's factor a to be 0·5 (rather than 0·3, as the writer considers the friction is increased due to the greater relative velocity between air and grain in a rising pipe, contrary to the opinion of some engineers) then:

material friction $\Delta P = 0·18 \times 0·5 \times 25/1·0 = 2·25$ in. Hg

The total pressure drop is therefore 6·1 in. Hg.

An experimental measurement carried out by the writer under these conditions gave 9·8 in. Hg.

Such an underestimate emphasizes the uncertainties involved. The static head is too small since the ultimate velocity of the particles was used, whereas the particles are actually accelerating and the average velocity would be rather less, thus increasing the density. There are other probable losses at the intake nozzle which cannot be calculated. However, this calculation is instructive in showing the relative values of the most important contributions.

Subsequent to the vertical conveying in this experiment, the wheat was conveyed a short distance horizontally into the receiver (vacuum 11 in. Hg) from which it was discharged through a seal while the air passed through a cyclone and filter to the vacuum pump (vacuum 12·5 in. Hg). The displacement of the pump is therefore about 1,600 ft.^{3}/min (assuming the volumetric efficiency to be 85–90 per cent). The figure of 40 ft.^{3}/min displacement per ton/h of material is generally considered about the minimum for a ship or barge unloading plant which consists principally of a direct lift of the material with only a short horizontal conveying path, regardless of the scale of the plant. For designing purposes, 50 ft.^{3}/min per ton/h is usually taken to allow for unavoidable leakage and to provide a margin above any guaranteed conveying rate. Vacuums greater than 14–15 in. Hg are seldom used owing to the increased size of the pumps required to handle the rarified air. The pressure drop is therefore limited, and in long pipelines the ratio of air to material must be increased to keep the pressure drop within bounds. To convey wheat 500 ft. horizontally (with some bends) a pump displacement of 150–200 ft.³/min per ton/h would be required, and dense materials such as minerals which need higher velocities in the pipeline would require even higher figures.

Blowing systems are not so restricted and, with suitable choice of compressor, pressures up to 50 lb./in.² have been used. The limitation here is rather the large change of air volume as the absolute pressure changes between nozzle and discharge point. The actual air velocity should remain approximately constant so that the pipe diameter at the nozzle may become inconveniently small when high pressures are used, leading to difficulties in feeding the material into the pipeline at a high enough rate.

Recent experimental work on the pressure drop in a conveying pipe has all been on a rather small scale and therefore difficult to relate to the size and conveying rates required in commercial equipment. In many cases the test section used was too short to allow adequate acceleration. BELDEN and KASSEL [23] conveyed two sizes of catalyst particle vertically up ½ in. and 1 in. approximate diameter pipes to obtain data for catalyst transport in a fluidized bed catalytic cracking plant. They showed that vertical conveying without choking is possible with gas velocities only slightly greater than the terminal velocity of free fall of the particles, but, of course, under such conditions the static head of material is very high. VOGT and WHITE [24] conveyed a number of materials (sands, steel shot, clover seed and wheat) both horizontally and vertically in a ½ in. diameter pipe and correlated

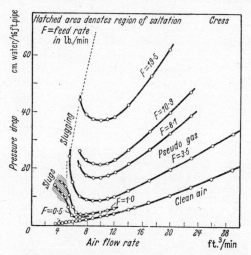

Figure 14. Experimental data on cress seed in 1 in. pipe. (*From* R. H. CLARK et al.[27], *by courtesy of* The Institution of Chemical Engineers)

their data by empirical graphical functions. Belden and Kassel point out that their data do not fit the Vogt and White correlation. PINKUS [25] conveyed two sizes of sand horizontally in a 1 in. pipe and attempted to correlate the data in terms of a friction factor (as used above for air pressure drop calculation). He also reported his data in terms of the Gasterstadt correlation, obtaining slightly curved lines with average values of a of 1·4 and 1·7 for small and large sand respectively.

HARIU and MOLSTAD[26] conveyed various sized sands in vertical pipes of $\frac{1}{4}$ in. and $\frac{1}{2}$ in. approximate diameter, and correlated their data with a similar friction factor. They calculated the length required for acceleration and found that a large part of their apparent friction was due to acceleration in the test section. CLARK, CHARLES, RICHARDSON and NEWITT[27] have reported extensive tests in a 1 in. diameter horizontal pipe and have made a largely empirical correlation of their data.

The general form of the pressure drop as a function of air velocity and conveying rate for cress seed is shown in *Figure 14*, taken from the paper by Clark *et al.* In vertical conveying, the static head which decreases with increasing velocity emphasizes the minimum in the curves. The aim of any practical conveyor is to use the velocity corresponding to the minimum pressure drop in the curves provided there is no danger of choking.

Practical Pneumatic Conveyors—The correct estimation of the pressure drop is only one factor of a pneumatic conveyor. The design of the associated equipment can make all the difference between a trouble-free conveyor and one which requires constant adjustment to ensure steady conveying or which requires excessive maintenance.

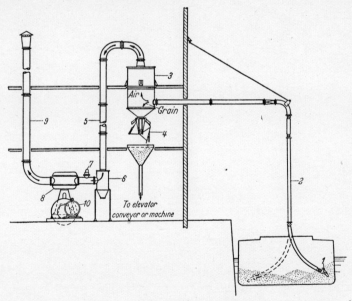

Figure 15. Arrangement of simple pneumatic suction conveyor. (By courtesy of Simon Handling Engineers Ltd.)

The basic features of a suction conveying plant are shown in *Figure 15*, and the following paragraphs comment on the details of such a plant and the analogous features present in a blowing plant.

(i) *Nozzles.* The essential feature is a means to regulate the rate of material intake so that the designed maximum material–air ratio cannot be exceeded, hence avoiding chokes. A suction nozzle to take material from a pile consists of a central conveying pipe with its end partially shrouded by an outer concentric sleeve. Air passes down the annular space to mix with the material as it enters the conveying pipe, when the nozzle is working with its end buried several inches in the material. The material–air ratio is regulated by adjustment of the projection of the sleeve to the end of the conveying pipe and if necessary by restricting the inflow of air at the top of the sleeve. This type of nozzle which is illustrated in *Plate X* is suitable for all materials which are not so poorly flowing that the hole created by removal of material is not immediately filled up by local collapse of the pile. Vibrators and mechanically driven agitators and prodders can be attached for dealing with very difficult materials. A curved nozzle which is more convenient for the operator to handle is used for free-flowing materials like grain. No outer sleeve is used, but the extra air necessary enters through an adjustable aperture part way up the nozzle.

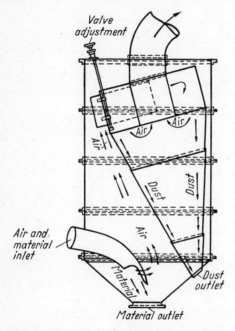

Figure 16. Receiver with internal cyclone. (*By courtesy of* Simon Handling Engineers Ltd.)

In taking material from a bunker or hopper, the material flow must be regulated by the size of the hopper outlet, controlled by a slide plate, by a special feeder or, in the case of a blowing system, the rotary seal used to prevent blow-back into the hopper can be used as a feeder. The material then drops into a longitudinal slot in the upper part of the horizontal conveying pipe. Some protection against choking can be given by an internal lip projecting into the conveying pipe along the edge of the slot to prevent a sudden surge of material completely filling the pipe.

Plate VIII. Drag chain conveyor using a single strand of chain. (By courtesy of Simon Handling Engineers Ltd.)

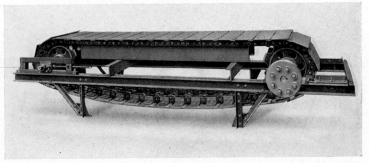

Plate IX. Apron conveyor with flat trays. (By courtesy of Simon Handling Engineers Ltd.)

Plate X. Sleeve type of intake nozzle. (By courtesy of Simon Handling Engineers Ltd.)

[To face page 424

(*ii*) *Bends.* The outer curves of bends are particularly subject to wear by abrasion of material. Either the pipeline should be designed so that bends can be easily replaced, or the outer curve should be made as a renewable wearing plate.

(*iii*) *Separation of Material and Air.* In a suction system the material is deposited into a receiver as a primary separator, while the dusty air passes through a cyclone and if necessary a filter before reaching the pump. The cyclone can in fact be inside the receiver as shown in *Figure 16*, saving space and also cost since the cyclone can be made of thinner metal as it does not have to withstand full atmospheric pressure externally. The discharge of material from the cyclone must pass through a separate seal. Some recently constructed plants for handling chemical powders omit the cyclone and fit filter units (porous ceramic or plastic tubes) in the upper part of the receiver. The filter cake is dislodged periodically by a short reverse air blow and falls out with the main bulk of material, making a compact unit and eliminating an extra seal. In blowing plants separation must take place at the delivery end of the pipeline. If this is an open area rather than a closed bunker, an excessive dust hazard may be created unless a cyclone is fitted. Such a cyclone must handle the full conveying rate of the plant and, although itself inefficient, will substantially reduce the amount of dust raised as a cloud.

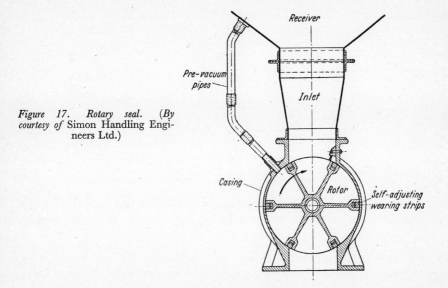

Figure 17. Rotary seal. (*By courtesy of* Simon Handling Engineers Ltd.)

(*iv*) *Discharge of Material.* The rotary seal in its many variations is in general use. As shown in *Figure 17* the rotor contains pockets which carry the material round from the receiver (under vacuum) to an outlet at atmospheric pressure. An alternative widely used in grain conveyors is the tipping seal (*Figure 18*) in which two chambers oscillate and are alternately filled and emptied. Both the rotary seal and tipping seal have air pipes to equalize the pressure in the pockets before material is allowed to enter, and both have spring-loaded bars to provide an efficient air seal against the casing. Rotary seals can be easily damaged by hard lumps or foreign bodies, whereas the tipping seal is self-clearing and springs in the driving rods absorb the drive motion during the jammed period. Rotary seals are widely used in blowing and combined suction–blowing plants at the feed-in to the blowing line.

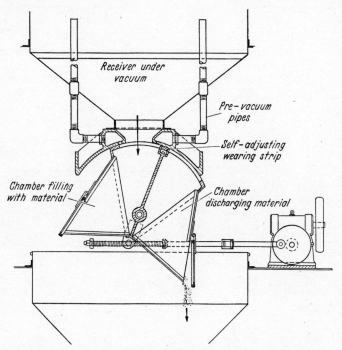

Figure 18. *Tipping seal.* (*By courtesy of* Simon Handling Engineers Ltd.)

(*v*) *Exhausters and Compressors.* Positive displacement types are used in preference to high pressure centrifugal machines since the air quantity is much less dependent on compression ratio. With suction plants the machine must be immune to the dust remaining in the air after separation of the material.

Dust Collecting Plants

In order to gather particles from an area where a dust cloud is being raised due to rough handling of a dusty material, such as at the feed or discharge of a conveyor, a hood is connected to a suction pipe so that an inward airflow induces the dust particles into the suction pipe before they have time to settle. The air velocity necessary at the mouth of the hood means that very large quantities of air are required relative to the amount of dust conveyed although the velocities are much less than those used in high concentration pneumatic conveyors. With such a low material–air ratio the increase in overall pressure drop due to the pressure of material is small and Gasterstadt's factor can be used with more confidence. However, heavy steel piping is not necessary for such light duty conveying and the pipe work is usually fabricated from light gauge steel sheet with lapped joints. To account for the greater apparent roughness, friction factors of about one-and-a-half times those given in *Table 6* should be used in calculating the pressure drop.

Several hoods can be connected to the same main conveying line provided the pressure drops in each branch pipe are matched and that at the same time the conveying air velocity is maintained. A typical application of a dust-collecting hood is shown in *Plate V* at the feed-on to a belt conveyor.

The overall pressure drop through the system is usually so small that a single stage centrifugal fan is adequate for operation. It is preferable, of course, to separate the dust from the air with a cyclone or filter before the air passes through the fan, but this requires an airtight vessel to contain the separated dust until it is convenient to close down the system for emptying, or else a mechanically driven seal is necessary to remove the dust continuously from the cyclone outlet. An alternative often used on lightly loaded systems is to allow the dust to pass through the fan and to be separated in a cyclone in the exhaust. The discharge from the cyclone is then at atmospheric pressure. A fan for this duty must be made specially robust to withstand the abrasion and it is usually of the paddle-bladed type. Dust collecting and ventilating systems are fully discussed in treatises by ALDEN[28] and by MADISON[29].

HYDRAULIC CONVEYING

For many years the hydraulic transport of solids in pipes has been successfully applied in a few specialized fields such as dredging, land reclamation, and the winning of some minerals. One important installation is that at Copper Cliff, Ontario, where 4 million tons of tailings a year are transported by pipeline over a distance of $7\frac{1}{2}$ miles. Another pipeline $5\frac{1}{2}$ miles long is being used to transport 1 million tons of fine coal a year to a power station at Carling in France. This pipe is of 15 in. diameter and is fed by five 400 h.p. centrifugal pumps. Even for comparatively short distances, however, the power requirements are large, especially when large solids are to be moved: as an example, in the construction of the Beauharnois Power Canal in Canada one 1,500 h.p., 18 in. pump was used to deliver 120 tons/h of rock crushed to 5 in., through 2,000 ft.

Existing installations for hydraulic transport utilize special centrifugal pumps whose impellers are capable of passing the largest pieces likely to occur in the mixture. Sometimes the solids and water in the desired proportions are fed to a hopper connected to the suction branch of the pump. The pump then delivers the mixture into the pipeline at a pressure sufficient to overcome the friction resistance and any static head. Where the solids are to be recovered, screening or other separating plant is used at the delivery of the pipeline. In suction dredging, prior mixing before the pump is not possible and it is necessary to rely on the skill of the operator to keep the suction mouthpiece close enough to the river bed to entrain a large quantity of material without starving the pump of water.

There is considerable interest today in the possibilities of hydraulic transport in other fields, such as the moving of coal in bulk. Earth and debris-moving plants are often required on temporary sites and the simplicity of hydraulic transport is then the major factor in determining its widespread use. For permanent installations, however, such as the feeding of coal to a power station, reliability and economic factors are relatively more important. For this reason more precise knowledge of the mechanism of the flow in pipes is needed so that reliable estimates can be made of the power consumption and the size and cost of the plant required for major projects. Large-scale laboratory and plant studies are now in progress to provide the basic information on plant operation and the design of new machinery.

The outstanding characteristics of hydraulic transport systems are the large power required and the high capacity of pipes of quite modest sizes. For instance, the pressure drop along a 12 in. pipe conveying coal may exceed 300 ft./mile, but the capacity is of the order of 250–300 tons/h of 3 in. or 4 in. coal. The compactness of the conduit gives hydraulic transport great advantages for conveying solids in confined spaces or where there are obstructions making it impossible to route directly. The ease of hydraulic 'loading' and 'unloading' is also important.

Permanent pipelines working with reasonable prospect of safety must be carefully designed and controlled to ensure stability in operation. To a large extent this depends on the combined characteristics of the pump and the pipeline and on the method of control of water and solids feed. Automatic control by instruments measuring pressure, velocity or concentration seems desirable for large-scale installations working continuously. It seems clear that with soft solids such as coal, a reasonable life of machinery and pipes can be expected, but with hard materials like sand, abrasion is severe and fairly frequent replacement of parts may be necessary. The fundamental problem in hydraulic conveying is the evaluation of the forces exerted on the individual solid particles by the fluid, and the relation of these forces to the actual particle behaviour and the resultant energy losses. These forces are a consequence of the motion of the fluid surrounding each particle and arise, in particular, from fluctuations in velocity due to turbulence. The turbulent eddies are constantly lifting the particles into the fluid stream. The problem is further complicated since the presence of suspended particles affects the fluid turbulence.

There is a distinction between homogeneous and heterogeneous transport. The former is characteristic of suspensions of particles of diameter less than about 30 μ; the suspension flows like an homogeneous fluid, but begins to settle under laminar flow conditions. The critical value of the Reynolds number for the transition from streamline to turbulent flow is the same as for liquids, provided that the density and viscosity terms relate to the suspension. It is then generally possible to use the normal friction chart to calculate head losses for turbulent flow through pipes.

Heterogeneous transport is characteristic of coarser particles. At high velocities it may take the form of fully suspended flow, though the concentration in a vertical plane is not uniform; the particles travel with a velocity only slightly less than that of the liquid. At lower velocities particles tend to collect at the bottom of the pipe and form a stationary bed. There are well defined intermediate regions, in which movement of particles at the interface occurs (saltation), or in which the bed as a whole may slide forward at a much lower velocity than the liquid.

Experimental Work and Correlations

By far the most extensive experimental work in hydraulic conveying has been carried out in France by DURAND and his co-workers[32, 33]. They used particles up to 1 in. diameter in concentrations up to 22 per cent in horizontal and vertical pipes varying in diameter from $1\frac{1}{2}$ to 22 in. They examined the effect of the roughness of the pipe and the elasticity of the solid particles and developed methods for measuring spatial concentration and distribution of particles in the pipe. Durand classified the solid phase as follows:

(a) Materials which give suspensions behaving as homogeneous fluids. These include clay and fine ash, with particle sizes up to about 25 μ.
(b) Intermediate materials, including silt, with particles in the size ranging from 25 to 50 μ.
(c) Materials giving rise to heterogeneous suspensions where the concentration is no longer uniform over the cross section. They found that this group could be subdivided into fine particles (50 μ to 0·2 mm) which obey Stokes' law when settling in a fluid, a transition range (0·2 to 2 mm), and coarse particles which obey Newton's law of settling.

They suggested that, as this classification which relates only to graded material corresponds to that normally made for particles settling in the gravitational field, the velocity of the fluid relative to the particles in hydraulic conveying is approximately equal to the terminal falling velocity of the particles.

Durand *et al.*[32] found that all their results could be correlated with reasonable accuracy by the expression:

$$\frac{i-i_w}{Ci_w} = 121\left\{\frac{gD(s-1)}{V^2} \cdot \frac{W}{\sqrt{[gd(s-1)]}}\right\}^{1\cdot 5}$$

where i, i_w are the hydraulic gradients due to the suspension and pure water;
C is the volumetric concentration;
d, D are the diameters of the particle and the pipe;
V is the fluid velocity;
W is the terminal falling velocity of a particle in water;
g is the acceleration due to gravity;
and s is the ratio of the density of the solids to that of the liquid.
(This term was introduced by WORSTER[34] to enable the results of materials of different densities to be correlated.)

WORSTER and DENNY[30] have made experiments with 3 in., 4 in. and 6 in. pipelines, using coal and gravel, and have published the simple empirical correlation reproduced in *Figure 19*. The abscissa $i_w/(s-1)$ can be written as $fV^2/2Dg(s-1)$, where f is the ordinary friction coefficient (not varying greatly) and the correlation is therefore closely related to Durand's expression given above.

NEWITT et al.[31] have made extensive experiments with a 1 in. pipeline using materials with a range of specific gravities from 1·18 (Perspex) to 4·1 (manganese dioxide) and size range from 240 mesh sand to $\frac{3}{16}$ in. coal and gravel. With many of these materials they were able to show the transition between homogeneous flow of fully suspended particles (at the higher velocities) and heterogeneous flow with the particles tending to settle out (at lower velocities).

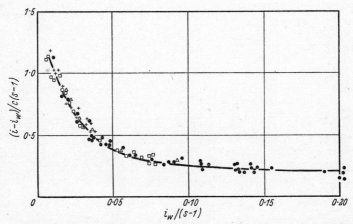

Figure 19. Curve of $(i-i_w)/C(s-1)$ plotted against $i_w(s-1)$. (From R. C. WORSTER and D. F. DENNY, by courtesy of The Institution of Mechanical Engineers)

Practical Considerations

All the experimental data indicate that there is a minimum value of total pressure drop in the pipeline, depending on the concentration, as the velocity is varied. From the practical point of view, therefore, the velocity

Table 8. *Minimum Velocities for Reasonable Concentrations of Coal or Gravel*

Pipe diameter in.	Minimum speed ft./sec	
	Gravel	Coal
1	3	1½
3	7	3½
6	10	5
9	13	6¼
12	15	7¼
18	17½	8¾

corresponding to the minimum pressure drop will be the most economical, provided there is no danger of choking. The velocities that correspond to the minimum in the friction curves increase as the square root of pipe diameter and of the density under water $(s-1)$. Approximate values are given in *Table 8* for reasonable concentrations of 20 or 25 per cent of large coal or gravel.

In an emergency it is possible for tightly packed solids to move along a parallel straight pipe, but in flowing around a pipe bend the packed bed must shear internally. In doing this there is a dilatation and the solids are very likely to seize and stop the flow. It is usually at a bend or restriction that a pipeline carrying solids will choke and it may be desirable to provide means of emptying the pipe at such danger spots. The maximum size of solids that can be passed through a pipe is not easy to fix. At low concentrations and high velocities single lumps almost as big as the pipe will even pass safely around pipe bends. It seems that the essential requirement at moderate speeds is that the sliding bed of solids should leave sufficient free space above it for the largest particles to turn over freely. Clearly, therefore, the maximum size of solids being transported depends, at these speeds, on the concentration in the pipe. For instance, in a 6 in. horizontal pipe a rather low speed for handling coal is about 5 ft./sec. At this speed the concentration present in the pipe is about 28 per cent when the delivered concentration is 20 per cent. Since the voids of a closely packed bed of coal are about 50 per cent, the coal bed occupies a depth of slightly more than half the pipe diameter, leaving only 3 in. of clear space above it. In this case the maximum safe size of coal would be about 2 in. At higher concentrations or lower speeds this would be reduced. The rule that the maximum size of solids is limited to one-third of the pipe diameter is only applicable at moderate concentrations if the speeds are low.

Methods of Pumping Solids

Low-pressure Systems—The simplest method of pumping solids is to mix them with the desired amount of water in a tank and then pass them through a centrifugal pump, the passages of which are wide enough to take the largest particles. If the pump speeds and solids feed rate are constant there is no difficulty in maintaining a uniform flow in the pipe. This method is widely used for conveying fine material, and one notable scheme is that at Carling, where five centrifugal pumps in series convey colliery tailings to a power station over a distance of 5½ miles (GILBRAT and CHENIN[35]).

It is usual with solids pumps to pressure-grease the internal bearings and to flush the glands continuously with clean water, but the real problem is the erosion of the volute and impeller and it is this factor which restricts their use. For instance, when handling solids the pump speed is limited and the maximum head available is about 150 ft./stage, so that a single pump is sufficient only for comparatively short distance transport unless staging is employed. Even with fine coal, erosion of high-speed pumps may be serious.

High-pressure Systems—Long distance horizontal transport or vertical lifts from deep mines necessitate a very high pressure at the entry to the pipe system, or boosters along its length. This can be achieved by any of the following methods.

1. The use of several centrifugal pumps in series.
2. The solids and water may be passed together through a positive-displacement pump. The problem of valves for such a pump is a serious one, for their lift or opening must be sufficient to allow the largest particles to pass freely through, and they are also subject to damage by the particles on the sealing surfaces. So far as is known this method has not been used for solids larger than $\frac{1}{8}$ in., but there is the possibility of designing a special valve for a pump to handle much larger solids than this with an efficiency better than that of the corresponding centrifugal pump.
3. The solids may be fed mechanically into the pipeline after the water has been raised to the necessary pressure. Several designs of feeding machine have been tested and these will be more fully described later. It is the essence of all such machines that they release water to atmospheric pressure at the same volume rate as the solids feed, and this, together with leakage water, must be pumped back into the pipe. The pressure of the system at this point is at its highest and the repumping of the displaced water which may be heavily laden with fine material presents a problem. Design of suitable centrifugal pumps for this duty may be difficult and, in many instances, the size and cost of cleaning plant for this purpose would be prohibitive. A positive-displacement pump may often be a better solution.

System Cycles—Hydraulic transport systems may be divided into two categories, closed cycle and open cycle, according to whether the liquid conveying the solids is recirculated and used again or not. Open cycles are applicable only where there is ample supply of liquid near the inlet end of the pipeline, and means of disposing of the liquid at the discharge end of the pipe. The conveying liquid will be contaminated by fine particles and, in an open cycle system, it may be difficult to dispose of, or uneconomic to clean the liquid. A typical application of the open cycle is the lifting of coal from a mine where water is available at the pit bottom.

In the closed cycle the water used to convey the solids is returned after separation at the discharge end and used again. In this way, both the provision of a continuous supply of water and its subsequent disposal after contamination are avoided. However, the fines that would otherwise accumulate in the system must be withdrawn at a rate sufficient to keep their concentration at an acceptable level. A closed cycle would be essential for lifting coal or other mineral from a dry mine, and for systems using liquid media other than water.

Types of Solids-feeding Machine—Feeders may be divided into two categories, those in which the only moving parts of the machine are valves, and those in which the moving parts themselves convey the solids from a region of low pressure (hopper) to a region of high pressure (delivery line). In practice, however, the problems of both types are almost identical. Except for the lock-hopper feeder, machines for feeding solids other than coal into high-pressure pipelines have not been developed.

One design of the first type, a lock-hopper feeder made by a commercial company for the Ministry of Fuel and Power, consists of two pressure vessels each communicating alternately with a hopper and a discharge vessel as shown in *Figure 20*. As an indication of size, the outer vessel is 8 ft. in diameter and 15 ft. high; the main valves are elliptical and measure 24 in. by 18 in. The throughput of the machine will be determined, not so much by the time cycle of events, but rather by the rate at which coal can fall through the valves against the up-flowing displaced water. With 3 in. coal the particle velocity through the valves is unlikely to exceed 1 ft./sec, which corresponds to a throughput of about 100 tons/h.

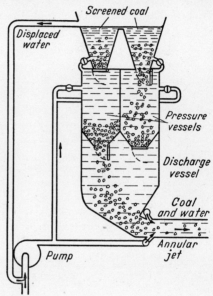

Figure 20. Lock-hopper feeder for coal. (From R. C. WORSTER and D. F. DENNY, by courtesy of The Institution of Mechanical Engineers)

The main point in favour of the lock-hopper feeder is that the leakage from the pipe to the hopper can be very small. This is a very important factor in the efficiency of the transport system. The valves move only after the pressures on either side are balanced, so that the power needed is small and erosion of the surfaces is minimized. Very large vessels are unsuitable for high pressures, and it may be necessary to use small vessels with a correspondingly short time cycle. The control system is complex, there being four coal valves and at least four water valves to be operated in their correct sequence.

Reciprocating Feeders—In the double-acting reciprocating feeder a plunger containing two rectangular pockets is reciprocated so that at each end of the stroke one pocket lies beneath a hopper and is filled with coal, whilst the other pocket falls opposite the pipeline into which its load of coal is discharged. The delivery from a single machine of this design is interrupted for a short time during each stroke because of the lap necessary to restrict axial leakage from the pressure pipe to the hopper. This interruption can be avoided either by allowing a continuous bypass of water round the feeder, which lowers the delivered concentration, or by intermittent bypass of water through radial holes in the plunger.

Rotary Feeders—From a design standpoint, continuous motion is advantageous because the drive is so much simpler and the inertia of the moving parts is less than with the corresponding reciprocating machine. To minimize inertia is important in regard to jamming of the machine by metal objects that may fall into the hopper. On the other hand, from a hydraulic point of view, the pockets have less chance to fill and empty when moving continuously than when reversing at each end of a reciprocating stroke. The choice of arrangement of rotary feeders is wide, though in practice many designs cannot be considered for high-pressure applications on account of their inherently high leakage.

Degradation of Coal Travelling along Pipes—Most present-day methods of moving coal do, to a certain extent, break it up. In hydraulic transport, the sliding of the layer of coal along the bottom of the pipe must cause some damage both to the coal and to the pipe; in particular, at bends or bad joints the whole layer of coal suddenly turns over. The top layer of coal moves along in a hopping motion at a speed which is little smaller than that of the water, while the sliding layer underneath may be moving at only a half or a third of this speed. Impacts must take place between coal moving at different speeds and between coal and the pipe walls. These shocks not only abrade the large coal and reduce its market value but also produce very fine particles which then create an effluent problem.

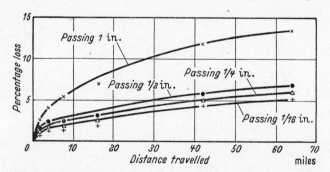

Figure 21. Degradation of 1 in. coal in hydraulic transport at 9·1 ft./sec. (*From* R. C. WORSTER *and* D. F. DENNY, *by courtesy of* The Institution of Mechanical Engineers)

In hydraulic transport installations the coal would usually pass once through a pump or feeder and then travel for perhaps several miles along the pipeline. It may be expected, therefore, that the damage to the coal done in the pipeline would predominate over other damage except for very short pipe lengths. In laboratory installations the length of the pipe is limited to 100 or 200 ft., so that a large proportion of the damage takes place in the pump and feeder and it is impossible to separate the effect due to the pipeline. Recirculating the coal round an experimental installation does not help in this problem because of the many passes then made through the pump or feeder.

Worster and Denny[30] have measured the degradation of coal experimentally in a toroidal pipe arranged in a vertical plane on the periphery of a large wheel. This is about three-quarters filled with water containing an appropriate concentration of coal. When rotated, the coal and water remain approximately stationary in the moving pipe and the relative motions are similar to those arising in ordinary hydraulic conveying. The pipe was of square section with a Perspex face.

Several series of tests on degradation of 1 in. coal were made with this machine at speeds of 7·5, 9·1 and 11 ft./sec for periods corresponding to distances up to 70 miles. The coal samples to be tested were put in the machine, run for a specified distance, and then removed: after drying, they were screened and weighed. A particular set of results is shown in *Figure 21*, from which it will be seen that the greater part of the attrition products are well below $\frac{1}{4}$ in. The degradation of this particular coal is tolerable—about 1 per cent loss of weight in the first mile followed by a gradually diminishing rate of loss with increasing distance travelled.

If degradation is defined as the attrition products below $\frac{1}{4}$ in. it has been found that, very roughly, at each speed the total loss increases in about the same proportion as the square root of the distance travelled. This diminishing rate is reasonable because as the coal becomes more rounded there are fewer corners to be knocked off.

Over a range of speed from 7·5 to 11 ft./sec the loss rate increased approximately as the cube of the speed. Within this range of conditions, and for this particular coal, the degradation seems to be given approximately by:

$$\text{Total loss} \propto (\text{speed})^3 \times (\text{distance})^{\frac{1}{2}}$$

This speed relation suggests that most of the damage is not done by the dragging of the bottom layer along the pipe but by impacts between 'flying' pieces of coal and perhaps also between flying coal and the side walls. This is another important reason for running a coal pipeline at the lowest safe speed.

It was observed in the ring-pipe test apparatus that when the doors in the front Perspex face of the pipe were not replaced exactly flush (causing a recess about 0·015 in. deep) the damage to the coal was greatly accelerated. This sort of irregularity is likely to happen at each pipe joint and it may be desirable, in a long pipeline carrying large coal, to use special pipe joints.

Degradation by Machinery—It is more difficult to estimate the damage done by other hydraulic devices such as centrifugal pumps or feeders. Centrifugal pumps used in the laboratory to circulate coal–water mixtures around pipe systems are presumed to be responsible for the greater part of the degradation occurring in these tests. In two or three hours' testing with a pump head of the order of 30–50 ft. $\frac{1}{2}$ in. coal may be reduced to about $\frac{1}{4}$ in. together with the production of a great quantity of fines. During this time the coal would have made about 200 passes through the pump so that it may be guessed that the degradation for a single pass through a centrifugal pump is of the order of $\frac{1}{4}$ or $\frac{1}{2}$ per cent for a pump working at this speed. At higher pump heads this would be greater, since the pump speed increases as the square root of head. If the relation between pump speed and coal degradation is also a cubic law it would seem that a similar pump generating a head of about 200 ft. might cause a degradation loss of about 4 per cent. Great differences between one pump and another may be expected: for instance, when a rubber-lined pump was used for the same duty as the iron one referred to above, the damage to the coal was reduced to about one-half.

Velocities in a coal feeder are low and independent of the head against which it works, so that damage caused by impact will therefore be very slight. On the other hand, there is the chopping associated with the valve action of the machine. It has been estimated that in a reciprocating machine only one piece of coal out of a hundred is broken by chopping, so the damage is quite slight.

Abrasion and Wear of Plant—In some instances where hard particles have been transported pneumatically, a very short life of only a few days has been reported for pipe bends. In hydraulic gravel-handling plant (where the velocities are about double those required for coal transport) the pipe may last a year or more if it is rotated periodically so that the wear is distributed around its periphery. Pipe bends may last four to six months before they require patching. Centrifugal-pump impellers made of abrasion-resisting steel may need replacing in a year, even though their speeds are generally low. A rubber lining to a pump or pipeline completely eliminates wear, but it is liable to be cut by large sharp flints or pieces of metal. The use of rubber lining is usually restricted to gravel no larger than about $\frac{3}{4}$ in. and precautions must be taken to keep out pieces of metal. Because coal is less sharp and less dense the limiting size of coal for rubber-lined pumps might be raised to about 2 in., but there is no experience on this point since the centrifugal pumps used in washeries are usually of low head and are of simple iron construction.

The damage to a pipeline when transporting coal will be much less than when transporting gravel. The coal pipeline at Hammersmith, which was in use for more than 10 years, appeared to have lost 0·1 in. of its wall thickness at the bottom, but it is not known how much coal was transported through this pipe or how often it was allowed to stand empty and corrode. The pipe could have been turned round into two other positions, therefore a real working life of at least 10 years may be expected for a coal-transport pipe. Pipe bends are likely to have a shorter life and there is the danger that erosion at misfitting pipe joints may be important.

Other workers (BERGERON[36]) in this field, all report a rapid increase in erosion with speed, usually as the cube of particle speed. The character of the wear depends on the angle of attack, and with sliding or scraping, the rate of wear increases almost in proportion as the particle size. It also increases with solids concentration. Naturally it also depends on the curvature of the wall and on the density and hardness and irregularity of the particles. With normal impact the phenomena are rather different, for instance, particle size is said to have little effect.

REFERENCES

[1] IMMER, J. R., *Materials Handling*, McGraw-Hill, New York, 1953
[2] HUDSON, W. G., *Conveyors and Related Equipment*, Chapman & Hall, 3rd edn, 1954
[3] 'Pallets for Materials Handling suitable for transport by road and rail', *Brit. Stand. Instn*, B.S. 2629:1955
[4] JORDAN, D. W., 'Adhesion of Dust Particles'. Paper in the Physics of Particle Size Analysis, *Brit. J. appl. Phys.*, Suppl. No. 3 (1954) S.194
[5] 'Dust in Industry', *Soc. chem. Ind., Lond.* Paper read at the Conference in Leeds. Sept. 28–30, 1948
[6] *H.M.S.O. Factory Acts 1948*, Sects 28, 47
[7] DALLA VALLE, J. M., *Micromeritics, the Technology of Fine Particles*, Pitman, 1948
[8] GRAY, W. S., *Reinforced Concrete Water Towers, Bunkers, Silos and Gantries*, Concrete Publications Ltd., London, 1953
[9] 'Steel Roller Conveyors', *Brit. Stand. Instn*, B.S. 2567:1955
[10] 'Rubber conveyor and elevator belting', *Brit. Stand. Instn*, B.S. 490:1951
[11] BEALING, E., *Mech. Handl.*, 37 (1950) 355
[12] HETZEL, F. V., and ALBRIGHT, R. K., *Belt Conveyors and Bucket Elevators*, John Wiley, 3rd edn, 1941
[13] WOODLEY, D. R., *Mech. Handl.*, 38 (1951) 61, 111
[14] WOODLEY, D. R., *Mech. Handl.*, 38 (1951) 484
[15] WOODLEY, D. R., *Mech. Handl.*, 38 (1951) 149
[16] WOODLEY, D. R., *Mech. Handl.*, 38 (1951) 324
[17] WOODLEY, D. R., *Mech. Handl.*, 38 (1951) 232
[18] WOODLEY, D. R., *Mech. Handl.*, 38 (1951) 392
[19] WRIGHT, D. K., *Heat. Pip. Air Condit.*, Oct.–Nov. (1945) 577
[20] CRAMP, W., and PRIESTLEY, A., *The Engineer*, 137 (1924) 34, 64, 89, 112
[21] SEGLER, G., 'Pneumatic Grain Conveying', *Nat. Inst. Agric. Engng*, 1951
[22] GASTERSTADT, H., *Z. Ver. dtsch. Ing.*, 68 (1924) 617
[23] BELDEN, D. H., and KASSEL, L. S., *Industr. Engng Chem.*, 41 (1949) 1174
[24] VOGT, E. G., and WHITE, R. R., *Industr. Engng Chem.*, 40 (1948) 1731
[25] PINKUS, O., *J. appl. Mech.*, 19 (1952) 425
[26] HARIU, O. H., and MOLSTAD, M. C., *Industr. Engng Chem.*, 41 (1949) 1148
[27] CLARK, R. H., CHARLES, D. E., RICHARDSON, J. F., and NEWITT, D. M., *Trans. Inst. chem. Engrs, Lond.*, 30 (1952) 209
[28] ALDEN, J. L., *Design of Industrial Exhaust Systems*, The Industrial Press, New York, 1948
[29] MADISON, R. D. (Ed.), *Fan Engineering*. Buffalo Forge Co., New York, 1949
[30] WORSTER, R. C., and DENNY, D. F., *Proc. Instn mech. Engrs, Lond.*, 1955 (to be published)
[31] NEWITT, D. M., RICHARDSON, J. F., ABBOTT, M., and TURTLE, R. B., *Trans. Instn chem. Engrs, Lond.*, 33 (1955) 93
[32] DURAND, R., and CONDOLIOS, E., *Compte Rendu des deuxième Journées de l'Hydraulique*, Société Hydrotechnique de France, Paris, June, 1952
[33] DURAND, R., *Proc. Minn. Hydraul. Conv.*, 1953
[34] WORSTER, R. C., *Proceedings of a Colloquium on the Hydraulic Transport of Coal in London*, 5th and 6th Nov., 1952, National Coal Board
[35] GILBRAT, R., and CHENIN, F., 'The transport and utilisation of colliery tailings at the Emile Huchet Power Station', *Instn elect. Engrs*, Pap. S 1630, Feb. 24, 1954
[36] BERGERON, P., *Compte Rendu des deuxième Journées de l'Hydraulique*, Société Hydrotechnique de France, Paris, June, 1952

15

SAMPLING, MEASURING AND GAUGING OF SOLIDS

E. J. SEBESTYEN

In very nearly every branch of industry, where material in bulk is handled and processed, it is not only desirable but almost invariably essential to ascertain the quantities of material handled, stored and used for processing or shipped from the premises to the consumers. Moreover, in many cases the rate of flow of the various materials and processes must be controlled to a limit appropriate to the requirements of the process in question and the nature of the material or materials. In addition, in many instances, not only has the rate of flow of the different materials used to be controlled, but also the respective ratio of one material to another must be maintained more or less accurately at a predetermined proportion and, sometimes it may be necessary to ensure that the whole stream of materials is stopped, should one of the materials run out or even if the pre-set ratio should be disturbed for some reason. Similarly, it may be desirable to limit the operation to a prearranged quantity (batch) automatically. In this group may be included the range of automatic sacking-off weighers and packing machines.

Obviously the manifold nature of the materials which are handled in the various chemical and food industries as well as the multitude of special requirements in their processing, has resulted in exceedingly numerous and varied designs of apparatus and machinery, which, broadly speaking, may be divided into two main groups:

(*i*) Machines and apparatus which check, register and/or control the *weight* of the materials which they handle, and

(*ii*) Machines, apparatus and devices which control, check and/or register the *volume* of materials which pass through them.

Before describing in detail and analysing the purpose of the different types of weighing and gauging (measuring) devices and machinery, another closely related subject should be briefly dealt with.

SAMPLING OF MATERIALS

This section is concerned only with those materials which are granular and more or less free flowing and which can be handled in bulk. Therefore it is proposed to examine the procedure and practice of sampling with this kind of material only and not to discuss the sampling of fluids and gases, nor that of masses of solids, such as pig-iron.

It is hardly possible to over-emphasize the importance of sampling. By the operation of sampling we understand the withdrawal of small quantities from the bulk of the material handled or stored, which quantities should be as closely representative of the whole and as similar in structure and consistency as may be practicable under the given circumstances. The object of sampling is to check, by appropriate tests, the chemical or physical properties, or both, of the material in question in order to decide its inherent value and its ultimate use. Simple as it may be thought, sampling of bulk material—unless of an exceptionally homogeneous nature and character—requires a great deal of skill and thorough knowledge of the materials, otherwise the sample withdrawn will not correspond closely enough with the average consistency and properties of the bulk. It must be borne in mind that heterogeneous materials may consist not only of particles of varying size and specific weight, but also of a widely differing chemical consistency and, consequently, of differing physical characteristics. Also, as they tend to segregate when in movement, in transit or when stored in silos and bunkers (during discharge) or in piles, very considerable 'grading', that is incongruity of the various parts and layers, may occur. Therefore it is evident that when sampling such materials the utmost care must be taken to incorporate a fair proportion of all parts.

Manual Sampling

Samples may be withdrawn as required, either from static material or material in movement, and can be done either manually or by means of mechanical sampling devices. Manual sampling of material in *movement* can be carried out, broadly speaking, in two ways:

(*a*) withdrawing, from time to time, the whole stream of material for a very short period;
(*b*) withdrawing continuously or periodically a small proportion of the material in movement.

In both cases the sample or samples obtained may be much larger than is required for testing and, therefore, it is customary to reduce it to the required size. This operation should be done in accordance with certain rules which may vary according to the type of material and local conditions, but, in general, once the principle of the method of sampling has been established it should be strictly adhered to so that uniformity in sampling may be achieved as nearly as possible.

Samples should be taken at regular intervals either by withdrawing the whole stream of material flow for a short period, or by diverting part of the stream. The latter method will yield reliable, homogeneous samples only if the material is fairly uniform in structure, *i.e.* particle size, and chemical consistency. In both cases the samples withdrawn may be far too bulky to be used directly to carry out any required tests and consequently should be reduced in size. There are a number of ways to perform this by no means simple task, all of which, unfortunately cannot be described here. However, by way of illustration, two typical methods are described for reducing a large sample to a handier size. The first method, known as 'coning and quartering', consists in taking the bulk sample, which has been withdrawn from the mass of material to be sampled, and piling it up into a heap or cone. This is subsequently flattened into a circular shape and divided into four, more or less equal, parts of which two diagonally opposite

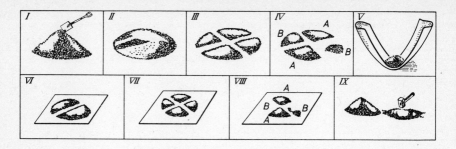

Figure 1. 'Coning and quartering' operation. (By permission from 'Chemical Engineers' Handbook', by John H. Perry. Copyright, 1950. McGraw-Hill Book Co., Inc.)

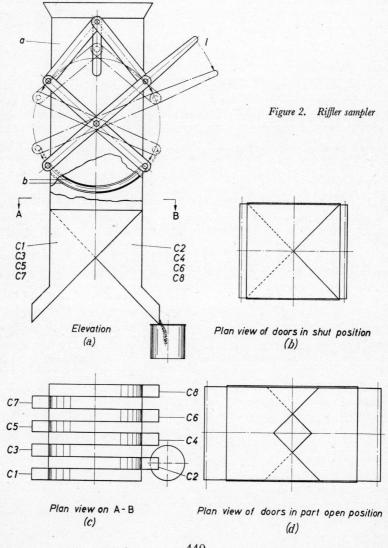

Figure 2. Riffler sampler

ones are rejected, whilst the remaining two quarters are piled up into a cone again, this time, perhaps, on a sheet of paper. This cone is then flattened out and the same procedure carried out. Should the original sample contain any large lumps, these may be broken up with a hammer, but care should be taken that the broken fragments are evenly distributed. This coning and quartering process, which is illustrated in *Figure 1* should be repeated until the sample is reduced to the size required for testing.

The second method described here makes use of a typical mechanical device which divides the sample with a view to reducing its size. For this purpose a mechanical apparatus called the Riffler Sample-Divider is used and one form of this device is shown in elevation in *Figure 2* (*a*) [shown in plan view, slides closed and open, in *Figure 2* (*b*), (*c*) and (*d*) respectively]. The apparatus [*Figure 2* (*a*)] consists of a small square bin or hopper (*a*) into which the sample to be divided is poured. The bottom of this hopper is formed by two radial slide valves (*b*) which both have a V-shaped edge. When these slides (*b*) are opened by quickly depressing lever *L*, the sample is discharged into a number of hoppered compartments [indicated by C_1 to C_8 in *Figure 2* (*c*)]; thus the original sample is divided into eight fairly equal parts of which one or two can be selected and used as the reduced sample.

It should be mentioned that the British Standards Institution as well as other Standards Institutions have standardized and described, and are recommending, certain methods of sampling for commodities such as coal, which are used in very large quantities, but there is no generally accepted standard method of sampling for the great majority of products.

Mechanical Sampling

Mechanical Withdrawal of Samples—When large quantities of a certain fairly uniform material are conveyed it is usually advantageous to carry out the task of withdrawing a sample at regular intervals by mechanical means. It is impossible to describe here all types of mechanical sampling equipment, of which a great variety is available, but it seems useful to say that whilst most mechanical samplers are designed only to withdraw material from the mass, some of these will break up large lumps and also discard the major part of the withdrawn material and retain only a representative portion for test purposes. In other cases when only one physical characteristic of the handled material has to be tested, the sampler will withdraw a certain quantity periodically, perform the testing operation, record its result, and return the sample to the main stream of material. A sampler of this type, marketed under the name of Litrograph, is shown in *Figure 3*. This is widely used in flour mills, maltings, breweries, *etc.*, and it takes samples of grain at regular intervals, measures and records their loose bulk density and then returns the samples to the bulk. It operates as follows: the main stream of grain enters at *A*, is spread evenly by a balanced gate *B* and divided by a saddle into two parts. In the path of one part is a rotating cylinder *C*, provided with a cylindrical pocket *D* of which the volume can be adjusted to contain, say, one litre of grain, by moving a piston *E* into the required position by means of an adjusting screw *F*. The pocket *D* fills with grain as the cylinder *C* rotates, and if the speed of rotation is one revolution per minute, a sample of one litre volume will be withdrawn every minute. This sample will be discharged via the hopper *G* into a weigh bucket *H* which is supported on a balance arm *J* provided with an indicator *L*. The indicator shows on the scale *M* the weight of one litre of grain and at the same time registers this weight on the recorder *O* by

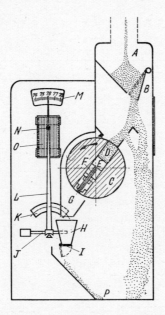

Figure 3. Litrograph

means of a pen *N*. In order to moderate the swinging movement of the indicator, a damper *K* is connected to it. Once weighing has taken place, the sample of grain is discharged from the weigh bucket through the gate *I* and joins the main stream of grain leaving the apparatus at the outlet *P*.

This apparatus has its limitations, of course, but where simple recording of the loose bulk density of the material only is required, as for example in flour mills, breweries, *etc.*, it has proved very useful indeed. On the other hand, when samples have to be tested in a laboratory or kept for further reference, another type of sampler may be mentioned as a typical apparatus. In *Figure 4* we show the Mercier sampler, which is widely used in flour mills, chemical and food processing plants, bakeries, *etc.*, for the purpose of withdrawing small samples at regular intervals and keeping these until all the samples withdrawn during a given period, say, an 8 hour shift, can be removed for examination in the laboratory. It consists of two self-contained units, the Timer and the Sampler. The latter can be fitted into any spout or duct in which the material to be sampled flows and it is housed together with the timer in a metal cabinet which can be locked to prevent interference. A sample scoop is provided which withdraws the sample from the stream of material at fixed intervals determined by the setting of the timer and deposits it in a small container. A certain number of these small containers are arranged in a turntable, which is rotated in such a way that one empty container is in position to take a sample every time the scoop withdraws one. The containers are closed by a lid and the samples are retained in the sealed cabinet until collected for examination.

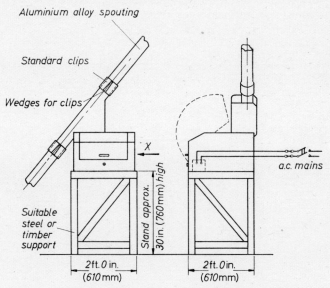

Figure 4. Mercier sampler. (By courtesy of Henry Simon Ltd.)

Mention is made, without illustrating or describing in detail, of the automatic sampling system developed by Birtley Ltd. which incorporates a hammer mill, the purpose of which is to break up lumps in the samples before reducing the size of the sample to the desired quantity.

MEASURING AND GAUGING OF SOLIDS

Intermittent Weighing Machines

It has been shown that the purpose of sampling is to ascertain the quality of the material received, handled, processed, stored and despatched. In addition to the quality, that is, the physical and chemical properties of the various materials, it is of no less importance to ascertain the quantities, *i.e.* to check and record the weight of the material or materials.

Broadly speaking the weighing of materials in bulk can be effected in two ways, either intermittently or continuously. The former method for the intermittent weighing of batches of material can be achieved either by manually operated, semi-automatic, or automatic weighers.

Weighbridges, for instance, which are used extensively to ascertain the weight of material in railway wagons, road vehicles, *etc.*, should be considered as coming in the intermittent category.

Weighbridges

A weighbridge, as its name implies, consists of a bridge or platform capable of receiving an entire vehicle including its load. A typical design is shown in *Figure 5*. Weighbridges should be arranged in such a position that incoming vehicles can easily pass through them by entering the platform on one side and leaving it on the opposite side. When the weight of the vehicle is known (as is usually the case with railway wagons) the net weight

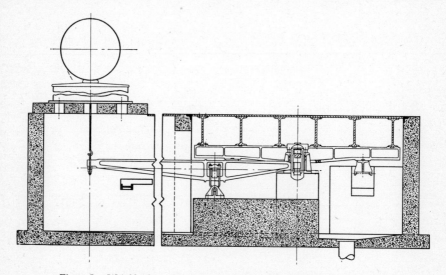

Figure 5. Weighbridge. (By courtesy of Simon Handling Engineers Ltd.)

of the material is obtained by deducting the weight of the vehicle from the weight indicated by the weighbridge. In the case of lorries *etc.* the weight of the vehicle is not usually known or may vary considerably and therefore it is necessary to weigh both empty and loaded in order to ascertain the weight of material delivered. This may be carried out by the same weigher or, where the intensity of traffic renders it worthwhile, a second weigher may be used. It should be emphasized that the positioning of the weighbridge or weighbridges requires much care and experience as obviously the easier the access to and from these, the less time is wasted, and consequently a larger number of vehicles can be handled.

Dormant Weighers

For bulk materials arriving in batches and being of homogeneous nature with variations in consistency only, it is often desirable to ascertain the weight of each batch separately. In these cases the so-called 'Dormant' weighers present many advantages. As can be seen in *Figure 6* these weighers consist of a weighbridge on which a hopper or bin is arranged which can be filled from another bin or hopper above it (usually called a garner bin). The material can be discharged from the weighing bin through a gate valve arranged at the apex of the conical bottom of the bin. The

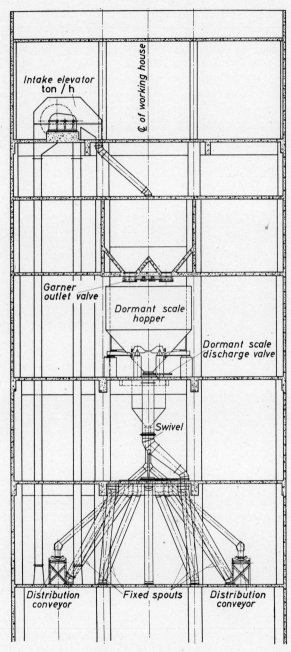

Figure 6. Dormant weigher arrangement. (By courtesy of Simon Handling Engineers Ltd.)

operation of these weighers is extremely simple and merely requires the opening and closing of the garner bin gate and the weighing bin gate and these operations can be facilitated by using mechanical or electrical devices or pneumatic cylinders. The weighers can be adapted to print tickets showing the exact weight of each batch handled. It is evident from *Figure 6* that each batch can easily be directed to any desired destination in accordance with requirements (that is variations in quality or variety, *etc.*). It should be noted that although this weigher is not fully automatic, it requires little or no manual handling and the duty of the operator is mainly to select the right destination for the product. These weighers, of course, can be situated in any convenient position and the material received can be conveyed to and from them by any suitable combination of conveyor, elevator, *etc.* Usually the operator is in touch by telephone or other signalling device with the actual intake plant and sometimes also with the storekeeper, thus ensuring maximum flexibility of the operation.

Automatic Bucket Weighers

Automatic bucket weighers of bulk materials are machines which check and register automatically in batches the weight of material passing through them. This type of machine normally consists of an equal-armed beam, on

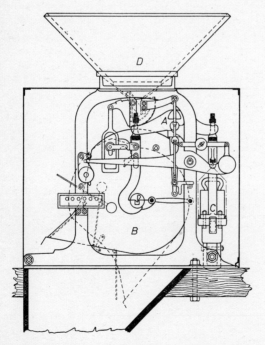

Figure 7. Chronos weigher. (By courtesy of Reuther & Reisert, K.-G., Germany)

one end of which is suspended a bucket or hopper into which batches of the bulk material flow or are fed, and on the other end the weight box carrying the weight corresponding to the batch of material. The first automatic batch weigher for free-flowing, dry, granular materials such as wheat, was patented in 1881 by Reuther & Reisert, of Hennef/Sieg, Germany, and marketed under the name of 'Chronos' weigher (*Figure 7*). It consists of a hopper D in which a feed mechanism with two feed gates is arranged and the equal-arm beam A carries the weight scale C on one end and the weigh bucket B on the other. The weigh bucket, which is carried in a fulcrum, fills with material, and as it does so its centre of gravity is gradually displaced to the extent that when the predetermined quantity of material has flowed into the bucket, the feed gate closes, and the catch which has been retaining the bucket in position is released, thus causing the bucket to tilt and discharge its contents (as shown by dotted lines in *Figure 7*).

In Great Britain three firms, Richard Simon of Nottingham, W. & T. Avery Ltd. of Birmingham, and Henry Simon Ltd. of Cheadle Heath, share the merit of having developed and of still producing automatic weighers based on the same principle of batch weighing, but differing from the 'tilting' bucket weigher by the design of the bucket itself.

The difference between a typical British automatic weigher for free-flowing granular material and the continental type is that the former has a bucket that does not tilt when discharging but is emptied through a gate arranged at the bottom, whilst the weigh bucket on the continental weigher tilts. The tilting action of the latter type of weigher causes a shock every time it tips, and the resulting vibration is transmitted through the frame to the bunker or hopper and building. Today, the larger size 'Chronos' weighers incorporate a very ingenious damping device which absorbs the trepidations caused by tilting, without interfering with the precision of the weighing. On the other hand, 'Chronos' and other continental makers have adapted the bottom discharging bucket for non free-flowing materials like flour and offal.

A typical British weigher is shown in *Figure 8* (*a*) and (*b*) and from the latter it can be seen that the weigher comprises an equal-armed beam (1) pivoted on two main fulcrum knife edges (2). From the knife edges (3) and (4), situated on the two ends of the beam (1) are suspended the weigh bucket (5) and the weight box (6). Hinged on the bucket is the bottom gate (7) which is kept closed by a counterweight (8). When both the weigh bucket and the weight box are empty they should be in perfect balance.

When the weight, corresponding to the batch, is placed in the weight box, the latter sinks under the load and tilts the beam (1), which in turn causes the feed gate to open. As the bucket fills with material it gradually sinks, causing the feed gate to close to the extent that the flow of material is reduced to a dribble. This dribble feed will complete the total weight required for the batch, causing a further downward movement of the beam, which, when again in the horizontal position, will completely close the feed gate and also release the catch holding the bottom discharge gate of the bucket, this gate is then forced open by the weight of the material, thus discharging the bucket content. As soon as the material is discharged the gate closes and the weigh bucket is lifted by the weight box and the sequence starts again.

It should be noted that if the weight placed in the weight box corresponds exactly to the weight of the batch of material required, each weighing would be slightly overweight (although the weigh bucket and empty weight box are in perfect balance) owing to the small amount of material which is in mid-air between the gate and the bucket at the moment of equilibrium, when the feed gate closes. This amount of material 'in transit' must be compensated for, and this can easily be achieved by means of a lever (9) carrying a sliding weight (10) which can be adjusted appropriately to compensate for the excess weight due to the material in transit. Adjustment of the sliding weight on the compensating lever will be required every time a change in the consistency of the material to be weighed occurs, as this would obviously have a slight effect on the quantity in transit and therefore adversely affect the accuracy of the weighing.

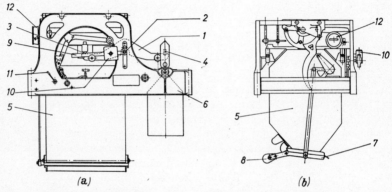

Figure 8. *Simon automatic weigher.* (*By courtesy of* Henry Simon Ltd.)

The balance of the weighing machine when loaded or unloaded can be tested quickly, merely by operating a small control handle (11). It is mechanically impossible for the feed gate and the weigh bucket discharge gate to be open simultaneously and therefore no material can pass through the weigher without being recorded by the tip or totalizing counter (12).

The larger size weighers are provided with residue weighing mechanism and ticket printing devices for weighing and recording odd quantities left over. Moreover, a 'feed control' device can be incorporated as a further refinement, which serves the purpose of preventing the feed gate from opening unless the reserve hopper above the weigher contains enough material for a full weighing. This not only reduces the generation and blowing out of dust at the feed gate, but also renders the weighing operations more accurate as the material in transit will tend to be more uniform in quantity.

The operation of these weighers is entirely automatic, and for free-flowing material, as a rule, requires no power as the weighers are activated by the weight of the material. For non free-flowing materials a feeding mechanism, consisting of a screw conveyor, vibrating shaker or band conveyor is required to ensure an even flow of material. A weigher of this sort is shown in

Plate I, and apart from the feed mechanism the design differs very little from the weigher shown in *Figure 8* (*a*) and (*b*). In the case of large size weighers it is the practice to use a small electric motor to bring the residue weighing and balance testing mechanisms into operation. Automatic weighers can be fitted with pre-set counters, such as a device by means of which the weigher will stop operating after a predetermined number of weighings have been made. The recording of the number of weighings and/or the total of the actual weight of material passing through the weigher can be effected not only locally but can be repeated at any distance from the weigher by means of electromagnetic repeat counters.

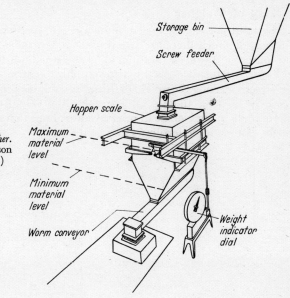

Figure 9. Richardson weigher. (By courtesy of Richardson Scale Company, U.S.A.)

It should be mentioned here that whilst free-flowing materials and also some not so fully free flowing can be handled successfully by the weighers described above, materials which tend to stick to the walls of the hoppers and weigh buckets cannot be handled where accuracy of weighing is essential. Several special types of automatic weighers have been developed recently, of which the Richardson differential weigher shown in *Figure 9* is a characteristic example. The main feature of this weigher being that it fills up with, say, about 150 lb. of material and, when the feed gate closes, the discharge gate opens and exactly 100 lb. of material are discharged, so that approximately 50 lb. of material will remain in it. The process then starts afresh by making up the remaining 50 lb. to 150 lb. again. By this ingenious arrangement even the stickiest materials can be weighed with reasonable accuracy.

The purpose of an automatic weigher, as its name implies, is to weigh and record automatically a material passing through it. Besides this prime object the automatic weigher can be used advantageously to control the rate of flow and the blending of various materials before, during, or after any process, but these applications of the automatic weigher are dealt with later on, when describing other apparatus—usually known as feeding mechanisms—having the same object, namely the gauging of more or less free-flowing and also sticky solids.

The bucket type of automatic weigher has another very important function, that of packaging free-flowing solids in sacks, bags, and containers of various types and sizes, made of textiles (such as jute, cotton, canvas, *etc.*), paper and cardboard, and a number of other materials.

Sacking-off Weigher

The most commonly used type in this group is the so-called sacking-off weigher, of which there are two distinct types: the *net weigher*, for weighing the content of the sack (not including the weight of the sack itself) and the *gross weigher* which weighs the material plus the sack. As a rule, the former is used for checking the weight of the more valuable products, whilst the latter is widely used for those products where the value of the material itself is approximately equivalent to or even lower than that of the sack, weight for weight (for example fertilizers and cement).

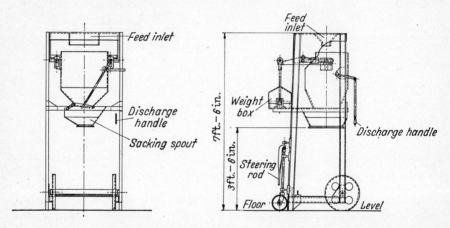

Figure 10.   Net sacking-off weigher.   (*By courtesy of* Richard Simon & Sons, Ltd.)

The net sacking-off weigher, shown in *Figure 10*, differs very little in its design from the automatic weighers shown in *Figure 8* and *Plate I*, and described above, except that the hopper under the weigher is provided with a sacking-off spout and a manually operated mechanism which prevents the weigh bucket discharging until the operator replaces the filled sack with an empty one and actuates the locking mechanism. On most of these weighers the replacement of the full sack with an empty one is carried out *after* the weigh bucket has discharged the material, but weighers are also available where several hoppers are fitted underneath the weigher in a kind of turntable which is timed to rotate at given intervals in such a way that a hopper with an empty bag is in position under the weigher every time the weigh bucket discharges, and the full sack is then turned round and lifted manually from the sacking-off spout and replaced by an empty one. This type of weigher-packer, not illustrated, is used mainly for packing flour, cereal foods, *etc.*, into jute, cotton or paper bags, or cartons.

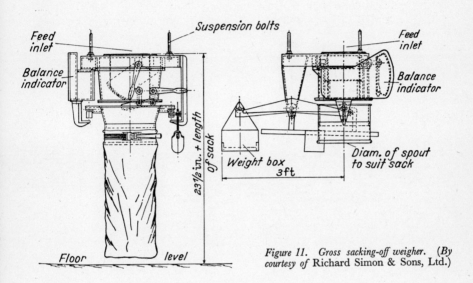

Figure 11. *Gross sacking-off weigher.* (*By courtesy of* Richard Simon & Sons, Ltd.)

The design of the gross sacking-off weigher (*Figure 11*) differs from the net weigher, in the first place by the fact that the weigh bucket is replaced by a small entraining hopper ending in the sacking-off spout to which the bag is attached, thus eliminating both the hopper under the weigh bucket and the whole discharge gate mechanism, so that the sack itself becomes virtually the weigh bucket, and in the second place, that the feed gate mechanism—instead of the discharge gate—is provided with a locking device, which prevents it from opening until the bag or sack is replaced, or until the operator manipulates the release lever. It should be noted that the output capacity of the gross sacking-off weigher is lower than that of the net sacking-off weigher (assuming that both are filling sacks of identical size with the same material) because the weigh bucket of the net weigher starts filling up immediately after discharging the previous load, whereas the weigh hopper of the gross weigher cannot start receiving a new load until the full sack is replaced and, consequently, time is lost.

A great variety of designs are available of both these types, but with little or no alteration in the basic principle. Two automatic sacking-off weighers usually known as 'valve-packers' and widely used for packing free-flowing products into self-sealing paper bags are described below. *Figure 12* shows a packer of this type, known as the Bates-Packer, and incidentally, the original of this ingenious self-sealing paper sack and packer design, which is a gross weigher, since the weight of the sack is included. Referring to *Figure 12* the free-flowing material, for example cement, is fed through a rotary seal *A* into a chamber in which a beater *B* breaks up any lumps which may be in the material. The material then drops and is caught by a thrower-wheel *C* which expels it through a nozzle *E* projecting into the paper sack to be filled. The sack is held in position by a clamp operated by a lever *D* and it rests on a platform *G* which is connected via a knife edge with one end of an equal-armed beam *J*, to the other end of which is

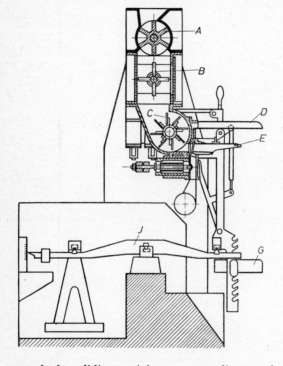

Figure 12. 'Bates' sacking-off weigher (gross). (From 'Hütte des Ingenieurs Taschenbuch', *by courtesy of* Wilhelm Ernst & Sohn Verlag, Berlin)

$A =$ *rotary feeder*
$B =$ *disintegrator*
$C =$ *thrower wheel*
$D =$ *sack platform lever*
$E =$ *feed nozzle*
$G =$ *sack platform*
$J =$ *balance beam*

attached a sliding weight corresponding to the gross weight of the sack. As a rule two nozzles and weighing mechanisms are connected to one filling mechanism. The operator has only to slip the paper bag over one nozzle and start the filling operation by pulling or pushing a lever. When the bag is filled to the correct weight, the nozzle is automatically closed by a slide and the bag can be removed. The ingenious 'valve' on the bag closes as soon as the bag is removed from the nozzle by the operator. It is not considered necessary to describe here the functioning of the paper 'valve' as it has been in general use for several years and it is assumed that the reader is familiar with it. In some modern machines now available, the full bags are removed, or fall off automatically as soon as the filling and weighing operation is complete, and then slide down a chute to a band conveyor to be transported wherever required. The capacity of some of these machines, although operated by only one man, attains approximately 1,000 bags of 100–110 lb. each per hour. The power requirements of these packers vary, of course, according to the material handled, the make, and the throughput capacity, but in general they are very low.

Figure 13 shows a further development of the type of weigher packer described above and which is noteworthy for two main reasons, namely (*a*) that it weighs the *net* weight of the material, and (*b*) that it can be used advantageously to pack granular, and in some cases even fragile, products such as rice and granulated sugar.

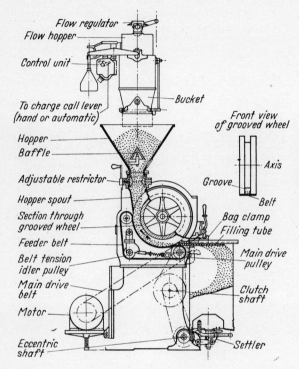

Figure 13. St. Regis sacking-off weigher (net). (By permission from 'Chemical Engineers' Handbook', by John H. Perry. Copyright 1950. McGraw-Hill Book Co., Inc.)

As can be seen in *Figure 13*, an ordinary weigh bucket type of automatic weigher is arranged above the packer, and the weigher incorporates a charge or discharge control unit which is operated either manually or automatically. The batch of material discharged by the weigher falls into a hopper and then, with the rate of flow controlled by a baffle and adjustable restrictor, it passes into a hopper-spout which feeds the endless feeder or thrower belt. The feeder or thrower belt is driven by an electric motor and wound round part of the periphery of a grooved pulley. The groove of the pulley, which is covered by the belt, forms a duct in which the material is accelerated by the friction of the high velocity belt (being in the region of 5,000 ft./min), and then expelled through the filling tube which projects into the bag through the self-sealing valve. Once the quantity of material discharged by the weigher has been filled into the bag, the operator detaches the clamp and lifts the bag on to a conveyor for transporting towards its destination. In the case of some materials a shaker or 'settler' device is provided to impart a gentle vibration which compacts the material in the bag. Usually two filling tubes are fitted so that the operator can slip a fresh bag on one tube whilst another bag is being filled through the other one. In most cases the operation by which the full bag is released also changes the flow from one tube to the other and simultaneously imparts an impulse to the control unit which releases the weigh bucket discharge gate. The power requirement of this type of weigher packer is again low since the weigher unit itself requires no power because it is operated by the weight of the material and only the belt and the settler (if fitted) need to be driven.

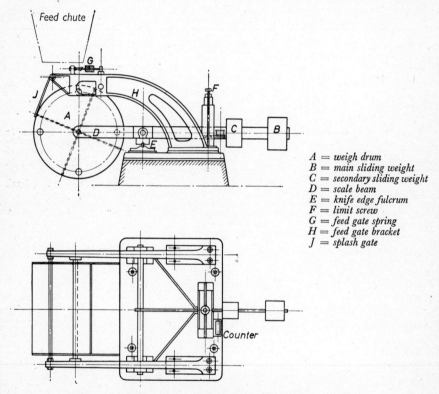

A = weigh drum
B = main sliding weight
C = secondary sliding weight
D = scale beam
E = knife edge fulcrum
F = limit screw
G = feed gate spring
H = feed gate bracket
J = splash gate

Figure 14. Edgar Allen weigher. (By courtesy of Edgar Allen & Co. Ltd.)

Mention must be made here of yet another type of weigher which may be considered an intermediary step between the intermittent and continuous type of weigher (see *Figure 14*). It is made by Edgar Allen & Co. of Sheffield and consists of a drum, divided into four equal compartments, and which can freely rotate on its centre spindle in one direction only. The drum is suspended through the spindle on one end of a scale beam, to the other end of which two sliding weights are attached. The material to be weighed should be fed at a steady rate of flow from a conveyor, elevator, *etc.*, into the uppermost compartment of the drum. When at rest, one of the sliding weights is set to balance the weight of the drum, the other to correspond to the required weight of material for each weighing. The action of the machine is as follows: gradually, as the top compartment of the drum fills, its weight overcomes the balance weight and the beam dips and releases a catch which permits the drum to revolve one quarter turn, so that it discharges its contents and consequently the original weight of the drum is restored causing the balance weight to return the beam instantaneously to the filling position so that the next compartment can be filled and the sequence repeated. Each time the drum makes a quarter turn, a tip counter is actuated, and since the standard weight of each weighing can be ascertained, the total weight of the material passing through the weigher can be obtained easily by multiplying the standard weight by the number of tips.

The whole apparatus is mounted on a heavy base plate, all parts are very solidly constructed, and it is most suitable for checking the weight of coal or the output of a rotary kiln in a cement works, that is, in cases where a certain latitude in the accuracy of the weighing is permissible.

Load Cells

Before closing consideration of the group of intermittent automatic weighers a description is given of a comparatively new device which will possibly gain increasing application in weighing static material in bulk. This new device is called the 'load cell' of which there are two distinct types, (*i*) the electric load cell, and (*ii*) hydraulic and pneumatic load cells. The principle of the electric load cell is based on the fact that the structure of any material is subjected to a strain under compression or tension load. If a fine wire filament is attached, either bonded or unbonded, to a metal block or cylinder, the resulting strain or minute distortion caused when this cylinder is subjected to compression will create a stress in the filament (known as a strain gauge), which affects its electrical resistance in direct proportion to the load or tension exerted on the metal block or cylinder. If the filament or gauge is connected to an electrical bridge circuit in such a way that one branch is formed through a pair of cells to be used as a weigher, and the other branch through a pair of cells to be used to balance the first pair when not loaded, a d.c. voltage passing through the bridge will vary proportionately to the alteration in resistance of the gauge. These variations in the electric current can be amplified, maintaining the exact linear relationship between the compression or tension load, and can be used to indicate on a dial or recorder, which can be calibrated to give a direct reading, the weight of the load exerting pressure or tension on the metal cylinder (or block) of the load cell. These load cells, which are surprisingly small (a cell rated for 10 ton capacity measures only approximately 4 in. high × 4 in. diameter), need not have any moving parts, can be applied easily under tanks, bunkers, hoppers or weighbridges (for railway wagons or lorries) and it is claimed that the accuracy is in the region of $\frac{1}{4}$ per cent between full rated load and 25 per cent of the total capacity, and that good sensitivity is maintained for loads under 25 per cent right down to zero. Cells of up to 100 ton capacity are available and obviously several cells can be arranged and their readings totalized.

The second type of load cell is the hydraulic load cell (pneumatic load cells are very similar in design and as they are used mainly for comparatively light loads their application is restricted to measuring stresses, torques, breaking strengths, *etc.*, rather than weighing, and therefore are not described here). The hydraulic cell is, in fact, a combination of cylinder, piston and diaphragm, in which friction has been eliminated so far as is technically possible. A thin layer of hydraulic fluid is kept between the piston and the diaphragm and this fluid is connected via an appropriate pipe to a pressure gauge or recorder which can be calibrated for direct reading of weight. The applications are similar to those of the electric load cells, and the hydraulic cells are said to be more accurate and sensitive than the electric ones.

A variation of the hydraulic load cell is the hydrostatic weigher which consists of a number of 'capsules' which carry the load, *e.g.* a hopper or bin, and transmit its weight to a cluster of smaller 'capsules'. This cluster of capsules is directly connected to a weight on an unequal armed beam, which operates a micro-switch or mercury switch, both when the bin is empty and when it is full (with the predetermined weight). In the first place the micro-switch or mercury switch operates a mechanism which opens the feed gate and closes the outlet of the weigh-hopper or bin, in the second case it reverses the operation, *i.e.* it opens the outlet and closes the feed gate. The feed gate and the outlet slide may be operated by hydraulic or pneumatic cylinders, which are controlled by electrical impulses received from the micro or mercury switches. This type of weigher is used in mines, cement works, *etc.* where accuracy of ± 1 per cent is quite acceptable, although manufacturers claim an overall efficiency of ± 0.5 per cent over a working period of a few hours (usually one shift of 8 hours).

CONTINUOUS WEIGHING MACHINES

Consideration is now given to the features and technical characteristics of the second group of automatic weighers—continuous weighing machines—serving the purpose of checking and recording the weight of materials moving or being moved in a continuous stream.

The first weighing machine of this type was developed by Samuel Denison & Sons Ltd. of Leeds, and was introduced under the name of the Blake-Denison Integrating Weigher. It is still used and its fundamental principle has been adapted with some modifications in detail by several other makers both in the United Kingdom and abroad. The advantages of these continuous weighing machines ensure that their application in certain conveying plants is the most economical way to check and record the weight of the materials handled both with regard to initial costs and operating expenses.

Plate II illustrates a modern continuous belt weigher made by W. & T. Avery Ltd. of Birmingham and which can be applied to a conveyor belt in order to check and register the weight of the material as it is conveyed, irrespective of whether the conveyor is being newly installed or is already in existence, or whether it is running horizontally or on an incline. The method of operation of this type of weigher is as follows. A section of the conveyor is incorporated in the weigher in such a way that six to eight sets of troughing idlers are carried in a weighing frame suspended by means of four rods supported on a weigh lever which is connected to the indicator cabinet, the cabinet being placed in a suitable position, for example, above the belt conveyor. The return strand of belt of the conveyor is diverted by two jockey pulleys to drive a totalizing pulley which is connected to the indicator cabinet by a vertical shaft. Incorporated in the indicator cabinet is an integrating mechanism which computes the momentary load, that is, the weight of material on the weighing section of the belt and the belt speed, and gives the total weight of material which passes through the machine. In addition, it is possible to incorporate a graphic recorder which will show the rate of flow of the material at any given time.

Basically, the same principle as that of the belt weigher described above can be used to combine the weighing of the material, with the possibility of controlling the rate of flow of the material within very close limits. This type of belt weigher, shown in *Figure 15*, is usually called a constant weight feeder and is not, as a rule, incorporated into a conveyor belt, but consists of a short belt conveyor which is driven on the discharge end. The belt is usually not troughed, but is flat, and one idler is suspended under the belt in such a way that it is depressed in direct proportion to the weight of

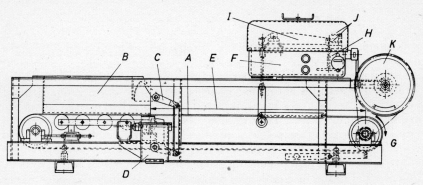

Figure 15. Constant feeder. (By courtesy of Richard Simon & Sons Ltd.)

A = *weigher band*
B = *feed hopper*
C = *feed gate*
D = *amplifier box*
E = *skirt board*
F = *integrator box*
G = *band pulley*
H = *totalizing counter*
I = *balance beam*
J = *integrator*
K = *driving pulleys*

material conveyed by the belt and it transmits the variations in load to the computer. The computer receives and converts these variations in load, combined with the belt speed, into a direct reading of the weight of the material passing along the belt; at the same time controlling the feed mechanism, which may be a simple radial gate or a power-driven feeder (*e.g.* screw conveyor, vibrating tray, *etc.*). The controlling of the feed mechanism may merely require the opening or closing of the radial gate, or it may entail varying the screw conveyor speed, or again, varying the number of oscillations or the amplitude or the frequency of the vibrations of the feeder, any of which actions result in increasing or reducing the rate of flow of material through the machine. Additionally, coloured electric lights may be used to signal any fluctuation in the rate of flow. This type of machine, which will be described later when dealing with the present-day technique of blending, is claimed to be very sensitive and to ensure a constant rate of flow within very close limits.

Before closing this section mention should be made of a recently developed continuous weigher for free-flowing powdery materials marketed under the name of 'Massometer' by Wallace & Tiernan Ltd. of London, which,

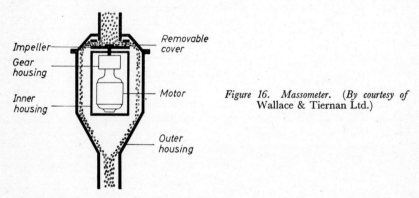

Figure 16. *Massometer*. (*By courtesy of* Wallace & Tiernan Ltd.)

although its throughput capacity is comparatively low, appears to have interesting features for certain applications. The operating principle is described hereafter with reference to *Figure 16*. The stream of material to be measured falls on a rotating impeller driven by a synchronous speed motor which is supported in bearings and is therefore free to move about the axis of the motor shaft. The torque on the motor housing which is created through accelerating the particles of the stream of material by the tangential velocity of the impeller, is proportional to the mass rate of flow. This torque is measured by a pneumatic prime relay which has an output directly proportional to the rate of flow of the material and can be recorded either in pounds of material per minute, or, additionally, can be integrated in a totalizing counter. The main feature of this machine is that it can be advantageously used to detect stoppage and fluctuation of flow rate as well as for the control of batches.

Blending

It was stated in the introduction to this section that in many cases both the rate of flow and the proportion of one material to another must be controlled. The degree of accuracy required of this control may vary considerably, depending on the nature and value of the materials used and the process in question. Therefore, in many instances, both control of the flow rate and blending of the ingredients may be achieved with comparatively simple and inexpensive apparatus, especially when recording of the weight

Plate I. Simon automatic weigher with feeder.
(*By courtesy of* Richard Simon & Sons Ltd.)

Plate II. Belt conveyor weigher (Avery). (*By courtesy of* W. & T. Avery Ltd.)

[*To face p. 458*

Plate III.   Exact measurer.   (*By courtesy of* Henry Simon Ltd.)

Plate IV.   Velofeeder.   (*By courtesy of* Henry Simon Ltd.)

[*To face page 459*

is not required and control of the rate of flow can be based on the volume rather than the weight of the material or materials. This type of control unit is called a gauging mechanism, or, more frequently, a measurer or feeder.

There are, of course, a very great number of these measurers or feeders, but they can be divided into about a dozen distinct types according to their basic principle of operation, and a brief description of some of these in more general use is given below.

The most commonly used device is shown in *Figure 17* and consists of a revolving roller which entrains material proportionately to its speed of revolution and to the 'gap' between the roller and the gate which can usually be controlled within certain limits. The roller can be provided with teeth, grooves, flutes, or indentations, or can be replaced by a pocket-wheel resembling the one shown in *Figure 18*. In some instances two rollers, sometimes running at different speeds, are used with the object of spreading more evenly the material of which they are controlling the rate of flow. The type of control device shown in *Figure 18*, usually called a measurer, is widely used for free-flowing granular materials, and in some cases the speed of rotation can not only be varied, but also the volume of the pockets increased or decreased.

A typical measurer where the volume of the pockets can be altered is shown in *Plate III*. The type of feeder shown in *Figure 19* consists of a tray which is moved either horizontally or on an incline and by its movement causes the material to flow at a rate proportional to the speed and stroke of the movement and to the variation in the gap of the opening. In this category may be included the modern vibrating feeders, of which a typical example, known as the 'Velofeeder', is shown in *Plate IV*. Vibrating feeders in which the vibrations are caused by electromagnetic means have gained considerable importance owing to the simple way in which remote control of the rate of vibration, and thus of the rate of flow, can be arranged. *Figure 20* illustrates the plunger type of feeder in which the rate of flow depends on the length and number of strokes of the plunger and it is used for less free-flowing materials. *Figure 21* shows a feeder, known as the rotating plate feeder, which is much used in cement works, *etc.* and consists of a rotating horizontal or slightly inclined disc which is placed with its centre under the hopper or bunker from which the material is to be fed at a given rate. This rate can be varied by altering the speed of rotation of the disc or by varying the angle or projection of the scraper blade fitted on the periphery of the disc. A feeder very similar in appearance, but differing from the former by the fact that instead of a revolving disc it has a number of revolving arms or blades which scrape the material from a 'shelf' or plate, is shown in *Figure 22*. The capacity variation of this type of feeder is very limited. *Figure 23* shows a device called an apron feeder or plate-band feeder which is used mostly for feeding heavy materials such as coal and ore, generally containing a large proportion of big lumps, and *Figure 24* shows a typical Redler conveyor feeder which, combined with other variations of the Redler En-Masse Conveyor, can be used as a versatile feeder or mixer.

Figure 17. Feed roll

Figure 18. Rotary feeder (measurer)

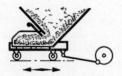

Figure 19. Shaker feeder

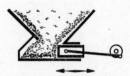

Figure 20. Plunger feeder

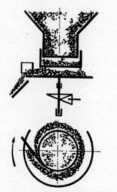

Figure 21. Turntable feeder

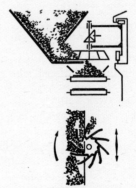

Figure 22. Wheel and shelf feeder

Figure 23. Apron feeder

(*From* 'Hütte des Ingenieurs Taschenbuch', *by courtesy of* Wilhelm Ernst & Sohn Verlag, Berlin)

It is regretted that lack of space makes it impossible to describe more of these measurer and feeder devices, but the few described above will give a general idea of the main types in use. It should be noted that many combinations of measurers and feeders are possible and are used to ensure regular flow and mixing of a great variety of materials at correctly proportioned rates. A limitation of all of these devices lies in the inherent fact that they gauge volume only, and therefore any variation in the loose bulk density of the material or materials may cause appreciable quantitative fluctuations. Whenever a greater degree of accuracy is required in the

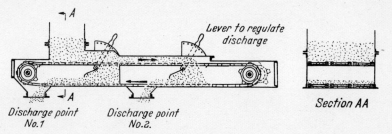

Figure 24. Redler feeder. (*By courtesy of* Redler Conveyors Ltd.)

blending of materials there is no doubt that preference should be given to weighers. Automatic weighers of either the intermittent, *i.e.* weigh bucket type or the continuous belt type, can easily be arranged in groups, interlinked mechanically or, for preference, electrically, to give simultaneous or continuous discharges at pre-set proportions. A great variety of combinations is possible, but to describe these would appear to be beyond the scope of the present article. Mention is made, however, that whilst the automatic weighers of the weigh bucket type need to be set by adjustment of fixed weights on the weigh-beam of the machine itself, other types of hopper or bucket weighers have been developed by several firms, which are finding ever-increasing application. The main feature of these types of weighers is that they can be set by remote control from a central control station and rely upon electrical signals for information regarding the weight of material delivered to the weigh hopper. Moreover, they can be set to receive and weigh batches of different materials of widely varying weights and, therefore, with these weighers blends can be formed to most complicated formulae set on a control panel. *Figure 25* shows a typical layout of such a system, although a comparatively simple one, in which two weighers are arranged, each receiving in a sequence pre-set on a panel, a certain number of materials of pre-set weight, and automatically stopping the feed and switching over to the next product, until the whole sequence, corresponding to the desired formula, is completed, when the total charge or batch in the hopper will be discharged into mixing worms or batch mixers. The blending sequence can be repeated either automatically or by manual control.

It should be noted that the worm conveyors shown feeding the weighers in *Figure 25* are frequently replaced by any type of feeder which is particularly suitable for the product to be handled. Feeders of the types shown in *Plates III* and *IV* and in *Figure 19* are some of those often used in connection with the 'Select-O-Weigh' system, which is one of the best known 'blending by weight' systems, and has been developed by the Richardson Scale Company and Henry Simon Ltd. The choice of the most appropriate feeder in connection with the weigher is of the utmost importance as obviously the weigher and feeder must be 'tuned' to each other—and to the product handled—in order to obtain the most efficient operation and most accurate blending. The extra cost involved in selecting the most appropriate feeders and weighers rather than cheaper and less accurate ones is

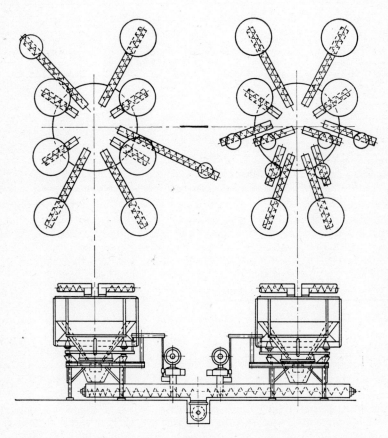

Figure 25. Select-O-Weigh system. (*By courtesy of* Richardson Scale Company, U.S.A.)

usually very small in comparison with the continuous losses incurred owing to inferior equipment. Some slight margin of error is inevitable with any commercial weighing system, but it is well worth while giving most serious consideration to the question of keeping this margin of error as low as possible.

In conclusion, perhaps a few words may be said of the maintenance of weighers and feeders in general. One of the most important points in this respect is cleanliness. Automatic weighers and feeders which handle dusty material—and the vast majority of free-flowing products contain some fines which will become airborne when poured—are much impaired in their operative accuracy by dust which settles on knife edges, toggles, levers and weight boxes, quite apart from the nuisance the dust causes to the staff and neighbourhood. Therefore, most weighers are usually connected to a dust-collecting system and are also provided with appropriate hoods or covers. Nevertheless, it is almost invariably necessary to remove the dust accumulating on weighers, and the frequency of these periodical cleanings depends on the effectiveness of the dust aspirating system as well as the nature of the product handled. Repair of weighers, which are usually robust, is infrequent, but it cannot be sufficiently emphasized that it should be carried out only by competent and experienced personnel. It is also important to realize that to neglect the prompt repair of weighers does not pay—a weigher which is out of order, even to a small extent, can cause very costly losses.

BIBLIOGRAPHY

ZIMMER, G. E., *The Mechanical Handling and Storing of Materials*, Technical Press Ltd., London, 1932

HANFFSTENGEL, G., *Die Förderung von Massengütern*, Julius Springer, Berlin, 1908

PERRY, H., *The Chemical Engineers Handbook*, McGraw-Hill, New York and London, 3rd edn., 1950

Hütte des Ingenieurs Taschenbuch, Wilhelm Ernst & Sohn Verlag, Berlin

Modern Materials Handling, Materials Handling Laboratories Inc., Boston, U.S.A.

16

CYCLONES

R. F. HEINRICH and J. R. ANDERSON

GENERAL PRINCIPLES

ONE of the oldest and most widely used methods of separating air or gas-borne dust is by the cyclone, either in its simplest form or in the more recently developed nested assemblies of small diameter cyclones. Cyclones possess fundamental advantages; in good efficiencies, if the dust particles are not very fine; low initial and operating costs, together with a stable working resistance; and, to a certain extent, self-adjusting characteristics which give them wide use for recovering a variety of industrial dusts where the fractions below 10 μ are only a small percentage of the total dust.

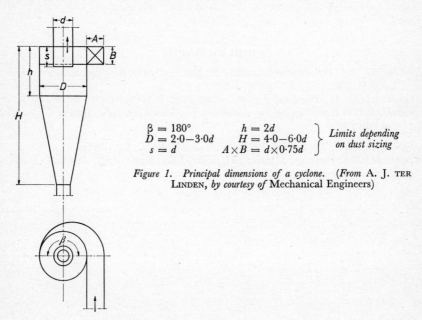

$\beta = 180°$ $\qquad h = 2d$
$D = 2 \cdot 0 - 3 \cdot 0 d$ $\qquad H = 4 \cdot 0 - 6 \cdot 0 d$ $\qquad$ Limits depending on dust sizing
$s = d$ $\qquad A \times B = d \times 0 \cdot 75 d$

Figure 1. Principal dimensions of a cyclone. (From A. J. TER LINDEN, by courtesy of Mechanical Engineers)

The separation of dust from its carrying media is caused by the centrifugal force set up in a vortex given to the gas on tangentially entering the cylindrical cyclone body (*Figure 1*). The dust is thrown out of the gas on to the inner surface of the cyclone and then spirals down to the dust outlet at the bottom where it settles out in the slowly moving air in the hopper, on to the pile of dust in the bottom. The gas vortex continues its spin into the hopper, then moves up and through the cyclone outlet at the top. From this it would appear an easy task to design a cyclone, but the flow pattern is complex and extremely difficult to measure, especially in the unstable and highly turbulent core in the centre of the cyclone.

A considerable amount of basic research has been carried out and readers interested in the theory of the cyclone and predicted efficiencies are referred to references 1–6.

There are many variations in cyclone proportions, as shown by the number of proprietary makes available. It will be obvious that a large diameter cyclone designed to handle wood chips, whilst giving almost 100 per cent efficiency on this material, would be totally inadequate to recover a chemical process dust containing nothing larger than 10 μ diameter. Conversely, a small diameter cyclone designed to give 90 per cent efficiency on dusts down to 10 μ would completely choke up if fed with wood shavings. In chemical engineering practice the majority of applications for cyclones involves handling the finer dusts demanding so-called 'high efficiency' cyclones of between 1–4 ft. diameter, used either singly or in multiple units, depending on the volume to be treated. A typical cyclone, according to TER LINDEN[2], would have proportions as given in *Figure 1*, and results of investigations showing the effect of a change in each of these dimensions on the cyclone efficiency are available.

Apart from the physical dimensions of a cyclone, the efficiency varies with volume, gas density, gas viscosity, particle size, particle shape, particle density, particle surface texture, dust concentration, agglomeration characteristics and natural or acquired electrostatic charge, so that prediction of efficiency is complex.

Handling Capacity

The gas volume which cyclones of a standard and sound design will handle at different temperatures and with pressure drops up to approximately 6 in. w.g., measured from the inlet flange to the outlet flange, can be calculated with reasonable accuracy according to the following formula:

$$V = K\sqrt{HT} \text{ ft.}^3/\text{min} \qquad \ldots \ldots (1)$$

where V (ft.^{3}/min) = gas volume per cyclone
K (———) = constant typical for the cyclone
H (in. w.g.) = pressure drop across the cyclone in inches water gauge
T (° F + 460) = absolute temperature of the gas passing the cyclone

Example

$$K = 5, \ H = 2 \text{ in.}, \ T = 60° \text{ F} + 460 = 520° \text{ abs.}$$

then:

$$V = 5\sqrt{2 \times 520} = 162 \text{ ft.}^3/\text{min}$$

If, on the other hand, the gas volume, the gas temperature, and the pressure drop have been measured, then the design constant K can be easily calculated according to the formula:

$$K = \frac{V}{\sqrt{HT}} \qquad \ldots \ldots (2)$$

For different gas volumes the pressure drop across the cyclone can be calculated according to the formula

$$H = \left(\frac{V}{K}\right)^2 \times \frac{1}{T} \text{ in. w.g.} \quad \ldots (3)$$

Table 1 gives the design constant K for reverse flow vane type cyclones, *Figure 2*, with 6, 9, 12, 16 and 24 in. inner tube diameters and shows also the approximate gas volumes which a single tube can handle at pressure drops of 1, 2, 3 and 4 in. w.g.

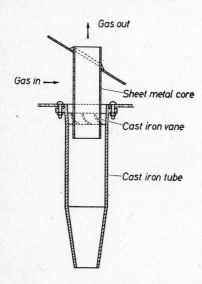

Figure 2. General view of multiclone. (*By courtesy of* Western Precipitation Corp., Los Angeles)

The design constant K, of course, depends upon the design of the vane, the length of the cyclone body and the inner diameter of the gas outlet tube, and can vary to quite a large extent according to the standards and the suitability of design of the different cyclone manufacturers.

Table 1. Performance of Reverse Flow Vane Type Cyclones

Inner Diameter of cyclone in.	Design constant K	Gas volume (ft.³/min) per tube at gas temp. 60° F and different pressure drops in in. w.g.			
		1 in. p.d.	2 in. p.d.	3 in. p.d.	4 in. p.d.
6	5	144	162	197	226
9	12	274	388	475	541
12	20	455	645	790	900
16	35	800	1,130	1,380	1,580
24	80	1,820	2,580	3,150	3,600

Reverse flow cyclones with a scroll or tangential gas inlet, as shown in *Figure 1*, give mostly a K which may be approximately 10–15 per cent higher compared with vane type cyclones, but this type seems to be more prone to abrasion at the gas inlet if the gas contains a large amount of dust particles with a diameter over 15 μ, especially when the particles are very hard and abrasive, as for example, sand, slag, coke, pyrites dust, *etc.*

To avoid excessive erosion and too high pressure loss in the cyclone, the gas velocity in the tangential gas inlet or in the vanes should normally not exceed 50–70 ft./s.

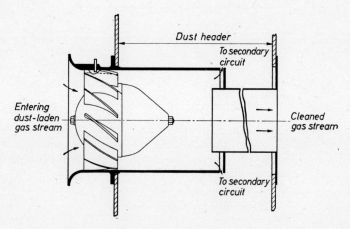

Figure 3. Arrangement of straight flow cellular dust collector. (By courtesy of Davidson & Co. Ltd., Belfast)

Straight flow cyclones as shown in *Figure 3*, especially the laminar flow type[7], allow design constants K which may be 25 to 35 per cent higher than the K values given in *Table 1* for reverse flow vane type cyclones of the same inner diameter.

As the inlet gas velocity to the cyclone is limited, for the reasons mentioned above, the tangential gas velocity in a cyclone with a large diameter will be mostly not higher than the tangential gas velocity in a cyclone with a small diameter.

COMPARISON OF TYPES OF CYCLONES

As the centrifugal forces, which in the rotating gas stream of the cyclone are exerted on a dust particle suspended in this gas stream, work according to the law:

$$\frac{v^2}{r} \qquad \ldots \ldots (4)$$

where v (ft./s) = gas velocity
r (ft.) = radius on which the dust particle rotates

the efficiency of a cyclone separator decreases when the diameter of the cyclone increases, in both cases the tangential gas velocity v remaining the same.

This decrease in efficiency with large diameter cyclones is shown in *Figure 4*. In this case tests were carried out with dust from a wet process cement kiln[8] using cyclones of 9, 16, 24 and 126 in. diameter. At a constant pressure drop of 2·5 in. w.g. across the cyclones, the following efficiencies were obtained:

Table 2. Variation of Cyclone Efficiency with Diameter

Cyclone diameter in.	Efficiency per cent
9	96·7
16	92·5
24	88·2
126	57·5

For a given dust in a given set of conditions it is possible for a manufacturer to carry out tests and to plot a grade efficiency curve for a particular cyclone. The following figures and curves show the dimensions of typical low, medium, high, and very high efficiency cyclone separators[5], and also the grade efficiency curves which in the average can be obtained with dust of a specific gravity of approximately 2·7 g/cm^3.

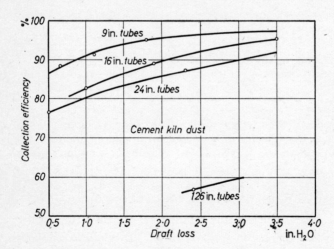

Figure 4. Collection efficiencies of cyclonic collectors with various-sized tubes on cement dust. (From EVALD ANDERSON, by courtesy of Chemical & Metallurgical Engineering)

Figure 5 shows a fan-scroll type collector with a 7 ft. diameter and the grade efficiency curve for this type of separator is given in *Figure 6*.

A high throughput cyclone with 4 ft. diameter is shown diagrammatically in *Figure 7* and its grade efficiency curve in *Figure 8*. In *Figure 10* a high efficiency cyclone with 3 ft. 6 in. diameter is illustrated and *Figure 9* shows its grade efficiency curve. *Figures 11* and *12* show a small highly efficient 6 in. diameter multicyclone assembly and its grade efficiency curve. Such curves would give some indication of what overall efficiency to expect when handling a different dust grading from that tested, but the actual efficiency on a different grading may well be a much higher or lower per cent depending on the many variables affecting efficiency as given in the preceding paragraph.

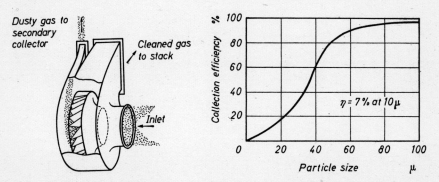

Figure 5. Scroll type collector

Figure 6. Grade efficiency curve for scroll collector (7 ft. diam.).

(*From* C. J. STAIRMAND *and* R. M. KELSEY, *by courtesy of* Chemistry & Industry)

It would be invidious to say that one particular make or type is better or worse than another; each generally has applications where its performance may be better than another and one is advised to check that the manufacturer has had practical experience of the material in question, operating under similar conditions, before fully accepting any claims on expected or guaranteed recovery figures.

MULTICYCLONES

Since efficiency in a cyclone is primarily dependent on centrifugal force, it follows that the smaller the diameter of the cyclone body, the greater will be the separating force for a given inlet velocity. This has led to the development of the so-called 'multiclones', 'tubular collectors' or 'cellular type dust extractors', to name three from the large number now available.

This type of cyclone is constructed with body diameters of 2 in. up to approximately 12 in. Some employ a tangential inlet whilst others use a radial vane ring to impart the swirl to the gases on entering the cyclone body. The vane ring exponents claim an advantage in that the gas and dust are uniformly distributed around the circumference of the tube by causing the gas to flow through a number of small entrances evenly spaced around the tube. They also claim that at higher inlet dust concentrations and with a large amount of dust particles of over 15 μ in the inlet gas, the effect of abrasion on the walls of the cyclone in the inlet area is considerably less with the vane type because the incoming gas stream is divided by the vanes into many smaller jets so that a very high dust concentration cannot easily build up at the walls near the gas inlet as can occur with cyclones having one or two tangential inlets only.

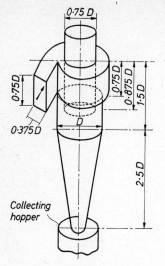

Figure 7. High throughput cyclone

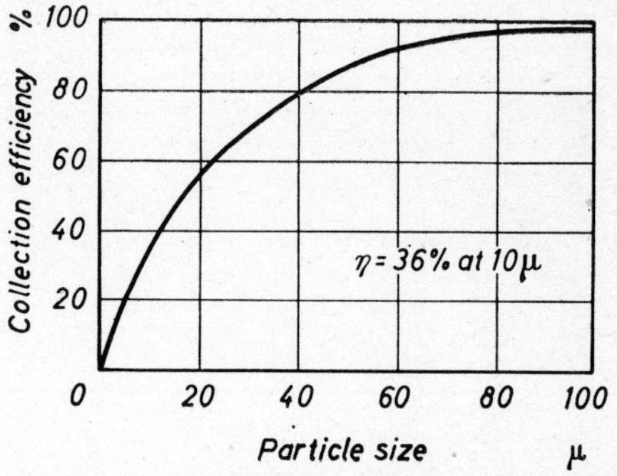

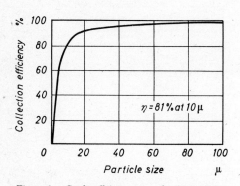

Figure 8. Grade efficiency curve for high throughput cyclone (4 ft. diam.)

Figure 9. Grade efficiency curve for high efficiency cyclone (3 ft. 6 in. diam.)

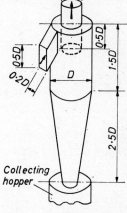

Figure 10. High efficiency cyclone

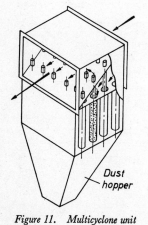

Figure 11. Multicyclone unit

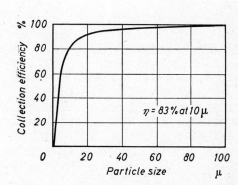

Figure 12. Grade efficiency curve for multicyclone (cells 6 in. diam.)

(*Figures 7–12 from* C. J. STAIRMAND *and* R. M. KELSEY, *by courtesy of* Chemistry & Industry)

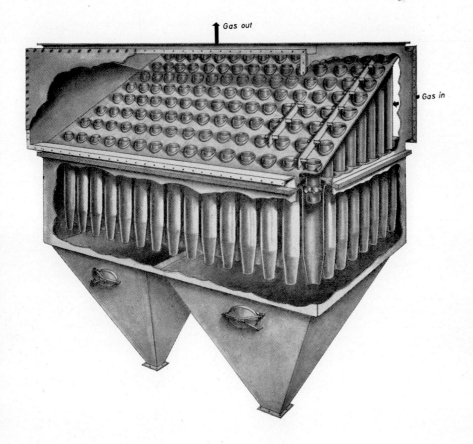
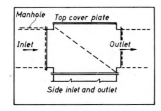
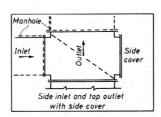

Plate I. *Phantom view of multicyclone assembly.* (*By courtesy of* Western Precipitation Corp., Los Angeles)

Undoubtedly the biggest advantage in small tubular collectors is the great saving in space compared with orthodox cyclones. The small diameter tubes, handling anything between 100 to 2,000 ft.^{3}/min each, depending on size and water gauge, demand nested assemblies of many units to handle large volumes, and some authorities consider any efficiency gain in using small tubes is more or less nullified by having to use such assemblies, with their associated gas distribution and dust collection problems, but of course these problems can be solved by proper design.

A typical reverse flow assembly of multicyclones is illustrated in *Plate I* which also shows the compactness and versatility of the layout. Dusty gas enters at the far side and the cleaned gas leaves vertically upwards at the top; the layout can easily be rearranged to permit the outlet to be in the near side.

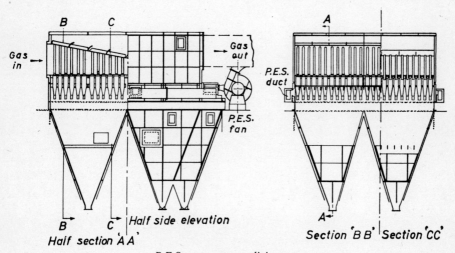

P.E.S. = *pressure equalizing system*

Figure 13. Centicell collector. (*By courtesy of* James Howden & Co. Ltd.)

In such a nested assembly it is usual to arrange for adjacent rows to have the rotation of the whirls or vortices in opposite directions so that the vortices from the cyclone dust outlets do not act against each other and upset the flow pattern. A further improvement is claimed if a small depression is given to the common hopper to equalize any difference in pressure at the bottom openings of the cyclones, thus preventing inter-cell flow by the use of a small fan handling approximately 3–5 per cent of the main flow; such a system is shown in *Figure 13*.

Sometimes difficulties arise owing to severe erosion of the cyclone tube at the bottom end where the conical outlet end begins to decrease the diameter of the tubes. This erosion can be avoided (see *Figure 14*) by using a bottom plate *A* formed as a helix with a central hole *B*. The dust and gas leave the cyclone through the annular space *A* between the helix bottom plate and the cyclone tube end *C*, the dust falling out into the hopper whilst the gas swirl returns to the cyclone body through the central hole B. This slotted bottom plate can be so arranged that it begins to vibrate violently under the action of the turbulent gas whirl at the bottom of the tube, and in doing so hits a stop-bar D welded to the bottom end of the tube. Here vibration also helps very efficiently to overcome the clogging of the tube outlet ends.

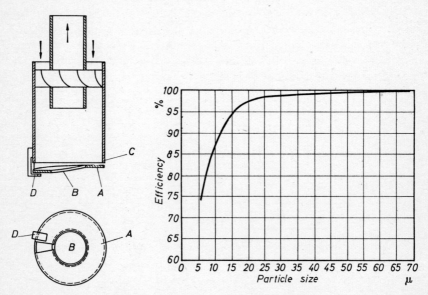

Figure 14. *Vibrating end plate on reverse flow cyclone*

Figure 15. *Typical grade efficiency curve for straight flow multiclone type cyclone*

The other development on nested multiclones concerns the so-called 'straight flow' as distinct from reverse flow discussed above. The straight flow type has a considerably higher handling capacity, size for size, but does not have the high efficiency obtainable with the reverse flow type.

A typical example of a straight flow cyclone is shown in *Figure 3* and the particle size–efficiency curve is given in *Figure 15*. Comparing these efficiencies with *Figure 16* shows the reduction compared with the reverse flow type.

The adoption of the term 'straight flow' is obvious. Reference to *Figure 3* shows the dust-laden gas stream entering the spinner vane which imparts an intense swirl to the gas, throwing the dust outwards centrifugally where it is carried to the outlet end of the cell and is assisted through an annular slot into the dust header by a secondary flow of gas, whilst the main volume of gas from which the dust has been extracted passes through the clean gas outlet under the action of the induced draught fan.

The use of a secondary air flow is essential to maintain a good efficiency and calls for a booster fan which would handle some 10–25 per cent of the main gas volume. This percentage skim-off carries a considerably higher dust concentration, which is recovered in the secondary collector system where reverse flow cyclones of either the small tubular type or a few relatively small diameter orthodox scroll inlet type are used. The secondary

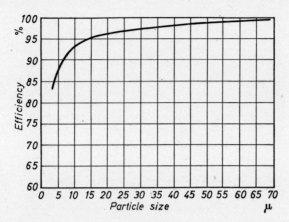

Figure 16. *Typical grade efficiency curve for reverse flow multiclone type cyclone*

air flow from the booster fan is then returned to either the inlet to the main cyclone assembly, where the dust loss from the secondary circuit can be re-treated, or into the cleaned air behind the main cyclone assembly.

Relative Costs

The following figures for a stoker fired boiler installation give some idea of the relative numbers of cyclones and cost for a given volume, between a reverse flow cyclone assembly and the alternative of straight flow with reverse flow secondary system, the grits containing only 2·5 per cent smaller than 10 μ.

Table 3. *Comparison between Reverse Flow and Straight Flow Cyclones*

Volume 75,000 ft.³/min	Water gauge in.	Number of units	Cyclone diameter in.	Cost
Reverse flow assembly	2	320 cyclones	7	£x
Straight flow assembly Reverse flow secondary system	2 3	210 cyclones 39 reverse flow cyclones	6 7	£0·5x

One important feature with straight flow assemblies is that the efficiency curve is relatively flat over a wide range below the design capacity. This characteristic is given by the secondary flow fan handling a constant volume irrespective of the reduction in the main flow. A secondary fan system to draw off, say, 15 per cent of the main flow at design rating will, of course, handle 30 per cent of the main flow under half load conditions, so the increase in efficiency given by the higher percentage of secondary flow offsets the drop in efficiency due to the lower water gauge loss set up by the main flow at half load.

GAS FLOW IN CYCLONES

Many efforts have been made to improve efficiency and these have been directed towards better control of the dust particles when once they are separated from the stream, or towards increasing the centrifugal force exerted upon the particles.

Nearly all mechanical separators are characterized by turbulent gas flow; that is, the flow of the gas stream is agitated or violent and contains many local eddy currents and whirls which in part are opposed to or detract from the main gas flow.

It has been found that the efficiency of a given device is improved if the flow can be maintained as nearly as possible in a streamline or laminar condition which can be defined as non-turbulent. This is possible only when there are no localized aberrations of the stream flow which produce transient changes in direction and velocity of portions of the gas stream, causing losses and friction in the stream.

It is known that if gas flows in a tube or between two parallel spaced surfaces the flow will be laminar up to a Reynolds number of approximately 2,000 and then rapidly becomes turbulent. Now it has been found [7] that by rotating gas streams under the influence of centrifugal forces laminar gas flow can be obtained even at a Reynolds number of up to approximately 10,000.

By using laminar instead of turbulent gas flow the efficiency of the cyclone can be considerably increased, and at the same time the pressure drop across it can be decreased very noticeably, while handling the same gas volume as under turbulent conditions.

THEORETICAL COLLECTION EFFICIENCY

By mathematical analysis it can be shown that the theoretical collection efficiency of a cyclone or centrifugal type dust collector with turbulent flow follows an exponential law, the collector efficiency being that fraction of the entering dust which is separated from the dust stream and retained within the collector. This exponential law may be expressed generally by the statement that in the centrifugal type collector, under given uniform conditions of gas flow, each unit area of collecting surface in succession, from the gas inlet toward the gas outlet, collects the same percentage of suspended particles of any given size in the gas stream flowing over this unit surface. This law of collection holds true only if the centrifugal force acting on the dust particles is always, on an average, of the same magnitude throughout the length of the collector, and therefore this condition is assumed to exist in order to give a common basis for analysis of various collectors, although minor variations in the centrifugal separating force may occur in actual practice. Expressed mathematically, the collection efficiency of a turbulent flow centrifugal type dust collector is as follows:

$$\Upsilon = (1 - e^{-\omega F}) 100 \text{ per cent} \qquad \ldots \ldots (5)$$

where Υ (%) = collection efficiency
 e (–) = base of naperian logarithms
 ω(m/s) = radial migration velocity of the particles within the collector measured in a direction perpendicular to the collecting surface
 F(m²/m³/s) = the specific collecting area expressed in square metres per cubic metre of gas per second
 m (–) = metre
 s (–) = seconds

According to the above expression collection efficiency can be calculated, strictly speaking, only for particles of any one size; but for practical purposes a given calculation may be assumed to apply to particles within a narrow size range. The collection efficiency for each one of a number of such ranges of particle sizes can be calculated separately and then the several efficiencies averaged in order to obtain the overall efficiency when the stream carries dust particles over a wide range of sizes. For example, to determine the overall collection efficiency for particles ranging in size from 0 to 10 μ, the efficiency for each of ten equal fractions of one micron size range can be calculated individually and then averaged.

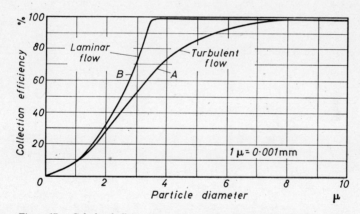

Figure 17. *Calculated efficiency curves for turbulent flow and laminar flow cyclones*

For particles of 10 μ and less in diameter, the collection efficiency for particles of given sizes, as determined by the above formula, when plotted appears as curve *A* in *Figure 17* where the collecting chamber is assumed to have a diameter *D* of 0·125 m and a specific collecting area *F* of 1·41 m²/m³/sec. The specific gravity of the particular dust in this case is 2·6 g/cm³.

It can be demonstrated mathematically that the collection efficiency Υ of a similar centrifugal type collecting unit in which the gas flow is laminar follows the law:

$$\Upsilon = (\omega F) 100 \text{ per cent} \qquad \ldots (6)$$

where ω and F have the same meanings as before. Assuming a collecting unit of the same physical characteristics as before, the calculated collection efficiency for particles of given sizes in a centrifugal type dust collector in which the flow is laminar is plotted as curve B in *Figure 17*.

It will be noted that at the upper range of particle sizes the maximum theoretical collection efficiencies both converge toward 100 per cent, whereas the theoretical possible collection efficiency under laminar flow conditions is substantially greater for particle sizes in the range of 2–7 μ with a maximum difference at about 3·5 μ. For example, for a particle size of 0·04 mm or 4 μ, the collection unit with turbulent flow collects only about 73 per cent, whereas with laminar flow the same unit collects approximately 99 per cent of the particles. It will be understood that these calculated efficiencies are ideal and can never be reached in actual practice, nevertheless, they can be approached and substantial improvements made in collection efficiencies by properly controlling the character of the gas stream flow.

The character of fluid flow can be predicted or determined in a satisfactory manner from the Reynolds number, which is widely used for this purpose. The Reynolds number is the well-known dimensionless constant, being a function of the dimensions of the fluid duct (or of a body around which the fluid is flowing), and the velocity, density and viscosity of the fluid, as expressed in the following formula:

$$R_e = \frac{v d \varrho}{\eta} = \frac{v d}{K} \qquad \ldots (7)$$

In the present case the fluid is a gas and the symbols have the following meanings:

v (cm/s) = gas velocity between the guiding surfaces
d (cm) = distance between the guiding surfaces
ϱ (g/cm^3) = specific gravity of the gas
η (g/cm) = absolute viscosity of the gas in poises
$K\left(\dfrac{\eta}{\varrho}\right)$ = kinematic viscosity of the gas

There are several factors in centrifugal type separating units that tend to stabilize flow conditions keeping them laminar rather than turbulent, as long as the Reynolds number is below approximately 10,000. The whirling gas stream tends to move in a spiral path without being diverted into local eddies. The comparatively high centrifugal acceleration acting on the stream is probably the reason for this behaviour. Also, the dust suspended in the stream increases its actual viscosity by an indefinite amount so that flow is more nearly laminar than for the same gas without any suspended particles, other conditions remaining the same. Friction and roughness at the bounding surface increase turbulence. Hence, by reducing friction and keeping all surfaces smooth and streamlined, normal tendencies toward turbulence are reduced and flow is stabilized as laminar at a higher Reynolds number than would otherwise be possible. By taking full advantage of these minor factors flow can be kept predominantly laminar in a centrifugal separating unit at values of about 10,000 or less for the Reynolds number produced by improved design.

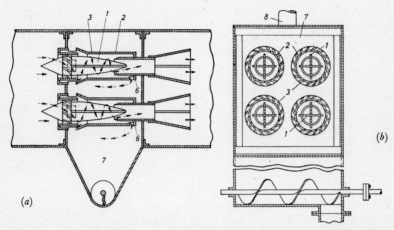

Figure 18. (a) *Sectional elevation of laminar flow cyclone assembly;* (b) *end view laminar flow cyclone assembly*

Assume a typical conventional centrifugal separating unit of circular cross section and a radius of 6·25 cm, through which a stream of air is moving under a head within the normal pressure drop range of 1·5 to 4·0 in. of water column across the separating unit. If the moving stream of air has a linear velocity of 1,500 cm/sec, and a temperature of 200° C, the density ϱ is 0·000742 g/cm³ and the absolute viscosity η is 0·00026 poises, giving a kinematic viscosity of 0·350. Substituting these values in the formula for the Reynolds number

$$R_e = \frac{vd}{K} = \frac{1{,}500 \times 6{\cdot}25}{0{\cdot}350} = 27{,}000$$

approximately. Gas flow in a dust collector having these characteristics is always turbulent.

By comparison consider the flow conditions in a separating unit designed for laminar flow as shown in *Figure 18*. Here core member (1) occupies a position at the centre of cylindrical tube (2), confining free gas flow to an annular zone bounded on the outside by the internal surface of tube (2) and having its inner boundary determined by the outer edges of radial vanes (1). Assume that the minimum distance between the wall of tube (2) and the outer edges of core member (1), which occurs at the point of maximum diameter of core member (1), is 1·25 cm or one tenth the diameter of the separating tube which has been assumed in the previous example to have an internal diameter of 12·5 cm. If the gas stream has the same velocity (1,500 cm/sec) as it moves through a separating unit (3), as illustrated in *Figure 18*, at a temperature of 200° C, then the Reynolds number may be calculated as follows:

$$R_e = \frac{vd}{K} = \frac{1,500 \times 1\cdot 25}{0\cdot 350} = 5,400 \text{ approximately.}$$

Under these conditions the gas flow may be assumed to be stabilized in a laminar condition, or at least a predominantly laminar condition, since laminar flow exists for most of the distance between the core and tube.

Laboratory tests with a collecting unit of this type demonstrate that with a given pressure drop across the collecting unit, for example 3 in. of water, the gas volume that can be passed through a centrifugal collecting unit having laminar flow is about 25–30 per cent higher than the gas volume which can be passed through a unit with the same size collecting tube (2) but having turbulent flow. Apparently this condition results from the fact that the gas stream, under turbulent conditions, uses up a substantial amount of energy in useless work. This energy goes to maintain the many internal eddies or local agitations in the main stream and does not help to maintain the main gas stream. When laminar gas flow without local agitations can be established, the energy required to maintain the gas flow at a given rate through the separating unit is substantially decreased; and conversely, for a given pressure drop across the tube a greater volume of gas can be passed through the collection unit per unit time.

A particular advantage of the form of core (1) illustrated is that it is believed to establish a desirable degree of laminar flow with a minimum frictional resistance to flow. Apparently in each of the four quadrants of the cross-shaped core, gas swirls [as at (4) in *Figure 19*] are set up by the frictional drag of the main stream rotating in the annular zone around the core. The swirl in each quadrant rotates at its outer surface in the same

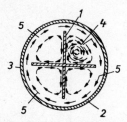

Figure 19. Gas flow paths in laminar flow cyclone core

direction as the main stream so that there is little, if any, frictional drag between the main stream and the localized swirl in each quadrant. The effect is to cause the main gas stream (5) to roll around and on the swirls (4) as if rolling on roller bearings. The result is stability of the main gas flow with a minimum of friction or energy loss by virtue of contact with core (1).

The gas swirls in the quadrants of the core have little, if any, movement in the direction of the axis of the separating tube. This fact lends stability to the main gas flow and helps maintain its laminar character. Any dust that enters these swirls is thrown out in a short time because the centrifugal force applied to it while in the swirl is relatively high.

In a centrifugal dust collector of the type discussed, a small fraction of the total gas stream travels along the wall of each separating tube (2) to the end where it leaves the tube through dust outlet (6). This part of the gas assists in discharging separated dust from the collecting tube and carrying it into the dust fall chamber (7). This type of circulation is maintained by fan (8) (*Figure 20*) withdrawing gas from chamber (7) through duct (14). Since the dust content of this gas is relatively high it is necessary to pass the air through some type of secondary collector, as indicated at (9) in *Figure 20*.

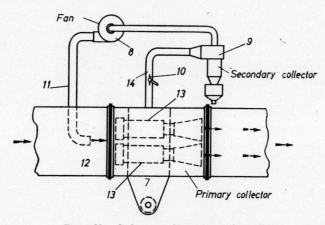

Figure 20. *Cyclone assembly with secondary collector*

The fraction of air so withdrawn is customarily in the neighbourhood of 10–15 per cent of the main gas stream, the quantity being regulated by any suitable means, such, for example, as damper (10) in duct (14). After withdrawal from chamber (7) it may be discharged to the atmosphere or reintroduced into the gas stream by duct (11) which discharges it to the main inlet duct (12) upstream from separators (13). As a result, this fraction of the total gas stream is continually recirculated.

In the present instance, the recirculation of a portion of the gas stream is utilized to assist in controlling and determining the Reynolds number of the collecting units. It will be noted from formula (7) that the Reynolds number of a particular construction decreases with an increase in the gas viscosity; and the viscosity of the gas stream can be increased by increasing the concentration of dust particles suspended in it. Hence, by recirculating through the cyclone a portion of the gas and with it some of the dust previously extracted from the hopper, the dust content of the stream delivered to the inlets of the several separators (13) can be increased over the concentration of dust otherwise appearing in the gas stream.

Extensive tests in the laboratory have proved that the efficiency of a well-designed cyclone with turbulent gas flow can be calculated with a reasonably good accuracy according to the formula:

$$\varUpsilon_{Tu} = (1 - e^{-\omega F}) \; 100 \; (\text{per cent}) \qquad \ldots \ldots (8)$$

and the efficiency of a cyclone with predominantly laminar gas flow can be calculated according to the formula:

$$\varUpsilon_{La} = (\omega F) \; 100 \; (\text{per cent}) \qquad \ldots \ldots (9)$$

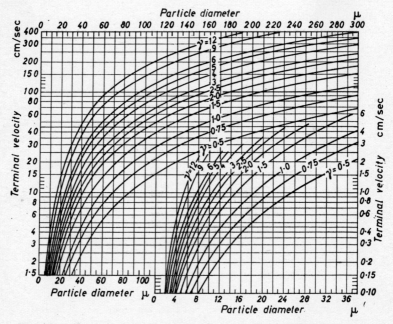

Figure 21. Terminal velocity of dust particles in air (20° C, 760 mm mercury) for different particle sizes and different specific gravities (according to A. Winkel)

This calculation can be made quite easily if the chart *Figure 21* is used. The chart shows the terminal falling velocity w_1 of dust particles in air of 20° C, 760 mm mercury, having a specific gravity over the range of 1 to 12 g/cm³ under the influence of gravitational acceleration of g. If the average centrifugal force which is acting in a cyclone on the dust particles is known in x times g (981 cm/s⁻²) then the migration velocity ω_2 (cm/s) of a dust particle of a known diameter and a known specific gravity will be, in air of 20° C, 760 mm mercury:

$$\omega_2 = \omega_1 . x . g \, (\text{cm/s})$$

If the viscosity of the gas η_2 passing the cyclone is different to that of the air η_1 at 20° C, then the migration velocity of the dust particle in the cyclone will be

$$\omega_2 = \omega_1 . g . x . \frac{\eta_1}{\eta_2} (\text{cm/s})$$

As the specific collecting area

$$F(m^2/m^3/sec)$$

of the cyclone is known, the efficiency of the cyclone on each particle size, say in steps of 1 μ, can now be assessed according to the formulae (8) and (9), and by adding up the different grade efficiencies thus calculated, the total efficiency of the cyclone can be found. As it would be difficult to calculate the average centrifugal force, $\omega_1 . x . g$ active in the cyclone because of the turbulent gas flow, the easiest method would be to find it by tests made with a dust feed of known particle size distribution by measuring the gas volume passing the cyclone, the pressure drop across the cyclone, and the grade efficiency. From these figures the virtually active centrifugal force in the cyclone at the given gas volumes can be calculated easily.

After the average centrifugal force acting in a certain type of cyclone at different gas throughputs has once been established, it is comparatively easy to calculate the total expected efficiency for the cyclone when it handles dusts with a different particle size distribution and gases with a different viscosity.

POWER CONSUMPTION OF CYCLONE PLANT

The power input to the fan motor is in the range of 0·25 kW per 1,000 ft.³/min per inch pressure drop if the efficiency of the fan is approximately 60 per cent.

Example
>Gas volume = 10,000 ft.³/min
>Pressure drop for cyclone inclusive of ducting, *etc.* = 10 in. w.g.
>Total power input $P = 10,000 \times 10 \times 0.25 = 25$ kW.

Plant Cost

Depending on the size, efficiency and design of the cyclone, the approximate cost[5, 10, 11] for such plant is in the range of 1s. to 3s. per installed cubic foot per minute gas capacity with total charges (including capital and running costs) ranging from 0·01d. to 0·02d. per 1,000 ft.³ of gas cleaned.

For particle size of various dusts and data concerning dust generated in various industrial processes readers are referred to the figures and references in the section on Electro-Precipitation.

PRACTICAL APPLICATIONS

The cyclone has been successfully applied to hundreds of thousands of industrial dust collection plants embracing almost all types of dusts, and it would be outside the scope of this article to detail every application; however, readers may like to know of the most recent application on open cycle gas turbines.

Cyclone Ash Separators on Open Cycle Gas Turbines [9]

Up to the present, the only methods used for separating from the combustion gases material likely to damage the blades of an open cycle gas turbine have been dust collectors of the cyclone type.

Elementary theory suggests very high cleaning efficiencies for cyclones, but is unable to allow for secondary flows which seriously disturb the performance.

Some work to elucidate the factors influencing the design of cyclone type dust separators for use under conditions of temperature and pressure in an open cycle gas turbine has been carried out by the British Coal Utilisation Research Association.[9] Most work has taken the form of experiments to determine the collection efficiency of various designs of cyclone at atmospheric pressure and temperature, using a prepared dust of known size grading, but some experiments with the same dust have been carried out with two different designs of cyclone at pressures up to 4·5 atm and temperatures up to 1,400° F (760° C). These experiments showed that collection efficiency for a given collector and type of dust was dependent on the ratio of gas velocity at the inlet to the absolute viscosity of the gas.

In addition, tests on three cyclones of one design built in three sizes have indicated that the dust collection efficiency is dependent on $v/\eta r$, where v is the gas velocity at the inlet, η the absolute viscosity of the gas, and r the radius of the cyclone body, and that the pressure loss is proportional to the inlet velocity head. The implication of this is that to obtain a given cleaning efficiency with the minimum pressure drop through the cyclones, a large number of small cyclones rather than a small number of larger cyclones should be used. However, the use of a large number of small cyclones introduces problems in the design of both the ducting to distribute the flow between them equally, and the arrangements for extracting the dust. Because of the great need for simplicity in equipment which has to operate at turbine inlet temperature and of the need to minimize the volume of gas contained in the dust collectors to facilitate control of the engine, a high gas throughput for each collector is desirable. The gases leaving most conventional forms of cyclone dust collectors do so with considerable swirl energy. Investigations have shown that to decrease the overall pressure drop by recovering some of the swirl energy in the form of pressure without also affecting the collection efficiency adversely is difficult, but even if the fitting of means for recovering swirl energy has no beneficial effect on the collection efficiency–pressure drop relation, it may still be justified if it increases the throughput of the collector for a given pressure drop.

The types of cyclone dust collector which have been studied have included: (1) reverse flow cyclones with an assembly of vanes in an annulus at the entry to impart swirl to the gases; (2) a tangential entry type in which the gases enter tangentially to the main body of the cyclone through a narrow slot extending over a large proportion of the length of the cyclone and leave through both ends, the dust being carried out in two 'blowdowns' situated in the walls at each end; and (3) reverse flow cyclones of the long cone type in which the gases enter through a small, nearly square, tangential entry at the top and leave through an axial tube in the top, the body of the cyclone being part cylindrical and part conical. The indications from dust collection efficiency determinations at atmospheric pressure are, that for cyclones of the same diameter the long cone cyclone gives the highest efficiency of the three for a given pressure drop, and the vane entry collector the lowest, but the long slot tangential entry collector gives the highest throughput for a given diameter and amount of space occupied. The higher throughput of the long slot tangential entry collector is in part due to the fitting, to the exit pipes, of simple conical diffusers which, for a given flow, reduce the overall pressure drop by about 40 per cent without significantly reducing the collection efficiency.

With dust collectors working at turbine inlet temperature, there is a danger that accumulations of lightly sintered dust may form in the collector and that these accumulations may periodically break away and choke the dust blowdown lines which must be of small bore to prevent dust settling out in normal operation. Passing gases from a cyclone type of combustor through different collectors has shown that accumulations of dust are less likely to form in the long slot tangential entry collector than in collectors with a vane type entry, and that keeping the blowdown pipes separate and fitting restrictors in them, so that a pressure difference is thrown across an incipient blockage, is very beneficial in keeping the lines clear.

Future Trends in Gas Cleaning

The experience so far obtained seems to indicate that with open cycle gas turbines burning solid fuel, abrasion of turbine blades by ash particles can be avoided by the use of cyclone type dust separators of high efficiency, but the development of equipment which will give the required efficiency without imposing an excessive pressure drop or entailing a high loss in 'blowing down' the dust, which will be free from troubles due to choking, abrasion or thermal expansion and which, at the same time, will not be excessively complicated or costly, will require much ingenuity.

It is possible that centrifugal dust separators with mechanically driven walls may be employed, possibly being incorporated in the turbine itself, as in this way the effects of boundary-layer flows—one of the main reasons for inefficiency in static cyclones—can be reduced considerably. In regard to deposition, there is considerable scope for making better cleaners than at present exist, but it is unlikely that any practical form of gas cleaning system will eliminate this and its incidence is likely to depend greatly on the characteristics of the fuel being burnt and on many other factors. Techniques for removing deposits by washing or other means need development.

REFERENCES

[1] ROSIN, P., RAMMLER, E., and INTELMAN, W., *Z. Ver. dtsch. Ing.*, 76, 1932, 433–7
[2] TER LINDEN, A. J., *Proc. Instn mech. Engrs, Lond.*, 160 (1949) 224–51
[3] SHEPHERD, C. B., and LAPPLE, C. E., *Industr. Engng Chem. (Industr.)*, 31 (1939) 972–84
[4] STAIRMAND, C. J., *Trans. Instn chem. Engrs, Lond.*, 29, No. 3, 1951
[5] STAIRMAND, C. J., *Chem. & Ind. (Rev.)*, 1955, 1324–30
[6] BARTH, W., Brennstoff-Warme-Kraft, Bd. 8, H.i, 1956, pp. 1–9
[7] British Patent 713,930
[8] ANDERSON, E., 'Effect of Tube Diameter in Cyclonic Dust Collectors', *Chem. metall. Engng*, vol. 40, No. 10, October 1933
[9] BATTOCK, W. V., HURLEY, T. F., and VOYSLY, R. G., *Proc. Jt Conf. Combustion*, 1955

17

ELECTRO-PRECIPITATION

R. F. HEINRICH and J. R. ANDERSON

HISTORY

THE earliest reference to electrostatic precipitation shows that the Italian, Beccaria, was the first to be credited with observing, in 1771, the phenomena that occur when an electric discharge takes place in a gas entrained smoke. In 1824 Hohlfeld of Leipzig found that when a charged wire was held in a smoke-filled bottle the smoke rapidly cleared and a deposit collected on the bottle, an experiment which was repeated independently by Guitard in 1850 and by Lodge in 1880; it was not until 1904, however, that serious study of these phenomena was undertaken by Walker and Lodge in Britain, by Moeller in Germany, and by Lewis, Davidson, Marscher, Dion, Blake, and especially Cottrell, Schmidt and others in the United States.

PRINCIPLES OF ELECTRO-PRECIPITATION

The basic principle of electro-precipitation is the removal of solid or liquid particles from a gaseous carrying medium by giving the particles an electric charge and precipitating them on to a receiving surface in an electric field.

When a wire of small cross section, which at a few inches spacing is surrounded by a metallic cylinder (*Figure 1*), is charged to a high potential, an intense electric field is created in its immediate vicinity. The air or gas surrounding the wire always contains approximately 200 to 1,000 electrons or ions per cubic centimetre which are continually created by light rays, cosmic rays or atomic radiation. Any electron or ion present in the intense electric field near the surface of the wire is so greatly accelerated that on striking a molecule it can set free another electron: both electrons are again accelerated, striking other molecules and setting free more electrons, so that in the end an eruption of negative and positive ions is brought about and so-called shock ionization takes place in a small belt around the wire. At a certain distance from the charged wire the field strength diminishes to a point at which electrons are no longer sufficiently accelerated to release further electrons by shock ionization; at this distance the corona discharge belt ends. If the wire is at negative potential then the positive ions migrate to the wire where they give up their charge on touching it. The negative ions migrate to the positively charged receiving electrode.

The discharge belt is visible as a bluish glow around the discharge electrode wire, and if closely observed, the belt can be seen to consist of numbers of separate glow points of somewhat unstable intensity; transportation of ions takes place in the dark space beyond the belt. In order to maintain the discharge fresh electrons must be produced at the wire: photo-electric effects may be involved, but the main source appears to be electron emission from the wire under the bombardment of the oncoming positive ions. The ion concentration in the discharge belt is in the region of approximately $10^8/cm^3$.

On their path from the discharge to the receiving electrode the gas ions and any free electrons collide with suspended dust or mist particles in the gas by which they are adsorbed to a certain amount so that the particles become electrically charged. Under the action of the electrical field between the electrodes the charged particles then migrate to the receiving

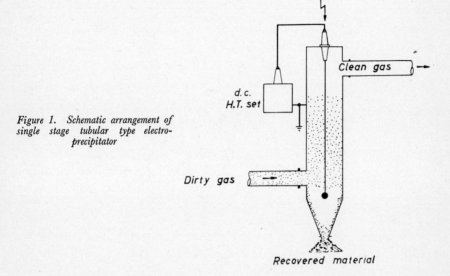

Figure 1. Schematic arrangement of single stage tubular type electro-precipitator

electrode where they are precipitated in a more or less thick layer and give up their charge in contact with the receiving electrode. Liquids drain off the receiving electrode easily, whereas the dust layer has to be shaken off the receiving electrodes periodically by vibration. The dust then falls by its own weight into a hopper arranged underneath the electrodes from which it can be removed periodically or continuously, *Figure 1*.

Discharge electrodes are normally at negative potential, because negative corona discharge is more stable than positive and the applied voltage can be maintained at a higher value as indicated in *Figure 13*. Positive corona is used, however, in two stage precipitators used for air cleaning purposes in order to minimize the generation of ozone.

The corona current obtained at a given voltage depends on the spacing of the electrodes, the radius of curvature of the discharge electrodes and the ion mobility, *Figure 22*, of the gas. The ionic current in a concentric discharge system is given by the equation

$$i = \frac{2b}{R^2 \ln \dfrac{R}{a}} (U - U_a) U \qquad \ldots \ (1)$$

where i = corona current per centimetre of discharge electrode (stat. A/cm)
  b = mobility of ions (cm/sec/stat.V/cm)
  U = applied voltage (stat. V)
  U_a = corona starting voltage (stat. V)
  R = radius of receiving electrode cylinder (cm)
  a = radius of discharge electrode wire (cm)

It will be seen that the corona current depends mainly on the applied voltage, the radius of the receiving electrode cylinder, and the radius of the discharge wire. The difference between the applied voltage and the corona starting voltage also appears in the equation: the latter depends on the corona starting field strength, which can only be arrived at experimentally, though an empirical equation has been given by Whitehead.

$$E_a = 30 \cdot 5 \left(\frac{1 + 0 \cdot 305}{\sqrt{a}} \right) \text{kV/cm} \qquad \ldots \ldots (2)$$

where E_a = corona starting field strength (kV/cm). In an undistorted electric field of coaxial cylinders (a round discharge electrode in the centre of a tubular receiving electrode as shown in *Figure 1*), the field strength at the inner cylinder is given by the equation

$$E = \frac{U}{a \ln \frac{R}{a}} \qquad \ldots \ldots (3)$$

and the required voltage is accordingly

$$U_a = 30 \cdot 5 \left(1 + \frac{0 \cdot 305}{\sqrt{a}} \right) a \ln \frac{R}{a} \qquad \ldots \ldots (4)$$

It will be seen that decreasing the radius of the wire reduces the field gradient required, and that the voltage necessary to produce this gradient increases approximately in proportion to the increase of radius. In practice it is undesirable to operate precipitators at very high voltages, and discharge electrodes are therefore kept as thin as possible consistent with adequate mechanical strength.

The characteristics of the gas on the corona current is seen in the variation of ion and electron mobility. For instance, pure nitrogen absorbs electrons only to a very small extent. The discharge current flowing between the electrodes is therefore mostly transferred by electrons which have a very high mobility, and not by nitrogen ions which have a much lower mobility. The corona current in nitrogen is therefore quite high at a comparatively low voltage at the discharge electrode. On the other hand, there are so-called 'electro-negative' gases which absorb electrons very readily, for example, oxygen, the current being much reduced because it is transported mainly by slow-moving negative ions. The characteristics of air are about midway between these two extremes between which there are many variations. The above formulae, of course, apply only to dust-free gases (see later comment under 'Corona Quenching Effect').

The charging and precipitation of the particles can be accomplished either in one stage or in two. In a single stage precipitator the discharge electrodes not only set up the corona and ionization leading to charging of the particles, but also make use of the applied voltage as the source of potential to produce the electric field for the precipitation of the charged particles. This type of precipitator is most widely used as its construction is relatively simple and a common applied voltage is used for the two functions of charging and precipitation, as shown diagrammatically in *Figure 1*.

In a two stage precipitator these two functions are carried out separately and more effectively. The sole purpose of the first stage is to produce corona discharge and ionization of the air or gas, and thereby to charge the particles: in the second stage, which is specially constructed to prevent corona formation, the sole purpose is to create a high field strength in order to precipitate the charged particles on the receiving electrodes. The two stage precipitator is of more complex construction and usually demands two separate values of applied voltage, but with certain types of dust it is the only means of achieving high collecting efficiencies. Moreover, it is electrically the most efficient type in that less current is used; current is consumed only in the ionizing stage and consumption is limited to what is necessary for effective ionization. The second stage or 'static zone' theoretically consumes no current, its duty being merely to supply the field with a sufficient voltage gradient between the electrodes to precipitate the charged particles.

Corona Quenching Effect (so-called Corona Suppression)

Under conditions of dust and fume being present in the gases handled by a precipitator, the average mobility of the charge carriers, *i.e.* ions and dust particles, is lessened to quite a noticeable amount. As the number of dust or fume particles increases and approaches that of the free ions (10^8 cm^3) the carrier mobility will be reduced to that of the fine fume particles which adsorb all the available ions. Under these conditions the theoretical discharge current, as calculated according to equation (1), will be reduced to only a fraction, and in the case of very fine or dense fumes, even to a few per cent.

This may be illustrated in a typical case of a four zone precipitator handling zinc oxide fume. In this electro-precipitator the first zone passed only 2·5 mA with very little precipitation, but already noticeable agglomeration; the second zone passed approximately ten times as much, that is, 25 mA; the third zone passed 32 mA which is about a thirteen-fold increase; whilst the fourth and last zone, where the fume concentration was reduced by approximately 95 per cent, passed around 52 mA, an increase of about twenty times as much. The overall efficiency of this plant was in the region of 99 per cent. This clearly shows how the corona quenching effect can be overcome by correct design and the use of zonal treatment where the voltage and current can be individually adjusted to suit the conditions in each zone.

The corona quenching effect is found in most smelter fume precipitators and also in mist precipitators where the process produces extremely fine mist in high concentration. This effect applies under certain conditions to modern power station flue dust precipitators which have to handle the gases from large boilers with very high temperatures in the centre of the combustion chamber: and on cyclone, wet bottom or slag tap type furnaces.

Under such high temperature conditions, silica, silicates, and alkalies are volatilized[1, 2] and condense into extremely fine fumes of the order of $0 \cdot 1$ μ diameter on cooling below the condensing point. The corona quenching effect of these fine fumes can also be overcome by the use of several precipitator zones in series.

TYPES OF ELECTRO-PRECIPITATOR

Dry Precipitators

Dry precipitators are used for handling dust, fumes, *etc.*, in a dry state, and can be constructed for either horizontal (*Plate I*) or vertical gas flow. The precipitated particles are removed from the electrodes by mechanical rapping devices.

Irrigated Precipitators

Irrigated precipitators are generally similar to the dry type, but they include some means of continuously wetting the receiving electrodes so that the water and precipitated particles form a sludge which drains off the electrodes without mechanical aid.

Mist Precipitators

Mist precipitators (*Plate II*) are used for the removal of particles such as acid mist and tar fog, *Figure 2*. They are usually self-cleaning, as the precipitated particles are liquid and drain off the receiving electrodes.

CONSTRUCTION

General

In size and shape as well as in details and materials of construction, electro-precipitators vary widely according to the purposes for which they are used. For instance, in a precipitator for removing tar fog at a gasworks, or acid mist in a sulphuric acid plant, the gas may be treated in a relatively small metal vessel, whereas the dust precipitators at a power station[3], where large volumes of flue gas must be handled, may have reinforced concrete treatment chambers of relatively vast dimensions (*Plate III*). The choice of constructional material is also influenced by the temperature of the gas treated, the necessity to withstand corrosion, and similar factors.

Discharge Electrodes

Since the function of the discharge electrodes is to give a corona emission, it is necessary to use electrodes of small cross section or to provide them with sharp edges to promote the discharge. At the same time, the electrodes must be mechanically strong enough to withstand high temperatures,

Plate I. General arrangement of a plate type precipitator

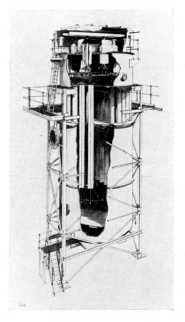

Plate II. Construction of the tubular type precipitator

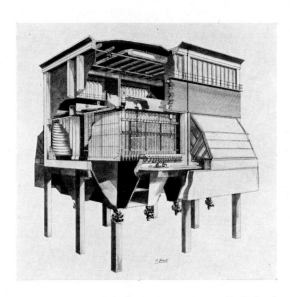

Plate III. Construction of plate type precipitator with reinforced concrete shell

Plate IV. Dust overhang on chute electrodes. Gas flow is from right to left

Plate V. Gas flow past chute electrodes, showing swirls set up behind the projecting fins

Plate VI. Behaviour of conductive dust on chute electrodes (left) and flat plate electrode (right)

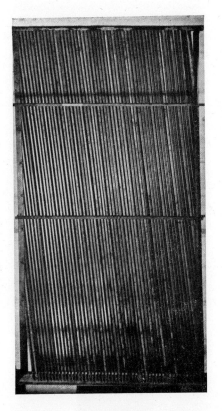

Plate VII. Chute electrode assembly for horizontal gas flow

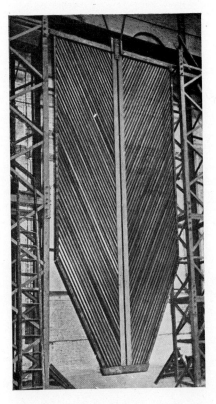

Plate VIII. Chute electrode assembly for vertical gas flow

Plate IX. Point-plate arrangement used for producing discharges which are the subject of the subsequent illustrations

Plate X. Back discharge from a layer of sulphur dust

Plate XI. A similar back discharge from a layer of sulphur dust but with additional spots of less intense back discharge

Plate XII. Back discharge from holes drilled into insulating paper

corrosion, and fatigue arising from mechanical rapping or vibration caused by varying electrical forces in the electrical field. They must be insulated from earth, and must be so arranged that they remain central between the receiving electrodes without any risk of swinging. In all circumstances they must be kept clean in order to maintain the necessary discharge.

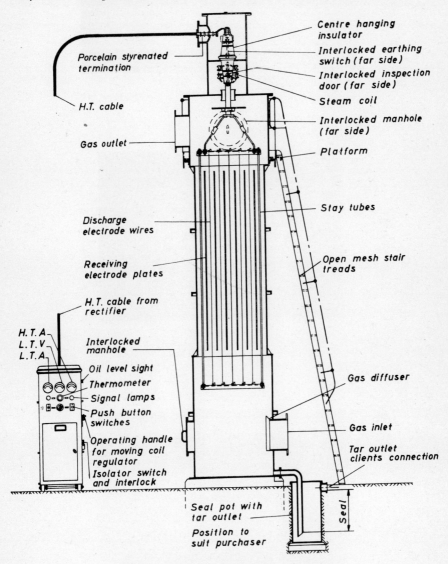

Figure 2. General arrangement of plate type detarrer

To meet these requirements in dry precipitators the discharge electrode assemblies must be light in weight so as to reduce mechanical stresses in the insulators on which they are supported; they must also be sufficiently rigid to remain central and taut, and so arranged that each discharge electrode can be effectively rapped.

In wet precipitators the requirements are less stringent since rapping is not used, the electrodes being either self-cleaning or needing only occasional flushing with water to keep them free from build-up.

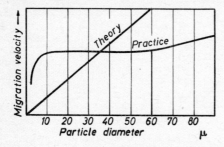

Figure 3. Relation between particle diameter and migration velocity

In practice one may expect with normal construction and applied kV an emission of up to 5 mA per 10 ft. length of discharge electrode when discharging in air, the current dropping to 0·1–2·0 mA per 10 ft. length when operating in dusty gases. This is equivalent to 5–100 mA per 1,000 ft.² of receiving electrode surface.

Receiving Electrodes

It is in the design of receiving electrodes that the highest skill and experience of the specialist are demanded in putting forward the most suitable type for the purpose in view.

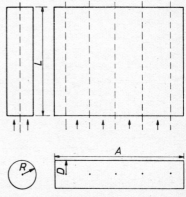

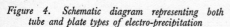

Figure 4. Schematic diagram representing both tube and plate types of electro-precipitation

Simple tubes or flat plates can be used in cases where the gas velocity through the field of treatment can be kept low enough to avoid erosion of the precipitated particles, *i.e.* blowing the particles off the electrodes and back into the gas stream; electrodes of these types have, however, a restricted field of application, for the gas velocity is the limiting factor, and to keep it below the critical point of approximately 2 to 4 ft./sec would in many cases involve enlarging the cross section of the precipitator to such dimensions that it could not be fitted into the available space. Among the alternative forms of construction used in appropriate cases are corrugated plates, rod-curtain and screen type electrodes, and electrodes of the graded resistance and pocket types.

A spectacular improvement in collection efficiency for a given receiving electrode area was brought about by the introduction of the so-called 'high duty' chute type[4] receiving electrode, which collects particles in such a way that they fall into the hopper in a quiescent zone undisturbed by the eroding effect of the gas stream; in fact the gas stream actually helps to keep the collected particles within the confines of the quiescent zone.

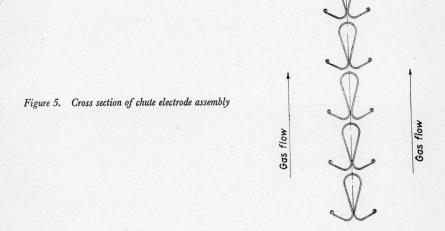

Figure 5. Cross section of chute electrode assembly

The chute type electrode element is a symmetrical rolled section of specially designed dimensions. The sections are assembled, as shown in Figure 5, to make up panels giving the required receiving area. These panel assemblies constitute the gas passes, and the projecting fins with their beaded edges form dust channels within which the precipitated dust is protected from the eroding effect of the gas stream. The electrodes are normally mounted at an angle to the vertical, so that the dust remains in contact with the channels while it slides down them to the collecting hopper. The risk of dust particles being blown off and re-entrained in the gas stream is thus reduced to a minimum.

Method of Dust Collection

The action of the chute electrode differs markedly from that of normal electrodes.

Figure 6 shows dust being precipitated on to a flat plate electrode in the ordinary way. The outer surface of the dust layer collected on the plate is of negative polarity owing to the continuous arrival of particles negatively charged by the discharge electrodes, but the inner dust surface attains the positive polarity of the collecting electrode with which it is in contact. When the electrode is rapped the particles adjacent to it are therefore repelled back into the gas stream and have to be precipitated all over again.

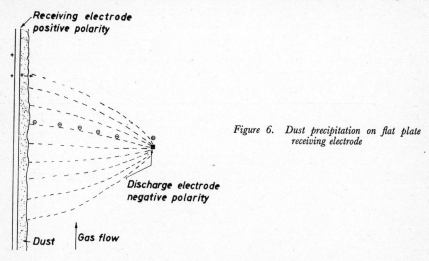

Figure 6. *Dust precipitation on flat plate receiving electrode*

The chute electrode prevents this undesirable phenomenon by causing the dust to be initially precipitated on the beaded edges of the projecting fins, where a dust overhang builds up on the lee side of the fins, as shown in *Figure 7* and *Plate IV*. The dust is thus kept a little distance from the

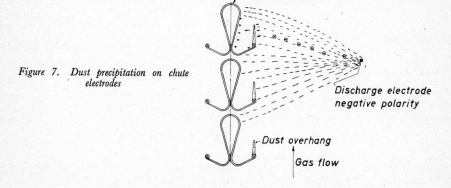

Figure 7. *Dust precipitation on chute electrodes*

main electrode surface, and as it has only a small area of contact with the electrode and is usually only semi-conductive, it retains some of its negative charge. When the electrode is rapped, the dust readily breaks away and retains enough negative charge to be attracted into the collecting channel instead of being repelled. Shielded from the eroding action of the gas stream, it then slides down the inclined channel into the collecting hopper. The formation of the dust overhang is aided by the small secondary gas swirls which occur behind the fins, as seen in *Plate V*. These swirls also help to draw the dust into the collecting channel when the dust overhang is loosened by rapping.

It is well known that conductive particles are not firmly precipitated on to flat plate electrodes, and that carbon particles in particular, being highly conductive, are specially prone to be repeatedly precipitated and repeatedly repelled and re-entrained, so that they tend to 'hop' or 'creep' through the precipitator without being retained. That this does not happen with chute electrodes is shown in *Plate VI*, which illustrates a test in which conductive dust was spread over an electrode consisting of a flat plate on the right and chute elements on the left; when voltage was applied to the discharge electrodes the conductive particles on the flat plate were violently repelled, while those in the channels of the chute elements were unaffected. The explanation lies in the dimensions of the chute electrodes; the projecting fins reduce the electrical field in the dust channels to an intensity too low to repel those particles which change their polarity on touching the receiving electrodes.

Construction of Chute Electrodes

Chute electrodes are equally effective for horizontal or vertical gas flow. *Plate VII* shows a receiving electrode assembly for horizontal flow, in which the inclined electrodes deliver the dust into a channel on the right of the assembly for discharge into the collecting hopper; *Plate VIII* shows an assembly for vertical flow, in which the two sets of inclined electrodes deliver the dust into central channels.

A feature of construction is that all the elements are suspended by their upper ends with their lower ends resting on spacers; in the type of assembly shown in *Plate VII* the spacer bars at the bottom are rapped, so that each element receives an effective rapping blow; in the assembly in *Plate VIII* the spacers are carried on the central channels, and the same effective rapping result is achieved by applying the rapping blow to the channels. This method has been found in practice to be far superior to orthodox methods by which comparatively large masses of material have to be vibrated.

Performance of Chute Electrodes

A test was carried out on a flue dust precipitator operating with a pulverized fuel fired boiler, half the precipitator being equipped with chute electrodes and the other half with pocket electrodes, both types having the same receiving area. For the purpose of the test, only one precipitation zone was used out of the two zones in each half of the precipitator. The tests on the pocket electrodes gave an average dust removal efficiency of 94·9 per cent, while the tests on the chute electrodes gave an average efficiency of 97·9 per cent; in other words the chute electrodes reduced the dust outlet burden by more than half. In terms of size, this means that chute electrodes would allow a plant for a given duty to be 30 per cent smaller; alternatively, it means that in a plant of given size the use of chute electrodes provides a substantial margin of performance, which is often important with new boiler designs.

Another existing vertical flow precipitator using a well established type of receiving electrode was rebuilt with chute electrodes. Formerly the removal efficiency was reduced after a short time by heavy dust build-up, which blanked off the slots in the receiving electrodes, and puffs of dust could be clearly observed when the electrodes were rapped: when chute electrodes were substituted they were found to be free from choking and almost self-cleaning, whilst the absence of dust puffs under rapping supported the claim that chute electrodes reduce particle re-entrainment.

With the advantages inherent in the design, the high duty chute electrode stands far ahead of other types and gives a higher efficiency in a plant of given size or a smaller and less costly plant for a given efficiency.

In a precipitator for a given set of conditions a certain area of receiving electrode surface is necessary, and if the whole of the area is to remain active an effective rapping system is essential. The receiving electrode assemblies must therefore be neither too massive nor too rigidly restrained in their mountings, failing which, only a limited area near the source of rap will be kept clean and deposits progressively build up on the remoter surfaces.

FACTORS INFLUENCING DESIGN

The most important factors which influence precipitator design are as follows.

Type of Material—Chemical analysis (content of alkali, mineral matter, carbon, *etc.*) together with physical characteristics and ability to agglomerate.

Resistivity—Electrical resistivity of the material under operating conditions.

Particle Size Range—The size analysis and the shape or shapes of the dispersoids.

Concentration—The dust, fume, mist or fog concentration, expressed in grains per cubic foot, at operating conditions.

Gas Composition—Moisture content and percentage of SO_2, SO_3, CO_2, CO, H_2, N_2, O_2, *etc.*

Gas Temperature—The expected range of operating temperatures of the gas to be treated.

Gas Volume and Pressure—The volume per minute and the pressure of the gas to be treated, and the range of variation expected.

EFFICIENCY OF COLLECTION

Efficiency Required

The efficiency of collection required has obviously a fundamental influence on the design of the plant[5] especially as regards its size. The basic function of particle collection in electro-precipitators is an exponential expression theoretically derived by DEUTSCH[6] and others[7] of the form

$$y = 1 - e^{-k} \qquad \ldots \ldots (5)$$

where y = efficiency, e = Napierian log base, k = factor

This equation, experimentally deduced by Anderson and Horne, has often been verified by tests[7, 8, 9]. The correlation of the factor k to the efficiency y is indicated in the following table:

y	60%	80%	90%	95%	97·5%	98%	99%
k	0·9	1·6	2·3	3·0	3·7	3·9	4·6

The factor k contains, among other factors assumed constant, the geometrical dimensions of the precipitator and is therefore a proportionality factor of the size. For instance, it can be seen from the figures given above that a precipitator for a collection efficiency of 97·5 per cent will be about four times as large as one for an efficiency of 60 per cent, the factor k having a value of 3·7 in the former case and 0·9 in the latter.

The problem may be put in another form: how much larger must a precipitator be in order to halve the outlet particle burden, *i.e.* to increase a given efficiency by half the difference between that efficiency and 100 per cent?

Table 1

Increase of efficiency y %		Increase of plant size (i.e. in value of k) %
from 60·0	to 80·0	78·0
,, 80·0	,, 90·0	43·5
,, 90·0	,, 95·0	30·5
,, 95·0	,, 97·5	23·0
,, 98·0	,, 99·0	18·0
,, 99·0	,, 99·5	15·0

It will be seen from *Table 1* that the higher the initial efficiency, the smaller the percentage increase in plant size to halve the outlet burden; but it is apparent from the preceding table that halving the burden always demands the same absolute increase in plant size, since the values of the factor k have the same difference of 0·7 for efficiency changes from 60 to 80 per cent, from 80 to 90 per cent, and from 90 to 95 per cent and so on. This reasoning is, of course, only theoretical and could only be applied if the dust being precipitated consisted of particles all having the same size of, say, 10 μ which in practice is seldom the case. If the dust consists of a large range of particles having various diameters, then a precipitator giving a very high efficiency of, say, 99·5 per cent, must be built larger than indicated above because small dust particles have a lower migration velocity in the electrical field than larger ones and it takes a longer time to precipitate the small particles with the same efficiency as the larger particles (see *Figure 3*).

Geometrical Dimensions of Precipitator as Contained in Exponent k

The constant k has been expressed in various terms according to practical requirements, as for instance:

(6) $k = wF$

(7) $k = w\dfrac{2L}{Rv}$

(8) $k = w\dfrac{L}{Dv}$

(9) $k = w\dfrac{SL}{Qv}$

(10) $k = w\dfrac{SL}{G}$

$k(-)$ = exponent
w(m/sec) = migration velocity
F(m²/m³/sec) = specific collecting surface
L(m) = length of collecting electrode in direction of the gas stream
v(m/sec) = gas velocity
R(m) = radius of tube
D(m) = spacing between discharge and collecting electrodes
S(m) = active circumference
Q(m²) = cross sectional area
G(m³/sec) = volume rate of flow

It will be seen that the constant k consists of two terms, namely, the factor w and a set of factors designating geometrical dimensions and gas velocity v. The most general of the above expressions is (6). It has become usual to regard w, which has the dimensions of a velocity, as the average migration velocity of the particles towards the collecting electrode. F has the dimensions of an inverse velocity, but if suitably enlarged as proposed[10], it attains a rational meaning:

$$F(\text{sec/m}) = F(\text{m}^2\text{sec/m}^3) = F\left(\dfrac{\text{m}^2}{\text{m}^3/\text{sec}}\right) \quad \ldots \ldots (11)$$

The last of these dimensions suggests that F constitutes a 'specific collecting surface', that is, a certain collecting surface per unit volume of gas treated per second.

Expression (7) is familiar in the design of tube type precipitators, L being the active length of the collecting surface in the direction of the gas stream and R the radius of the tube (see *Figure 4*). For plate type precipitators, the expression (8) is substituted, D being the distance between the discharge and collecting electrodes. Expression (8) is more general in character and disregards the cross sectional shape of the individual gas passages. This formula is well suited to the calculation of precipitators using tubes of other than circular section, S being the total active circumference and Q the total free cross sectional area. This method of expressing the constant k explains the difference between expressions (7) and (8).

For circular tubes the ratio S/Q is

$$\frac{S}{Q} = \frac{2\pi R}{R^2 \pi} = \frac{2}{R}$$

In the case of plates extending over a width A (see *Figure 4*) and having a spacing of $2D$ (remembering that D is by definition half the distance between adjacent plates)

$$\frac{S}{Q} = 2\frac{(A+2D)}{(A.2D)} \simeq \frac{1}{D} \text{ if } A \gg D$$

which is true in most industrial applications. Expression (9) easily resolves the confusion that often arises as to when the factor 2 appears. In expression (10) Qv has been replaced by the equivalent G, the volume rate of flow per second. This expression readily solves the frequent problem of what circumference is required for a given volume rate of flow and a standard length of collecting surface.

Migration Velocity w

Having dealt with the second of the terms appearing in the expression $k = wF$, detailed consideration will now be given to the factor w, the average migration velocity of the particles in the electric field. Assuming ideal conditions, the migration velocity of a particle is proportional to its charge and the electric field strength.

(12) $w \propto q\, Ec$

  q (stat. coulomb) = particle charge
  Ec (stat. V/cm) = field strength

The charge acquired by a particle can be calculated by assuming two different mechanisms to be involved[11]: the charging of particles by ions in an electric field of field strength E_i and the charging of particles due to the thermal movement of the gas ions. In the first case the maximum charge n_m on a particle will be:

(13) $n_m \varepsilon = f E_i r^2$

$n(-)$	= number of elementary charges
ε (stat. coulomb)	= elementary charge
E_i (stat. V/cm)	= field strength
$f(-)$	= factor $\left(1 + 2\dfrac{\delta-1}{\delta+2}\right)$
$\delta(-)$	= dielectric constant
r (cm)	= radius of particle.

This formula is commonly used for particles over about $2\,\mu$ diameter. For smaller particles the charge due to the thermal movement of the ions will be predominant, although both mechanisms will apply. For this second case AHRENDT and KALLMANN[12] have developed an equation, and LADENBURG[11] has shown that for practical purposes the following equation is a good approximation:

(14) $\quad n\varepsilon \simeq 2.10^6 r$

Further formulae for the charging of particles are given in a paper by WALKER[13].

The migration velocity is further determined by Stokes' law, modified in that the gravitational field is replaced by the electric field.

(15) $\quad w = \dfrac{n\varepsilon E_c}{6\pi \eta r} \quad \eta$ (poise) = absolute viscosity.

Expression of the charge by means of equations (13) and (14) gives respectively:

$$w = \frac{fE_i E_c r}{6\pi \eta} = \frac{fE^2 r}{6\pi \eta} \quad \ldots (16)$$

$$w = \frac{E_c 10^6}{3\pi \eta} \quad \ldots (17)$$

For larger particles to which equation (16) applies, the migration velocity is proportional to the radius of the particles. It is also proportional to the square of the field strength, since in a conventional electro-precipitator the charging or ionization field strength E_i is the same as the field strength E_c in the collecting zone.

The field strength around a thin wire emitting corona discharge in a cylinder is given by the equation:

$$E = \sqrt{\frac{2i}{b} + \left(Ea \cdot \frac{a}{d}\right)^2} \quad \ldots (18)$$

$$Ea = 103 + \frac{31 \cdot 8}{\sqrt{a}} \text{ (stat. V/cm)}$$

where i (stat. A/cm) = current per unit length of discharge wire
b (cm/sec/stat. V/cm) = mobility of ions
Ea (stat. V/cm) = corona starting field strength
a (cm) = radius of discharge wire
d (cm) = distance from centre of discharge wire to point in the field

For thin wires this reduces essentially to:

$$E = \sqrt{\frac{2i}{b}} \quad \ldots (19)$$

In the case of plate type precipitation the field strength, as pointed out by Troost[14], is

$$E = \sqrt{\frac{4i}{b}} \qquad \ldots (20)$$

when the distance between wires is approximately the same as the receiving electrode plate spacing.

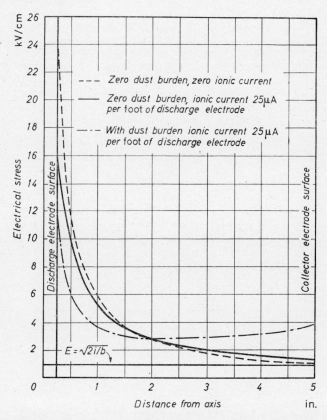

Figure 8. *Field distribution in a tube type precipitator.* (From H. J. Lowe and D. H. Lucas, *by courtesy of* The British Journal of Applied Physics)

If dust is introduced into the space between electrodes a space charge is set up which levels out the field strength near the discharge electrode and increases it near the collecting electrode; according to Lowe and Lucas[15] the field strength is then given by the equation

$$E = \frac{1}{d}\left[\left(\frac{i}{bf^2T^2} + C_o^2\right)\exp(2fTd) - \frac{2i}{fTb}\left(d + \frac{1}{2fT}\right)\right]^{\frac{1}{2}} \qquad \ldots (21)$$

where T = total surface area of dust per cubic cm of gas (cm²/cm³)
C_o = constant of integration (stat. V)

and *Figure 8* shows field gradient curves for various conditions as given by them.

Calculation of the field strength according to equation (19) gives the lowest value, corresponding to the field strength at the receiving electrode produced by the static field from the discharge electrode without any dust burden. With ionic current flowing, the field strength near the receiving electrode is slightly increased, but the increase becomes pronounced when there is dust burden as well as ionic current flow.

From (19), (20) and (16), it follows that w is proportional to i.

From the foregoing formulae the efficiency of a precipitator can be calculated, on the important assumption that the particle size distribution is known since efficiency under given conditions is different for each size of particle. Such calculations have been made, and the theoretical results have been compared with the results of tests under laboratory conditions[8, 9]. The close agreement observed suggests that it is unnecessary to introduce the effect of the so-called electrical wind developed in corona discharge into collection formulae applied to precipitators of present-day design. It should, however, be noted that particle migration velocities much higher than those expected from calculation have been actually measured by LADENBURG[11], a fact which he attributes to the electrical wind. The action close to the electrode surface, however, could not be clearly observed and it appeared that dust particles came to rest or reversed their direction of motion. No definite statement of the real effect of the electrical wind was made therefore, nor was it indicated how such an effect should be embodied in calculations of precipitator efficiency.

Table 2

Particle diameter (sp. gr. 2·7) μ	Free falling velocity cm/sec	Migration velocity in electric field of 3,000 V/cm cm/sec	Migration velocity N times greater than free-falling velocity N
0·1	0·00008	2·8	35,000
0·2	0·00032	2·8	8,700
1	0·008	2·94	370
2	0·032	5·88	183
10	0·80	29·40	32·6
20	3·20	58·80	18·3
40	13·20	117·60	8·9

It is apparent from the efficiency formula that a given efficiency requires a larger specific collecting surface for small particles than for large ones, and it follows that when extrapolating performance data from one precipitator to another working under similar conditions, consideration must be made for particle size distribution. For the same reason, different values of w will be obtained when assessing this factor in a precipitator designed for, say, 90 per cent efficiency and one designed for 99 per cent efficiency under identical conditions. The average migration velocity w will be lower in the latter case because more of the finer particles will be precipitated, see

Figure 3. Nevertheless, the introduction of the migration velocity w into industrial practice in the early stages of electro-precipitation made it possible to compare different installations on a common basis.

Table 2[16] gives an idea of the comparison between the theoretical migration velocities of charged particles in an electric field and the free-falling velocity of the particles in the gravity field.

Precipitation under Actual Working Conditions

Theory and practice seem to agree provided that two extremely important factors are not involved, namely (*i*) the re-entrainment of particles already collected, and (*ii*) the effects of highly insulating layers of precipitated particles. Under such ideal conditions the migration velocity w corresponds to its real meaning, *i.e.* the speed with which a particle of a certain size moves towards the collecting electrode. If particles of different sizes are involved, their several migration velocities may be averaged to give a mean migration velocity. In practice, however, one or both of these disturbing factors are usually present, and it is for this reason that electro-precipitation becomes an art demanding thorough experience.

Re-entrainment of Particles

If dust is poured from a window at a height above ground, it is obvious that little of it will reach the ground perpendicularly below the window in even the slightest wind. Conditions obtaining in electro-precipitators are very similar; the collecting electrodes are 10–20 ft. high, dust is collected from top to bottom of them, and when the dust is dislodged by rapping it is highly unlikely that all of it will reach the hopper without some re-entrainment. If the gas velocity is high, re-entrainment may also occur during the actual precipitation process due to erosion of particles from the collecting surface. It is also obvious that re-entrainment must decrease the efficiency of a precipitator. Its direct result is a reduced migration velocity w because w is the only variable in the efficiency formula which takes account of decreased efficiency under otherwise identical conditions. In the event of re-entrainment w no longer has its literal meaning but is 'degraded' to a mere operation factor. It is still useful, however, to call this factor 'migration velocity' because the term creates a mental image more easily appreciated than a mere numerical factor. Apart from this it obviously remains true that the reduced migration velocity represents the average velocity with which particles are precipitated, allowing for those which are re-entrained.

The extent to which dust is re-entrained can be reduced appreciably by adopting the correct design for the collecting electrodes. Spectacular improvements were achieved in the early 1920's when so-called high duty electrodes were invented and introduced[7]. The design of these electrodes arose from the problem of cleaning the flue gases from lignite-fired boilers, the gases passing through an existing flue at a velocity of about 4·6 m/sec (14 ft./sec). It was found impossible to obtain the guaranteed efficiency with the wire mesh screens which were used as collecting electrodes, but the problem was solved by the development of electrodes which collected the particles in such a way that they could fall into the hopper through a quiescent zone undisturbed by the eroding effect of the gas stream. With wire mesh electrodes w had a value of the order of 4 cm/sec at about 15 per cent overall efficiency, whereas with the design of high duty electrodes finally adopted, the value of w rose to 51 cm/sec at about 95 per cent overall efficiency under otherwise identical operating conditions. It is interesting to note that the theoretical migration velocity calculated for the prevailing conditions and particle size distribution was 65 cm/sec. From this it is evident that, whilst re-entrainment can be considerably reduced, the theoretical migration velocity still differs from the velocity calculated from the measured efficiency and physical dimensions of the precipitator.

Precipitator designs can only be based on values of w found in actual tests. Designers nowadays know from long experience what values of w can be used under given operating conditions and with given types of collecting electrodes, so the effects of re-entrainment are no longer a serious problem, but developments are nevertheless in progress with the aim of improving the performance of collecting electrodes by still further reducing the rate of re-entrainment. It has also been claimed[17] that efficiency can be improved by rapping the electrodes with only very slight blows and by adjusting the frequency and amplitude of these blows very carefully, but it still remains to be seen how far the apparent migration velocity w can be increased in this way above the values obtained with orthodox methods of rapping.

THE RESISTIVITY OF SEMI-CONDUCTORS AS A FUNCTION OF TEMPERATURE

The electrical conductivity of semi-conductors is, in general, partly electronic, as in metals, except that the number of free electrons is greatly reduced, and partly ionic, resembling electrolytic conduction. In the case of salts such as lead sulphide, the conductivity is entirely electronic, whereas in halides such as lead chloride, it is entirely ionic. In either case, however, it is found that above 800° F or 900° F, semi-conductors of either kind obey a law of the type

$$\varrho = Ae^{a/T}$$

where ϱ is the resistivity, T is the absolute temperature, and A and a are constants. In fact, this law is often obeyed fairly accurately down to temperatures as low as 400° F. This equation indicates that the resistivity decreases very rapidly as the temperature increases. Thus, it may be 100 times as great at 500° F as it is at 700° F.

The Effect upon Resistivity of Moisture Adsorption at Low Temperatures

In the case of loosely compacted dusts such as the ones under consideration, there is another effect which becomes prominent below 400° F. As the temperature drops below this value it becomes possible for molecules of water vapour to attach themselves to the surface of the particles of most dusts. As the temperature continues to drop, more and more water molecules attach themselves, and this gives each dust particle a conducting surface. Assuming that the dust is in a moist atmosphere, this surface conduction increases rapidly as the temperature drops, until at ordinary room temperature it predominates entirely over the internal volume conductivity. Hence, in a lightly compacted dust at room temperature, the apparent resistivity is nearly always as low as 10^8 Ω.cm or less, although there are some exceptions like sulphur dust which has an apparent resistivity of 10^{12} Ω.cm or more at room temperature. Presumably this obtains because sulphur is unusually water-repellent and does not adsorb water molecules. Another plausible explanation is that even if water molecules are adsorbed by the sulphur dust particles, there may be no ionizing salt or impurity present to make this adsorbed layer conducting, as is ordinarily the case with most other dusts.

Resistivity Behaviour in General

To summarize, then, most dusts have a low apparent resistivity of the order of 10^8 Ω.cm at room temperature because they adsorb moisture which forms a conducting layer on the surface of the particles. As the temperature rises, more and more of this moisture is driven off the particle surfaces and consequently the apparent resistivity rises to a high value such as 10^{12} Ω.cm or more, in the vicinity of 200–300° F. As the temperature is further increased the resistivity again falls rapidly to low values such as 10^8 Ω.cm in the region of 700° F, because this is the general law followed by most semi-conductors, and at these high temperatures the electrical current flows through the dust particles and not just over their surfaces. This general behaviour is characteristic of most of the materials such as cement dust, P.F. boiler grits, lead fume, oil refinery catalyst, plaster dust, *etc.* commonly collected in precipitators.

Effects of Electrical Resistivity of the Collected Dust Layer

SPROULL and NAKADA[19] point out that the resistivity of collected particles covers the range of 10^{-3} to 10^{14} Ω.cm. This extremely wide range of 10^{17} has important effects in electro-precipitation. The full range of resistivity values can be divided into three:

(*i*) resistivity below about 10^4 Ω.cm
(*ii*) resistivity between 10^4 and 5×10^{10} Ω.cm
(*iii*) resistivity above 5×10^{10} Ω.cm

The border between (*i*) and (*ii*) is not yet clearly defined and no investigations seem to have been published. Generally speaking, range (*i*) covers the very low resistivity values, at which the collected particles too readily give up their original charge acquiring the polarity of the collecting electrode and being then repelled from it and re-entrained in the gas stream. In the case of carbon black, with a resistivity as low as 10^{-3} Ω.cm, it is virtually impossible to precipitate any significant amount of particles because the carbon particles only touch the receiving electrodes for a very short time and then are repelled again towards the discharge electrodes so that they dance backward and forward between the electrodes causing these fine particles to agglomerate. These larger agglomerates leaving the precipitator are then separated from the gas stream in a cyclone in which the gas velocity is kept so low that the fluffy agglomerates are not broken up again. It is also well known that carbon particles in flue gases tend to 'hop' or 'creep' through the precipitator if the collecting electrodes are flat plates without means for preventing re-entrainment. Semi-conductive electrode surfaces help to prevent the adverse effects of low resistivity, but even so, precipitators for dealing with boiler flue grits having a carbon

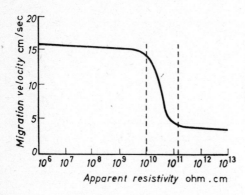

Figure 9. *Typical curve showing variation in migration velocity with resistivity of precipitated material*

content substantially above 25 per cent of the total dust content must be designed to proportions larger than precipitators for handling gases with lower carbon contents. In general, resistivities of these low values seldom occur in industry nowadays. Range (*ii*) resistivity, *i.e.* between 10^4 and 5×10^{10} Ω.cm, seems most suitable for electro-precipitation as the particles do not give up their charge too readily nor do the adverse effects of high resistivity dust layers arise. Range (*iii*) resistivity above about 5×10^{10} Ω.cm presents difficult problems in design. It is hardly possible to predict the performance of precipitators dealing with dust of this resistivity range; any guarantees of efficiency can only be based on previous practical experience of similar installations. The curve shown in *Figure 9* presents a general picture of the way performance of a precipitator passes through a transition from good performance, from below 10^{10} Ω.cm, to a poor performance above 10^{11} Ω.cm.

Dust Resistivity Above 5×10^{10} Ω.cm

As early as 1918 Wolcott observed what has been termed 'back discharge' or 'back ionization', a phenomenon which he related to high resistivity of the collected dust layer which would not allow the particles to give up their charges. In 1928 the occurrence of back discharge was described with special reference to the cleaning of blast furnace gas[7]. A detailed description of the phenomenon and some investigations into its effects were published by MIERDEL and SEELIGER[20] in 1935, while more recently Sproull and Nakada[19] published a comprehensive study of the electrical resistivity of dusts at various temperatures and various gas moisture contents. Back ionization seems likely to occur if the resistivity of the precipitated dust is above about 5×10^{10} Ω.cm. In the case of negative discharge electrodes positive ions are emitted from the dust layer of the collecting electrodes; the discharge can be readily observed and has been photographed[21].

The point-plate arrangement used in preparing the photographs which are reproduced in this article is shown in *Plate IX*. Sulphur dust was spread over the plate and *Plate X* shows the discharge from the point and the much heavier back discharge from a number of individual spots of the dust layer, while *Plate XI* shows a similar phenomenon but with additional spots of less intense back discharge. The condition seen in *Plate XII* was produced in a different way, a flat piece of insulating paper with minute holes pricked in it being placed on top of the plate; back discharge was produced from these holes.

The discharge out of the dust layer of positive ions neutralizes negatively charged dust particles, and also considerably reduces the spark-over voltage. Stable operation of the precipitator can only be maintained at low voltages, resulting in a very low discharge current. Since the migration velocity is proportional to the discharge current, as shown by expression (16), the migration velocity is greatly reduced if the precipitator can only be operated at a low current input. If, for instance, the normal current is 0·2 mA/m² of collecting electrode with a design migration velocity ω of 10 cm/sec, then ω will drop to approximately 2·5 cm/sec, if the current can only be maintained at 0·05 mA/m², in which case the attainment of a given efficiency will require a precipitator four times as large. With dust of very high resistivity it may also happen that the current consumption becomes very high at comparatively low voltages. In this case a very heavy back discharge covering practically the whole surface of the dust-coated receiving electrodes can be seen in the form of a bluish discharge emanating out of the dust layer. In this case an electric charge builds up on the surface of the dust layer, which layer has a very high resistivity, because the corona current is too great to be drained off entirely through the dust layer to the metallic receiving electrode. A high field strength builds up across the dust layer from the surface charge to the receiving electrode. Shock ionization takes place in the gas which fills the interstices of the dust layer. The negative ions created in the interstices migrate to the receiving electrode whereas the positive ions migrate into the space between the electrodes. These positive ions neutralize the negatively charged dust particles suspended in the gas stream and also the negative ions emanating from the corona zone near the discharge wire. As a result, the efficiency of the precipitator drops considerably and the applied voltage has to be reduced until the discharge current becomes so small that no harmful surface charge can build up on the dust layer covering the receiving electrodes[20].

It is hardly possible to give a direct relation between the applicable current and the dust resistivity, but an approximation can be obtained by considering the voltage developed on the exposed side of the dust layer. In fact, the system should be regarded as a capacitor with a leakage resistance. Sproull and Nakada have derived a formula for this case, but they show that under normal conditions the full equation reduces essentially to Ohm's law.

(22) $V = x \cdot \varrho \cdot \chi$

V (V)   tension
x (cm)   thickness of dust layer
ϱ (Ω/cm)   = resistivity
χ (A/cm^2) = current density.

The current density is normally in the range of 0.2 mA/m^2 or 2×10^{-8} A/cm^2. If the thickness of the dust layer is taken as 0.1 cm the following voltages and field strengths are developed across the dust layer at various resistivities:

Resistivity Ω.cm	Voltage kV	Field strength kV/cm
10^{11}	0.2	2
10^{12}	2	20
10^{13}	20	200

The gases contained in the pores of the dust layer are likely to break down electrically at about 20 kV/cm or less and give rise to back discharge. It may be noted that this calculation suggests that trouble due to highly non-conductive dust layers may be expected at resistivities of the order of 10^{12} Ω.cm, but experience shows that trouble may occur at resistivities between 10^{10} and 10^{11} Ω.cm. Information is very scarce, and a more clearly defined boundary may be found if resistivities are measured under actual operating conditions. It is also possible that the electric field in the pore of the dust layer is far from homogeneous so that breakdown occurs more readily.

As regards the mechanism of back discharge, it is unlikely that it is associated with projections of the dust layer. Mierdel and Seeliger[20] made a test by producing corona discharge in a point-plate arrangement in which the plate was covered with dust. When a needle was pushed through the dust layer from below, back ionization was not observed until the needle protruded 2–4 mm. In another test, however, they showed how back discharge develops from a smooth surface; tiny holes were drilled in an insulating plate, and when this was placed on top of the metallic plate of the point-plate arrangement, back discharge was clearly observed (see also *Plate XII*) indicating that the gas in the holes had broken down electrically under high stress.

High Resistivity Dusts in Industrial Practice

Mineral dusts and metal oxides usually have a high electrical resistivity. In some cases the process from which they result conditions the dust sufficiently for successful precipitation, but when no conditioning occurs (especially with metal oxides from smelter plants) electro-precipitation is difficult and migration velocities as low as 3 cm/sec must be assumed when designing the precipitator. In the absence of conditioning during the process of dust production, artificial conditioning must be resorted to in order to reduce resistivity. A good example is provided by the extensive investigations made by Guthmann[17] into the moisture content of blast furnace flue gases required to ensure efficient electro-precipitation of the flue dust. The peak value of the resistivity of a number of dusts, particularly metallurgical fumes, occurs in the range 200° to 300° F and electro-precipitation of such dusts is therefore best carried out by operating the precipitator outside this temperature range. Another way of reducing the adverse effects of back discharge is to allow time for the electrical charges to drain through the dust layer. Half-wave rectification or pulse energizing[22, 23] has been found superior under bad conditions to a constant unidirectional current supply, but it will be realized that such methods cannot be as satisfactory as a constant-rate energy supply when such a supply can be maintained.

The flue dust from pulverized fuel-fired boilers has recently had to be added to the category of dusts which are troublesome because of high resistivity. The cause of the difficulty was not always realized at first. During the past three decades many hundreds of boiler flue dust precipitator installations have been built and have given excellent results. In some cases special precautions were needed owing to the high carbon content of the dust, but generally plants were designed for a certain value of w which was confirmed by long experience. In some recent installations, however, the expected efficiency could not be obtained, mainly owing to the fact that coal combustion in modern high efficiency boilers is so nearly complete that the carbon content of the dust may be less than 2–3 per cent. The result is that most pulverized fuel boiler dust is today a highly insulating material similar to that produced in cement works when the kilns are fed with a dry raw material. It is known that precipitators for dry process cement kilns must be designed for migration velocities of sometimes half those appropriate for the earlier boiler flue dust precipitators. Carbon content is not, however, the only significant factor. White, Roberts and Hedberg[24] point out that the SO_3 in flue gases helps to reduce resistivity of the dust. Other constituents of boiler flue gas, and especially moisture and water soluble alkali salts, also exert some influence. Hitherto it has not been found possible to predict the resistivity of boiler flue dust under operating conditions; one type of coal may cause no trouble in the precipitator, while another may result in a considerable loss of efficiency.

The precipitator designer faces two conflicting problems. On the one hand he wishes to maintain his reputation by building plants which can cope with the worst conditions likely to arise and by giving guarantees which he is confident of fulfilling; on the other hand, clients naturally tend to favour the cheapest quotation, and one which does not involve elaborate guarantee conditions. Consider a case in which tenders are invited for a boiler flue dust precipitator to give 99 per cent efficiency. A designer, aware of the risk of trouble with highly non-conductive dust, will design for a migration velocity which takes account of bad conditions, and will tender for a precipitator requiring a collecting surface of, say, 60,000 ft.² Another designer, basing his design on previous experience of less efficient boilers, may tender for a precipitator with only 40,000 ft.^{2} of collecting surface. The smaller plant may achieve its 99 per cent efficiency on certain coals, but on others it may reach no more than 96 per cent. The difference between 96 and 99 per cent efficiency may seem relatively trifling, but a moment's reflection will show that an efficiency of 96 per cent allows four times as much dust to escape up the chimney compared with the escape at an efficiency of 99 per cent. It is therefore clear that the interests of both client and designer are best served by relating performance guarantees to the resistivity of the dust to be dealt with. If high resistivity is not expected the client need not pay for an unduly large precipitator, but in case of doubt a bilateral assessment of the risk is advisable. It may prove desirable to build the plant with three precipitation zones in series, one zone being initially left without electrodes, which can be added subsequently if the desired efficiency is not achieved owing to operating conditions which cannot be economically improved by conditioning the gas. If, however, high dust resistivity is expected at the outset the client should obviously safeguard himself by accepting a plant of adequate size for the worst conditions.

Measurement of Dust Resistivity

The approximate resistivity of a dust can be measured by placing dust between two electrodes of known surface area and measuring the thickness and resistance of the dust layer; this gives a very rough value. The mechanical handling of the dust when placing it between the electrodes does not reproduce the actual precipitation process. Sproull and Nakada[19] have accordingly devised an apparatus called the 'race track' whereby a quantity of dust is circulated in suspension in an air stream through closed ducting; the circuit includes a small precipitation section consisting of a point-plate arrangement. The air can be raised to high temperatures and the moisture content adjusted to various values. This apparatus enables the actual gas conditions to be simulated to a certain degree. The dust layer is precipitated in the same way as in an actual installation, and its resistivity is measured by lowering a plate on to the surface of the dust and measuring the current flowing at a certain voltage; the specific resistance is then calculated from the measured thickness of the dust layer and the area of the plate lowered on to it.

Since actual gas conditions can hardly be reproduced in the laboratory, dust resistivity tests are envisaged[21] in which a small precipitation chamber will receive gas and dust directly from the inlet duct of an actual precipitator. The chamber is designed to allow measurement of the dust resistance and the thickness of the precipitated layer. Care is necessary to keep the dust layer at the same temperature as that prevailing in the actual precipitator. For exact measurements of dust resistivity it is necessary to follow the dust sampling procedure recommended in B.S.893/1940. In large ducts multi-point isokinetic sampling is specified.

Resistivity measurements in conjunction with guarantee specifications will also have to follow the above procedure. It may seem cumbersome to make a full dust sampling test in order to assess the actual dust resistivity in addition to such a test to assess the gravimetric efficiency of the precipitator, but at present no simpler method of resistivity measurement which would ensure relevant results can be suggested. It should, however, be borne in mind that cases of high resistivity which cannot be foreseen with reasonable accuracy are comparatively rare, and that only in these cases would it be necessary to measure the dust resistivity in order to prove that guarantees had been fulfilled.

It has been found possible to calculate the efficiency of electro-precipitators correctly from basic assumptions and given design data, and under what can be considered ideal conditions electro-precipitation can be regarded as a science. In most practical applications, however, the process is disturbed by various secondary factors, foremost of which are those of re-entrainment and the electric resistance of the collected dust layer. It is these factors, the effects of which cannot be included in design calculations, which make electro-precipitation a matter of practical experience.

GAS CONDITIONING

Most industrial gases contain enough SO_2, SO_3, moisture, *etc.* to allow stable corona emission to take place from the discharge electrodes and to bring the electrical resistivity of the particles within the range of 10^4–10^{10} Ω.cm so that the precipitated particles can give up their charge on reaching the receiving electrodes, but should the gas and the entrained particles be deficient in these respects it becomes necessary to condition the gas in dry precipitators.

As previously explained with dusts of high resistivity[19, 24] of, say, $10^{11}\Omega$.cm or more, charges are built up on the precipitated dust layer and back discharge occurs. In these circumstances a conditioning agent is necessary to reduce the inherent high resistivity of the particles and so allow them to give up their charge. The conditioning agents generally used are steam, water, or sulphuric acid mist[17, 24, 25].

PLANT ARRANGEMENT
Single Zone and Multi Zone Precipitators

For a given set of conditions and a given efficiency an electro-precipitator requires a certain area of receiving electrode surface. Whether this surface is installed in a single chamber or subdivided between two, three or four chambers, depends on several considerations.

For vertical gas flow a single treatment chamber is easily provided, but complications arise if two chambers are installed in series one above the other. For horizontal flow the design is simplified, since each chamber has its own receiving hopper, and any number of chambers or treatment zones can be arranged to give series treatment to the entrained particles.

Single zone, vertical flow precipitators are widely used, (*a*) for treating relatively small volumes of gas, (*b*) for materials which are easily precipitated, (*c*) when high efficiency is only obtainable at so low a gas velocity that re-entrainment of precipitated particles does not occur during the normal

10 to 12 ft. of travel through the treatment zone, and (*d*) when horizontal flow is prohibited by lack of space.

Multi zone precipitators are in most cases preferable owing to their greater flexibility and other advantages not obtainable with single zone treatment. One advantage of subdivision into zones is that each zone can be provided with its own H.T. electrical equipment operating at its optimum voltage. Thus, the first treatment zone, which handles the full particle burden, is maintained at a voltage that gives electrical stability of operation under the given conditions: the subsequent zones have separate H.T. sets operating at higher voltages on the reduced particle burden, making it possible to precipitate efficiently the finer particles which escape precipitation in the first zone.

Subdivision has the further advantage that if for any reason spark-over occurs, the resulting loss of efficiency is confined to one portion of the plant.

In dry precipitators a third advantage of subdivision is that the rapping applied to the receiving electrodes of each zone can be separately adjusted. In a two zone precipitator, for instance, the electrodes of the inlet zone, which retain about 85 per cent of the dust burden, need about seven times as much rapping as those of the outlet zone, which retain only about 12 per cent of the original dust burden. For maximum collecting efficiency the receiving electrodes should be rapped clean when (*a*) the deposited layers become thick enough for the particles to agglomerate without redispersal when falling, but not so thick that back ionization can occur; (*b*) when the electrical clearances are so reduced by dust build-up as to cause internal flash-over; and (*c*) when erosion and consequent re-entrainment of nodules begin to occur. If, in the example of a two zone precipitator quoted above, these considerations demand rapping of the inlet zone once every 4 min, then the outlet zone will need rapping once every 28 min: if the two zones were combined into one, any compromise in frequency of rapping would necessarily be inadequate for the electrodes at the inlet and excessive for those at the outlet.

GAS DISTRIBUTION

In all but very small precipitators it is usual to divide the gas flow into two separate streams, using two separate casings in large plants and a single casing with a division wall in smaller ones.

This division of the gas stream provides a measure of safety, for in the event of breakdown the affected zone can be isolated until the plant can be shut down for inspection: in the case of a four zone precipitator three zones will be left in operation with only a slight reduction in overall efficiency, and in cases where the process is continuous, the affected part of the plant can be isolated for inspection and the whole of the gas passed through the remaining part with a drop in efficiency of 10 per cent or so.

Uniform gas distribution across the active section is essential in all precipitators, and in most plants it demands deflectors, splitter plates or some other form of baffle device. Each individual case must be considered on its merits, and suitable baffles must be included to guard against a drop in designed efficiency.

On some large plants using vertical flow, it is necessary to divide the casing up into a number of parallel compartments, each compartment fitted with a controlling damper to apportion the flow into each chamber before it is possible to distribute the gases evenly across the active precipitation section. Such an arrangement may well add up to several inches of water gauge, thereby reducing the I.D. fan margin and increasing the running costs.

A good gas distribution is essential if the design efficiency is to be obtained and with good design the water gauge drop from the inlet to the outlet flanges of a precipitator should not exceed 0·5 in. but a uniform gas distribution in large plants should not be obtained at the expense of high resistance to gas flow.

ELECTRO-PRECIPITATORS COMBINED WITH MECHANICAL COLLECTORS

Although electro-precipitators can usually be designed for any required efficiency, there are cases where a combination of precipitator and mechanical precollector provides the most effective arrangement.

Multicyclone arresters are widely favoured for such cases, because their layout is best suited to the requirements of the precipitator ducting. Multicyclones can be expected to give efficiencies ranging from about 60 to 85 per cent according to the fineness of the dust handled: thus, if an overall efficiency of 99 per cent is required, the precipitator following the cyclones must have an efficiency ranging from 97·5 per cent (for 60 per cent cyclone efficiency) down to 93·5 per cent (for 85 per cent cyclone efficiency): if the overall efficiency required is 97 per cent, the corresponding figures for the precipitator will be 92·5 per cent down to 87 per cent. Precipitators with high duty chute electrodes can be designed without difficulty for such low efficiencies, since their performance is stable and predictable at gas speeds far above the critical velocity which would cause re-entrainment of the precipitated dust in plants with orthodox tube or flat plate electrodes.

Combined plants usually merit consideration when the dust concentration is high, or where, as in the cement industry, the coarser dust fractions, which contain only a low percentage of alkali as Na_2O and K_2O, can be used again in the process concerned. These plants are also used in the manufacture of carbon black, but in this case precipitation precedes and does not follow mechanical collection, the function of the precipitator being mainly to agglomerate the carbon flakes so that they can be more effectively collected in the cyclones. Every application of a combined plant must be separately considered on its merits, since the operating costs are higher than those of precipitation alone, owing to the pressure drop in the mechanical collector and the attendant increase in fan horsepower.

H.T. ELECTRICAL EQUIPMENT

The H.T. electrical equipment must be capable of giving a voltage high enough to set up corona discharge with whatever type of electrode design and spacing is adopted, the usual values ranging from 30 to 75 kV or more, with standard capacities from 30 to 350 mA, but larger sets are often supplied when required for special cases.

Electro-precipitation requires high tension direct current electricity and a transformer-rectifier unit is therefore necessary to convert the normal alternating current supply.

Three types of equipment are mainly used today:
1. The electronic valve.
2. The metal rectifier.
3. The mechanical rectifier.

The Electronic Valve Rectifier

The electronic valve rectifier has found wide use in America where an average valve life of 20,000 h is obtained and up to 30,000 h is not uncommon. This type has not found wide use in Great Britain since equivalent valves have not been developed, therefore valve rectifiers have only been used in small pilot plants or on atmospheric air cleaning precipitators where only 12 kV or so is required.

The Metal Rectifier

The metal rectifier has been used extensively in Great Britain for some 30 years and is being adopted more and more in the U.S.A. and on the Continent. The use of selenium elements has ousted the original copper oxide rectifier, but with the rapid development of germanium and silicon rectifiers, these will probably in time be used in high voltage rectifier sets.

The main advantages of the metal rectifier are (*a*) its silent operation, (*b*) total enclosure in oil without any moving parts, (*c*) requires no radio and television suppression devices, (*d*) is extremely simple to operate from a remote position, (*e*) requires no polarity indicator, (*f*) does not generate nitrous oxide.

The Mechanical Rectifier

The mechanical rectifier used with electro-precipitators is simply a commutator adapted for high voltages, and it takes the form of an insulated disc with contacts on its periphery, or an insulated arm with two contact blades with slip rings, rotated by a synchronous motor so arranged that the high voltage peaks of the a.c. wave form are selectively commutated, the H.T. current is picked up by jumping across an air gap from the a.c. contacts arranged around the insulated disc or insulated arms.

Mechanical rectifiers are comparatively cheap, essentially rugged and crude, and cannot come to much harm under short circuits. They are, however, noisy, they generate toxic gases, attract dirt and require regular maintenance on bearings, tips and shoes; they also require stringent equipment to minimize radio and television interference.

In general, no hard and fast rule can be laid down on which type is best, any difference in the rectifier wave form and mean current passed between the different types is always taken into account by the electro-precipitator makers when putting forward a specific design for a particular duty, *Figure 10*.

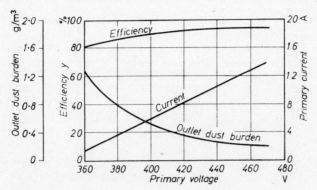

Figure 10. *Precipitation efficiency versus primary current and voltage*

A recently revived system of supplying high voltage power to electro-precipitators was introduced in America in 1952[26] based on the use of pulse principles and techniques originally conceived in Germany[22, 23, 27]. This system is designed so that both the duration and frequency of the current pulses supplied to the electro-precipitator are subject to precise adjustment and control, which then makes it possible to supply optimum electrical energization and hence maximum electro-precipitator efficiency.

Automatic Voltage Control

In applications where the gas conditions, dust burden, temperature and other factors are subject to cyclic or at times violently fluctuating conditions, it is very desirable to install an automatic device that will permit the applied kV to remain as high as possible (see *Figure 11*); without such a device the H.T. sets under manual control of an operator are apt to be run well under the optimum setting with a consequent loss in efficiency.

There are a number of devices in present-day use, some of which automatically raise or lower the voltage on the electrodes to the point where either momentary flash-overs occur or the maximum rating of the set is being passed.[30] Such methods rely on a current-sensitive relay to raise or lower the applied kV.

Other types employ electronic equipment to monitor the number of momentary flash-overs to a predetermined figure before lowering the voltage for a brief period before attempting to raise it again. Another method employs transductors; by varying the voltage drop across the transductor the input voltage to the H.T. transformer and rectifier is varied, the control being on the d.c. saturating circuit of the transductor which is monitored by relays sensitive to spark-overs in the precipitator.

The use of such automatic devices undoubtedly maintains precipitator performance at its peak and they should be installed on all plants where the operating conditions are likely to vary.

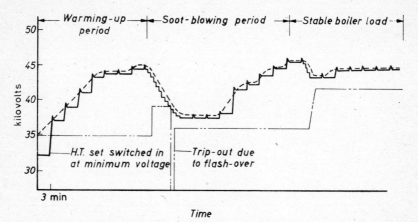

Figure 11. Graph of applied voltage. The dotted curve shows the variation of the flash-over voltage as found during operation: it is assumed that the conditions vary quickly. The continuous curve shows how, under automatic voltage control, the applied voltage closely follows the optimum curve: in order to illustrate the action of the control clearly, it is assumed that its operation is relatively coarse in relation to the changes in operating conditions. The chain-dotted curve shows how the applied voltage might be adjusted under manual control

Automatic Reclosing Device

This equipment is now included as standard on the majority of H.T. sets supplied today and as its name implies, automatically recloses the H.T. set contactor in the event of a short circuit inside the precipitator. The device is usually designed to reclose for a pre-set number of attempts before permitting the H.T. set to trip out finally and give audible warning that a fault exists.

CAPITAL AND WORKING COSTS

Capital Costs

In view of the many variables affecting precipitator designs and materials of construction, it is only possible to give an indication of capital cost; the following figures are British prices as ruling in the first half of 1956:

(a) Dusts easy to precipitate, *i.e.* dusts with resistivities up to 10^{10} Ω.cm with sizing up to 30 per cent smaller than 10 μ: 4s. to 8s. per cubic foot gas per minute at temperature.

(b) Dusts not so easy to precipitate, *i.e.* resistivity above 10^{10} Ω.cm and/or sizing up to 60 per cent smaller than 10 μ: 12s. to 20s. per cubic foot gas per minute at temperature.

(c) Difficult materials with high resistivity embracing fume and mists: 20s. to 60s. per cubic foot gas per minute at temperature.

The prices include the precipitator structure with access, the terminal points being inlet and outlet flanges and dust hopper outlet flanges, all electrical equipment and high tension house, together with complete erection, commissioning and testing for efficiency.

The range of prices assumes a plant of at least 6,000 ft.^{3}/min, with an efficiency range between 93 and 98 per cent, plants smaller than this capacity may cost more, lower efficiencies would reduce the cost per cubic foot per

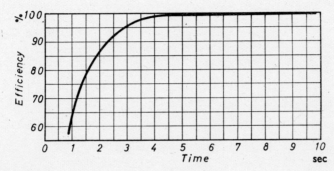

Figure 12. Relationship between collection efficiency and time of treatment

minute, whilst efficiencies higher than 98 per cent would increase the cost, generally the greater the volume the lower the cost per cubic foot per minute.

Working Costs

Here again, due to the very wide range of plants which can be supplied to meet diverse duties, working costs vary enormously, the power consumed may be anything between 10 and 180 kW per 100,000 ft.^{3}/min of gas treated. Likewise, maintenance and replacement costs may be a few pounds per

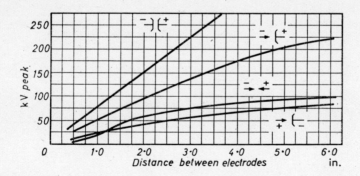

Figure 13. Sparking voltage between different gap arrangements in air at 70° F at normal pressure. No dust on electrodes

annum up to a few hundreds of pounds per annum, again depending on the size of plant, type of precipitator, nature of material collected, type of rectifier used and other factors. Readers are referred to references 28 and 29 for two interesting papers which also include costs for other forms of gas-cleaning equipment.

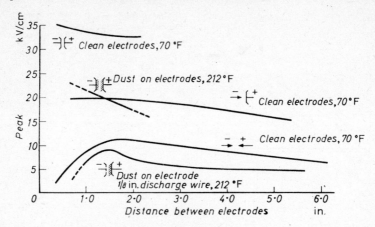

Figure 14. Sparking voltage between plane–plane, point–plane and $\frac{1}{8}$ in. wire–plane gaps in air at 70° F and 212° F at normal pressure, cement dust on electrodes, H.T. source, mechanical rectifier

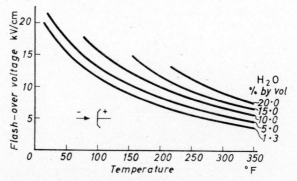

Figure 15. Flash-over voltage between point–plate arrangement in air at different temperatures and at different water vapour content of the air

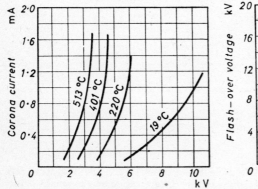

Figure 16. Corona discharge current in a tube type precipitator at different temperatures and voltages

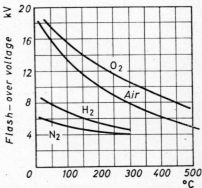

Figure 17. Flash-over voltage of different gases at different temperatures in a tube type precipitator

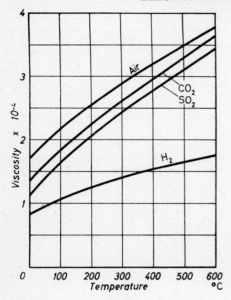

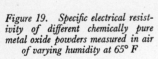

Figure 18. Viscosity of various gases at different temperatures

Figure 19. Specific electrical resistivity of different chemically pure metal oxide powders measured in air of varying humidity at 65° F

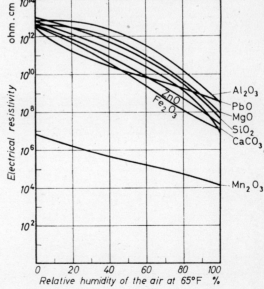

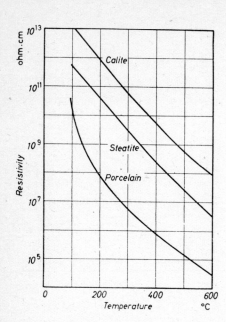

Figure 20. Specific electrical resistivity of different ceramic materials at different temperatures

Figure 21. Typical curves showing resistivity as function of temperature at various percentages of moisture

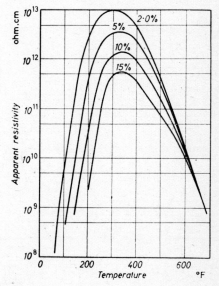

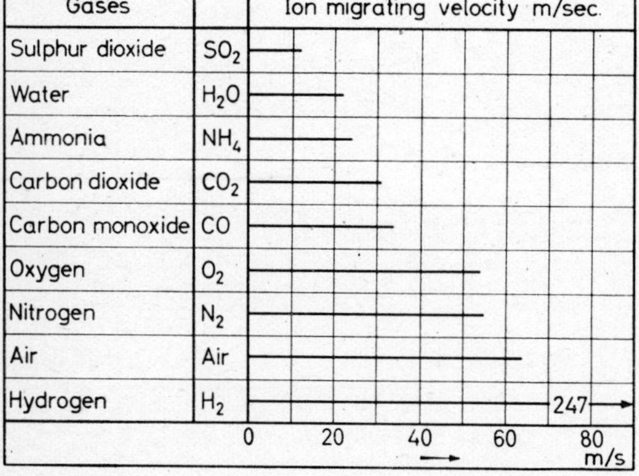

Figure 22. Migrating velocity of negative ions of various gases in an electrical field of 3 kV/cm

Gases		Ion migrating velocity m/sec.
Sulphur dioxide	SO_2	
Water	H_2O	
Ammonia	NH_4	
Carbon dioxide	CO_2	
Carbon monoxide	CO	
Oxygen	O_2	
Nitrogen	N_2	
Air	Air	
Hydrogen	H_2	247

GRAPHICAL PRESENTATION OF SOME OF THE VARIABLE FACTORS IN ELECTRO-PRECIPITATOR DESIGN AND THEIR EFFECT

A series of graphs has been prepared (*Figures 12–20*) showing some of the various factors in electro-precipitator design and their effect.

In general, the curves indicate the complexity of the many variables affecting electro-precipitator design, especially at gas temperatures above 400° C. The migration velocity on which design is based is limited by the flash-over voltage of the precipitator, and this, in turn, is approximately inversely proportional to the absolute temperature. Thus, for example, if the gas temperature is raised from 100° C (373° abs.) to 267° C (540° abs.) the flash-over voltage is approximately halved, thereby greatly reducing the migration velocity of the dust particles. The increase in temperature will also increase the gas viscosity, thus reducing still further the migration velocity; on the other hand, by increasing the moisture content of the gas the spark-over voltage will be increased considerably. The electrical resistivity of the precipitated material alters considerably with changes in temperature (*Figure 21*) and usually runs through a peak value between 200° and 500° F; however, it does not follow that the same migration velocity can be expected when the resistivity is the same value but at different temperatures, the one on the rising characteristic and the other on the falling characteristic portion of the temperature–resistivity curve.

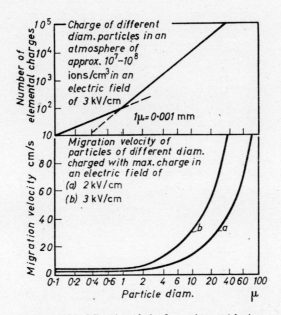

Figure 23. Migration velocity for varying particle size

Migrating velocity of negative ions of various gases in an electrical field of 3 kV/cm is shown in *Figure 22*. It will be seen that the velocity of hydrogen ions is approximately ten times higher than that of water vapour, which explains why the breakdown voltage in gases containing a high moisture content is so much higher than hydrogen or other dry gases.

The curve in *Figure 23* gives the theoretical migration velocity of dust particles with a diameter of 1–100 μ at field strengths of 2 and 3 kV/cm showing how the migration velocity for particles larger than approximately 1 μ goes up in direct proportion to the diameter, and *Figure 24* shows how certain dusts, notably silica dioxide, maintain a very high electrical resistivity at percentage moisture contents five times higher than the percentage necessary to reduce the resistivity on other dusts.

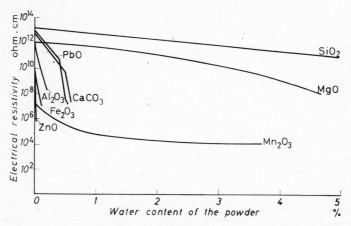

Figure 24. *Specific resistivity of different chemically pure powders depending on the moisture content of dust*

SOME APPLICATIONS OF ELECTRO-PRECIPITATION

The following list gives the principal industrial mists and dusts, *etc.*, which are collected by electro-precipitation in various industries.

Metallurgical Industry

    Alumina from calcining kilns.
    Arsenic from roasting concentrates.
    Recovery of most metallic oxides, *e.g.* those of tin, copper, lead, barium, strontium, silver, gold, antimony, cadmium, bismuth, zinc, selenium, *etc.* from smelting and sintering processes, and of molybdenum, vanadium and tungsten from 'Thermite' processes.

Cement Manufacture

    Dust from cement kilns, coal dryers, clinker mills, and raw material dryers and crushers.

Coal Industries

    Coal dust from dryers, grinders, briquetting plants, *etc.*

Mining Industries

    Dust from dryers, crushers, *etc.*
    Lime and limestone dust, bauxite, phosphate rock dust, sand, *etc.*

Gas Industry
    Tar fog.
    Oil mist.

Chemical Industry
    Acid mist recovery (sulphuric acid, phosphoric acid, *etc.*).
    Pyrites dust, arsenic, selenium from roaster gases, borax, chalk dust, clay dust, soap dust, sodium carbonate.
    Calcium arsenate, dolomite dust, fertilizer dust, iron ore dust, iron oxide.
    Gypsum and barium sulphate recovery.
    Elemental sulphur mist.
    Dust from dryers, crushers, *etc.*
    Removal of dust, arsenic, selenium, elemental sulphur, and SO_3 mist from SO_2 gases from pyrites or blende roasters.

Electricity Generation
    Pulverized fuel flue dust from power station boilers.
    Stoker grits from power station boilers.
    Dust from coal dryers and coal mills.
    Coke dust.

Paper Making
    Sodium sulphate recovery from waste gases of black liquor recovering furnaces.
    Removal of dust, arsenic, selenium, elemental sulphur and SO_3 mist from SO_2 gases from pyrites or blende roasters.

Fertilizer Industry
    Calcium cyanide dust.
    Phosphate dust.
    Potash dust.
    Dust from hot pyrites roaster gases.
    SO_3 mist, selenium, arsenic, sulphur and iron oxide from cold SO_2 gases.

Food Industries
    Processed dust recovery, *etc.*

Petroleum Industry
    Catalyst recovery.
    Acid mist from concentration plants.

Steel Industry
    Flue dust from blast furnace and cupola furnace gas.
    Dust from ferro-manganese blast furnaces.
    Dust from sintering and hot scarfing machines.
    Dust from foundries and sand-blasting plants.
    Iron-ore dust from crushers, dryers and calciners.
    Dust from open-hearth furnaces.
    Tar from coke oven gas and bituminous coal producer gas.
    Dust from coke-fired gas producers.

Carbon Black Manufacture
   Agglomeration of carbon black flakes.
   Lamp black and acetylene black.

Miscellaneous Industries
   Process dusts from lathes, grinders, linishers, *etc.*

DATA REQUIRED FOR DESIGN PURPOSES

To enable the fullest consideration to be given to the type and size of plant required, clients are asked by manufacturers to supply the following data wherever possible:

1. Type and capacity of plant for which precipitator is required.
2. Site of plant.
3. Is the problem one of (*a*) nuisance elimination, or (*b*) material recovery?
4. What is the process (in detail) which produces the gas to be treated, or the material to be recovered, in the precipitator?
5. Operating temperature and pressure of gas at precipitator inlet.
6. Volume per minute (at operating temperature and pressure) of gas to be treated.
7. Moisture content of gas under operating conditions.
8. Dew point of gas.
9. Gas analysis.
10. Entrained particle burden at operating conditions, expressed in grains per cubic foot or grams per cubic metre.
11. Chemical analysis of dust, mist or fume.
12. Size analysis of dust.
13. Give (*a*) required efficiency of recovery, or (*b*) maximum permissible residual particle burden per unit volume of gas.
14. Is gas, dust or fume corrosive? If so, what is the corroding agent?
15. Is plant experience available regarding corrosion–resisting materials?
16. Give phase, cycles and voltage of available electricity supply.
17. Give fullest possible drawings or other particulars of space available for precipitator, and indicate limiting factors, *e.g.* ground conditions, existing structures, *etc.*

EXAMPLES OF TYPICAL INSTALLATIONS

Power Plants

The removal of entrained solids from flue gases coming from pulverized fuel fired boilers is one of the major applications of electro-precipitators. This is primarily because in the burning of the pulverized fuel a very heavy concentration of fine particles (see *Figure 25*) results which cannot effectively be removed from the gas stream, except by electro-precipitation.

The solid particles which are essentially the ash content of the coal, together with some 2–10 per cent carbon in ash, consist mainly of the oxides, silicates, and sulphates of aluminium, calcium, and iron. The particles from a modern boiler with present-day standards of coal grinding

Table 3. Generation of Dust, Mist, etc. in Various Industrial Processes

Type of plant	Source of dust, mist or fume	Concentration in gases at working temperature gr/ft.^{3}	Weight generated as percentage of finished product % or plant product
Metallurgical furnaces	Steelworks blast furnaces	4·5–17·0 measured at inlet of main dust catcher	4·0–16·0
	Sintering machines, lead and tin blast furnaces and reverberatory furnaces	1·3–9·0	3·0–12·0
	Converters for copper refining	2·6–4·3	3·0–6·0
	Brass melting furnaces	0·4–2·2	2·0–4·0
Chemical plants	Pyrites roasters	1·1–2·2	3·0–6·0 of burnt pyrites
	Pyrites and zinc blende flash roasters	9·0–26·0	20·0–60·0 of burnt material
	Zinc blende roasters	2·2–6·5	6·0–15·0 of burnt blende
	SO_2 and H_2SO_4 mist from cold roaster gases	5·0–20·0	1·0–5·0
	Sulphuric acid concentrators	1·7–4·5	0·5–1·5
	H_2SO_4 mist from wet catalyst sulphuric acid plants	20·0–50·0	99·0
Coal treatment plants	Lignite dryers	5·0–11·0	6·0–12·0
	Dryers for bituminous coal	4·5–9·0	3·0–5·0
	Internal exhaust systems for mills and conveyors	9·0–22·0	2·0–3·0
	Gasification and distillation of bituminous coal	0·9–15·0 tar and oil mist	4·0–6·0
Cement manufacture	Rotary kilns, wet process	2·2–4·5	6·0–10·0
	Rotary kilns, dry process	3·5–9·0	8·0–20·0
	Limestone dryers	9·0–35·0	8·0–25·0
	Cement mills and internal exhaust systems	9·0–22·0	3·0–6·0
Dryers for various materials	Dryers and calciners for aluminium hydroxide	13·0–45·0	10·0–25·0
	Dryers for potassium and sodium salts	2·2–9·0	3·0–8·0
	Spray dryers for soap, etc.	4·0–13·0	5·0–15·0
	Dryers for limestone, gypsum and sand	2·2–22·0	4·0–20·0
Electric furnaces	Carbide furnaces	0·4–0·9	1·0–2·0
	Aluminium furnaces	0·2–0·7	0·5–1·5
	Phosphorus furnaces	0·9–4·0	0·5–2·5
	Steel furnaces	1·0–4·5	1·5–7·0
Boiler plants	Stoker and grate fired boilers	0·6–2·2	2·0–5·0 of coal input
	Pulverized fuel fired boilers	2·5–18·0	2·0–20·0 of coal input
Paper manufacture	Paper mills using sulphate process sodium salts precipitated from waste gas of black liquor recovery furnaces	2·0–5·0	3·5–6·5 of pulp output
Carbon black	Oil-burning furnaces	10·0–30·0	100

are extremely fine and often approach more than 60 per cent smaller than 10 μ in diameter. The dust concentration in the gases varies considerably, depending on the ash content of the coal, the type of boiler, and its burning equipment and other factors, and it is not unusual to find concentrations up to 18 gr/ft.³ of gas at n.t.p. in the coals used in Great Britain today.

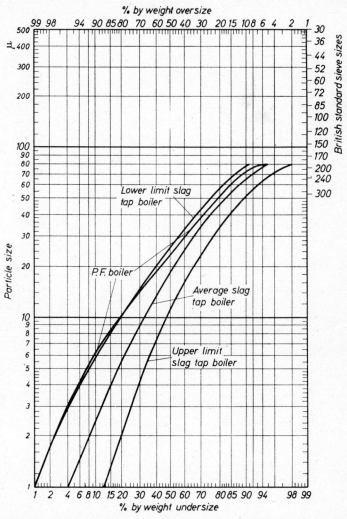

Figure 25. *Typical particle size distribution of pulverized fuel fired boiler flue dust*

With the present measures proposed by the Clean Air Act, the Central Electricity Authority, being the largest user of pulverized fuel fired boilers, will, as always, demand a high standard of dust collection, and probably an outlet emission of not more than 0·2 gr/ft.³ at n.t.p., which requires overall collection efficiencies of up to 99 per cent.

To meet these high efficiencies, either straight electro-precipitators, or a combined mechanical collector, usually nested assemblies of small diameter cyclones, followed by an electro-precipitator, can be installed.

The tendency today for larger boilers with higher temperatures and the need to burn low grade coals, produces an entrained burden containing metal oxide fumes not experienced in boilers of smaller capacity. This state often results in a dust having a very high electrical resistivity with associated difficulties in effective precipitation. These difficulties can, of course, always be overcome by designing a precipitator large enough to ensure that guaranteed efficiency will be attained even under the worst possible conditions which could be expected.

The magnitude of the task can be realized when it is considered that a boiler rated at 1·4 million pounds of steam per hour, may pass on 1 ton of dust every 4–5 min for recovery by the dust collector.

Disposal of such vast quantities of dust presents a problem, but a process has recently been developed to use recovered dust to make into bricks, slabs, and other forms of material for use in building.

Table 4. Typical Performance Data of Electro-Precipitators for Power Stations

Type of plant	Gas flow rate ft.^{3}/min at temp.	Dust concentration gr/ft.^{3} at temp.		Collecting efficiency %	Power consumption kW/100,000 ft.^{3}/min
		Inlet	Outlet		
Pulverized fuel fired boilers	161,850	5·72	0·071	98·67	19·6
Pulverized fuel fired boilers	144,231	4·76	0·027	99·43	22·3
Refuse burning boilers	50,000	7·3	0·252	96·6	23·8
Lignite stoker fired boilers	235,300	0·7–0·874	0·00743–0·01616	98·15	20·4
Lignite pulverized fuel fired boiler (hammer mills)	942,000	2·015	0·0698	96·5	6·8

Coal Industry

Electro-precipitators have been applied very successfully in many hundreds of plants on the Continent for cleaning the gases from coal processing plants. They are now being installed in Great Britain, America and other countries at an increasing rate. Typical performance data are appended in *Table 5*.

Iron Blast Furnaces

Production of steel requires the use of large quantities of combustible gas generated from the blast furnace, and this gas must be cleaned of its entrained burden before being used to fire the stoves, boilers, soaking pits, open-hearth and other steel plant equipment. In a typical blast furnace installation three stages of gas cleaning are usual. First, the removal of the large dust particles in a mechanical collector, next, scrubbing of the gas by wet washing, thus reducing the dust and fume concentration to around 0·1–0·3 gr/ft.³ at n.t.p., and finally, an irrigated type electro-precipitator is used which reduces the dust and fume content to 0·01–0·005 gr/ft.³ at n.t.p. or less, enabling the gas to be used on the stoves, *etc.*, without choking the burners.

Table 5. Coal Industry

Type of plant	Gas flow rate ft.³/min at temp.	Dust concentration gr/ft.³ at temp.		Collecting efficiency %	Power consumption kW/100,000 ft.³/min
		Inlet	Outlet		
Lignite rotary type steam dryer	17,050	15·32	0·1225	99·25	5·1
Lignite rotary type steam dryer	15,300	7·96	0·0394	99·40	4·76
Lignite plate type steam dryer	14,700	3·42	0·0272	99·20	4·25
Combustion gas lignite dryer	24,700	6·25	0·0481	99·50	8·5
Lignite mill dryer	23,580	10·93	0·1445	98·67	13·6
Lignite conveying system de-dusting	12,050	23·8	0·121	99·40	3·4
Bituminous coal tube type steam dryer	25,300	7·09	0·0355	99·50	11·9
Bituminous coal conveying system de-dusting	6,480	9·75	0·0656	99·30	49·3
Bituminous coal/coke grinding plant	2,825	6·02	0·0245	99·59	51·0

Average power consumption for blast furnace gas cleaning precipitators is of the order of 0·6–0·8 kW per 100,000 ft.³/min, the power required for the water circulating pumps being of the order of 8 kW per 100,000 ft.³/min with around 400 gal. of flushing water per 100,000 ft.³/min. for the cleaning of the electrodes.

Open-hearth Gas Cleaning

Electro-precipitators have been used successfully on a commercial scale in open-hearth plants in America since 1951 and are applicable either to cold or hot metal furnaces. The cleaning of open-hearth stack gases involves removal of a heterogeneous mixture of metallic oxides from the flue gases of oil or gas fired furnaces. As much of the material in suspension is actually fume, the particle size is extremely small, also the dust concentration varies considerably over the operating cycle. The gases leaving the furnaces are at a high temperature, therefore, for effective treatment, cooling and conditioning of the gases is necessary to reduce the temperature from around 1,200° F down to 400°–550° F at the precipitator inlet. This can be done by utilizing waste heat boilers and/or spraying with water. The gases after treatment in the precipitator pass to the fan and stack to atmosphere.

Iron Ore Sintering Gas Cleaning

Another successful commercial application of electro-precipitators is the cleaning of gases from sintering machines handling iron ore or flue dust in steel plants. Dust originating in the sintering of the iron ores is carried out of the wind boxes by the gases and transported to the dust-collecting equipment where it is recovered, permitting the clean gases to discharge into the atmosphere, thus minimizing local pollution.

Cupola Gas Cleaning

In normal foundry practice, pig and scrap iron, together with coke and limestone, are charged to a cupola. Blast air is blown into the cupola through tuyeres located near the bottom, the air essentially burns the coke which provides the heat for melting the iron, bringing about a refinery operation incident to this class of work. The gases resulting from the burning of the coke contain finely-divided solid particles of varying chemical composition, together with fumes volatilized from the charge.

With the cold blast system, the cupola stack is usually enclosed and the gases drawn off to the precipitator after cooling and conditioning. With the hot blast system, air preheaters are used to preheat the blast air to the cupola and the combustion gases furnish the heat in the air preheaters. The cupola stack is enclosed and the gases are first drawn off to the pre-heaters, then sent to the precipitator after cooling and conditioning.

Manufactured Gas Cleaning

For the production of coke oven gas, carburetted water gas and producer gas, it is necessary to remove the suspended tar and oil from these gases to prevent contamination of subsequent products, to eliminate operating difficulties such as plugging, and to provide a clean gas for distribution and use.

Coke Oven Gas

It is usual in coke oven plants for a precipitator to be located after the exhausters and before the ammonia saturators, removing up to 99 per cent or more of the tar and oil before the gas enters the saturators. Also, in order to ensure as clean a gas as possible for burning at the ovens and to eliminate any possibility of burner choking, another precipitator is often installed to remove any traces of tar and oil from the gas which is passed to the ovens for under-firing. The use of precipitators in coke oven plants has become almost standard practice to clean the gas. In most cases where the gases are cooled and washed to recover the free ammonia, one de-tarrer is generally used operating at approximately 20° C. In some instances hot gas cleaning is practised at 80°–85° C to give a practically moisture-free tar. This is followed by an indirect cooler and de-tarrer for light oils at 20°–25° C, the oil and water separate easily in settling tanks.

Carburetted Water Gas

In the case of gas plant where carburetted water gas is manufactured, the suspended tar and oil is removed before the purifiers by electro-precipitators. In this type of plant the precipitator is usually located after the coolers and exhausters, and handles the gas at temperatures around 20° C. The gas, after cleaning, passes to the purifiers for sulphur removal; again, efficiencies of 99 per cent or higher are common on this application.

Producer Gas

Precipitators have been used for the removal of suspended matter from producer gas after this gas has passed through water scrubbers and before its distribution to the mains for use as fuel.

Table 6. Coal Gas Industry

Type of plant	Gas flow rate ft.³/min at temp.	Dust concentration gr/ft.³ at temp.		Collecting efficiency %	Power consumption kW/100,000 ft.³/min
		Inlet	Outlet		
Peat gas producer	2,650	2·32	0·00351	99·85	119·0
Cracking plant for natural gas	5,120	0·0976	0·000875	99·20	204·0
Producer gas from lignite briquettes	7,650	16·4	0·0875	99·47	110·5
Producer gas from semi-bituminous lignite	28,250	12·47	0·0437	99·7	102·0
Shale-gas cleaning plant	20,000	17·4	0·0026	99·9	153·0
Coke oven town gas cleaning	1,825	10·5	0·00437	99·9	153·0
Coke oven town gas cleaning	1,350	7·35	0·00131	99·9	272·0
Coke oven gas cleaning	8,230	12·17	0·0342	99·8	127·5
Oil carburetted water gas cleaning	7,360	2·06	0·0171	99·2	238·0
Tar carburetted water gas cleaning	2,358	4·37	0·0219	99·5	306·0

Oil Refinery Application

To produce high octane gasoline by the fluid catalytic cracking process, oil and powdered catalyst are brought together in a reactor to effect the desired cracking. In the process, carbon forms on the catalyst and reduces its efficiency. Consequently, the 'spent' catalyst is sent to a regenerator, where the carbon is burned off and the regenerated catalyst is fed back into the reaction system.

The regenerator flue gas, which carries large quantities of catalyst dust in suspension, passes first through cyclone separators (usually located in the top of the regenerator), then to a waste heat boiler where the temperature is reduced to 400°–600° F, and finally passes to the precipitator, in which the very fine residual catalyst particles are recovered. In addition to recovering valuable catalyst, the precipitator also serves a community relations function by minimizing atmospheric pollution.

Paper and Pulp Mill

Electro-precipitators are used in soda and sulphate pulp mills to recover valuable sodium salts from gases leaving black liquor recovery furnaces. The large savings resulting from this operation are such that the precipitator pays for itself in a relatively short period of time.

Electro-precipitators designed for this purpose have steel reinforced tile shells to withstand corrosion, unique inlet and outlet gas connections to control distribution, and a wet bottom arrangement which enables the collected salts to be continuously returned to the black liquor flow almost immediately after collection.

Black liquor is sprayed into the recovery furnace where it is burned to release heat for steam generation and to recover chemicals later used in the pulping process. In this burning process some of the sodium salts are volatilized and pass out of the furnace with the flue gas. This volatilized material condenses into solid fume particles when the gases are subsequently cooled, and is recovered in a precipitator located after disc or tower evaporators. The dust collected on the electrodes is removed by rapping.

At the same time, black liquor from the multiple effect evaporators is pumped into the wet bottom of the precipitator, where it receives the recovered salts. The flowing black liquor is agitated to keep the dust particles in suspension or solution, and the liquor level is automatically controlled. The black liquor moves from the precipitator bottom to the disc or tower evaporator for further concentration, and then to the furnace.

Table 7. Electro-Precipitators in the Paper Industry

Type of plant	Gas flow rate ft.^{3}/min at temp.	Dust concentration gr/ft.^{3} at temp.		Collecting efficiency %	Power consumption kW/100,000 ft.^{3}/min
		Inlet	Outlet		
Pyrites roaster 25 ton/day	5,570	1·398	0·0205	98·5	115·7
Pyrites roaster 36 ton/day	8,230	1·79	0·0157	99·1	93·5
Acid mist from sulphur burning furnace 7·5 ton/day	1,470	2·99	0·0179	99·4	136
Sulphuric acid mist following cooling tower	2,530	5·33	0·0267	99·5	161·5
Sulphuric acid mist following cooling tower	2,940	3·29	0·0306	99·1	132·5
Black liquor burning plant	78,000	1·234	0·058	95·3	45

Efficiency of removal usually ranges from 90 to 98 per cent, dependent upon design conditions, and recoveries range from 100 to 150 lb. of chemicals per ton of pulp.

Acid Applications

In contact sulphuric acid plants roasting sulphur-bearing ores on multiple hearths, flash roasters, turbulent layer or similar types of furnaces, there are three separate applications of precipitators for cleaning the gases:

1. The removal of dry dust and fume from the hot gases (400°–800° F) after they leave the furnace and before they reach the acid scrubbers.

2. The removal of sulphuric acid mist and solid impurities from the cooled gases (90°–100° F) after the gases have passed through the acid scrubbers and before they go to the converters.

3. The cleaning of the tail stack gases after the absorber towers, in order to prevent the discharge of acid mist to atmosphere.

The first two applications actually involve the purification of the gases so that they can be efficiently and economically handled in the acid plant equipment; consequently, high efficiency of removal is required. The third application is primarily one of preventing atmospheric pollution during normal and abnormal operation of the acid plant.

The precipitators used to clean the hot gases and remove dry dust and fume are usually of the plate type and constructed of steel or brick, whilst those for the removal of acid mist and other impurities are of the tube or plate type and constructed of lead to withstand corrosion.

Table 8. Electro-Precipitators in the Chemical Industry

Type of plant	Gas flow rate ft.^{3}/min at temp.	Dust concentration gr/ft.^{3} at temp.		Collecting efficiency %	Power consumption kW/100,000 ft.^{3}/min
		Inlet	Outlet		
Pyrites roaster 29 ton/day	6,480	0·95	0·0162	98·3	136
Pyrites roaster 35 ton/day	7,660	0·525	0·00157	99·7	163
Blende roaster	9,140	2·23	0·0328	98·5	87
Arsenic and sulphuric acid mist removal	8,250	1·18	0·0000022	99·99	145
Tail gas for sulphuric acid concentration	12,380	5·38	0·022	99·6	61
Elemental sulphur fume from H$_2$S combustion plant	2,530	11·2	0·0875	99·2	289

Precipitators for the removal of solids have also been utilized for cleaning the gases in chamber acid plants, in which case they are located ahead of the Glover Tower. In this type of plant it is necessary to handle the gases at high temperatures and specially designed alloy construction has been used successfully.

Precipitators have also been extensively applied in acid concentrating plants for cleaning the gases before their discharge to atmosphere.

Carbon Black

A large portion of the carbon black used as a reinforcing agent in synthetic rubbers is made in 'furnace black' plants. Here the black is produced by burning great volumes of natural gas, as in America, or oil under controlled conditions in specially designed furnaces. Precipitators are widely used in this field, where they employ a principle not generally used in other applications. The electro-precipitator not only precipitates but also agglomerates the carbon particles so that those not collected in the precipitator can be recovered in specially designed cyclone collectors located after it. The carbon particles leaving the furnaces are extremely fine, and under the influence of the electric field they combine to form small clusters which are collected in relatively low velocity cyclones following the precipitators.

If it is necessary to remove the small amount of black remaining in the gases after they pass through the cyclone a special wet scrubber, precipitator, or a bag filter can be used for this final cleaning.

Cement Mill Applications

One of the first applications for precipitators was the recovery of dusts from cement kiln gases. Since the first installation many years ago, the electro-precipitator has been generally accepted by the cement industry as the most efficient dust recovery equipment for kiln gases. In recent years the precipitator has also been utilized for the recovery of dust from the ventilating air of finish mills and their auxiliary equipment, as well as from

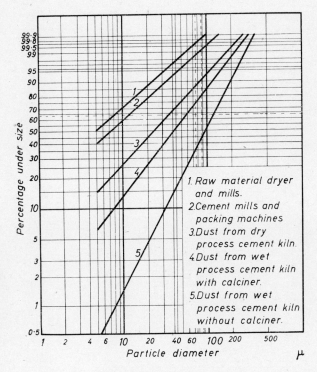

Figure 26. *Typical dust size analyses for various sources*

stone dryer combustion gases (*Figure 26*). In addition to eliminating air pollution problems, the installation of high efficiency precipitators enables the recovery of valuable material for further processing and use. In some cases it is necessary to separate the large particles containing a low alkali from the fine particles consisting mostly of Na_2O and K_2O fume. When this condition exists, an integral combination mechanical-electro-precipitator is used for this separation, material with the low alkali can be added to the kiln feed without spoiling the cement.

Precipitators also offer many operational advantages. Where high temperatures and small particles are factors in a dust recovery problem (and both of these are factors in the cement industry applications), use of electro-precipitators has been found to be the most practical and efficient solution to the problem. Electro-precipitators maintain high collection efficiencies ranging from 90 per cent to almost 100 per cent, even when treating gases bearing minute particles.

Table 9. Electro-Precipitators in the Cement Industry

Type of plant	Gas flow rate ft.^{3}/min at temp.	Dust concentration gr/ft.^{3} at temp.		Collecting efficiency %	Power consumption kW/100,000 ft.^{3}/min
		Inlet	Outlet		
Rotary kiln dry process 520 ton/day	80,000	9·0	0·084	99·06	15·3
Lepol rotary kiln dry process 470 ton/day	74,750	2·75	0·032	98·85	5·1
Rotary kiln wet process 350 ton/day	85,400	9·26	0·029	99·68	14·5
Rotary kiln with calciner, wet process, 350 ton/day	85,400	4·82	0·105	98·2	17
Vertical kiln	73,600	0·78	0·021	97·3	15·3
Raw material dryer	21,200	21·3	0·052	99·75	79·9
Cement mill	14,100	22·3	0·037	99·8	76·4
Packing machine	8,840	16·4	0·048	99·7	79·9

Precipitator design also permits the handling of hot gases from the dry process kilns, and also from the wet process kilns. Draft loss through a precipitator is less than 0·5 in. of water, and the power required for operation is nominal.

Table 10. Mineral Earth and Salts Processing

Type of plant	Gas flow rate ft.^{3}/min at temp.	Dust concentration gr/ft.^{3} at temp.		Collecting efficiency %	Power consumption kW/100,000 ft.^{3}/min
		Inlet	Outlet		
Bauxite dryer 180 ton/day	11,800	6·73	0·021	99·69	19·4
Bauxite calcining and processing kiln 220 ton/day	25,900	2·185	0·026	98·8	30·6
Alumina calciner with multicyclone precleaner 45 ton/day	8,950	129·5	0·013	99·99	78·8
Potassium chloride dryer	17,100	3·5	0·035	99·0	32·3
Fullers earth dryer	17,650	1·88	0·00875	99·54	35·7

Metallurgical Applications

In addition to the metallurgical applications described in the steel and acid sections, precipitators are also used to treat gases coming from smelting and refining furnaces, such as roasters, reverberatory furnaces, ore or dust sintering machines, converters, and other equipment. Valuable materials such as gold, silver, cadmium, lead, *etc.*, are economically recovered. In recent years still other metallurgical applications have been developed and these include cleaning gases from aluminium reduction pots, cleaning gases from scarfing machines, and the cleaning of gases from electric furnaces in the steel industry.

Table 11. Non-Ferrous Metallurgical Industry

Type of plant	Gas flow rate ft.³/min at temp.	Dust concentration gr/ft.³ at temp.		Collecting efficiency %	Power consumption kW/100,000 ft.³/min
		Inlet	Outlet		
Vertical blast furnace: lead ore	5,880	5·25	0·0289	99·5	37·4
Vertical blast furnace: lead ore	9,410	2·76	0·0656	97·5	32·3
Rotary kiln processing: zinc ores	4,710	17·25	0·193	98·90	34·0
Rotary kiln processing: zinc ores	7,360	5·74	0·0267	99·53	27·2
Vertical blast furnace: tin ores	5,700	2·145	0·0149	99·29	30·6
Vertical blast furnace: tin ores	2,120	2·98	0·0425	98·70	40·8
Vertical blast furnace: antimony ores . .	3,650	1·64	0·00307	99·80	71·4
Copper converters . .	8,530	1·97	0·0591	97·00	11·9
Rotary kiln for nickel bearing iron ores . .	26,500	12·03	0·0285	99·76	45·9

Phosphorus Manufacture

In the manufacture of elemental phosphorus by electric furnaces, it is essential to remove all dust and fume carry-over from the gases in order to get a pure phosphorus and to prevent choking of the mains and condensers.

Gases leaving the electric furnaces contain dust and fume, mostly silica, and other elements from the phosphate ore. The electro-precipitators used treat the gases between 300° and 360° C and special precautions are taken to prevent condensation of the elemental phosphorus at all times during the operation of the plant. Dust removal from the precipitator hoppers is carried out dry and the dust can be disposed of in this state or, in order to avoid dust nuisance, in the form of a sludge, after it has been mixed with water.

Atmospheric Air Cleaning

The principles of electro-precipitation are also widely used for the cleaning of atmospheric air and are an essential part in the air conditioning of large stores, theatres, cinemas and public buildings. This application uses a two stage precipitator operating at comparatively low voltages with reduced electrical clearances since the entrained atmospheric concentration is exceedingly small compared with industrial dust concentrations. Such units can give up to 99 per cent recovery efficiency by weight, but due to the very low dust concentration, the equipment is usually offered to give 85–90 per cent efficiency by a blackness test method. This method of test consists of passing measured quantities of air through filter papers and comparing the time required to obtain an equal density of stain on the papers on both the dirty and clean sides of the equipment. In other words, if a certain density of stain was attained in 5 min on the inlet to the air cleaner and 100 min are required to obtain a similar density of stain on the outlet side of the precipitator, then the efficiency by this method would be 95 per cent.

The air cleaning type of precipitator construction has also been applied to the removal of bacteria and mould bearing spores and providing practically sterile air for pharmaceutical processes where antibiotics, surgical dressings, *etc.* are manufactured and packed.

GENERAL APPLICATIONS

In addition to the applications described in general terms above, precipitators are also used to treat gases coming from smelting and refining furnaces, as listed in the field of applications. Precipitators have also been successfully applied to the removal of odours, if the odour is due to particulate matter, and to the recovery of solvent mist passing over from enamelling drying ovens.

REFERENCES

[1] PRACHT, P., *Energie No. 9*, 15th Sept., 1955, 323–7
[2] SCHÄFF, K., *Z. Ver. dtsch. Ing.*, Vol. 98, No. 2 (1956) 4750
[3] HEINRICH, D. O., *Elect. Rev., Lond.*, 19th Nov., 1954
[4] *Brit. Pat.* 716,622
[5] HEINRICH, D. O., *Engng Boil. Ho. Rev.*, June, 1953
[6] DEUTSCH, W., *Ann. Phys. Lpz.*, 16 (1933) 588; *Z. Phys.*, 34 (1933) 448
[7] HEINRICH, R. F., *Dissert. techn. Hochschule*, Berlin, 1930
[8] MIERDEL, G., *Z. tech. Phys.*, 1932, 564
[9] KALASCHNIKOW, S., *Z. tech. Phys.*, 1933, 267
[10] HEINRICH, R. F., *Unpublished Notes*
[11] LADENBURG, R., *Der Chemie-Ingenieur*, Vol. I, Pt. IV (1934) 31
[12] AHRENDT, P., and KALLMANN, H., *Z. Phys.*, 35 (1926) 421
[13] WALKER, E. A., and COOLIDGE, J. E., *Heat. Pip. Air Condit.*, March, 1951, 107
[14] TROOST, N., *Proc. Instn elect. Engrs*, Vol. 101, Pt. II, No. 82 (1954) 369–89
[15] LOWE, H. J., and LUCAS, D. H., *Brit. J. appl. Phys.*, Suppl. 2 (1953) 540
[16] HEINRICH, R. F., *Brennst.-Wärme. Kr.*, I, 8 (1949) 195
[17] GUTHMANN, K., *Dissert. techn. Hochschule*, Berlin, 1932
[18] SCHMIDT, W. A., SPROULL, W. T., and NAKADA, Y., *A.I.E. & M.E.*, New York, 1950
[19] SPROULL, W. T., and NAKADA, Y., *Industr. Engng Chem.*, 43 (1951) 1350
[20] MIERDEL, G., and SEELIGER, R., *Trans. Faraday Soc.*, 1936, 1284
[21] LAKEY, J. R. A., and BOSTOCK, W., *Trans. Inst. Chem. Engrs*, Vol. 33, No. 4 (1955) 261
[22] *Ger. Pat.* 562,891
[23] *Ger. Pat.* 638,700
[24] WHITE, H. J., ROBERTS, L. M., and HEDBERG, C. W., *Mech. Engng*, Nov., 1950, 873
[25] WHITE, H. J., *Air Repair*, Vol. 3, No. 2 (1953) 79
[26] WHITE, H. J., *A.I.E.E.*, Pt. I, Nov., 1952, No. 3
[27] *Amer. Pat.* 2,000,007
[28] STAIRMAND, C. J., *J. Inst. Fuel*, Feb. 1956
[29] DALLAVALLE, J. M., *Chem. Engr*, Nov., 1953, 177–83
[30] *Brit. Pat.* 726,556

INDEX

ACCO dirt extraction gear, 204
Acid,
    cresylic as frothing agent, 218
    mist recovery, 521
    sulphuric, in electro-precipitators, 529
Aerofloats, flotation reagents, 215
Aggregates, mixtures, 363
Aggregation and destabilization, 252
Air,
    aspirator, 352
    cleaning, 485
    conditioning, 533
    corona current in, 486
    dust separation from, 464
    elutriation by, 32, 125, 340, 352
    flow selection, 337
        references, 361
Airslide, handling of powders, 384, 391
Akins classifier, 293, 290
Alkyl dithiocarbonates as flotation agents, 215
Alkalis, depressors for flotation, 219
All-flotation, 235
Alum, batch sedimentation costs, 276
Alumina, handling and storage, 384
Ammonium sulphate, handling data, 384
Anhydrite, dry grinding, 74
Anthracite,
    grinding, 93
        dry, 75, 75
        jigging, 193
Antimony, grinding of sulphide, 88
Apatite, flotation treatment, 210
Apron conveyor, 413
Apron feeder, 460
Asbestos,
    fibre, handling data, 384
    screening, 160
Asphalt, grinding, 88
Aspirator, air, 352
Atritor mills, air flow selection, 360
Atritor pulverizer, 88
Attrition mills, 116
    provender milling, 122
    special applications, 116
Avery belt weighers, 456

Bahco centrifugal dust classifier, 347
Bakelite, grinding, 93

Ball mills,
    cascading, 103
    cataracting, 103
    centrifuging, 103
    circulating load, 107
    cylindrical, 69, 71
    dry grinding, 74
    dynamics, 102
    efficiency, relative, 19
    gold ore grinding, 69
    grindability tests, 15, 107
    grinding by, 13, 65, 102, 107
    Hardinge, 65, 69, 70, 71, 74, 76
    lead–zinc ore grinding, 70, 71
    load, circulating, 107
    Marcy, 66, 69, 71, 72
    mixing of solids, 370, 376
    operating data, 69
    power–throughput relationship, 97
    solids, mixing, 370, 376
    speed, 63, 66
    states of motion, 103
    tests, 6, 7, 13, 15, 16, 108
    throughput–power relationship, 97
    Tricone, 76
    U.S. Bureau of Mines, 16
    wet grinding, 69
Banbury mixers, 370, 377
Barvoys system of coal washing, 322
Barytes,
    flotation treatment, 210
    grinding, 80, 83, 93
Bates packer, 451
Baum jig, 202
Bauxite,
    crushing, 59
    handling data, 384
Beater mixers, 375
Beet sugar, sedimentation, 264
Belt conveyors, 391
    capacities, 395
    cleaning, 400
    discharge of materials, 397
    drive, 398
    fabrics for, 394
    feeding, 396
    incline of, 398
    speed, 395
    storage of materials, 384

INDEX

Belt weighers, 456
Belting, fabrics for, 394
Bendelari jig, 197
Bins, storage of materials, 387
Bird centrifugal classifier, 299
Blake crushers, 49
Blake–Denison integrating weigher, 456
Blanket filtration, clarification and thickening, 270
Blast furnaces, gas cleaning for, 525
Blende, separation from galena, 242
Boilers, pulverized fuel fired, dust from, 507
Bolting cloth, 141
Borax, grinding, 88
Bowl classifiers, 288, 304
Boylan cone classifier, 290
Bradley–Poitte roller mill, 84
Brewing, sampling, 441
Bucket elevators, 402
Bucket weighers, 446
Bunker Hill classifier, 298
Bunkers,
  construction, 387
  load cells, 455
  overfilling, 389
  storage of materials, 382, 387
Buss table, 179
Butchart table, 183

Calcite, flotation, particle size limits, 222
Calcium fluoride, handling data, 384
Calcium phosphate, grinding, 88
Caldecott machinery,
  cone classifier, 289
  thickening cones, 265
Callow apparatus,
  cone classifier, 290
  flotation cell, 231
  thickening cone, 266
Calomel, grinding, 93
Calumet classifier, 296
Camphor oil, frothing agent, 218
Carbon, electro-precipitation, 503
Carbon black, electro-precipitation, 530
Carborundum, dry grinding, 74
Cascade cells, froth flotation, 229
Cascade mill, 65
Cascading in ball mills, 103
Casein as flotation depressor, 219
Cataracting in ball mills, 103
Caustic soda sedimentation processes, 265
Cement,
  dust,
    collection, 511
    generation, 523
  electro-precipitation, 520, 531
  grinding, 77, 300
  handling data, 384
  milling, electro-precipitation, 531
  weighing, 455

Centicell collector, 471
Centri-cleaner, 300, 310
Centriclone, 300
Centrifugal classification, 289
Centrifuging,
  ball mills, 103
  flour sieving by, 153
  hydraulic, 299
  screening by, 150
  velocity separation by, 161
Cereals, packing, 451
Cerussite, flotation, 219
Chain conveyors, carrying types, 413
Chalk,
  crushing, 59
  grinding, 80, 83, 88
Chance coal washing system, 319
China clay, handling data, 384
Chronos weigher, 446
Chutes, handling of materials, 390
Clarification,
  batch processes, 256
  coal, in long tube, 278
  collective subsidence, 257
  limitations in settling, 252
  line settling, 257
  mechanics of, 256
  phase subsidence, 257
  pump behaviour during, 258
  unit capacity, 259
Clarifiers,
  area determination, 281
  Clariflocculator, 270
  cones, 265
  costs, 274
  dilution tests, 279
  Dorr multifeed, 271
  green liquor, redesign, 283
  Hardinge, 267
  maintenance, 275
  power consumption, 274
  RapiDorr, 271
  recausticizing, redesign, 283
  references, 285
Clariflocculator, 270
Classification,
  aqueous, 287
  centrifugal, 289
  drainage, 287
  factors affecting, 289
  granular material, 128
  grinding in closed circuit, 287
  hydraulic, centrifugals, 299
  leaching, 287
  pneumatic, 289
  thickening and, 290
  washing, 287
  wet, 287
    references, 310

INDEX

Classifiers,
  air, fan separators, 350, 355
  air-flow, 337, 348
  Akins, 290, 292
  Allen cone, 289
  ancillary equipment, 309
  aqueous, operating characteristics, 304
  Bahco centrifugal dust, 347
  Bird, 299
  bowl, 288, 304
  Boylan cone, 290
  Bunker Hill, 298
  Caldecott cone, 289
  Callow cone, 290
  Calumet, 296
  centrifugal, 299
    advantages, 344
    air, 344
    dust, 347
    fan separators, 355
    helical flow of air, 356
    spiral air flow, 358
  comparison of, 288
  cone,
    Allen, 289
    double, 359
    operating characteristics, 304
  construction materials, 310
  countercurrent, 293
  cyclones, 288, 300
  Deister Concenco, 297
  Deister cone baffle, 297
  Delano, 298
  design and performance, 303
  Dorr, 290, 291, 292, 294, 295, 307
  DorrClone, 300
  drag, 290, 295
  dust, Bahco centrifuge, 347
  Esperanza, 290
  Evans, 296
  feeding, 309
  free settling, 295
  functions, 290
  Geco, 292
  Hardinge, 290, 293
  hindered settling, 295
  hydraulic, 295, 304
  hydrocyclones, 300
  Hydrorotator, 298
  hydroscilator, 298, 304, 307
  hydroseparators, 288
    operating characteristics, 304
    pumping, 309
    size considerations, 306
    wet classification, 288
    working conditions, 293
  launder vortex, 296
  mechanical, 290
  Multideck, 295
  non-centrifugal, 349

Classifiers (cont.)
  non-hydraulic, 289
  non-mechanical, 289
  Nordberg–Wood, 290
  Pellett, 298
  performance,
    comparisons, 307, 308, 309
    design, 303
  rake, 291, 292
    characteristics, 288, 290, 291, 292
  references, 310
  Richards, 296
  Richards–Janney, 298
  sand washers, operating characteristics, 304
  screw washers, 294
  Spitzkasten, 289
  spiral, 288
  washing, 294
  Wemco, 290
  wet screen, 288
  Wuensch cone, 290
Chromium, dry grinding of ore, 75
Clay, grinding, 80, 87, 88
Climate and storage, 385
Coal,
  ball mill tests, 6, 7, 13, 15, 16, 108
  bituminous, jig washing, 202
  classification, in long tube, 278
  cleaning, 210
  coarse, flotation, 223
  compression tests, 5
  crushing, 7, 58
  degradation during hydraulic conveying, 434
  dense medium, washing of, 312
  dry grinding, 74
  drying, 520
  dust extraction, 520
  feeders, 432
  flotation, 216
    creosote as collector, 216, 221
    depressors, 219
    Ekof method, 244
    froth, 237
    particle size limits, 222
    pulp density, 224
    reagent requirements, 217
    selective, 243
    Simcar-Geco cell, 227
    vacuum, 233
  froth flotation, 237
  grinding, 80, 87, 106
  handling data, 384
  hydraulic conveying, 428
  jig washing, 196, 202
  lock-hopper feeder, 433
  middlings, separation, 321
  processing plant, gas cleaning, 525
  pulverizing for boiler firing, 95

INDEX

Coal (cont.)
  sampling, 441
  screening, 156
  sedimentation, calculation of equipment size, 277
  separation, 179, 312
  sink-and-float analysis, 312
  soft, stress–strain curves, 3, 6
  superclean, production, 241
  vacuum flotation, 233
  washing,
    Barvoys system, 322
    Chance system, 318, 319
    dense media plant, 315
    desanding, 317
    Dutch State Mines system, 323
    Humboldt-Deutz process, 332
    Link belt process, 333
    magnetic medium cleaning, 318
    Nelson Davis separator, 328
    references, 336
    Ridley-Scholes process, 318, 326
    separating bath, 315
    Simcar process, 329
    S.K.B. process, 334
    thickeners, 317
    Tromp process, 318, 324
  weighing, 455
Coke,
  cyclones as separators, 467
  dry grinding, 74, 75
  handling data, 384
Colloids, flocculation, 253
Colloid mills, solid mixing, 376
Colloids separation, 253
Compression tests, 3, 6
Compressors, 426
Compound millings, 122
Conditioners, Denver, 234
Conditioning tanks, 233
Cone classifiers, operating characteristics, 304
Cone crushers,
  capacities, 55
  overload precautions, 54
  principles, 48, 53
  product analysis, 55
  sizes, 55
Conveyors,
  abrasion, 390
  apron, 413
  belt, 382, 391
  capacity, 390
  chains, 412
  concentration of materials, 390
  corrosion, 390
  cyclones, 425
  drag chain, 410
  En masse, 414
  flight, 410
  grain, discharge of, 425

Conveyors (cont.)
  gravity, 390
  Hapman, 416
  hydraulic, 427
  inclined, 396
  pneumatic, 418
  roller, 391
    storage of materials, 382
  run-around, 413
  site limitations, 390
  slat, 382, 413
  suction, 419
    bends, wearing at, 425
    nozzles, 424
  system cycles, 432
  vibrating, 417
  wire belt, 413
  worm, 406
  zipper, 401
Cooley jig, 195
Copper,
  grinding of oxide, 88
  ore grinding, 72, 73
  sedimentation, 264
  separation data, 181, 183
Corn cutters, usage, 126, 127
Corn, winnowing of, 338
Corona quenching effect, 487
Corona suppression, 487
Corrosion,
  conveyors, 390
  electro-precipitators, 488
  sedimentation units, 272
Costing, sedimentation units, 273
Countercurrent decantation,
  definition, 275
  calculations, 277
  flow sheet, 275
  industrial applications, 275
  operating data, 276
  washing tray units, 269
Creosote, flotation reagent, 216, 221
Cresylic acid, frothing agent, 218
Crushers,
  Blake, 49
  coal, 7, 58
  cone, 48, 53
  Dodge, 151
  efficiency relative, 19
  gyratory, 19, 48, 52
  Hadsel, 65
  hammer, 48
  impact, 5, 19
  jaw, 3, 14, 48
  maintenance of, 49, 52
  ring hammer, principles, 59
  roll, principles, 48, 56
  rotary, 52
  special applications, 109
  swing hammers, 48, 58

INDEX

Crushing,
  ball mill tests, 6
  bauxite, 59
  chalk, 59
  component theory of comminution, 13
  compression tests, 3, 6
  crack formation, 21
  crushing roll tests, 6
  efficiency of machines, 19
  equipment, 48
  impact tests, 3, 6
  Kick's law, 1
  mills, 3
  power requirements, 61
  principles, 1
    references, 22
    summary, 20
  Rittinger's law, 1
  rod mills, 60
  roller mills, design, 109
  size distribution, 8, 9, 12
  small-scale plant, 61
  sizing analysis, 24
  stress–strain curves, 3
  tertiary, 60
Cryolite, grinding, 93
Cyanide as flotation depressor, 219
Cyclones,
  abrasion problems, 303
  air flow pattern, 346
  ash separators, 481
  Centri-cleaner, 300, 310
  Centriclone, 300
  Centicell collector, 471
  comparison of types, 467
  conveyors, 425
  costs, 481, 473
  dust extractors, 469
  efficiency, 468
    theoretical, 474
  erosion of tube, 471
  fan-scroll type, 468
  gas flow in, 474
  gas recirculation, 479
  general principles, 464
  handling capacity, 465
  handling of materials, 425
  Hetl and Patterson, 300
  high throughput, 470
  hydraulic, coal washing, 323
  Kreba, 302
  laminar flow, efficiency, 475
  multiclones, 469
  multiple, 302
  operating characteristics, 304
  power consumption, 481
  practical applications, 481
  references, 483
  reverse flow, 359
    vane type, 466

Cyclones (*cont.*)
  straight flow, 467, 472
  turbulent flow, efficiency curves, 475
  types, comparison, 467
  wet classification, 288
  Whirlcone, 300

Decantation, continuous countercurrent,
  definition, 275
  calculations, 277
  flow sheet, 275
  industrial applications, 275
  operating data, 276
  washing tray units, 269
Delano classifiers, 298
Densludge thickener, 284
Desiltor, bowl, operating characteristics, 304
Destabilization and aggregation, 252
Detarrers, 489
Diesel oil as flotation agent, 216
Discharge, back, 504
Disintegrators,
  bar type, 90
  speed, 64
Dithiophosphates as flotation agent, 215
Dodge crushers, 51
  Hydroseparator, 294
  multifeed clarifiers, 271
  sand washers, 294
  sizers, 296
  thickeners, 266
  TM multiple cyclone, 302
Dorrclone, 300
  pumps for, 310
Dough mixers, 370, 377
Drag chain conveyors, 410
Drag classifiers, 290, 295
Drainage, separation, 287
Drewboy separator, 329
Drum mixers, 370
Dust,
  classification, Bahco centrifuge, 347
  collection, 419, 491
    centrifuge type unit, 478
    efficiency, 494
    fan-scroll type separator, 468
    multicyclones, 511
    plants, 426
    steel industry, 521
    theoretical efficiency, 474
    tubular, 469
  definition, 24
  electro-precipitation, 485
  extractors, cellular type, 469
  flue, analysis, 523
  generation in industry, 523
  health hazards, 384
  high resistivity, industrial practice, 506
  mineral, resistivity, 506

INDEX

Dust (*cont.*)
  precipitation, 484, 491
  presence in gases, 487
  removal, 488
  resistivity, 502, 508
  separation, 464, 469
Dutch State Mines system of coal washing, 323

Edgar Allen weigher, 454
Electro-precipitation, 484
  acid applications, 529
  applications, 520
  automatic voltage control, 513
  back ionization, 504
  blast furnaces, 525
  carbon, 503
  carbon black manufacture, 530
  cement manufacture, 520
    mills, 531
  chemical industry, 521
  chute electrodes, 493
  coal industry, 520, 525
  corona current, 485
  dust collection, 491
  dust generation, 491, 523
  electronic valve rectifiers, 512
  fertilizer industry, 521
  food industry, 521
  gas, characteristics and, 486
    conditioning, 509
    distribution, 510
  installations, typical, 522
  metal rectifiers, 512
  metallurgical industry, 520, 532
  mining industries, 520
  oil refinery application, 528
  paper industry, 521, 528
  petroleum industry, 521
  plant arrangement, 509
  principles, 484
  references, 534
  steel-making, 521
  working conditions, 501
Electro-precipitators,
  construction, 488
  corona current, 485
  costs, 514
  design, 494, 507, 516, 522
  detarrer, 489
  dust collector, 491
  dust resistivity, measurement, 508
  electrical equipment, 512
  electrodes, 488
  electronic valve rectifiers, 512
  geometrical dimensions, 496
  iron ore sintering, gas cleaning, 526
  manufactured gas cleaning, 527
  mechanical collectors and, 511
  mechanical rectifiers, 512
  metal rectifiers, 512

Electro-precipitators (*cont.*)
  open-hearth gas cleaning, 526
  plate type, 488
    geometrical dimensions, 496
  reclosing of circuit, 514
  sulphuric acid in, 529
  tube type, 488
  types, 488
  typical installations, 522
  working costs, 515
Electron microscope, size analysis, 29
Elevators,
  bucket, 402
  grain, 402, 404, 406
Elmore jig, 195
Elmore-Vacuum process for froth flotation, 232
Elutriation,
  air, 32
    provender milling, 125
  sizing analysis, 25, 29
  velocity separation by, 161
  water, 31
Elutriators,
  air, 32, 352
    flow in, 340
  centrifugal, 342
  water, 31
En masse conveyors, 414
Energy losses in pulverizer, 101
Entoleter mixers, 375
Erosion, cyclones, 471
Esperanza classifiers, 290
Eucalyptus oil as frothing agent, 218
Evans classifier, 296
Exact measurer, 459
Exhausters, 426

Fahrenwald sizer, 296
Feed roll, 460
Feeders, 459
  apron, 460
  coal, 432
  lock hopper, 433
  maintenance, 463
  reciprocating, 433
  rotary, 460
  shaker, 460
Feldspar,
  dry grinding, 74, 75
  jig, 205
Ferraris table, 179
Fertilizers, electro-precipitation, 521
Filter, thickeners, 269
Filtration and powder handling, 425
Flight conveyors, 410
Floatability and wetting, 213
Flocculation,
  Clariflocculator, 270
  definition, 252

INDEX

Flocculation (*cont.*)
  flow diagram, 238
  sedimentation units combined with, 269
Flotation,
  apatite, 210
  cerussite, 219
  coal, 216
  creosote as collector, 216, 221
  depressors, 219
  Ekof method, 244
  Simcar-Geco cell, 227
  vacuum, 233
  froth, 209
    activating agents, 218
    aerofloats as reagents, 215
    agitation zone, 226
    applications, 235
    bibliography, 245
    Callow cells, 231
    cascade cells, 229
    cell requirements, 225
    coal, 237
    collectors, 214
    concentration zone, 227
    conditioning tanks, 233
    contact angle, 211
    Denver cell, 227
    depressing agents, 219
    Elmore-Vacuum process, 232
    floatability, 213
    frothing agents, 217
    Inspiration cell, 232
    limitations, 210
    machinery used, 225
    MacIntosh machine, 232
    mechanism, 209
    oils as reagents, 215, 218
    particle size effects, 222
    plant performance, 240
    pneumatic cells, 231
    principles, 210
    pulp density, influences, 224
    reagents, 214
      activators, 218
      collectors, 214
      depressors, 219
      quantities and effects, 221
      regulators, 219
      sulphidizers, 220
      use of, 220
    references, 244
    regulations, 219
    separation zone, 227
    Simcar-Geco cell, 227
    sub-aeration cells, 226
    sulphidizers, 220
    temperature control, 223
    uranium, 237
    wetting and floatability, 213
    xanthates as reagents, 215

Flotation (*cont.*)
  Kleinbentink, 231
  primary, 235
  secondary, 235
  selective, 242
  separation by, 161
  straight, 236
  vacuum, 232
Flour,
  milling, 118
    'break process', 113, 120
    flow design of mill, 119
    plansifters, 151
    preparation process, 120
    purifying, 121
    screening, 160
  packing, 451
  preparation process, 120
  sampling, 441
Flow,
  forced vortex, 347
  free vortex, 345
  sheets, continuous countercurrent decantation, 275
  streamline, 163
  turbulent, 165, 474
Fluidization,
  air, oscillatory screen and, 161
  handling of powdered materials, 383
  particle size analysis, 134
  screening with oscillatory motion, 160
Food, electro-precipitation in industry, 521
Foundries, gas cleaning, 526
Fountain mixers, 371
Foust jig, 196
Froth flotation, 209
  activating agents, 218
  aerofloats as reagents, 215
  agitation zone, 226
  applications, 235
  bibliography, 245
  Callow cell, 231
  cascade cells, 229
  cell requirements, 225
  coal, 237
  collectors, 214
  concentration zone, 227
  conditioning tanks, 233
  contact angle, 211
  Denver cell, 227
  depressing agents, 219
  Elmore-Vacuum process, 232
  floatability, 213
  frothing agents, 217
  Inspiration cell, 232
  limitations, 210
  machinery used, 225
  MacIntosh machine, 232
  oil as reagents, 215, 218
  particle size effects, 222

INDEX

Froth flotation (cont.)
  plant performance, 240
  pneumatic cells, 231
  principles, 210
  pulp density, influences, 224
  reagents, 214
    activators, 218
    collectors, 214
    depressors, 219
    quantities and effects, 221
    regulators, 219
    sulphidizers, 220
    use of, 220
  references, 244
  regulators, 219
  separation zone, 227
  sub-aeration cells, 226
  sulphidizers, 220
  temperature control, 223
  uranium, 237
  wetting and floatability, 213
  xanthates as reagents, 215
Frothers, 217
Fuel,
  oil as flotation agent, 216
  pulverized, 95
Fullers earth, grinding, 80, 83, 88, 93
Fume, removal, 488
Furnaces, open-hearth, gas cleaning, 526

Galena,
  flotation of, particle size limits, 222
  separation from blende, 242
Gas,
  carburetted water, cleaning, 527
  cleaning, 484, 526
    future trends, 483
  coke oven, cleaning, 527
  conditioning, 509
  corona current and, 486
  dust in,
    precipitation, 487
    separation, 466
  electro-negative, 486
  electro-precipitation and, 494, 509
  flotation reagents, 216
  flow in cyclones, 478
  producer, cleaning, 528
  turbulent flow in cyclones, 474
Gas turbines cyclone ash separators, 481
Gasoline, high octane, production, 528
Gauze, grit, 141, 143
Geco classifiers, 292
Glue, depressor for flotation, 219
Gold,
  jig washing, 200
  ore grinding, 69, 73
Grading,
  bibliography, 171
  coarse, machinery, 147

Grading (cont.)
  granular material, 128
  machines, 147
  particle analysis, 135
  references, 171
Grain
  conveyors, discharge of materials, 425
  handling, 402, 404, 406
  milling, 118
    flour, 118
    provender, 122
    purifying, 121
  sampling, 441
  separation, 349
  storage, 387
Granulators, 50, 61
Graphite,
  dry grinding, 74
  flotation treatment, 210
Gravel, hydraulic handling plant, 436
Gravity conveying, 390
Grindability,
  ball mill tests, 15, 107
  coal tests, 21
  comparison of tests, 18
  Hardgrove machine test, 17
  laboratory tests, 15
  Mohr's scale, 62
Grinding,
  anthracite, 74, 75, 93, 193
  antimony sulphide, 88
  asphalt, 88
  attrition, 63, 116
  Babcock and Wilcox 'E', 85
  Bakelite, 93
  ball mill, 13, 65, 102, 107
  barytes, 80, 83, 93
  batch processes, 106
  borax, 88
  Bradley-Poitte roller mill, 84
  'break system', 120
  cake breaker, 126
  calcium phosphate, 88
  calomel, 93
  cement, 77, 300
  chalk, 80, 83, 88
  chromium ore, 75
  classification of mills, 63
  clay, 80, 87, 88
  closed circuit, 287
  corn cutter, 126, 127
  dry,
    operating data, 74
    Raymond roller mills, 83
    rod mills, 79
  efficiency of machines, 19
  energy-to-surface ratio, 8
  equipment, 62
    applications, special, 109
    bibliography, 127

INDEX

Grinding (*cont.*)
  hematite, 80
  impact, 63, 124
    air flow, 360
    provender milling, 124
  Kek mill, 90
  Kick's law, 1
  Lopulco mill, 86
  machinery, special applications, 109
  mechanics, 97
  Micronizer, 92, 360
  millstones, special applications, 117
  Mohr's scale, 62
  pebble mills, 65
  pin disc mills, 89
  power losses, 101
  power requirements, 61
  principles, 1
    references, 22
    summary, 20
  provender, 122
  Rittinger's law, 1
  rod mills, 78
  roller mills, 80, 109
  selective effects, 106
  size distribution, 8
  sizing analysis, 24
  special applications, 109
    bibliography, 127
  stone grinders, 118
  swing hammer pulverizers, 91
  Tricone mills, 76
  tube mills, 77
  wet,
    ball mills in, 69
    operating data, 69
    rod mills, 79
  Wheeler fluid energy mill, 94
Grit,
  definition, 24
  gauze, 141, 143
Grizzlies, 148
Gross sacking-off weighers, 451
Gypsum,
  crushing, 58, 59
  grinding, 80, 83
Gyratory crushers,
  capacities, 53
  efficiency, relative, 19
  principles, 48, 52
  sizes, 53

Hadsel mill, 65
Halkyn jig, 192
Hammer crushers, principles, 48, 58
Hammer mills,
  high speed, 88
  provender milling, 124
  speed, 64
Hancock jig, 190

Handling,
  bucket elevators, 402
  conveying equipment, 390
  conveyors, 382
  dust collecting plants, 426
  dust removal, 384
  level measurement, 389
  solids, 380
    fluidization, 383
    references, 437
  weight measurement, 389
  zipper conveyor, 401
Hapman conveyor, 416
Hardgrove machine, grindability testing, 17
Hardness, Mohr's scale, 62
Harz jig, 194
Heat conservation, settling apparatus, 272
Hematite, grinding, 80
Heyl and Patterson hydrocyclones, 300
Holman table, 179
Hoppers, load cells, 455
Humboldt-Deutz process of coal washing, 332
Hursting mills, 122
  provender milling, 122
  special applications, 117
  under running, 117
Hydraulic conveyors, 427
  degradation of materials, 434
  experimental work, 429
  feeders, 432
  plant wear, 436
  practical considerations, 430
  system cycles, 432
Hydroclassifiers, 294
Hydrocyclones, classification by, 300
Hydrorotators, 298
Hydroscillator, 298
  operating characteristics, 304
  screen analyses, 307, 308, 309
Hydroseparators,
  operating characteristics, 304
  pumping, 309
  size considerations, 306
  wet classification, 288
  working conditions, 293
Hydrostatic weighers, 456
Hydro-treator, 270

Ilmenite, wet classification, 297
Impact mills,
  crushers, 5
  efficiency relative, 19
  energy and throughput, 99
  provender milling, 124
Impact tests, falling weight, 5
Inspiration flotation cell, 232
Ionization, back, 504, 506
Iron ore, sintering of, gas cleaning, 526
Ironstones, crushing of, 56

INDEX

James table, 179
Jaw crushers,
  Blake type, 49
  Dodge type, 51
  principles, 3, 48
  size-frequency curves, 4
Jeffrey diaphragm jig, 198
Jig washers,
  ACCO dirt extraction gear, 204
  air pressure and plunger stroke, 206
  artificial beds and sieve plates, 207
  Baum jig, 202
  Bendelari jig, 197
  classification, 188
  Cooley jig, 195
  cycle, 206
  Denver jig, 197
  dirt extraction, 203
  Elmore jig, 195
  Foust jig, 196
  feed rate, 206
  Feldspar jig, 205
  froth flotation, 236
  Halkyn jig, 192
  Hancock jig, 190
  hand operation, 188
  Harz jig, 194
  Jeffrey diaphragm jig, 198
  Neill jig, 205
  operating variables, 206
  Pan-American jig, 200
  plunger stroke and air pressure, 206
  product removal, 207
  Richards pulsator jig, 199
  sieve plate and artificial beds, 207
  separation in, 187
  typical, description, 190
  Vissac jig, 201
  water feed, 206
  Wilmot pan jig, 193
Jigging,
  dense medium, 207
  general considerations, 187
  gold, 200
  theory, 172

Kek mills, 90
Kick's law, crushing and grinding, 1
Kleinbentink flotation machine, 231
Kreba cyclone, 302

Leaching, separation, 287
Lead,
  separation data, 181, 183, 179
  wet classification, 297
Lead-zinc ore grinding, 70, 71
  size distribution, 73
Lime,
  dust extraction, 520
  handling data, 384
  storage, 389

Limestone,
  flotation treatment, 210
  grinding, 80, 87
Link belt process of coal washing, 333
Liquids,
  centrifugal sedimentation, 36
  electro-precipitation, 484
  liquid–solids,
    contact angle, 211
    flocculation, 252
    viscous, flow of, 163
Litrograph sampler, 441
Load cells, 455
Lopulco mill, 86

MacIntosh pneumatic machine, 232
Magnesite, grinding, 80
Magnesium, grinding of carbonate, 88
Maintenance,
  clarifiers, 275
  thickeners, 275
  weighing machines, 463
Malt, sampling, 441
Marcy ball mills, 66, 69, 71, 72
Massometer, 458
Materials,
  blending, 458
  feeders, 459
  handling, 380
  sampling, 438
  storage, 380
  weighing, 443
Mercier sampler, 442
Merco centrifugal separator, 299
Metallurgy, electro-precipitation in industry, 520
Micronizer pulverizer, 92
Microscopy, sizing analysis, 27, 29
Milling,
  'break process', 113
  compound, 122
  flour, 118
    'break process', 113, 120
    flow design of mill, 119
    plansifter, 151
    preparation process, 120
    purifying, 121
    screening, 160
  grain, 118
  provender, 122
Mills,
  ash content of product, 106
  Atritor, air flow selection, 360
  attrition, 122
    design, 116
    provender milling, 122
    special applications, 116
  ball,
    cascading, 103
    cataracting, 103

x

INDEX

Mills *(cont.)*
  ball *(cont.)*
    centrifuging, 103
    circulating loads, 107
    cylindrical, 69, 71
    dry grinding, 74
    dynamics, 102
    efficiency, relative, 19
    gold ore, grinding, 69
    grindability tests, 15, 107
    grinding by, 12, 65, 102, 107
    Hardinge, 65, 69, 70, 71, 74, 76
    lead–zinc ore, grinding, 70, 71
    load, circulating, 107
    Marcy, operating data, 69, 71, 72
    mixing of solids, 370, 376
    operating data, 69
    power and throughput relationship, 97
    solids, mixing, 370, 376
    speed, 63, 66
    states of motion, 103
    tests, 6, 7, 13, 15, 16, 108
    throughput–power relationships, 97
    Tricone, 76
    U.S. Bureau of Mines, 16
    wet grinding, 69
  cascade, 65
  cement, electro-precipitation, 520, 531
  clinker, dust extraction, 520
  colloid, solid mixing, 376
  crushing, 3
  flour,
    purifier, 121
    screening, 160
  flow diagram, 119
  grindability tests, 15, 107
  grinding,
    batch processes, 106
    classification, 63
    fineness and power, 98
  hammer, 88
    provender milling, 124
    speed, 64
  high speed, 88
  hursting, 117, 122
    provender milling, 122
    special applications, 117
  impact, 5
    crushers, 5
    energy and throughput, 98
    power–throughput relationship, 99
    provender milling, 124
  Kek, 90
  Lopulco, 86
  paper, electro-precipitation, 528
  pebble, 63, 65, 74
  peg, 88
  pin disc, 89
  power losses, 101

Mills *(cont.)*
  ring-roll,
    characteristics, 98
    efficiency, relative, 19
    power–throughput relationship, 98
  rod, 78
    crushing by, 60
    dry grinding, 79
    speed, 63
    uses, 48
    wet grinding, 79
  roller,
    air separation and, 82
    bearings, 114
    Bradley-Poitte, 84
    break, 121
    capacity, 116
    design, 109
    diagonal, 110, 112, 114, 125
    dry grinding, 83
    feed gate, 115
    flour milling, 120
    fluted rolls, 113
    horizontal, 110
    provender milling, 125
    Raymond, 80
    reduction, 121
    roll arrangements, 110
    special applications, 109
    speed, 63
    three high, 111
    uses, 80
    vertical, 110
  serrated disc, 88
  stamp, relative efficiency, 19
  swing hammer, relative efficiency, 19
  Tricone, 76
  tube,
    description, 77
    efficiency, 19
    power–throughput relationship, 97
    speed, 63
    throughput–power relationship, 97
  Wheeler fluid energy, 94
Millstones,
  construction, 124
  design, 117
  grinding action, 116
  provender milling, 122
  special applications, 117
Minerals, reduction, 48
Mining, dust extraction, 520
Mix-Muller, 371
Mixers,
  Banbury, 370, 377
  batch, 368
  beater, 375
  dough, 370, 377
  drum, 369, 370
  Entoleter, 375

XI

INDEX

Mixers (cont.)
  fountain, 371
  mullers, 370
  paddle, 374
  pan, 376
  ribbon, 369, 374
  Rotocube machine, 370
  sifters, 150, 370, 375
  spiral, 371
  spray, 372
  trough, 369, 373
  tumbling, 369, 370
Mixing,
  assessment, 364
  batch samples, 364
  continuous, 368
  convective, 367
  diffusive, 367
  machinery, 369
  mechanical action, 367
  shear, 367
  solids, 362
    references, 379
Morton dough mixer, 377
Mullers, mixing of solids, 370
Multiclones, 469
  diagram, 466
  relative costs, 473
  reverse flow, 472
  straight flow, 472
Multicyclones, 469, 470
  arrestors, 511
  relative costs, 473
  reverse flow, 472
  straight flow, 472

Neill jig, 205
Nelson Davis coal separator, 328
Net sacking-off weighers, 450
Nitrogen, corona current in, 486
Nordberg–Wood classifiers, 290
Nylon as screening media, 142

Oil,
  electro-precipitation in refinery, 528
  flotation reagent, 216, 218
  mist, electro-precipitation, 521
Ores,
  crushing, 56
  screening, 156
Overstrom table, 181
Oxides, flotation treatment, 209
Oxygen, corona current, 486

Packaging of free-flowing materials, 450
Packing machines,
  Bates, 451
  St. Regis, 450
  valve packers, 451
Paddle mixers, 374

Pallets, material handling, 381
Pan mixers, 376
Pan-American jig, 200
Paper milling, electro-precipitation, 528
Paraffin oil as flotation reagent, 216
Particles
  air flow selection, 337
  collection, theoretical efficiency, 475
  critical velocity, 162
  definition, 24
  drag forces, 168
  dust, cyclone as separator, 467
  electro-precipitation, 487
  equivalent diameters, relationship, 40
  flow of, 130
  mobility in granular bed, 128
  packing in beds, 129
  re-entrainment, 501
  resistivity, 502
  scale of segregation, 366
  separation of, 135
  solid mixing, 364
  size,
    analysis,
      agitation effects, 129
      air flow through beds, 134
      distribution, 8, 128
      fluidization, 134
      friction, effects, 129
      grading, 128
      horizontal oscillation, 130, 133
      irregularly shaped particles, 37
      mobility in granular bed, 128
      precipitators, efficiency, 500
      screening, 128, 135
      sieving, 128
      size and flow, 130
      storage of materials, 383
    distribution, 8, 12, 73, 107
    definitions, 24
    Gaudin's theorem, 10
  spherical, terminal velocity, 338
  surface area measurement, 43
  vertical oscillatory motion, 134
  terminal velocity, 162
  velocity of, 162
Pebble mills, 65
  dry grinding, operating data, 74
  speed, 63, 66
Pellett classifiers, 298
Permutit precipitator, 271
Petroleum industry, electro-precipitation, 521
Phosphate, grinding, 80, 87
Phosphorus, electro-precipitation, 533
Pilot plant,
  grinding, batch processes, 106
  precipitator equipment, 512
  separating units, 477
Pine oil as frothing agent, 218
Plansifters, 151

XII

INDEX

Plant, siting of, safety precautions, 385
Plat-O table, 181
Plunger feeder, 460
Pneumatic classification, 289
Pneumatic conveyors, 418
   abrasion of pipes, 436
   discharge of materials, 425
   nozzles, 424
   plant wear, 436
   rotary seal, 425
   tipping seal, 425
Porous masses, fluidization, 134
Powders,
   Airslide in handling units, 384
   conveying of, 391
   definition, 24
   fluidization in handling of, 383
   handling of, 383, 425
   measurement of, Massometer, 458
Power plants, electro-precipitation, 522
Power stations, stock-piling of supplies, 385
Precipitation, electrostatic, 484
   acid applications, 529
   applications, 520
   automatic voltage control, 513
   back ionization, 504
   blast furnaces, 525
   carbon, 503
   cement manufacture, 520, 531
   chemical industry, 521
   chute electrodes, 493
   coal industries, 520, 525
   corona current, 485
   dust collection, 491
   dust generation, 523
   electronic valve rectifiers, 512
   fertilizer industry, 521
   food industry, 521
   gas, characteristics, 486
      conditioning, 509
      distribution, 510
   geometrical dimensions, 496
   installations, typical, 522
   metal rectifiers, 512
   metallurgical industry, 520
   mining industries, 520
   oil refinery applications, 528
   paper industry, 521, 528
   petroleum industry, 521
   plant arrangement, 509
   principles, 484
   references, 534
   steel making, 521
   working conditions, 501
Precipitators,
   construction, 488
   corona quenching effect, 487
   dry, 488
   detarrer, 489
   dust collection efficiency, 494

Precipitators (*cont.*)
   electrodes, 489
   electrostatic,
      automatic voltage control, 513
      chute electrodes, 493
      construction, 488
      costs, 514
      design factors, 494, 507
      equipment, 512
      graphical presentation, 516
      mechanical collectors and, 511
      mechanical rectifiers, 512
      paper making, 521
      reclosing of circuit, 514
      types of, 488
      typical installations, 522
      working costs, 515
   fume, corona quenching effect, 488
   irrigated, 488
   mist, 488
   multizone, 509
   Permutit, 271
   plate type, 488
   single zone, 509
   tube type, 488, 499
   wet, electrodes, 489
Provender milling, 122
   attrition mills, 122
   cake breakers, 126
   corn cutters, 126
   elutriation, 125
   hammer mills, 124
   hursting mills, 122
   Impact grinder, 124
   millstones, 122
   roller stones, 125
Pulverizers, 48
   Atritor, 88
   boiler fuel, 95
   classification, 97
   energy losses, 101
   energy and throughput, 98
   fluid jet, 64
   Kek mills, 90
   mechanics, 97
      references, 108
   Micronizer, 92
   output measurement, 100
   pin-disc mills, 89
   power, characteristics, 98
   power–throughput relationship, 97
   Raymond Impax, 91
   special applications, 109
   swing hammer, 91
   throughput–power relationship, 97
Pumping of solids, 431
Pumps,
   centrifugal, 431
   diaphragm, sedimentation units, 272
   high pressure systems, 431

INDEX

Pumps (*cont.*)
  hydroseparators, 309
  sedimentation units, 272
Purifiers, flour mill, 121, 160
Pyritic ore, wet grinding, 72
Pyrites, flotation of, 222, 244

Quartz, surface energies, 22

Rake classifiers,
  spiral, 292
  wet classification, 288, 290, 291
RapiDorr clarifiers, 271
Raymond Impax pulverizer, 91
Raymond roller mills, 80
Redler feeder, 460
Resistivity, moisture adsorption effects, 502
Ribbon mixers, 369, 374
Richardson automatic weigher, 449
Ridley-Scholes process of coal washing, 326
Riffler sample divider, 441
Ring-hammer crushers, 59
Ring-roll mills,
  characteristics, 98
  efficiency, relative, 19
  power–throughput relationship, 98
Rittinger's law, crushing and grinding, 1
Rod mills,
  action of, 78
  crushing by, 60
  dry grinding, 79
  speed, 63
  uses, 48
  wet grinding, 79
Roll crushers, principles, 48, 56
Roller conveyors, 391
Roller mills,
  air separation and, 82
  bearing alignment, 114
  Bradley-Poitte, 84
  break, 121
  capacity, 116
  design, 109
  diagonal, 110, 112, 114, 125
    adjustments, 114
    double sided, 112
  dry grinding, 83
  feed gate, 115
  flour milling, 120
  fluted rolls, 113
  horizontal, 110
  provender milling, 125
  range of use, 80
  Raymond, 80
  reduction, 121
  roll construction, 112
  special applications, 109
  speed, 63
  three high, 111

Roller mills (*cont.*)
  vertical, 110
Rotary crushers, 52
Rotary feeders, 460
Rotocube mixing machine, 370

Sacking-off weighers, 450
Sampling,
  Birtley's apparatus, 443
  coning and quartering, 439
  granular materials, 438
  Litrograph, 441
  manual, 439
  mechanical, 441
  Mercier sampler, 442
  Riffler sample divider, 441
Sand,
  conveyance of, 423
  cyclones as separators, 467
  dust extraction, 520
  handling data, 384
  washing, Dorr machine, 294
Sawdust, handling data, 384
Select-O-Weigh machines, 462
Screening,
  agitation, threshold, 145
  asbestos, 160
  bars as media, 136
  bibliography, 171
  centrifugals, 150
  clogging by particles, 144
  coal, 156
  efficiency, 144
  fluidization, and oscillatory motion, 160
  granular material, 128
  grids as media, 136
  grindability tests, 16
  grizzlies, 148
  high speed rotary screens, 156
  horizontal oscillation, 133, 156
  horizontal rotary motion, 130
  machines, types, 147
  media, 135
  oscillatory motion,
    fluidization and, 160
    rotation and, 155
  particle analysis, 135
  perforated plates as media, 137
  plansifters, 151
  rate of, 144
  reels, 148, 150
  references, 171
  roll grizzly, 148
  rotary screens, 155
  rotary sifters, 150
  silk screens, 143
  sizing analysis, 27
  speed of machines, 146
  stationary screens, 148
  trommels, 148

INDEX

Screening (*cont.*)
  vertical oscillation, 134, 156
  vertical rotary motion, 156
  wedge wires as media, 136
Screens,
  British Standards, 138
  Hummer, 158
  high speed rotary, 156
  mechanically perforated, 137
  oscillatory motion, 156
  perforated, 137, 149
  rotary, vertical, 156
  Sherwen, 158
  stationary, 148
  woven textile, 141
  woven wire, 137, 149
Sedimentation,
  aggregation, 252
  ancillary equipment, 272
  batch processes, 260
  beet sugar, 264
  caustic soda in, 265
  centrifugal, in liquids, 36
  clarification,
    batch, 256
    collective subsidence, 257
    definition, 248
    limitations, 252
    line settling, 257
    mechanics, 256
    phase subsidence, 257
    pulp behaviour during, 258
    terminology, 250
  construction of units, 271
  corrosion of units, 272
  countercurrent decantation, 269
  data for design of units, 271
  definition, 248, 249
  density controls, 273
  destabilization, 252
  diagram showing process, 251
  equipment, 34, 265
    calculation of size, 277
    costs, 273
  flocculation, 252, 269
  heat conservation during, 272
  multiple units, 269
  new developments, 283
  overload alarm, 273
  practical considerations, 263
  principles, 33
  pulp characteristics, 256
  pumps, 272
  settling capacity, 259
  settling tests, 278
  sizing analysis, 25, 29, 34
  theory of, 254
  thickening,
    capacity of unit, 261
    definition, 248

Sedimentation (*cont.*)
  thickening (*cont.*)
    limitations, 252
    mechanics, 260
    terminology, 250
  tray type units, 269
  underflow lines, features, 273
  unit, materials for, 271
Semi-conductors, resistivity, 502
Separation,
  air elutriation, 352
  air flow, 337
  centrifugal type units, 476
  coal, 312
    Barvoy's system, 322
    Chance system, 318, 319
    data, 179
    dense media plant, 315
    desanding, 317
    Dutch State Mines system, 323
    Humboldt-Deutz process, 332
    Link belt process, 333
    magnetic medium cleaning, 318
    media stability, 314
    Nelson Davis separator, 328
    references, 336
    Ridley-Scholes process, 318, 326
    Simcar process, 329
    S.K.B. process, 334
    Tromp process, 318, 324
  density, flotation, 161
  dust in gas, 466
  fan, 355
  froth flotation, 209
  gas, 466
  grain, 349
  jigging, 172, 187
  principles, 177, 187
  settling rate, 161
  tabling, 172, 176
  velocity, elutriation, 161
  wet screen, 289
  wheat, 354
Separators,
  air, spiral flow, 347
  catch box, 339, 349
  centrifugal, 299, 340
  cyclone, air flow pattern, 346
  Drewboy, 329
  helical flow, 341
  hydroseparators, 288, 293, 304, 306, 309
  trajectory, 350
Settling,
  air movement in chamber, 339
  capacity of unit, 259
  clarification, 257
  clarifier dilution test, 279
  compression point determination, 278
  cones, 265
  gravity, 248

INDEX

Settling (*cont.*)
   intermittent tanks, 265
   laws, 161
   preliminary test, 278
   pulp characteristics, 256
   sedimentation, 254
   unit size determination, 280
Shaker feeders, 460
Sherwin electromagnetic vibrator, 184
Sieves,
   British Standards, 26
   gyratory, high speed, 156
   high speed gyratory, 156
   high speed rotary, 156
   rotary, high speed, 156
   silk bolting cloth, 141
   test, 140
Sieving,
   clogging of apertures, 144
   efficiency, 144
   feed and efficiency, 145
   flour, 153
   mechanical, 26
   particles, 128
   particle analysis, 135
   perforated plates, 137, 149
   rate of, 144
   sizing analysis, 26
   wire screws, 137, 149
Sifters,
   mixing of solids, 370, 375
   rotary, 150
Sillimanite, dry grinding, 74
Silos, 388
Simcar process of coal washing, 329
Simcar-Geco flotation cells, 227
Simon automatic weighers, 447
Size reduction,
   distribution laws, 8
   energy requirements, 7
   principles, 1
      references, 22
      summary, 20
   impact crushers, 5
   jaw crushers, 3
   relative energies, 2
Sizers,
   Dorrco, 296
   Fahrenwald, 296
   hydraulic, 295
   jet, 296
   operating characteristics, 304
Sizing analysis,
   agitation effects, 129
   definitions, 24
   elutriation, 25, 29
   equivalent diameters, relationship, 40
   fluidization, 134
   friction effects, 129
   grading, 128

Sizing analysis (*cont.*)
   irregularly-shaped particles, 37
   methods, 26
   microscopical measurement, 27, 29
   photo-extinction measurements, 45
   precipitation, 484
   principles, 24
   references, 46
   results of, 42
   screening, 128, 135
   sedimentation, 25, 29, 34
   Stokes' equation, 29
S.K.B. process of coal washing, 334
Slag, cyclones as separators, 467
Slat conveyor, 413
Slate, grinding, 74, 75, 83
Sludge collection, 270
Slurry, recovery, 238
Soda ash, handling and storage, 384, 389
Soil, analysis, 35
Solids,
   air affinity, 212
   ball mills as mixers, 376
   Banbury mixers, 370, 377
   batch mixing, 368
   blending, 362, 458
   closed circuit, 287
   colloids as solid mixing, 376
   continuous mixing, 368
   convective mixing, 367
   conveying equipment, 390
   diffusive solids, 367
   dough mixers, 370, 377
   drum mixers, 369, 370
   electro-precipitation, 484
   feeders, 459
   fountain mixers, 371
   gauging,
      bibliography, 463
      measuring and, 438
   handling of, 380
   high speed beater mixing, 375
   hydrophilic, 211
   hydrophobic, 210
   liquid–solids,
      contact angle, 211
      flocculation, 252
      sedimentation, 248
      separation, 248
   manual sampling, 439
   measurement,
      automatic weighers, 447
      belt weighers, 456
      bibliography, 463
      bucket weighers, 446
      Chronos weigher, 446
      constant feeder, 457
      dormant weighers, 444
      Edgar Allen weigher, 454
      gauging, 438

INDEX

Solids (cont.)
  measurement (cont.)
    load cells, 455
    packers, 450
    Richardson weigher, 449
    sacking-off weigher, 450
    weighing machines, 443
  mechanical sampling, 441
  mixing, 362
    ball mills, 370, 376
    bibliography, 379
    machinery for, 369
    references, 379
    rolls, 377
    texture, 365
  mullers for mixing, 370
  paddle mixers, 374
  apron mixers, 376
  pumping methods, 431
  ribbon mixers, 369, 374
  sampling, 438
  segregation scale, 366
  shear mixing, 367
  sifter mixers, 370, 375
  spiral mixers, 371
  spray mixers, 372
  storage of, 380
  trough mixers, 369, 373
  tumblers for mixing, 369, 370
Sphalerite, flotation of, 219, 222
Spiral classifiers, wet classification, 288
Spiral mixers, 371
Spitzkasten, classification, 289
Spray mixers, 372
Stamp mills, relative efficiency, 19
Stress–strain curves for soft coal, 3
Stag ball mills, 67
St. Regis packing machines, 450
Starch, depressor for flotation, 219
Steel making, electro-precipitation, 521
Steelworks, dust generation, 523
Stokes–Navier equations, 162
Storage,
  angle of repose of materials, 383
  bins, 387
  bulk density of materials, 382
  bunkers, 387
  coal, 385
  fork lift trucks, 382
  grain, 387
  indoor, 386
  materials in bulk, 385
  open air, 385
  safety precautions, 385
  silos, 387
  solids, 380
    references, 437
Sugar, belt conveyors, 394
Sulphides, flotation treatment, 209
Sulphidizers, flotation reagents, 220

Sulphur,
  flotation treatment, 210
  grinding, 83, 93
  removal, 527
Supersorter, 297
Swing hammer mills, relative efficiency, 19
Swing hammer pulverizers, 91

Tabling,
  Buss table, 179
  Butchart table, 183
  costs, 186
  Deister tables, 180
  dry, 185
  Ferraris table, 179
  Garfield, 179
  general considerations, 176
  head motion, 184
  Holman table, 179
  James table, 179
  operating variables, 183
  Plat-O table, 181
  product launders, 185
  Sherwin electromagnetic vibrator, 184
  sizes and capacity, 178
  slope of table, 184
  theory, 172
  types of table, 178
  wash water, 184
  Wilfley table, 176
Talc, grinding, 93
Tar fog, electro-precipitation, 521
Tar oil, cleaning, 527
Temperature,
  electrical resistivity and, 519
  flotation, effects on, 223
  resistivity of semi-conductors as function of, 502
Thickeners,
  Allen cone, 266
  Caldecott cones, 265
  Callow cone, 266
  chemical, redesign, 284
  clarifiers and, 266
  cone, 265
  construction materials, 271
  costs, 274
  design data, 271
  Dorr Densludge, 284
  Dorr mechanism, 266
  filter, 269
  Hardinge, 267
  Humboldt, 333
  hydroseparation and, 293
  invention of, 248
  maintenance, 275
  metallurgical, redesign, 284
  multiple units, 269
  power consumption, 274
  redesign, 284

INDEX

Thickeners (cont.)
  references, 285
  Torq, 267
  tray type units, 269
  washing type, continuous countercurrent decantation, 276
Thickening,
  capacity of unit, 261
  classification, 290
  equipment, 265
  intermittent settling tanks, 265
  limitations in settling, 252
  mechanics of, 260
  underflow, 285
Titanium, grinding, 80, 83, 93
Titanium dioxide,
  classification, 300
  grinding, 93
Torq thickeners, 267
Transportation, fork lift trucks, 381
Tricone mills, 76
Tromp process of coal washing, 324
Trommels, 149
Trough mixers, 369, 373
Trucks, fork lift, 381
Tube mills,
  grinding, 77
  efficiency, relative, 19
  power–throughput relationship, 97
  speed, 63
  throughput–power relationship, 97
Tumblers, solid mixing, 369, 370
Tungsten, separation data, 179, 181
Turntable feeders, 460

Uranium, flotation of, 236
U.S. Bureau of Mines ball mill test, 16

Velofeeder, 459
Vibrating conveyors, 417
Vissac jig, 201
Vacuum flotation, 232

Washers,
  Allis-Chalmers, 294
  Dorr, 294, 295
  dregs, redesign, 283
  jig, 188
  log, separation of materials, 294
  sand, operating characteristics, 304
  screw, 294

Washing,
  classifiers, 294
  separators, 287
Water, elutriation by, 31
Weighbridges, 443, 455
Weighing machines, 443
  Bates packer, 451
  belt weighers, 456
  Blake–Denison Integrating weigher, 456
  bucket weighers, 446
  Chronos weigher, 446
  constant feeder, 457
  continuous, 456
  dormant weighers, 444
  Edgar Allen weigher, 454
  hydraulic load cells, 456
  hydrostatic weigher, 456
  load cells, 455
  maintenance, 463
  Massometer, 458
  net sacking-off weighers, 450
  packers, 450
  Richardson automatic weigher, 449
  sacking-off weigher, 450
  Select-O-Weigh, 462
  Simon automatic weigher, 447
  St. Regis weigher, 453
  weighbridges, 444
Wemco classifiers, 290
Wet classification, 287
Wet grinding, 69, 79
Wetting and floatability, 213
Wheat,
  conveyors, 421
  handling data, 384
  separation, 354
Wheel-and-shelf feeder, 460
Wheeler fluid energy mills, 94
Wilfley table, 176
Wilmot hydrorotators, 298
Wilmot pan jig, 193
Whirlcone, 300, 302
Worm conveyors, 406
Wuensch cone classifier, 290

Xanthates, flotation reagents, 215

Zinc,
  separation data, 179
  wet classification, 179
Zipper conveyors, 401
Zircon sands, dry grinding, 74, 75